The Epithelial-to-Mesenchymal Transition (EMT) in Cancer

Special Issue Editor
Joëlle Roche

MDPI • Basel • Beijing • Wuhan • Barcelona • Belgrade

MDPI

Special Issue Editor
Joëlle Roche
Université de Poitiers
France

Editorial Office
MDPI AG
St. Alban-Anlage 66
Basel, Switzerland

This edition is a reprint of the Special Issue published online in the open access journal *Cancers* (ISSN 2072-6694) from 2017–2018 (available at: http://www.mdpi.com/journal/cancers/special_issues/emt).

For citation purposes, cite each article independently as indicated on the article page online and as indicated below:

Lastname, F.M.; Lastname, F.M. Article title. *Journal Name* **Year**. *Article number, page range.*

First Edition 2018

ISBN 978-3-03842-793-3 (Pbk)
ISBN 978-3-03842-794-0 (PDF)

Table of Contents

About the Special Issue Editor.   v

Joëlle Roche
The Epithelial-to-Mesenchymal Transition in Cancer
doi: 10.3390/cancers10020052   1

Julia Thierauf, Johannes Adrian Veit and Jochen Hess
Epithelial-to-Mesenchymal Transition in the Pathogenesis and Therapy of Head and
Neck Cancer
doi: 10.3390/cancers9070076   5

Monica Fedele, Laura Cerchia and Gennaro Chiappetta
The Epithelial-to-Mesenchymal Transition in Breast Cancer: Focus on Basal-Like Carcinomas
doi: 10.3390/cancers9100134   18

Yuliya Klymenko, Oleg Kim and M. Sharon Stack
Complex Determinants of Epithelial: Mesenchymal Phenotypic Plasticity in Ovarian Cancer
doi: 10.3390/cancers9080104   37

Antoine Legras, Nicolas Pécuchet, Sandrine Imbeaud, Karine Pallier, Audrey Didelot,
Hélène Roussel, Laure Gibault, Elizabeth Fabre, Françoise Le Pimpec-Barthes,
Pierre Laurent-Puig and Hélène Blons
Epithelial-to-Mesenchymal Transition and MicroRNAs in Lung Cancer
doi: 10.3390/cancers9080101   69

Nicola Gaianigo, Davide Melisi and Carmine Carbone
EMT and Treatment Resistance in Pancreatic Cancer
doi: 10.3390/cancers9090122   98

Trung Vu and Pran K. Datta
Regulation of EMT in Colorectal Cancer: A Culprit in Metastasis
doi: 10.3390/cancers9120171   115

Simon Grelet, Ariel McShane, Renaud Geslain and Philip H. Howe
Pleiotropic Roles of Non-Coding RNAs in TGF-β-Mediated Epithelial-Mesenchymal
Transition and Their Functions in Tumor Progression
doi: 10.3390/cancers9070075   137

Joëlle Roche, Robert M. Gemmill and Harry A. Drabkin
Epigenetic Regulation of the Epithelial to Mesenchymal Transition in Lung Cancer
doi: 10.3390/cancers9070072   152

Zheng Fu and Donghua Wen
The Emerging Role of Polo-Like Kinase 1 in Epithelial-Mesenchymal Transition and
Tumor Metastasis
doi: 10.3390/cancers9100131   166

Eve Duchemin-Pelletier, Megghane Baulard, Elodie Spreux, Magali Prioux, Mithila Burute,
Baharia Mograbi, Laurent Guyon, Manuel Théry, Claude Cochet and Odile Filhol
Stem Cell-Like Properties of CK2β-down Regulated Mammary Cells
doi: 10.3390/cancers9090114   181

Robert H. Blackwell, Kimberly E. Foreman and Gopal N. Gupta
The Role of Cancer-Derived Exosomes in Tumorigenicity & Epithelial-to-Mesenchymal Transition
doi: 10.3390/cancers9080105 . . . . **195**

Jianrong Lu and Anitha K. Shenoy
Epithelial-to-Pericyte Transition in Cancer
doi: 10.3390/cancers9070077 . . . . **206**

Elvira Forte, Isotta Chimenti, Paolo Rosa, Francesco Angelini, Francesca Pagano,
Antonella Calogero, Alessandro Giacomello and Elisa Messina
EMT/MET at the Crossroad of Stemness, Regeneration and Oncogenesis: The Ying-Yang
Equilibrium Recapitulated in Cell Spheroids
doi: 10.3390/cancers9080098 . . . . **219**

Dongya Jia, Mohit Kumar Jolly, Prakash Kulkarni and Herbert Levine
Phenotypic Plasticity and Cell Fate Decisions in Cancer: Insights from Dynamical
Systems Theory
doi: 10.3390/cancers9070070 . . . . **234**

About the Special Issue Editor

Joëlle Roche, received her PhD in 1985 at the University of Grenoble (France). This was followed by a position of Associate Professor at the University of Lyon (France) and a sabbatical at the University of Colorado in Denver (CO, USA), where she developed her interest in lung cancer after discovering the semaphorin SEMA3F, from a recurring 3p deletion. She was full Professor of Biochemistry and Molecular Biology at the University of Poitiers (France) from 1995 to 2016. During that period, she had a sabbatical at the Medical University of South Carolina (Charleston, SC, USA) from 2009–2011. She is presently Emeritus Professor at the University of Poitiers. Dr. Roche is interested in gene expression, chromatin remodeling and guidance molecules. Her work has focused on semaphorin expression and function in lung cancer, EMT and epigenetic regulation.

cancers

MDPI

Editorial

The Epithelial-to-Mesenchymal Transition in Cancer

Joëlle Roche

Laboratoire EBI, SEVE, UMR-CNRS 7267, Université de Poitiers, F-86073 Poitiers, France;
joelle.roche@univ-poitiers.fr

Received: 11 February 2018; Accepted: 13 February 2018; Published: 16 February 2018

The epithelial-to-mesenchymal transition (EMT) occurs during normal embryonic development, tissue regeneration, organ fibrosis, and wound healing. It is a highly dynamic process, by which epithelial cells can convert into a mesenchymal phenotype. However, it is also involved in tumor progression with metastatic expansion, and the generation of tumor cells with stem cell properties that play a major role in resistance to cancer treatment [1–3]. EMT is not complete in cancer cells, and tumor cells are in multiple transitional states and express mixed epithelial and mesenchymal genes. Such hybrid cells in partial EMT can move collectively as clusters, and can be more aggressive than cells with a complete EMT phenotype [4]. EMT is also reversible by the mesenchymal-to-epithelial transition (MET), thought to affect circulating cancer cells when they reach a desirable metastatic niche to develop secondary tumors. The EMT process involves the disruption of cell–cell adhesion and cellular polarity, remodeling of the cytoskeleton, and changes in cell–matrix adhesion. It is associated with improvement in migratory and invasive properties. In cancers, EMT inducers are hypoxia, cytokines, and growth factors secreted by the tumor microenvironment, stroma crosstalk, metabolic changes, innate and adaptive immune responses, and treatment with antitumor drugs. Switch in gene expression from epithelial to mesenchymal phenotype is triggered by complex regulatory networks involving transcriptional control with SNAI1 and SNAI2, ZEB1 and ZEB2, Twist, and E12/E47 among transcriptional factors, non-coding RNAs (miRNAs and long non-coding RNAs), chromatin remodeling and epigenetic modifications, alternative splicing, post-translational regulation, protein stability, and subcellular localization [5]. EMT is becoming a target of interest for anticancer therapy [6]. However, more knowledge about the role of EMT in metastasis, its control, and its reversion is necessary. Indeed, alternative modes of dissemination, colonization via a MET-independent pathway, and investigation of circulating cancer cells in the blood support a more nuanced view of the role of EMT and MET in cancer metastasis.

This current Special Issue entitled "Epithelial to Mesenchymal Transition in Cancer" sheds more light on EMT in several selected cancers, and the first six reviews describe the pathology, EMT inducers and their corresponding pathways, EMT involvement in metastasis and drug resistance, and therapeutic perspectives [7–12]. The first review by Thierauf et al. describes the most recent findings on the clinical relevance of a mesenchymal-like phenotype for head and neck cancer patients, including more rare cases of mucosal melanoma and adenoid cystic carcinoma [7]. Among classical EMT actors, the abnormal expression in head and neck squamous cell carcinoma (HNSCC) of both the pluripotency-associated transcription factor SOX2 (sex determining region Y-box 2) and of the Kallikrein-related peptidase 6 (KLK6) is discussed. This is followed by a review by Fedele et al. that recapitulates the main endogenous molecular signals involved in the acquisition of the mesenchymal phenotype in metastatic breast cancer, mainly basal-like and claudin-low subtypes that belong to the group of triple-negative breast cancer (TNBC), and their cross-talk with paracrine factors [8]. The importance of the extracellular matrix, paracrine mechanisms, and exosomes is underlined. In the next review, Klymenko et al. describe the EMT and the way in which epithelial ovarian cancer (EOC) metastasizes [9]. Unlike most epithelial malignancies which metastasize hematogenously, EOC metastasis occurs primarily via transcoelomic dissemination, and this exceptional microenvironment is highly permissive for phenotypic plasticity. Current knowledge on EOC heterogeneity in an EMT context is highlighted. Computational modeling of EMT is also presented. This review is followed by a description of EMT in non-small cell lung cancer (NSCLC) by Legras et al.—a cancer that is the major cause of cancer-related death in developed countries [10]. The authors report the role of the transcription factor TWIST1 in EMT in cells with epidermal growth factor receptor (EGFR) mutation and its association with low disease-free survival in EGFR-mutated lung adenocarcinoma patients. A large part of this overview is dedicated to EMT regulation by microRNAs. Next, EMT is described by Gaianigo et al. in pancreatic cancer (PC), which has a poor prognosis due to metastatic dissemination in the very first stages of tumor development mainly attributed to EMT [11]. The focus of this review is the involvement of EMT in treatment resistance. Among EMT actors, the authors give special attention to mutated RAS and the nuclear transcription factor yes-associated protein 1 (YAP1), one of the main transducers of the Hippo pathway. The authors also develop the importance of the large fibrotic tumor microenvironment called desmoplasia, which contributes to PC advancement, chemoresistance, and reduced drug delivery. The role of inflammation, microbiota, and treatment resistance related to EMT are also discussed. The review by Vu et al. summarizes EMT in colorectal cancer (CRC) [12]. In addition to classical transcription factors, the authors describe deeper PROX1 homeobox and forkhead box (FOX) transcription factors, their signaling pathways, several novel EMT inducers in colorectal cancers including neuropilin-2 (NRP2, a receptor for cell guidance molecule and growth factors), and downregulation of several microRNAs. Furthermore, the role of the EMT in circulating tumor cells (CTCs) is also discussed with the possible use of the mesenchymal markers as potential biomarkers for metastasis.

The following review by Grelet et al. focuses on specific aspects of EMT in cancers and cell lines, the molecular mechanisms regulating the TGF-β-induced EMT, and their implications in tumor metastasis [13]. The authors discuss the recent reports emphasizing the regulatory functions of non-coding RNA (microRNAs and lncRNAs, piRNAs, snRNAs, snoRNAs, circRNAs). In the second part of their survey, the authors provide experimental evidence of change in the expression of several tRNAs during TGF-β treatment of the human lung cancer cell line A549.

The third series of reviews addresses more specific mechanisms involved in EMT regulation [14–18]. First, Roche et al. present epigenetic regulation of EMT mainly in lung cancer with recent data on EZH2 (enhancer of zeste 2 polycomb repressive complex 2 subunit) that behaves as an oncogene in lung cancer associated with gene repression [14]. EZH2 is the catalytic subunit of the PRC2 (polycomb group PcG), which methylates lysine 27 of histone H3, inducing a repressive transcriptional mark. The authors mention that EZH2 inhibitors are under development with clinical trials ongoing. Next, Fu et al. describe the role of the Polo-like kinase 1 (PLK1)—a serine/threonine kinase—in EMT and tumor invasion [15]. Its overexpression correlates with increased cellular proliferation and poor prognosis in cancer patients. To its canonical role in almost every stage of cell division and cytokinesis execution, modulation of DNA replication, and cell survival, overexpressed PLK1 orchestrates EMT and associated events (such as invasion and therapeutic resistance) in tumor cells via the MAPK pathway by directly binding and phosphorylating CRAF (a member of the Raf kinase family), and through AKT or FoxM1-dependent pathways. The therapeutic use of PLK1 inhibitors is discussed. Next, Duchemin-Pelletier et al. present data about the role of CK2β, the regulatory subunit of the ubiquitous protein kinase CK2 [16]. In MCF10A mammary epithelial cells where CK2β expression was downregulated, the authors provide evidence that CK2β downregulation can promote the acquisition of characteristics commonly associated with the cancer stem cell phenotype. They demonstrate that a CK2β level establishes a critical cell fate threshold in the control of epithelial cell plasticity. Since EMT supports the induction stem-cell phenotype, identifying the CK2 substrates will improve the discovery of new specific markers for breast cell stemness. In the following review, Blackwell et al. describe exosomes that are important mediators of intercellular signaling and EMT, with resultant transformation of cancer cells to a more aggressive phenotype [17]. By concentrating proteins or RNA, exosomes may transform nearby cells, as an explanation for the "field effect" phenomenon. In addition, cancer-derived exosomes exert pro-angiogenic effects on epithelial cells. Then, Lu et al. review another interesting aspect of EMT events: the gain of multiple pericyte markers by cells undergoing EMT—called the epithelial-to-pericyte transition (EPT), possibly mediated by the serum response factor (SRF) [18]. These mesenchymal cells phenotypically and functionally resemble pericytes. They are attracted to blood vessels via paracrine PDGF signaling and associate to blood vessels through N-cadherin. EPT cancer cells associate with and stabilize blood vessels to fuel tumor growth, and EPT may promote therapy resistance to antiangiogenic agents. The authors propose a model for EPT regulation.

Next, Forte et al. address concerns about 2D cellular models to study EMT and strongly recommend in vitro tissue-derived cell spheroid models [19]. These three-dimensional (3D) cell culture systems—whose phenotype has been shown to be strongly dependent on TGF-β-regulated EMT/MET processes—present the advantage of exposing cells to more physiological conditions in a microenvironment where cell–cell and cell–matrix interactions are present, and recapitulate in vitro the hypoxic micro-environment of tissue stem cell niches and their formation. The authors explore the mechanistic correspondence in vivo and the possible pharmacological perspectives.

The last review by Jia et al. illustrates the concept of "landscape" previously defined by Waddington in the epigenetic field [20]. Each cell phenotype is considered as an "attractor" which is characterized by a unique gene expression pattern, and determined by interactions between multiple molecular players buffered against environmental fluctuations. Here, the authors illustrate how phenotypic plasticity in cancer cells enables them to acquire hybrid phenotypes during EMT. The perspective of this review is intended to encourage the exchange of ideas between cancer biologists and physicists interested in exploring the physics of biology.

In summary, this special issue covers EMT in several types of cancer, and reviews different actors in EMT regulation with potential therapeutic applications. More work is needed to advance this complex field with the best available models.

Conflicts of Interest: The author declares no conflict of interest.

References

1. Nieto, M.A.; Huang, R.Y.-J.; Jackson, R.A.; Thiery, J.P. EMT: 2016. *Cell* **2016**, *166*, 21–45. [CrossRef] [PubMed]
2. Lambert, A.W.; Pattabiraman, D.R.; Weinberg, R.A. Emerging Biological Principles of Metastasis. *Cell* **2017**, *168*, 670–691. [CrossRef] [PubMed]
3. Moustakas, A.; de Herreros, A.G. Epithelial-mesenchymal transition in cancer. *Mol. Oncol.* **2017**, *11*, 715–717. [CrossRef] [PubMed]
4. Jolly, M.K.; Boareto, M.; Huang, B.; Jia, D.; Lu, M.; Ben-Jacob, E.; Onuchic, J.N.; Levine, H. Implications of the Hybrid Epithelial/Mesenchymal Phenotype in Metastasis. *Front. Oncol.* **2015**, *5*. [CrossRef] [PubMed]
5. De Craene, B.; Berx, G. Regulatory networks defining EMT during cancer initiation and progression. *Nat. Rev. Cancer* **2013**, *13*, 97–110. [CrossRef] [PubMed]
6. Marcucci, F.; Stassi, G.; De Maria, R. Epithelial-mesenchymal transition: a new target in anticancer drug discovery. *Nat. Rev. Drug Discov.* **2016**, *15*, 311–325. [CrossRef] [PubMed]
7. Thierauf, J.; Veit, J.A.; Hess, J. Epithelial-to-Mesenchymal Transition in the Pathogenesis and Therapy of Head and Neck Cancer. *Cancers* **2017**, *9*. [CrossRef] [PubMed]
8. Fedele, M.; Cerchia, L.; Chiappetta, G. The Epithelial-to-Mesenchymal Transition in Breast Cancer: Focus on Basal-Like Carcinomas. *Cancers* **2017**, *9*. [CrossRef] [PubMed]
9. Klymenko, Y.; Kim, O.; Stack, M.S. Complex Determinants of Epithelial: Mesenchymal Phenotypic Plasticity in Ovarian Cancer. *Cancers* **2017**, *9*. [CrossRef] [PubMed]
10. Legras, A.; Pécuchet, N.; Imbeaud, S.; Pallier, K.; Didelot, A.; Roussel, H.; Gibault, L.; Fabre, E.; Le Pimpec-Barthes, F.; Laurent-Puig, P.; Blons, H. Epithelial-to-Mesenchymal Transition and MicroRNAs in Lung Cancer. *Cancers* **2017**, *9*. [CrossRef] [PubMed]
11. Gaianigo, N.; Melisi, D.; Carbone, C. EMT and Treatment Resistance in Pancreatic Cancer. *Cancers* **2017**, *9*. [CrossRef] [PubMed]

12. Vu, T.; Datta, P.K. Regulation of EMT in Colorectal Cancer: A Culprit in Metastasis. *Cancers* **2017**, *9*. [CrossRef] [PubMed]
13. Grelet, S.; McShane, A.; Geslain, R.; Howe, P.H. Pleiotropic Roles of Non-Coding RNAs in TGF-β-Mediated Epithelial-Mesenchymal Transition and Their Functions in Tumor Progression. *Cancers* **2017**, *9*. [CrossRef] [PubMed]
14. Roche, J.; Gemmill, R.M.; Drabkin, H.A. Epigenetic Regulation of the Epithelial to Mesenchymal Transition in Lung Cancer. *Cancers* **2017**, *9*. [CrossRef] [PubMed]
15. Fu, Z.; Wen, D. The Emerging Role of Polo-Like Kinase 1 in Epithelial-Mesenchymal Transition and Tumor Metastasis. *Cancers* **2017**, *9*. [CrossRef] [PubMed]
16. Duchemin-Pelletier, E.; Baulard, M.; Spreux, E.; Prioux, M.; Burute, M.; Mograbi, B.; Guyon, L.; Théry, M.; Cochet, C.; Filhol, O. Stem Cell-Like Properties of CK2β-down Regulated Mammary Cells. *Cancers* **2017**, *9*. [CrossRef]
17. Blackwell, R.H.; Foreman, K.E.; Gupta, G.N. The Role of Cancer-Derived Exosomes in Tumorigenicity & Epithelial-to-Mesenchymal Transition. *Cancers* **2017**, *9*. [CrossRef]
18. Lu, J.; Shenoy, A.K. Epithelial-to-Pericyte Transition in Cancer. *Cancers* **2017**, *9*. [CrossRef] [PubMed]
19. Forte, E.; Chimenti, I.; Rosa, P.; Angelini, F.; Pagano, F.; Calogero, A.; Giacomello, A.; Messina, E. EMT/MET at the Crossroad of Stemness, Regeneration and Oncogenesis: The Ying-Yang Equilibrium Recapitulated in Cell Spheroids. *Cancers* **2017**, *9*. [CrossRef] [PubMed]
20. Jia, D.; Jolly, M.K.; Kulkarni, P.; Levine, H. Phenotypic Plasticity and Cell Fate Decisions in Cancer: Insights from Dynamical Systems Theory. *Cancers* **2017**, *9*. [CrossRef] [PubMed]

Review

Epithelial-to-Mesenchymal Transition in the Pathogenesis and Therapy of Head and Neck Cancer

Julia Thierauf [1,*], Johannes Adrian Veit [2] and Jochen Hess [1,3]

[1] Section Experimental and Translational Head and Neck Oncology, Department of Otorhinolaryngology, Head and Neck Surgery, University Medical Center Heidelberg, Im Neuenheimer Feld 400, Heidelberg 69120, Germany; j.hess@dkfz-heidelberg.de

[2] Department of Otorhinolaryngology, Head and Neck Surgery, University Medical Center Ulm, Ulm 89075, Germany; johannes.veit@uniklinik-ulm.de

[3] Research Group Molecular Mechanisms of Head and Neck Tumors, German Cancer Research Center (DKFZ), Heidelberg 69120, Germany

* Correspondence: JuliaCara.Thierauf@med.uniklinik-heidelberg.de; Tel.: +49-6221 566757

Academic Editor: Joëlle Roche
Received: 6 June 2017; Accepted: 30 June 2017; Published: 3 July 2017

Abstract: Head and neck cancer (HNC) is one of the most prevalent human malignancies worldwide, with a high morbidity and mortality. Implementation of interdisciplinary treatment modalities has improved the quality of life, but only minor changes in overall survival have been achieved over the past decades. Main causes for treatment failure are an aggressive and invasive tumor growth in combination with a high degree of intrinsic or acquired treatment resistance. A subset of tumor cells gain these properties during malignant progression by reactivating a complex program of epithelia-to-mesenchymal transition (EMT), which is integral in embryonic development, wound healing, and stem cell behavior. EMT is mediated by a core set of key transcription factors, which are under the control of a large range of developmental signals and extracellular cues. Unraveling molecular principles that drive EMT provides new concepts to better understand tumor cell plasticity and response to established as well as new treatment modalities, and has the potential to identify new drug targets for a more effective, less toxic, and individualized therapy of HNC patients. Here, we review the most recent findings on the clinical relevance of a mesenchymal-like phenotype for HNC patients, including more rare cases of mucosal melanoma and adenoid cystic carcinoma.

Keywords: epithelial-to-mesenchymal transition; mesenchymal-to-epithelial transition; head and neck cancer; biomarkers

1. Head and Neck Cancer

Head and neck cancer (HNC) originates from the mucosal epithelia of the upper aero-digestive tract, and in the majority of cases is diagnosed as head and neck squamous cell carcinoma (HNSCC) of the oral cavity, the naso-, oro- or hypopharynx, the larynx, the paranasal sinuses, or nasal cavity [1]. Apart from tobacco and alcohol abuse, infection with high-risk types of human papilloma virus (HPV—in particular HPV16—has been established as an additional risk factor with a rising incidence in oropharyngeal SCC (OPSCC) [2]. HPV-related OPSCCs represent a distinct tumor entity with regard to cellular and molecular features and exhibit a favorable survival as compared to their HPV-negative counterparts [3].

Despite the implementation of interdisciplinary treatment modalities and improvements in early detection, surgical techniques, radiation therapy protocols, and chemotherapeutic regimes, the overall survival of advanced HNSCC has only marginally improved over the past decades and appropriate treatment remains a major challenge [4]. This is mainly due to the aggressive and invasive growth pattern as well as high resistance against available therapies, leading to loco-regional relapse and/or distant metastasis [5].

Mucosal melanoma (MM) and adenoid cystic carcinoma (ACC) of salivary glands belong to rare cases of head and neck cancer. MM originates from melanocytes of mucosal epithelia and makes up to 1% of all melanomas [6]. However, more than half of all cases can be found in the head and neck region (MMHN). There is compelling evidence highlighting molecular and clinical differences between cutaneous and mucosal melanoma in terms of tumor growth, metastasis, and pathogenesis [7]. MMHN is characterized by an infiltrative and local destructive growth pattern, and overall survival in these patients is more often limited by local recurrences rather than by distant metastases [8].

ACC originates from epithelial cells of the major or minor salivary glands of the head and neck, and is the second most common cancer of the salivary glands [9]. ACC is a neurotropic tumor with an infiltrative growth pattern preferentially along nerve fibers and a high tendency to local spread and early distant metastases, limiting therapeutic options with a curative intent of treatment [9].

2. Epithelial-to-Mesenchymal Transition in Cancer

Epithelial-to-mesenchymal transition (EMT) is a complex process in which epithelial cells lose their characteristic features and acquire a mesenchymal-like phenotype [10]. Phenotypic hallmarks of EMT are the loss of cell-cell junctions, loss of apical-basal polarity, and acquisition of migratory and invasive properties (Figure 1). It resembles a fundamental process in embryonic development and during wound healing. In a pathophysiologic context, cancer cells are able to reactivate the EMT program to gain new properties such as accelerated motility and treatment resistance [11].

A large range of developmental and growth factor signals can drive EMT by triggering genetic and epigenetic programs, which are under the control of or regulate a core set of transcription factors (EMT-TF) belonging to different families, including SNAI1/2, TWIST1, and ZEB1/2, among others (Figure 1) [12]. In the past, most studies relied on morphological alterations complemented by the detection of epithelial (e.g., E-cadherin) and mesenchymal markers (e.g., vimentin, fibronectin, N-cadherin) to assess an epithelial or a mesenchymal state as a simple binary decision (Table 1). However, more recent evidence has advanced and broadened the definition of EMT as a program with dynamic transitional states characterized by metastable intermediates (Figure 1) [13]. This concept is in line with the high degree of tumor cell plasticity, which has been described for most tumor entities (including HNC), and might be the main driver of intrinsic or acquired resistance to established treatment regimens like platinum-based chemotherapy or radiation.

Epithelial plasticity and mesenchymal conversion is often seen in cancer cells as they leave the primary tumor and disseminate to other parts of the body to colonize distant organs and form metastases, which is responsible for the vast majority of cancer-related deaths [13]. However, EMT-related invasion in combination with tumor cell dissemination is initiating, but is not sufficient for completion of the metastatic cascade [14]. For metastatic colonization, a reversal of EMT known as mesenchymal-to-epithelial transition (MET) supports tumor cell expansion as an important prerequisite for metastatic growth (Figure 1). In contrast to EMT, extracellular signals or cell intrinsic programs involved in the induction of MET have not been well characterized [13]. It is also worth noting that tumor cell dissemination and metastatic colonization does not exclusively rely on changes in cell identity related to the EMT program and its reversal, and other scenarios are also reasonable [15].

Table 1. Expression of different molecules and their participation in biological behavior and epithelial-to-mesenchymal transition (EMT). HNSCC: Head and neck squamous cell carcinoma.

Established EMT-Markers	Biological Behavior	Role in EMT/Tumorigenesis	Reference
Vimentin	Type III intermediate filament that is found in mesenchymal cells of various types	Marker of cells undergoing an epithelial-to-mesenchymal transition (EMT) during both normal development and metastatic progression	[16]
E-cadherin	Protein encoded by the CDH1 gene, also been designated as CD324, tumor suppressor gene	Loss of E-cadherin function/expression is implicated in cancer progression/metastasis, downregulation decreases the strength of cellular adhesion, resulting in an increase in cellular motility	[17]
N-cadherin	In embryogenesis, N-cadherin is the key molecule during gastrulation and neural crest development	Promotes tumor cell survival, migration, and invasion, and high levels are often associated with poor prognosis	[18]
Fibronectin	Many different cells are capable of incorporating plasma fibronectin into their extracellular matrix of any tissue	Cancer-associated fibroblasts (CAFs) are essential sources of increased extracellular matrix deposition and altered remodeling to pave the way for cancer cell invasion.	[19]
SNAI1/2	Snail superfamily of zinc-finger transcription factors, involved in cell differentiation and survival. Snail1: essential for gastrulation. Snail2: embryonic development	Snail1: Common sign of poor prognosis in metastatic cancer, and tumors with elevated Snail1 expression show high rates of treatment failure Snail2: Tumor metastasis promotes EMT through activation of SNAIL2 in HNSCC	[20]
TWIST1	Helix-loop-helix transcription factor, plays an essential and pivotal role in multiple stages of embryonic development	Promotes the formation of cancer stem cells and EMT, targeting TWIST1-related molecules significantly inhibits tumor growth and thus improves the survival of cancer patients	[21]
ZEB1/2	Zinc finger E-box binding homeobox 1/2, acts as transcriptional repressor	Dual role: (1) repressor for epithelial genes. (2) a transcriptional activator when associated with YAP (Hippo Pathway); also known to induce EMT in various cancers, but has also been linked to promote treatment failure in an EMT-independent manner	[22]

Markers Associated with EMT	Biological Behavior	Role in EMT/Tumorigenesis	Reference
PD-L1 and PD-1	PD-L1: cluster of differentiation 274(CD274) or B7 homolog 1 (B7-H1), 40kDa type I transmembrane protein, plays a role in suppressing the immune system during pregnancy, tissue allografts, autoimmune disease, and others, ligand of programmed cell death protein-1 (PD-1). PD-1: Cell surface receptor that plays a role in promoting self-tolerance by suppressing T cell inflammatory activity	Many tumor cells express PD-L; inhibition of the interaction between PD-1 and PD-L1 can enhance T-cell responses in vitro and mediate preclinical antitumor activity. This is known as immune checkpoint blockade	[23]
SOX2	Pluripotency-associated transcription factor SOX2 (sex determining region Y-box 2), essential during mammalian embryogenesis, adult tissue regeneration, and homeostasis	Identified as a lineage-survival oncogene in lung and esophageal SCC and recurrent copy number gain of chromosome 3q26, the gene locus encoding SOX2 represents a frequent alteration in HNSCC	[24]
KLK6	Kallikrein-related peptidase 6 (KLK6), Family of 15 secreted serine proteases with trypsin or chymotrypsin-like activity, encoded by a cluster of genes located on chromosome 19q13.3–13.4	Common feature for many human cancers, promising biomarker for early diagnosis or unfavorable prognosis. KLK6 can degrade components of the extracellular matrix and is implicated in tissue remodeling and induction of tumor-relevant processes such as proliferation, migration, and invasion	[25]
BMI1	BMI1 (B lymphoma Mo-MLV insertion region 1 homolog) has been reported as an oncogene by regulating p16 and p19	BMI1 deregulation is associated with enhanced migration, invasion, and poor prognosis in salivary adenoid cystic carcinoma	[26]
BDNF	Brain-derived neurotrophic factor (BDNF) acts on certain neurons of the central and the peripheral nervous system, helping to support the survival of existing neurons and encourage the growth and differentiation of new neurons and synapses	Elevated expression of the brain-derived neurotrophic factor (BDNF) and its receptor NTRK2 together with reduced E-cadherin expression is a common feature of salivary adenoid cystic carcinoma (ACC) and significantly correlated with invasion, metastasis, and poor prognosis of ACC	[27]
NTRK2	Receptor tyrosine kinase involved in the development and maturation of the central and the peripheral nervous systems through regulation of neuron survival, proliferation, migration, differentiation, and synapse formation and plasticity	NTRK2 levels are positively correlated with expression of the EMT-related protein S100A4 but negatively associated with E-cadherin levels	[27]

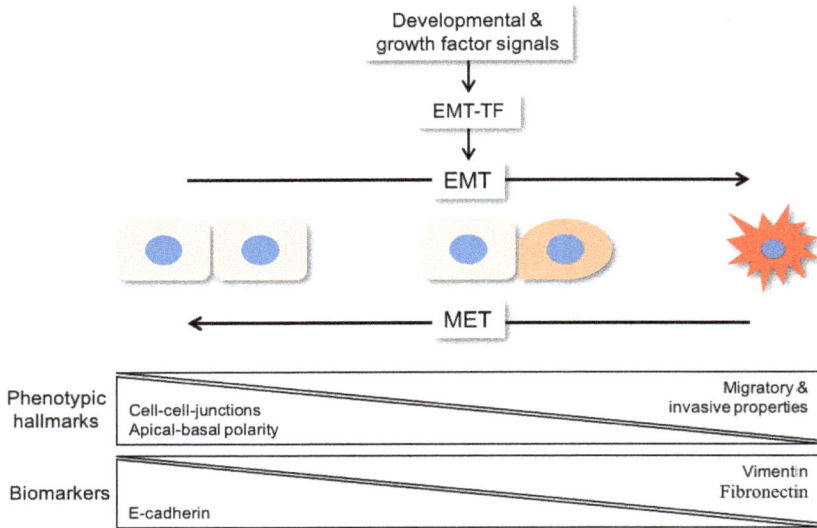

Figure 1. Scheme of EMT (epithelial-to-mesenchymal transition) as a program with dynamic transitional states, which are characterized by metastable intermediates. For metastatic colonization, a reversal of EMT known as mesenchymal-to-epithelial transition (MET) supports tumor cell expansion as an important prerequisite for metastatic growth. Phenotypic hallmarks of EMT are loss of cell-cell junctions, loss of apical-basal polarity, and acquisition of migratory and invasive properties. Those changes are induced by loss of E-cadherin and increased levels of biomarkers like vimentin and fibronectin. EMT-TF: EMT-transcription factor.

3. Epithelial-to-Mesenchymal Transition in HNSCC

The expression of well-established markers and key regulators of EMT as well as their potential impact on the pathogenesis and treatment of HNSCC has been reviewed in several recent publications [28]. Further support for the clinical relevance of EMT has been provided by global expression profiling studies, which confirmed the existence of a distinct patient subgroup with a mesenchymal-like gene expression signature in independent HNSCC cohorts [29]. As an example, Keck et al. identified HPV-related and non-HPV-related subgroups with a prominent immune and mesenchymal phenotype [30]. This inflamed/mesenchymal subtype was characterized by a prominent tumor infiltration with cytotoxic T-lymphocytes independent of the HPV status, suggesting a common mechanism to evade immune surveillance by activation of the EMT program. Numerous mechanisms by which cancer cells can escape attack by the immune system have been postulated, and several key regulators of the immune checkpoint control have been identified [31]. The availability of immune checkpoint inhibitors, such as antibodies against PD-L1 (programmed cell death ligand-1), PD-1, and CTLA-4, provides the unique opportunity to revolutionize the treatment of HNSCC [32]. In this context, it is also noteworthy that an increasing body of studies has demonstrated an association between EMT and elevated expression of PD-L1, but also other key regulators of immune checkpoint control in distinct tumor entities, including HNSCC [33]. Consequently, the identification of additional key regulators driving the EMT program could not only pave the way for the establishment of new therapeutic strategies to prevent tumor cell dissemination, but could also improve our understanding of molecular principles modulating the immune checkpoint control.

4. The Transcription Factor SOX2

The pluripotency-associated transcription factor SOX2 (sex determining region Y-box 2) is essential during mammalian embryogenesis, adult tissue regeneration, and homeostasis [24]. More recently, SOX2 has been identified as a lineage-survival oncogene in lung and esophageal SCC [34] and recurrent copy number gain of chromosome 3q26; the gene locus encoding SOX2 represents a frequent alteration in HNSCC [35]. As a large body of published studies demonstrates a correlation between high expression levels and poor prognosis, an elevated incidence of recurrence, and/or invasive and metastatic capacity of tumor cells, SOX2 has been considered as a potential therapeutic drug target [36]. However, high SOX2 expression is not uniformly an unfavorable risk factor for survival, and in gastric cancer as well as SCC of the lung and head and neck region, low levels have been reported to correlate with a higher risk for lymphatic metastasis and poor prognosis [37]. These controversial data indicate a high level of context dependency concerning the mode of action during carcinogenesis and the impact of SOX2 on clinical outcome.

In a recent study, global gene expression profiling unraveled a pattern of differentially expressed genes after SOX2 silencing in HNSCC cell lines with 3q amplification, which are related to cell motility, regulation of locomotion, and response to wounding [37]. As described above, these biological processes resemble the activation of EMT, indicating that SOX2 activity stabilizes the epithelial phenotype and prevents mesenchymal conversion of HNSCC cells. In line with this assumption, several well-established mesenchymal marker genes (e.g., VIM and FN1) were up-regulated after SOX2 silencing and are also elevated in primary HNSCC with low SOX2 expression according to public available data from TCGA (The Cancer Genome Atlas; Figure 2A). Moreover, 3q amplification and SOX2 copy number gain is a rare event in the mesenchymal-like subgroup of HNSCC patients [30].

It is also worth noting that for the initiation of somatic reprogramming of fibroblasts into induced pluripotent stem cells, the induction of MET is triggered by SOX2 in a complex with OCT4 [38]. Moreover, the secreted protein CTGF (connective tissue growth factor) has been shown to promote MET in HNSCC cells by inducing c-Jun—a component of the AP-1 transcription factor which activates the transcription of SOX2 and OCT4 [39].

In summary, these data support a model in which high expression of SOX2—as a consequence of copy number variation—but also other so-far less well characterized mechanisms contribute to the pathogenesis of HNSCC by promoting tumor cell proliferation and survival. However, its presence in advanced tumor stages might interfere with tumor cell plasticity and activation of mesenchymal transition by stabilization of the epithelial phenotype, including stemness-like traits (Figure 2C). Although cancer cells with reduced or no SOX2 expression might be more sensitive to systemic treatment with cytotoxic drugs, the mesenchymal-like phenotype potentiates their motility and invasive capacity, which makes an escape from local treatment by surgical resection or radiotherapy more likely. Future studies will be required: (i) to confirm this concept, (ii) to unravel relevant SOX2-related signaling and gene regulatory networks, and (iii) to address the question of whether detection of SOX2-negative tumor cells at the invasive front of primary HNSCC predicts the risk for treatment failure and subsequent loco-regional and/or distant relapse.

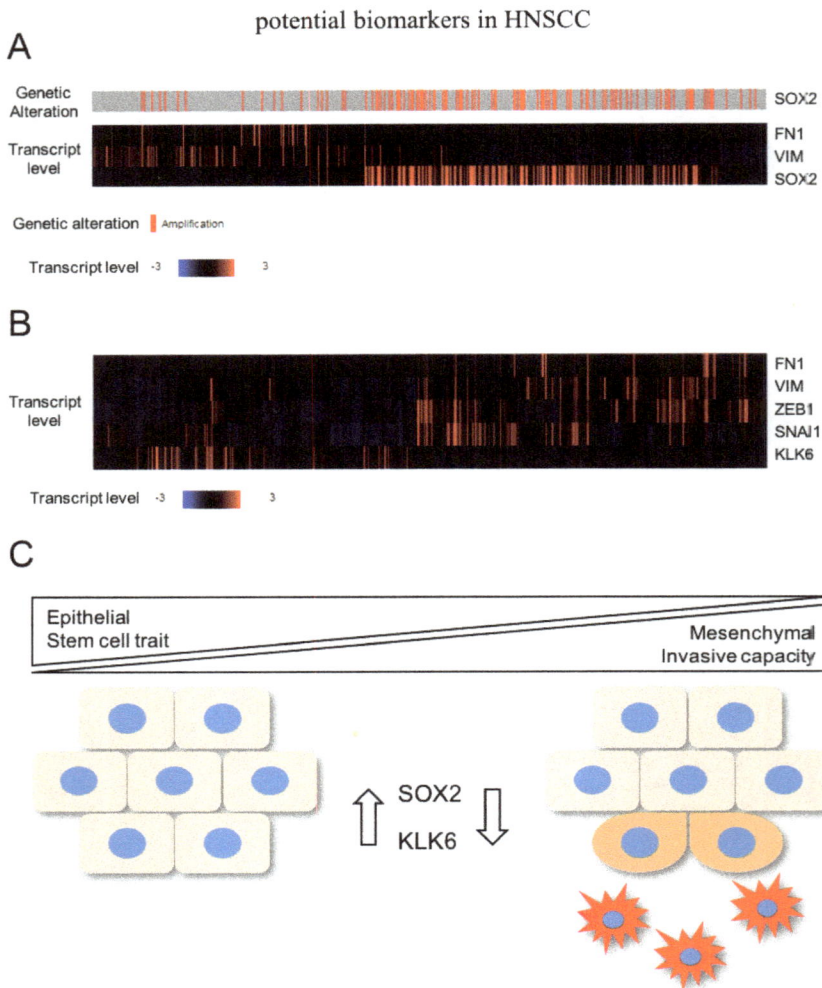

Figure 2. (A) Several well-established mesenchymal marker genes (e.g., VIM and FN1) are elevated in primary HNSCC with low SOX2 expression according to TCGA (The Cancer Genome Atlas; https://cancergenome.nih.gov) (B) Inverse expression pattern between KLK6 and mesenchymal markers as well as key regulators of EMT in primary HNSCC of the TCGA cohort; (C) High expression of SOX2 contributes—among others—to the pathogenesis of HNSCC by promoting tumor cell proliferation. In advanced tumor stages, SOX2 might interfere with tumor cell plasticity and activation of mesenchymal transition by stabilization of the epithelial phenotype including stemness-like traits. KLK6 (kallikrein-related peptidase 6).

5. The Kallikrein-Related Peptidase 6

The kallikrein-related peptidase 6 (KLK6) belongs to a family of 15 secreted serine proteases with trypsin or chymotrypsin-like activity, which are encoded by a cluster of genes located on human chromosome 19q13.3–13.4 [25]. Aberrant KLK6 expression is a common feature for many human cancers, and numerous studies evaluated KLK6 as a promising biomarker for early diagnosis or unfavorable prognosis [40]. KLK6 can degrade components of the extracellular matrix, and is implicated in tissue remodeling and the induction of tumor-relevant processes such as proliferation, migration, and invasion [41]. Despite numerous studies pointing to a critical role of KLK6 during neoplastic transformation and malignant progression, more recent data questioned its general tumor-promoting role and stressed the importance of considering its context-dependent function, as exemplified in breast and renal cancer [42].

In the context of HNSCC, a recent study provided experimental evidence that silencing of KLK6 activates the EMT program accompanied by a mesenchymal-like cell morphology as well as accelerated tumor cell migration and invasion [43]. In line with these findings, there is an inverse expression pattern between KLK6 and mesenchymal markers as well as key regulators of EMT in primary HNSCC of the TCGA cohort (Figure 2B). The clinical relevance of these findings was evident by the fact that low KLK6 protein levels in primary HNSCC serve as an unfavorable risk factor for progression-free and overall survival [43]. A critical role of KLK6 in regulating the transition between epithelial and mesenchymal phenotypes was already postulated by Pampalakis et al. [42] They demonstrated that KLK6 acts as a suppressor of tumor progression by promoting MET in breast cancer cell lines, suggesting common mechanisms of KLK6 function in breast cancer and HNSCC cells. However, the underlying mode of action and putative proteolytic downstream targets of KLK6 implicated in EMT and tumor cell plasticity remain to be elucidated.

Concerning the regulation of KLK6 expression, tumor-specific loss in breast cancer is at least in part mediated by epigenetic silencing due to DNA methylation [42]. It will be interesting to address the question of whether a similar mode of regulation also occurs in primary HNSCC with low KLK6 expression. As a consequence, inhibitors of DNA methyltransferases could be used as promising drugs to restore KLK6 expression in order to reverse the EMT program and prevent tumor cell plasticity and dissemination [44].

6. EMT-Like Phenotype in Mucosal Melanoma of the Head and Neck

Melanoma is an aggressive tumor arising from melanocytes, endowed with unique features of cellular plasticity. The high degree of phenotypic and functional diversity—which is also a characteristic trait of melanoma cells—is at least in part due to their capacity to reversibly switch between phenotypes with non-invasive and invasive potentials, and is driven by oncogenic signaling and environmental cues [45].

As melanocytes do not express a classic epithelial phenotype, the term EMT cannot be formally attributed to the progression or metastatic spread of melanoma. However, melanocytes exhibit a stable differentiated state via E-cadherin-mediated communication with keratinocytes, while melanoma cells progressively lose E-cadherin in favor of N-cadherin expression [46]. In addition, altered expression of EMT-TFs have been reported in numerous studies, but only recently has the mode of action—how these factors are regulated and orchestrate cellular plasticity in a non-epithelial context (e.g., melanoma)—been addressed in more detail [45]. As an example, Caramel et al. provided a comprehensive overview of the regulatory network of EMT-TFs in cutaneous melanoma with links to oncogenic transformation and tumor cell plasticity [47]. They confirmed prominent SNAIL2 and ZEB2 expression in normal melanocytes, and postulated that both act as tumor-suppressor proteins by activating an MITF-dependent (MITF = Microphthalmia-associated transcription factor) melanocyte differentiation program. However, upon activation of MEK-ERK signaling (e.g., as a consequence of oncogenic NRAS or BRAF mutations), the EMT-TF network undergoes a profound reorganization in favor of TWIST1 and ZEB1. This switch results in E-cadherin loss, enhanced invasion, and constitutes an independent factor of poor prognosis in patients with cutaneous melanoma [48].

Due to the small number of cases of mucosal melanoma patients, there is limited information available on the pathogenesis of this aggressive tumor entity that tends to form local recurrences and regional lymph node metastases in 21%, but a significantly lower number of distant metastases than cutaneous melanoma. Furthermore, there is neither a commercially available cell line for mucosal melanoma nor an established in vivo model for further analysis. Apart from the postulation by Hussein et al. and Dupin et al. that mucosal melanomas arise from melanocytes migrating to non-cutaneous organs after neural crest cells undergo EMT, reliable studies on the expression of classical EMT markers or EMT-TFs in mucosal melanoma are still missing [48].

With regard to the newly identified factors related to an EMT-like phenotype in HNSCC, the expression of KLK6 in cutaneous melanoma was analyzed by Krenzer et al. [41]. Although KLK6 was not detectable in cutaneous melanoma cells, a strong KLK6 protein expression was found in keratinocytes and stromal cells located adjacent to benign nevi, primary melanomas, and cutaneous metastatic lesions. These data suggested a paracrine function of extracellular KLK6 during neoplastic transformation and malignant progression. In our previous work on mucosal melanoma, we characterized MMHN for the expression of KLK6. Paraffin-embedded MMHN of 22 patients were analyzed by immunohistochemical staining, and results were correlated with clinical and pathological data. A positive KLK6 staining was observed in 77.3% (17/22) of MMHN cases, and in line with the situation in HNSCC, a high pattern was significantly correlated with favorable outcome concerning local recurrence-free survival ($p = 0.013$) [7]. The same cohort was used to analyze patterns of PD-L1 expression in MMHN. Interestingly, only 13% (3/23) of mucosal melanoma showed PD-L1 expression, while prominent PD-L1 staining was detected in 100% of tissue sections from a control group of cutaneous melanoma (n = 9) [49]. PD-L1 expression in mucosal melanoma was not correlated with age, sex, nor anatomical localization of the tumor. However, patients with PD-L1-positive mucosal melanoma had a significantly longer recurrence-free survival ($p = 0.026$).

7. EMT-Like Phenotype in Salivary Gland Malignancies

Common carcinomas of the salivary glands are classified as adenocarcinoma, mucoepidermoid carcinoma, and adenoid cystic carcinoma (ACC). ACC represents one of the most common malignancies of the salivary glands, and three distinct growth patterns have been identified, which differ in their clinical behavior. The cribriform and the tubular type are usually associated with a better clinical outcome as compared to the solid type of ACC [50]. It is also the most aggressive salivary gland tumor in terms of treatment failure, with a high rate of distant metastases and common perineural invasion [9]. Similar to mucosal melanoma, the availability of reliable preclinical models is limited [51], and most published studies on expression and regulation of EMT-related proteins rely on retrospective studies with FFPE (Formalin-fixed paraffin-embedded) tissue samples. However, reduced expression of E-cadherin in salivary ACCs could be described as compared to paraneoplastic normal salivary tissue [52]. SNAI2 and E-cadherin expression were negatively associated in a cohort of 115 salivary ACC. High SNAI2 and low E-cadherin expression were significantly correlated with perineural invasion [52]. Yi et al. investigated the expression of BMI1—a major component of the polycomb group complex 1 and a candidate stem cell marker—together with EMT-related proteins SNAI1, SNAI2, and E-cadherin in a cohort of 102 ACC patients [53]. They identified a positive correlation between high BMI1 levels and SNAI1 and SNAI2 overexpression. Moreover, high BMI1 levels indicated an unfavorable metastasis-free survival and served as a high-risk marker for salivary ACC. The association between deregulated BMI1 expression and clinical outcome was also observed in an independent study [26]. Elevated expression of the brain-derived neurotrophic factor (BDNF) and its receptor NTRK2 together with reduced E-cadherin expression is a common feature of salivary ACC and significantly correlated with invasion, metastasis, and poor prognosis of ACC patients [27]. The crucial role of the BDNF/NTRK2 axis was further supported by a more recent study demonstrating that NTRK2 levels are positively correlated with expression of the EMT-related protein S100A4 but negatively associated with E-cadherin levels [54]. Both NTRK2 and S100A4 were positively associated with perineural invasion, indicating that specific targeting of the BDNF/NTRK2 axis might represent a promising new therapeutic strategy for ACC patients.

8. Conclusions and Outlook

Tumor cell dissemination as enabled by EMT and followed by MET has been considered as a hallmark of metastasis [55]. However, alternative modes of dissemination, such as collective or cluster-based migration and invasion can exist and may not even exhibit an overt up-regulation of mesenchymal markers. Furthermore, disseminated cancer cells may undergo metastatic colonization via an MET-independent pathway. Together, the wealth of data acquired thus far support a more nuanced view of the role of EMT and MET in cancer metastasis. While in some cases these programs are critically important, in other scenarios EMT and MET may not play an important role, but more of permissive and potentially catalytic roles by regulating phenotypes that accelerate the processes necessary to escape and colonize [14]. Moreover, recent reports investigating circulating cancer cells in the bloodstream or employing genetic lineage-tracing have questioned a critical role of an EMT in metastasis formation. Hence, we need to better understand the molecular networks underlying the cell plasticity conferred by an EMT or a MET and its functional contribution to malignant tumor progression. Although the exact mechanisms of EMT remain complex, certain aspects and connections have become more evident. The loss of KLK6 in HNSCC seems to create a mesenchymal-like morphology and accelerated motility of tumor cells with an EMT-phenotype, identified by loss of E-cadherin and prominent induction of vimentin expression in HNSCC cells. The clinical relevance of these findings is supported by the fact that low KLK6 protein level in primary HNSCCs serves as an unfavorable risk factor for progression-free and overall survival [43]. Additionally, it has been shown that not only the loss of KLK6 but also the loss of SOX2 expression induces cell motility via vimentin up-regulation and is an unfavorable risk factor for survival of head and neck squamous cell carcinoma [37]. Low SOX2 expression seems to be an unfavorable risk factor for poor clinical outcome, and serves as a prognostic marker to identify HNSCC patients with high risk for treatment failure due to an invasive phenotype. On the other hand, KLK6 in MMHN seems to be highly overexpressed, but also correlated with a better clinical outcome [7]. Although the accuracy of this statement must be questioned given the small number of MMHN cohorts, the protective role of KLK6 has been described before in other tumor entities such as breast cancer, and therefore seems to be highly dependent on the microenvironment and the tumor entity. In contrast to previous results in HNSCC, SOX2 seems to promote invasion and tumor progression by creating a more mesenchymal phenotype in ACCs [56]. These controversial findings of well-described mesenchymal and epithelial markers as well as new mediators of EMT in three tumor entities with different progressive behavior underlines that the exact molecular mechanisms underlying the cross-linking between various EMT pathways in different HNCs remains to be fully elucidated. The main difficulty seems to be the quantification of "partial EMT" in each disease state and capturing these dynamic states. Partial EMT phenotypes reflect the enhanced plasticity of tumors and are important for understanding the progression of EMT and MET processes. However, this "moving target" presents a major challenge in rational drug design. Nevertheless, with fresh knowledge and the benefit of hindsight, certain principles have emerged. The exploration of EMT and MET mechanisms is destined to pave the way for future clinical implications. EMT markers could not only serve as biomarkers for chemotherapy response and survival prognosis, but also represent a targetable process, aiming to prevent metastasis and overcoming drug resistance. In different solid cancers, EMT has been linked to a generation of cancer cells with stem cell attributes of tumor initiation and resistance to chemotherapy [57]. In clinical tumor samples, EMT has been implicated with the acquisition of drug resistance. Moreover, the inhibition of EMT signal cascades by specifically tailored drugs has been tested in vivo. As a result, miR-506 sensitized EOC (epithelial ovarian cancer cells) to chemotherapy and inhibited EMT-mediated metastasis [58]. The group of Chiu et al. described FOXM1 as a critical regulator of an EMT, and were able to show that the combination of a FOXM1 inhibitor and cisplatin led to increased expression of EMT-related markers in chemoresistant cells [59]. In contrast to classical chemotherapy, these experimental approaches towards targeting EMT cascades in cancer aim to diminish or even abolish metastasis formation without doing harm to healthy cells.

In summary, we are facing an urgent need to characterize the distinct cellular states associated with EMT plasticity to identify targetable EMT components and create therapeutic strategies to effectively eliminate cells undergoing EMT, since this is ultimately leading to treatment failure. However, by further analyzing the already existing results and investigating the pathogenesis of EMT in different head and neck cancers, the knowledge on cellular and molecular principles of malignant progression can be improved, ultimately aiming to stratify cancer patients at high risk for poor therapy response and to provide applicable therapies targeting EMT processes. By exploring the complex underlying mechanisms which lead from tumor cell dissociation in the primary tumor through EMT towards the formation of metastasis (including MET), our findings might offer promising perspectives for the future.

Acknowledgments: We acknowledge funding of our research program by the German Cancer Aid (70112000 to JH), the German Research Foundation (HE5760/3-1 to JH), and Stiftung Tumorforschung Kopf-Hals (to JT).

Conflicts of Interest: The authors declare no conflict of interest.

References

1. Pai, S.I.; Westra, W.H. Molecular pathology of head and neck cancer: implications for diagnosis, prognosis, and treatment. *Annu. Rev. Pathol. Mech. Dis.* **2009**, *4*, 49–70. [CrossRef] [PubMed]
2. Gillison, M.L.; Chaturvedi, A.K.; Anderson, W.F.; Fakhry, C. Epidemiology of human papillomavirus—Positive head and neck squamous cell carcinoma. *J. Clin. Oncol.* **2015**, *33*, 3235–3242. [CrossRef] [PubMed]
3. Ang, K.K.; Harris, J.; Wheeler, R.; Weber, R.; Rosenthal, D.I.; Nguyen-Tân, P.F.; Westra, W.H.; Chung, C.H.; Jordan, R.C.; Lu, C.; et al. Human papillomavirus and survival of patients with oropharyngeal cancer. *N. Engl. J. Med.* **2010**, *363*, 24–35. [CrossRef] [PubMed]
4. Siegel, R.L.; Miller, K.D.; Jemal, A. Cancer statistics, 2016. *CA. Cancer J. Clin.* **2016**, *66*, 7–30. [CrossRef] [PubMed]
5. Agrawal, A.; Hammond, T.H.; Young, G.S.; Avon, A.L.; Ozer, E.; Schuller, D.E. Factors affecting long-term survival in patients with recurrent head and neck cancer may help define the role of post-treatment surveillance. *Laryngoscope* **2009**, *119*, 2135–2140. [CrossRef] [PubMed]
6. Prasad, M.L.; Jungbluth, A.A.; Patel, S.G.; Iversen, K.; Hoshaw-Woodard, S.; Busam, K.J. Expression and significance of cancer testis antigens in primary mucosal melanoma of the head and neck. *Head Neck* **2004**, *26*, 1053–1057. [CrossRef] [PubMed]
7. Thierauf, J.; Veit, J.A.; Lennerz, J.K.; Weissinger, S.E.; Affolter, A.; Döscher, J.; Bergmann, C.; Knopf, A.; Grünow, J.; Grünmüller, L.; et al. Expression of kallikrein-related peptidase 6 in primary mucosal malignant melanoma of the head and neck. *Head Neck Pathol.* **2016**. [CrossRef] [PubMed]
8. Thierauf, J.; Veit, J.; Döscher, J.; Theodoraki, M.-N.; Greve, J.; Hoffmann, T. Schleimhautmelanome des Kopf-Hals-Bereichs. *Laryngorhinootologie* **2015**, *94*, 812–818. [CrossRef] [PubMed]
9. Coca-Pelaz, A.; Rodrigo, J.P.; Bradley, P.J.; Vander Poorten, V.; Triantafyllou, A.; Hunt, J.L.; Strojan, P.; Rinaldo, A.; Haigentz, M.; Takes, R.P.; et al. Adenoid cystic carcinoma of the head and neck—An update. *Oral Oncol.* **2015**, *51*, 652–661. [CrossRef] [PubMed]
10. Lamouille, S.; Xu, J.; Derynck, R. Molecular mechanisms of epithelial-mesenchymal transition. *Nat. Rev. Mol. Cell Biol.* **2014**, *15*, 178–196. [CrossRef] [PubMed]
11. Marcucci, F.; Stassi, G.; de Maria, R. Epithelial-mesenchymal transition: A new target in anticancer drug discovery. *Nat. Rev. Drug Discov.* **2016**, *15*, 311–325. [CrossRef] [PubMed]
12. Gonzalez, D.M.; Medici, D. Signaling mechanisms of the epithelial-mesenchymal transition. *Sci. Signal.* **2014**, *7*, re8. [CrossRef] [PubMed]
13. Nieto, M.A. Epithelial plasticity: A common theme in embryonic and cancer cells. *Science* **2013**, *342*. [CrossRef] [PubMed]
14. Diepenbruck, M.; Christofori, G. Epithelial-mesenchymal transition (EMT) and metastasis: yes, no, maybe? *Curr. Opin. Cell Biol.* **2016**, *43*, 7–13. [CrossRef] [PubMed]
15. Ledford, H. Cancer theory faces doubts. *Nature* **2011**, *472*, 273. [CrossRef] [PubMed]

16. Satelli, A.; Li, S. Vimentin in cancer and its potential as a molecular target for cancer therapy. *Cell Mol. Life Sci.* **2011**, *68*, 3033–3046. [CrossRef] [PubMed]
17. Canel, M.; Serrels, A.; Frame, M.C.; Brunton, V.G. E-cadherin-integrin crosstalk in cancer invasion and metastasis. *J. Cell Sci.* **2013**, *126*, 393–401. [CrossRef] [PubMed]
18. Derycke, L.D.M.; Bracke, M.E. N-cadherin in the spotlight of cell-cell adhesion, differentiation, embryogenesis, invasion and signalling. *Int. J. Dev. Biol.* **2004**, *48*, 463–476. [CrossRef] [PubMed]
19. Wang, K.; Seo, B.R.; Fischbach, C.; Gourdon, D. Fibronectin mechanobiology regulates tumorigenesis. *Cell Mol. Bioeng.* **2016**, *9*, 1–11. [CrossRef] [PubMed]
20. Scanlon, C.S.; van Tubergen, E.A.; Inglehart, R.C.; D'Silva, N.J. Biomarkers of epithelial-mesenchymal transition in squamous cell carcinoma. *J. Dent. Res.* **2013**, *92*, 114–121. [CrossRef] [PubMed]
21. Morris, J.C.; Tan, A.R.; Olencki, T.E.; Shapiro, G.I.; Dezube, B.J.; Reiss, M.; Hsu, F.J.; Berzofsky, J.A.; Lawrence, D.P. Phase I study of GC1008 (Fresolimumab): A human anti-transforming growth factor-beta (tgfβ) monoclonal antibody in patients with advanced malignant melanoma or renal cell carcinoma. *PLoS ONE* **2014**, *9*, e90353. [CrossRef] [PubMed]
22. Smith, B.; Bhowmick, N. Role of EMT in metastasis and therapy resistance. *J. Clin. Med.* **2016**, *5*, 17. [CrossRef] [PubMed]
23. Patel, S.P.; Kurzrock, R. PD-L1 expression as a predictive biomarker in cancer immunotherapy. *Mol. Cancer Ther.* **2015**, *14*, 847–856. [CrossRef] [PubMed]
24. Sarkar, A.; Hochedlinger, K. The sox family of transcription factors: Versatile regulators of stem and progenitor cell fate. *Cell Stem Cell* **2013**, *12*, 15–30. [CrossRef] [PubMed]
25. Borgoño, C.A.; Diamandis, E.P. The emerging roles of human tissue kallikreins in cancer. *Nat. Rev. Cancer* **2004**, *4*, 876–890. [CrossRef] [PubMed]
26. Chang, B.; Li, S.; He, Q.; Liu, Z.; Zhao, L.; Zhao, T.; Wang, A. Deregulation of Bmi-1 is associated with enhanced migration, invasion and poor prognosis in salivary adenoid cystic carcinoma. *Biochim. Biophys. Acta* **2014**, *1840*, 3285–3291. [CrossRef] [PubMed]
27. Jia, S.; Wang, W.; Hu, Z.; Shan, C.; Wang, L.; Wu, B.; Yang, Z.; Yang, X.; Lei, D. BDNF mediated TrkB activation contributes to the EMT progression and the poor prognosis in human salivary adenoid cystic carcinoma. *Oral Oncol.* **2015**. [CrossRef] [PubMed]
28. Natarajan, J.; Chandrashekar, C.; Radhakrishnan, R. Critical biomarkers of epithelial-mesenchymal transition in the head and neck cancers. *J. Cancer Res. Ther.* **2014**, *10*, 512–518. [CrossRef]
29. De Cecco, L.; Nicolau, M.; Giannoccaro, M.; Grazia Daidone, M.; Bossi, P.; Locati, L.; Licitra, L.; Canevari, S. Head and neck cancer subtypes with biological and clinical relevance: Meta-analysis of gene-expression data. *Oncotarget* **2015**, *6*, 9627–9642. [CrossRef] [PubMed]
30. Keck, M.K.; Zuo, Z.; Khattri, A.; Stricker, T.P.; Brown, C.D.; Imanguli, M.; Rieke, D.; Endhardt, K.; Fang, P.; Brägelmann, J.; et al. Integrative analysis of head and neck cancer identifies two biologically distinct HPV and three non-HPV subtypes. *Clin. Cancer Res.* **2015**, *21*, 870–881. [CrossRef] [PubMed]
31. Sharma, P.; Hu-Lieskovan, S.; Wargo, J.A.; Ribas, A. Primary, adaptive, and acquired resistance to cancer immunotherapy. *Cell* **2017**, *168*, 707–723. [CrossRef] [PubMed]
32. Bauml, J.; Seiwert, T.Y.; Pfister, D.G.; Worden, F.; Liu, S.V.; Gilbert, J.; Saba, N.F.; Weiss, J.; Wirth, L.; Sukari, A.; et al. Pembrolizumab for platinum- and cetuximab-refractory head and neck cancer: Results from a single-arm, phase II study. *J. Clin. Oncol.* **2017**, *35*, 1542–1549. [CrossRef] [PubMed]
33. Lee, Y.; Shin, J.H.; Longmire, M.; Wang, H.; Kohrt, H.E.; Chang, H.Y.; Sunwoo, J.B. CD44+ cells in head and neck squamous cell carcinoma suppress T-cell-mediated immunity by selective constitutive and inducible expression of PD-L1. *Clin. Cancer Res.* **2016**, *22*, 3571–3581. [CrossRef] [PubMed]
34. Bass, A.J.; Watanabe, H.; Mermel, C.H.; Yu, S.; Perner, S.; Verhaak, R.G.; Kim, S.Y.; Wardwell, L.; Tamayo, P.; Gat-Viks, I.; et al. SOX2 is an amplified lineage-survival oncogene in lung and esophageal squamous cell carcinomas. *Nat. Genet.* **2009**, *41*, 1238–1242. [CrossRef] [PubMed]
35. The Cancer Genome Atlas Network. Comprehensive genomic characterization of head and neck squamous cell carcinomas. *Nature* **2015**, *517*, 576–582. [CrossRef]
36. Wuebben, E.L.; Rizzino, A. The dark side of SOX2: Cancer—A comprehensive overview. *Oncotarget* **2017**. [CrossRef] [PubMed]

37. Bayo, P.; Jou, A.; Stenzinger, A.; Shao, C.; Gross, M.; Jensen, A.; Grabe, N.; Mende, C.H.; Rados, P.V.; Debus, J.; et al. Loss of SOX2 expression induces cell motility via vimentin up-regulation and is an unfavorable risk factor for survival of head and neck squamous cell carcinoma. *Mol. Oncol.* **2015**, *9*, 1704–1719. [CrossRef] [PubMed]

38. Goding, C.R.; Pei, D.; Lu, X. Cancer: Pathological nuclear reprogramming? *Nat. Rev. Cancer* **2014**, *14*, 568–573. [CrossRef] [PubMed]

39. Hess, J.; Angel, P.; Schorpp-Kistner, M. AP-1 subunits: Quarrel and harmony among siblings. *J. Cell Sci.* **2004**, *117*, 5965–5973. [CrossRef] [PubMed]

40. Prassas, I.; Eissa, A.; Poda, G.; Diamandis, E.P. Unleashing the therapeutic potential of human kallikrein-related serine proteases. *Nat. Rev. Drug Discov.* **2015**, *14*, 183–202. [CrossRef] [PubMed]

41. Krenzer, S.; Peterziel, H.; Mauch, C.; Blaber, S.I.; Blaber, M.; Angel, P.; Hess, J. Expression and function of the kallikrein-related peptidase 6 in the human melanoma microenvironment. *J. Invest. Dermatol.* **2011**, *131*, 2281–2288. [CrossRef] [PubMed]

42. Pampalakis, G.; Prosnikli, E.; Agalioti, T.; Vlahou, A.; Zoumpourlis, V.; Sotiropoulou, G. A tumor-protective role for human kallikrein-related peptidase 6 in breast cancer mediated by inhibition of epithelial-to-mesenchymal transition. *Cancer Res.* **2009**, *69*, 3779–3787. [CrossRef] [PubMed]

43. Schrader, C.; Kolb, M.; Zaoui, K.; Flechtenmacher, C.; Grabe, N.; Weber, K.-J.; Plinkert, P.K.; Heß, J. 10 Kallikrein-related peptidase 6 regulates epithelial-to-mesenchymal transition and serves as prognostic biomarker for head and neck squamous cell carcinoma patients. *Oral Oncol.* **2015**, *51*, e30. [CrossRef]

44. Tam, W.L.; Weinberg, R.A. The epigenetics of epithelial-mesenchymal plasticity in cancer. *Nat. Med.* **2013**, *19*, 1438–1449. [CrossRef] [PubMed]

45. Vandamme, N.; Berx, G. Melanoma cells revive an embryonic transcriptional network to dictate phenotypic heterogeneity. *Front. Oncol.* **2014**, *4*, 352. [CrossRef] [PubMed]

46. Kuphal, S.; Bosserhoff, A.K. E-cadherin cell-cell communication in melanogenesis and during development of malignant melanoma. *Arch. Biochem. Biophys.* **2012**, *524*, 43–47. [CrossRef] [PubMed]

47. Caramel, J.; Papadogeorgakis, E.; Hill, L.; Browne, G.; Richard, G.; Wierinckx, A.; Saldanha, G.; Osborne, J.; Hutchinson, P.; Tse, G.; et al. A switch in the expression of embryonic EMT-inducers drives the development of malignant melanoma. *Cancer Cell* **2013**, *24*, 466–480. [CrossRef] [PubMed]

48. Dupin, E.; Le Douarin, N.M. Development of melanocyte precursors from the vertebrate neural crest. *Oncogene* **2003**, *22*, 3016–3023. [CrossRef] [PubMed]

49. Thierauf, J.; Veit, J.A.; Affolter, A.; Bergmann, C.; Grünow, J.; Laban, S.; Lennerz, J.K.; Grünmüller, L.; Mauch, C.; Plinkert, P.K.; et al. Identification and clinical relevance of PD-L1 expression in primary mucosal malignant melanoma of the head and neck. *Melanoma Res.* **2015**. [CrossRef] [PubMed]

50. Büchsenschütz, K.; Veit, J.; Schuler, P.; Thierauf, J.; Laban, S.; Fahimi, F.; Bankfalvi, A.; Lang, S.; Sauerwein, W.; Hoffmann, T. Molekulare Ansatzpunkte für systemische therapien adenoidzystischer karzinome im Kopf-Hals-Bereich. *Laryngorhinootologie* **2014**, *93*, 657–664. [CrossRef] [PubMed]

51. Phucharoen, J.; Ohta, Y.; Woo, J.M.; Eisele, D.W.; Tetsu, O. Genetic profiling reveals cross-contamination and misidentification of 6 adenoid cystic carcinoma cell lines: ACC2, ACC3, ACCM, ACCNS, ACCS and CAC2. *PLoS ONE* **2009**, *4*, e6040. [CrossRef] [PubMed]

52. Xie, J.; Feng, Y.; Lin, T.; Huang, X.-Y.; Gan, R.-H.; Zhao, Y.; Su, B.-H.; Ding, L.-C.; She, L.; Chen, J.; et al. CDH4 suppresses the progression of salivary adenoid cystic carcinoma via E-cadherin co-expression. *Oncotarget* **2016**, *7*, 82961–82971. [CrossRef] [PubMed]

53. Yi, C.; Li, B.-B.; Zhou, C.-X. Bmi-1 expression predicts prognosis in salivary adenoid cystic carcinoma and correlates with epithelial-mesenchymal transition-related factors. *Ann. Diagn. Pathol.* **2016**, *22*, 38–44. [CrossRef] [PubMed]

54. Shan, C.; Wei, J.; Hou, R.; Wu, B.; Yang, Z.; Wang, L.; Lei, D.; Yang, X. Schwann cells promote EMT and the Schwann-like differentiation of salivary adenoid cystic carcinoma cells via the BDNF/TrkB axis. *Oncol. Rep.* **2015**, *35*, 427–435. [CrossRef] [PubMed]

55. Jolly, M.K.; Ware, K.E.; Gilja, S.; Somarelli, J.A.; Levine, H. EMT and MET: Necessary or permissive for metastasis? *Mol. Oncol.* **2017**. [CrossRef] [PubMed]

56. Dai, W.; Tan, X.; Sun, C.; Zhou, Q. High expression of SOX2 is associated with poor prognosis in patients with salivary gland adenoid cystic carcinoma. *Int. J. Mol. Sci.* **2014**, *15*, 8393–8406. [CrossRef] [PubMed]

57. Mani, S.A.; Guo, W.; Liao, M.-J.; Eaton, E.N.; Ayyanan, A.; Zhou, A.Y.; Brooks, M.; Reinhard, F.; Zhang, C.C.; Shipitsin, M.; et al. The epithelial-mesenchymal transition generates cells with properties of stem cells. *Cell* **2008**, *133*, 704–715. [CrossRef] [PubMed]

58. Sun, Y.; Hu, L.; Zheng, H.; Bagnoli, M.; Guo, Y.; Rupaimoole, R.; Rodriguez-Aguayo, C.; Lopez-Berestein, G.; Ji, P.; Chen, K.; et al. MiR-506 inhibits multiple targets in the epithelial-to-mesenchymal transition network and is associated with good prognosis in epithelial ovarian cancer. *J. Pathol.* **2015**, *235*, 25–36. [CrossRef] [PubMed]

59. Chiu, W.-T.; Huang, Y.-F.; Tsai, H.-Y.; Chen, C.-C.; Chang, C.-H.; Huang, S.-C.; Hsu, K.-F.; Chou, C.-Y. FOXM1 confers to epithelial-mesenchymal transition, stemness and chemoresistance in epithelial ovarian carcinoma cells. *Oncotarget* **2015**, *6*, 2349–2365. [CrossRef] [PubMed]

cancers

Review

The Epithelial-to-Mesenchymal Transition in Breast Cancer: Focus on Basal-Like Carcinomas

Monica Fedele [1,*], Laura Cerchia [1] and Gennaro Chiappetta [2]

[1] CNR—Institute of Experimental Endocrinology and Oncology, 80131 Naples, Italy; cerchia@unina.it
[2] Dipartimento di Ricerca Traslazionale a Supporto dei Percorsi Oncologici, S.C. Genomica Funzionale, Istituto Nazionale Tumori—IRCCS—Fondazione G Pascale, 80131 Naples, Italy; chiappettagennaro@gmail.com
*   Correspondence: mfedele@unina.it; Tel.: +39-081-5455751

Academic Editor: Joëlle Roche
Received: 27 July 2017; Accepted: 28 September 2017; Published: 30 September 2017

Abstract: Breast cancer is a heterogeneous disease that is characterized by a high grade of cell plasticity arising from the contribution of a diverse range of factors. When combined, these factors allow a cancer cell to transition from an epithelial to a mesenchymal state through a process of dedifferentiation that confers stem-like features, including chemoresistance, as well as the capacity to migrate and invade. Understanding the complex events that lead to the acquisition of a mesenchymal phenotype will therefore help to design new therapies against metastatic breast cancer. Here, we recapitulate the main endogenous molecular signals involved in this process, and their cross-talk with paracrine factors. These signals and cross-talk include the extracellular matrix; the secretome of cancer-associated fibroblasts, macrophages, cancer stem cells, and cancer cells; and exosomes with their cargo of miRNAs. Finally, we highlight some of the more promising therapeutic perspectives based on counteracting the epithelial-to-mesenchymal transition in breast cancer cells.

Keywords: breast cancer; TNBC; EMT; tumor plasticity; molecular signaling; exosomes; miRNAs; $\alpha v \beta 3$; differentiation therapy

1. Introduction

Breast cancer is the most common cancer in women worldwide, and the fifth most common cause of death from cancer overall [1]. However, when we talk about breast cancer, as for most human cancers, we are referring to different tumors with respect to histopathological appearance, molecular alterations, presentation, and clinical outcome. According to most recent molecular classifications, breast carcinomas can be divided into at least six subgroups. These include normal-like (expression profile similar to noncancerous breast tissue); luminal A and B (generally estrogen receptor (ER)-positive tumors, with expression of epithelial markers; luminal B shows a higher Ki67 index and worse prognosis compared to luminal A); HER2 positive (overexpressing ERBB2 oncogene); basal-like (expressing basal cytokeratins and other markers characteristic of the myoepithelium of the normal mammary gland); and claudin-low (enriched in epithelial-to-mesenchymal transition (EMT) features, immune system responses, and stem cell-associated biological processes). Basal-like and claudin-low subtypes belong to the group of triple-negative breast cancer (TNBC), which are characterized by the lack of progesterone receptor (PR), ER and HER2 expression, and have high incidence of distant disease recurrence within three years of diagnosis, with a high frequency of visceral metastases [2]. A recent meta-analysis of a large cohort of TNBC cases allowed the subclassification of this group into at least four TNBC subtypes: luminal androgen receptor (LAR), mesenchymal (MES), basal-like immune-suppressed (BLIS), and basal-like immune-activated (BLIA) [3,4]. This subclassification is further supported by The Cancer Genome Atlas (TCGA) Program through mRNA, miRNA, DNA, and epigenetic analyses [5]. Considering the five main breast cancer subtypes, luminal A, luminal B, HER2, basal-like and claudin-low, a differentiation hierarchy that resembles the normal epithelial mammary developmental cascade has been proposed [6]. The claudin-low, which overlaps with the mesenchymal group, represents the most primitive tumors that are also the most similar to the mammary stem cells (MaSC). The following step in the mammary development is the luminal progenitor, which corresponds to the basal-like subtype. Then, a further development may lead a luminal progenitor/basal-like cell to the HER2 subtype, which represents the loss of the basal features and the acquisition of a luminal phenotype. Finally, the most differentiated groups are the luminal A and B subtypes [7]. Breast cancer patients with an undifferentiated phenotype similar to the normal MaSC have a worse prognosis compared with breast cancers with the more differentiated/luminal phenotype [6]. The process of dedifferentiation, which leads tumor cells to become increasingly more aggressive, is characterized in the last passage by an EMT process toward the claudin-low subtype. Indeed, the majority of death (90%) in breast cancer patients is caused by invasion and metastasis, two features related to the EMT [8]. The acquisition of EMT and stem cell-like features have been linked to each other [9,10], and have been associated with therapeutic resistance [5]. Indeed, breast cancer stem cells, which were originally isolated on the basis of the $CD44^{high}/CD24^{low}/Lin^-$ immunophenotype [11], may be generated from breast cancer cells through the induction of an EMT, and EMT markers are expressed in stem-like cells isolated from mammary glands [9]. Cancer stem cells (CSCs) represent a small subpopulation of the tumor identified in most human tumors, including breast cancer [11]. These cells have self-renewal and tumor-initiating capabilities, which are determinant for the metastasization process [12]. A group of transcription factors playing critical roles during embryogenesis are also critical in the process of de-differentiation of the cancer cells. They induce EMT through transcriptional control of E-cadherin and include SNAIL1/2, ZEB1/2, TWIST1/2, FOXC1/2, TCF3, and GSC [13]. Among them, SNAIL and TWIST are able alone, if activated, to induce a mesenchymal/CSC phenotype in human immortalized human mammary epithelial cells [9,10]. Moreover, TWIST1, FOXC2, SNAIL1, ZEB2, and TWIST2 are overexpressed in stem-like cells isolated from primary breast carcinomas compared with more differentiated cancer cells [9].

2. The Role of EMT in Basal-Like Carcinomas

The EMT program associated with malignancy, invasion, and metastasis, also called EMT type 3 to distinguish it from those related to embryogenesis (type 1) and tissue regeneration (type 2), leads to a loss of cellular adhesion, changes in the polarization of the cell and cytoskeleton, migration, intravasation, survival in the vascular system, extravasation, and metastasis [8]. Therefore, it is believed to be a critical step in the progression of cancer toward a metastatic disease, even if the role for EMT in breast cancer metastases has been the matter of a recent debate on Nature [14,15]. Furthermore, EMT confers stem cell features contributing to chemoresistance and poor outcome [9]. Indeed, whereas neoadjuvant chemotherapy is associated with high pathologic complete response rates in basal-like carcinomas, metaplastic breast cancers (MBCs), which are aggressive TNBC tumors mostly characterized by EMT, are usually also chemoresistant and associated with worse outcomes [16]. The claudin-low subset is closely related to the MBC group by transcriptional profiling. Indeed, they are both characterized by the low expression of GATA3-regulated genes and genes involved in cell-cell adhesion, while are enriched of stem cell and EMT markers. However, they show differences in the presence of *PIK3CA* mutations and are therefore considered two different TNBC subgroups, even though they may have related cellular origins [17].

An intriguing capacity of the EMT process is that it is potentially reversible at any time by simply changing the expression of key molecular components. Accordingly, recent studies have indicated that mesenchymal-to-epithelial transition (MET), the reverse program of EMT, is observed in fibroblasts during the generation of induced pluripotent stem cells [18,19]. Further studies have shown that reprogramming factors introduced in cancer cells are able to attenuate their malignancy by letting them regain epithelial properties by MET [20]. Changes in cell phenotype between the epithelial and mesenchymal states are parts of the tumor progression process that leads tumor cells to disseminate in metastases. EMT is required for acquiring capability to migrate and invade, while MET is required to colonize the metastatic sites [21].

This opens a potential challenge in that, by deeply dissecting all the pathways involved in the EMT program, we may discover new biomarkers and therapeutic agents for the most aggressive breast tumors. Indeed, different studies have shown that basal-like breast cancer, which is associated with mesenchymal features, is the most deadly subtype [6,22,23]. The acquisition of mesenchymal traits could be due to differences in the cells of origin, or the activation of oncogenes other than the paracrine induction of various EMT programs. However, how the mesenchymal phenotype is maintained is still a matter of intense investigation. There are both endogenous cell autonomous and exogenous non-cell autonomous signals concurring in the process of the EMT in breast cancer. The main endogenous pathways include those orchestrated by TGF-β, Notch, Wnt, Hedgehog, and receptor tyrosine kinases. Meanwhile, the exogenous signals include those coming from the extracellular matrix that act directly on the endogenous pathways, and those coming from the microenvironment, which act in a paracrine way. The latter includes the urokinase plasminogen activator system, the secretome of cancer associated fibroblasts, macrophages, cancer stem cells and cancer cells, and exosomes with their cargo of miRNAs (Figure 1). An integrated cross-talk among all these pathways, which adds further complexity to all of the process, has been observed.

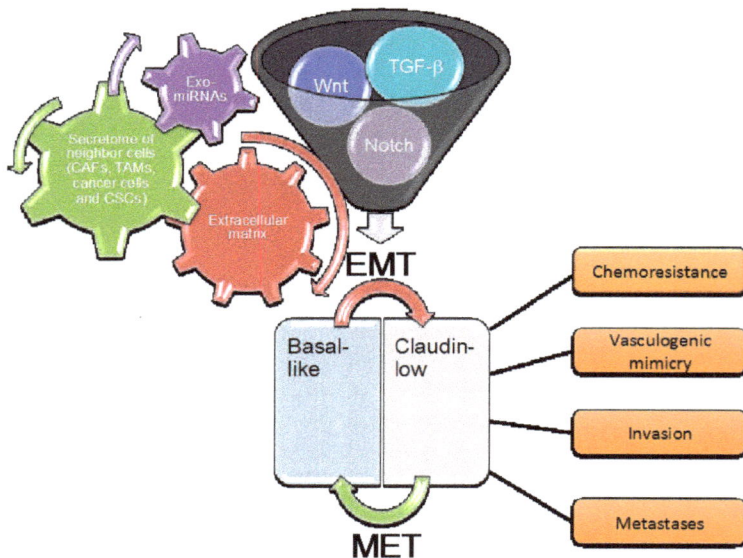

Figure 1. Contributing factors to the claudin-low phenotype and its impact on tumor behavior. The funnel encloses some of the main endogenous pathways involved in the epithelial-to-mesenchymal transition (EMT) process. The gear diagram indicates the various exogenous factors acting in a paracrine way on the endogenous EMT pathways. In the lower part of the scheme, the red upper arrow refers to the origin of claudin-low from basal-like carcinomas, whereas the green lower arrow depicts a possible reversion of the mesenchymal phenotype (mesenchymal-to-epithelial transition (MET) with the re-acquisition of the basal-like features. On the right, in orange, the cell functions resulted enhanced in the claudin-low phenotype as a consequence of the EMT induction.

3. Main Critical Endogenous Pathways of EMT in Breast Cancer Cells

Six main critical pathways may be activated by means of genetic/epigenetic alterations, paracrine stimulation from neighbor cells, or direct interaction with ECM components in breast cancer cells, regulating their transition to a mesenchymal state (Figure 2).

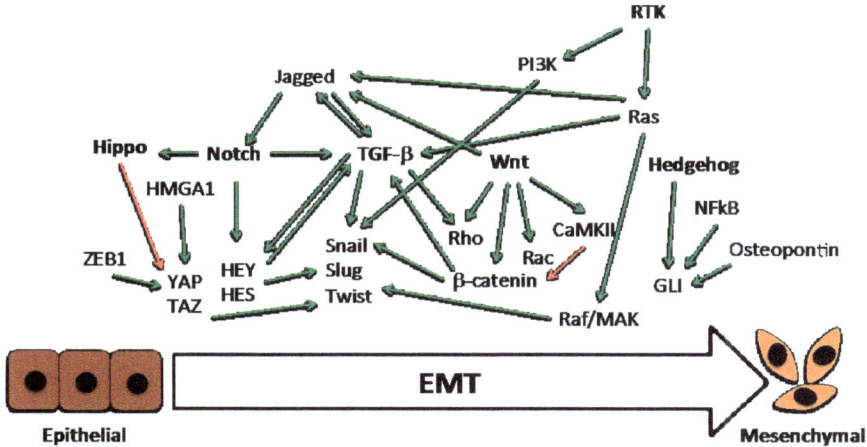

Figure 2. Schematic overview of the main critical endogenous pathways involved in breast cancer EMT. Green and red arrows indicate positive and negative regulation, respectively. The six pathways are in bold, and are further detailed in the text.

3.1. The TGF-β Pathway

TGF-β signaling has a crucial and dual role in breast tumorigenesis. In early tumorigenic lesion, it has a tumor suppressive role due to its ability to induce growth inhibition. However, as cancer progress, it promotes tumor progression and metastasis mainly through the induction of EMT [24]. The TGF-β family of growth factors can initiate and maintain EMT in different cellular contexts [25]. They bind to cell surface receptors (types I and II) and form tight complexes with members of the Smad protein family, leading to their phosphorylation [26]. Phosphorylated Smads associate with cytoplasmic Smad4 and translocate to the nucleus where Smad complexes control transcription of target genes [27]. Moreover, TGF-β may alter the cell surface protein complex structure and expression directly through its receptor complex, independently from nuclear gene regulation. Indeed, TGF-β ligand binding enables type II TGF-β receptor kinase, which is associated with occludin at tight junctions, to phosphorylate Par6 [28]. This protein-protein interaction is direct and independent of Smad proteins [28]. The phosphorylation of Par6 allows it to recruit Smurf1, which in turn leads to the ubiquitination and degradation of RhoA [28], a small GTPase family member responsible for stress fiber formation and for the maintenance of apico-basal polarity and junctional stability [29,30]. Other members of the Rho family, which regulate the cytoskeletal remodeling associated with the gain of cell motility, focal adhesions, and adherens junction formation; and with the interactions between adherens junctions and actin filaments, are also activated by EMT signaling events induced by TGF-β [31–34]. Thus, Rho family GTPases and their regulatory molecules (GEFs and GAPs) exert key roles in epithelial plasticity, and are crucial effectors of TGF-β-induced EMT [30].

The TGF-β pathway cross-talks with different pathways for the induction and maintenance of the EMT phenotype. In mammary epithelial cells, oncogenic Ha-Ras activation promotes EMT through autocrine production of TGF-β [35], and activation of the Raf/MAK pathway is required for metastatic features of EMT in vivo [36]. Another cross-talk of the TGF-β signaling in EMT has been described with the Notch pathway. On one hand, a subset of Notch target genes, including HEY1, HEY2, HES1, and HES5, and the Notch ligand Jagged1, are induced by TGF-β at the onset of EMT in a panel of epithelial cells from the mammary gland, kidney tubules, and epidermis. On the other hand, the silencing of HEY1 or Jagged1, as well as the chemical inactivation of Notch, inhibits TGF-β-induced EMT [37]. Finally, a TGF-β crosstalk with the Wnt pathway has been also described. In a mammary gland model, β-catenin transcriptional activity leads to the activation of autocrine TGF-β signals, which synergize with the Wnt signaling to maintain the mesenchymal phenotype [38].

3.2. The Wnt Pathway

The Wnt pathway plays a critical role in the development and progression of breast cancer [39]. In many human cancers, including breast cancer, it promotes cell migration and EMT through the stabilization of Snail [40]. Consistently, Wnt signaling up-regulates the transcription factors Slug and Twist [41,42]. The Wnt pathway is composed of two distinct branches: the canonical and non-canonical pathways. In the canonical pathway, Wnt protein binds to the cell surface receptor Frizzled, which forms a complex with the coreceptors Lrp5 and 6 to promote Axin binding to Dishevelled. This leads to stabilization and translocation to the nucleus of β-catenin. There, β-catenin interacts with transcription factors of the Tcf/Lef family to activate the expression of several genes [43]. β-catenin translocation after Wnt activation can also be mediated through the sequestration of GSK3, the protein that phosphorylates and destabilizes β-catenin, inside multivesicular endosomes [44]. The secreted Frizzled-related protein SFRP1, which is homologous to the extracellular cysteine-rich domain of Frizzled, prevents Wnt ligands from binding, thereby acting as a negative regulator of Wnt signaling [45]. Another secreted protein, DKK1, also inhibits the canonical Wnt pathway by binding directly to the Lrp5/6 coreceptors [46]. Both SFRP1 and DKK1 are frequently silenced by methylation in many human cancers, including breast cancer [47]. The non-canonical, β-catenin-independent Wnt pathways include the planar cell polarity pathway, which regulates cellular organization and polarity through cytoskeletal organization, and the Wnt/Ca^{2+} pathway, which results in a release of intracellular calcium regulation of cell adhesion and migration [48]. In the planar cell polarity pathway, Wnt binds the Frizzled and coreceptors ROR and Ryk. This activates Rho and Rac, which in turn activate Rho-associated protein kinase (ROCK) and c-Jun N-terminal kinase (JNK), respectively, leading to actin polymerization. In the Wnt/Ca^{2-} pathway, Wnt ligands bind to Frizzled, which interacts with G proteins and Dishevelled, thereby activating cGMP-specific phosphodiesterase or phospholipase C, which cause the release of intracellular calcium. Ca^{2+} activates CaMKII, which in turn regulates cell adhesion and migration through activation of nuclear factor of activated T cells (NFAT) [48]. This non-canonical Wnt pathway antagonizes canonical Wnt/β-catenin signaling by activation of Nemo-like kinase, which phosphorylates TCF4 and prevents the β-catenin–TCF4 complex from binding to DNA [49].

Using an orthotopic model of human breast cancer metastasis to lung, DiMeo et al. demonstrated that Wnt signaling is required for tumor formation and metastases, and is necessary for the capacity of cancer cells to self-renew and to maintain the dedifferentiated EMT state, thus providing a molecular link among self-renewal, EMT, and metastasis in breast cancer [50]. More recent studies have shown that the nuclear accumulation of β-catenin is required for TNBC development by controlling tumor-associated properties, such as migration, stemness, anchorage-independent growth, and chemosensitivity, thus suggesting that canonical Wnt signaling is a major driving force in breast cancer [51]. Other studies have also implicated the dysregulation of non-canonical Wnt signaling pathways in the highly metastatic behavior of TNBC cells and CSCs, specifically through aberrant JNK activation [52].

3.3. The Notch Pathway

The Notch ligand-receptor interaction system in vertebrates comprises four receptors (Notch1–Notch4) and five ligands from the Delta and JAG/Serrate (DSL) families: JAG1 and 2 and Delta-like (DLL)-1, 3, and 4 [53]. The interaction between ligand and receptor triggers a series of proteolytic cleavages that release the Notch intracellular domain (NotchIC), allowing it to translocate into the nucleus. Within the nucleus, NotchIC binds to the transcriptional repressor CSL, resulting in the derepression and coactivation of target genes, which regulate various cellular processes. Interestingly, in the development of cancer, Notch may act as either an oncogene or a tumor suppressor gene, depending on the tumor type [54]. In breast carcinoma pathogenesis, Notch signaling and its cross-talk with various pathways—Notch has been reported to be activated downstream of Ras and Wnt in the promotion of mammary tumors [55,56]—play an important role in cell growth, migration, invasion, angiogenesis, and metastasis [57]. Its activation correlates with poor prognosis and poor patient survival [58], induces EMT [59], and promotes the malignant features of breast cancer [60]. JAG1-mediated activation of Notch in breast epithelial cells induces EMT through the induction of Slug and the subsequent repression of the cell–cell adhesion protein E-cadherin, and EMT plays a crucial role in promoting metastases in tumor xenografts exhibiting ligand-induced Notch signaling [59]. Consistently, mammary-specific overexpression of constitutively active Notch1, Notch3, or Notch4 in mice leads to the formation of aggressive, metastatic breast tumors [61,62], and Notch signaling plays a crucial role in stemness [63]. On the other hand, it has been also recently reported that Notch3, inhibits EMT in breast cancer by activating the Hippo pathway, but Notch1 does not [64].

3.4. The Hippo Pathway

The Hippo tumor suppressor pathway consists of a large network of proteins that play important regulatory functions during organ development and regeneration. The core components of this network include a kinase and a transcription module. The kinase module includes the MST1/STK4 and MST2/STK3 protein kinases, the large tumor suppressor proteins LATS1/2, and the adaptor proteins SAV1 and MOB1A/B. This module contributes to the LATS1/2-dependent phosphorylation of yes-associated protein (YAP) and tafazzin (TAZ), which are members of the transcriptional module. The phosphorylation of YAP and TAZ represses their activity by creating 14-3-3 binding sites that cause their cytoplasmic accumulation and proteasome degradation [65]. In breast cancer cell lines, the phosphorylation and activation of YAP, which enhance cell motility and invasiveness, are dependent on the HMGA1-CyclinE2 axis [66]. YAP and TAZ activity have been shown to be increased in basal breast cancers that show a stem-cell-like phenotype [67], and YAP overexpression has been reported to promote the EMT of cultured breast cancer cells [68]. Furthermore, YAP and TAZ activity were increased in high-grade metastatic breast cancer specimens compared with low-grade non-metastatic breast cancer [67], and TAZ is required for the metastatic activity and chemoresistance of breast cancer stem cells [69]. Still in breast cancer, signal transduction from the metastasis suppressor leukaemia inhibitory factor receptor (LIFR) was shown to sequester and inactivate YAP. Therefore, the loss of LIFR expression could be one mechanism that results in YAP or TAZ hyperactivation during the metastasis of breast cancers [70]. Another mechanism might be the loss of E-cadherin, which causes YAP and TAZ derepression in metastatic breast cells [71]. More recently, the Hippo pathway inhibition has been shown to be required for the increased migratory and invasiveness ability of breast cancer cells in twist-mediated EMT [72]. Also, the EMT-inducing transcriptional repressor ZEB1 has been shown to directly interact with and activate the Hippo pathway effector YAP [73], and the E3 ubiquitin-protein ligase Itchy homolog (ITCH) has been shown to enhance EMT in breast cancer by negatively regulating LATS1, and therefore increasing YAP activity [74]. All together, these findings support a central role of the Hippo pathway in counteracting EMT and metastases in breast cancer.

3.5. The Hedgehog Pathway

A growing body of literature supports the role of the stem cell renewal Hedgehog (Hh) pathway in breast cancer [75]. The Hh pathway plays a key role in embryonic development, and regulates stem cell renewal and tissue homeostasis [76]. It involves a signaling cascade starting from the three secreted proteins Sonic (SHH), Indian (IHH), and Desert (DHH) Hedgehog, and the two trans-membrane receptors Patched and Smoothened. It then terminates with the activation of the three glioma-associated oncogene (GLI) transcription factors, GLI1, GLI2, and GLI3, which can function as either activators or repressors of transcription [77]. As for Wnt signaling, canonical and non-canonical pathways have been described for the Hh pathway, too. The canonical pathway is the above-described signaling that through Hh/receptor binding leads to GLI activation, whereas non-canonical Hh pathways are considered either a cellular response mediated by Patched and Smoothened but independent from GLI [78], or GLI activation independent from Hh ligand/receptor binding [79].

Evidence supporting the contribution of the Hh pathway to EMT in breast cancer has been reported different studies. Using a high throughput inhibitor screen, Colavito et al. identified the high expression of GLI1 as a critical determinant of breast cancer cell lines that have undergone an EMT [80]. Their work also showed the importance of the Hh pathway in the maintenance of CSC features, and uncovered a cross-talk between NFkB and GLI1 [80]. Other studies reported that the non-canonical activation of GLI1 by the inflammatory cytokine osteopontin or hypoxia results in the induction of EMT, drug resistance, and invasion capabilities in breast cancer cell lines [81,82]. Moreover, the development of mammary tumors by the conditional expression of GLI1 in experimental mouse models further supports the implication of the Hh pathway in EMT-mediated breast tumorigenesis [83].

3.6. Pathways Emanating from Receptor Tyrosine Kinases

Receptor tyrosine kinases (RTKs) have a crucial role as sustainers and effectors of EMT in a variety of tumors, including breast cancer [84]. The activation of RTKs occurs through homodimerization induced by ligand binding, or ligand independent mechanisms, including transactivation or heterodimerization with other RTKs or non-RTKs receptors [85]. Growth factors such as EGF, FGF, IGF, and PDGF, stimulate RTKs to initiate intracellular signaling (including those mediated by Ras, PI3K, Src, and ILK), which ultimately could promote the expression of EMT-inducing transcription factors such as Snail1/2, ZEB1/2, and Twist, contributing and/or regulating EMT [84]. Some RTKs, as PDGFRβ [3] and Axl RTKs [86], are emerging as mesenchymal/stem cell-specific markers in breast cancers. However, whether RTKs induces EMT or whether EMT induces receptor expression is still an open debate [87].

Importantly, many of the signaling cascades (including various branches of Mitogen-activated protein kinase, Rho-like GTPase, and PI3K/AKT pathways) induced by TGF-β, a primary inducer of EMT, are also induced by RTKs in response to ligand binding, and a complex cross-talk of oncogenic signaling has been implicated in EMT [88]. It has been reported that mammary cancer metastasis is strongly promoted by an autocrine PDGF/PDGFR loop, which is established as a consequence of TGF-β-induced EMT [89]. Also, RTK-dependent signaling has not only an established role in the induction of classical EMT transcription factors, it also regulates the deposition of several ECM components and integrin binding to ECM, thus activating intracellular cascades that mediate EMT (see below).

4. Role of the Extracellular Matrix

It has been recently shown that breast cancer cell lines representative of the mesenchymal/claudin-low subtype have the capability to undergo endothelial transdifferentiation forming spiderweb-like networks. This phenomenon is known as vascular mimicry (VM), which provides the blood supply for tumor growth and promotes metastasis with mechanisms distinct from classical angiogenesis [90]. The VM process is essentially dependent on cell-matrix interaction mediated by integrins, which are cell surface adhesion molecules representing the main receptors by which the cells bind to and respond to extracellular matrix (ECM) components. Among them, integrin αvβ3 expression strongly correlates with tumor invasion, EMT, and metastases of highly aggressive cancers [91,92]. Different RTKs have been shown to associate with αvβ3, thus promoting many aspects of tumor progression, including VM, migration, invasion, and metastases. In response to matrix, integrin αvβ3 forms a complex with the epidermal growth factor receptor (EGFR) on the surface of TNBC claudin-low MDA-MB-231 and BT-459 cell lines, which are crucial for VM [92]. This interaction allows integrin to adopt a conformation competent for binding to the ECM, which is required for VM. A similar role is played by the PDGFRβ [93], and other examples of cross-talk between integrins and RTKs, also based on a physical interaction among them, have been reported [91]. Indeed, the high expression of integrin αvβ3 has been recently shown to be a marker of breast carcinomas with stem-like features, and high resistance to tyrosin kinase inhibitors [94]. In many cases, the cross-talk between integrins and RTKs leads to the degradation or recycling of the receptor, thus regulating the engagement of matrix ligands [95]. It has been also reported that the association of αvβ3 with different RTKs, including PDGFRβ and VEGFR2, in the presence of ECM ligands, augments the ability of RTKs to respond to their growth factors, thus resulting in the induction of cell proliferation and migration [96]. Another component of the ECM, periostin (POSTN), is induced in breast cancer metastases, where it has been found to play a critical role in their development through the maintenance of CSCs [97]. To this aim, POSTN interacts with Wnt ligands, boosting Wnt signaling, which in turn control stem cell maintenance [97]. Also, the matrix metalloproteinases (MMPs), which degrade and modify the ECM as well as cell-ECM and cell-cell contacts, facilitating detachment of epithelial cells from the surrounding tissue, are upregulated in breast cancer, where they stimulate tumorigenesis, cancer cell invasion, and metastasis by activating EMT [98].

5. Paracrine Mechanisms

Cancer cells secrete proteins that modify the extracellular milieu, acting as autocrine and paracrine stimulatory factors, and have a relevant role in cancer progression [99]. This secretome, which is released by the cells via different pathways [100], contributes to EMT, the metastatic spreading of cancer cells, and the maintenance of CSCs. Also, cancer-associated fibroblasts (CAFs) assist tumor invasion and promote the oncogenic transformation of surrounding epithelial cells by secreting numerous pro-tumorigenic factors [99,101]. In breast cancer cells, CAFs promote aggressive phenotypes through EMT induced by paracrine TGF-β1 [102]. They may also originate by differentiation of bone marrow-derived mesenchymal stem cells, which migrate to the tumor site and contribute to the tumor microenvironment [103,104]. They also promote the aggressiveness of TNBC cell lines that are evaluated as capable of migrating, invading, and acquiring stemness markers [105]. CSCs themselves have their own secretome, which is different from that of the bulk tumor cells and their derived differentiated cancer cells. Different studies indicate a role for CSC-secreted TGF-β in the transformation of breast cancer cells to CSCs, and in the TGF-β-mediated metastasis of the cancer cells tissues [99]. Indeed, breast cancer cells have shown a gene signature that is consistent with the activation of TGF-β signaling. This signature includes the elevated expression of TGF-β and its receptors in CD44$^+$/CD24$^-$ CSCs compared with the CD44$^-$/CD24$^+$ non-stem cells [106]. Furthermore, in vitro treatment of human mammary epithelial cells with TGF-β has been shown to give rise to CD44$^+$/CD24$^-$ CSCs through induction of the EMT [9]. Also, it has been recently reported that in mammary glands, tumor CSCs activate CAFs via the paracrine activation of Hedgehog signaling, thus inducing the CAFs' secretion of factors that promote the expansion and self-renewal of CSCs [107]. Finally, a paracrine loop between tumor cells and tumor-associated monocytes and macrophages (TAMs) has been described in mammary tumors to allow tumor cell migration and CSC niche support [108,109]. In the latter case, the EMT program mediates the physical interactions of CSCs with TAMs by receptor-counter-receptor binding, thus activating signalings in CSCs that culminate with the secretion of cytokines sustaining the stem cell fate [109]. The reciprocal reprogramming of both the tumor cells and the surrounding cells and tissue structures not only guides invasion, it also generates diverse modes of dissemination [110]. Some of the factors that are necessary for the induction of different EMT pathways in breast cancer cells are secreted by cells that are in an epithelial state and fail to act in an autocrine way, but act in a paracrine way on neighbor cancer cells. Then, once cells have passed through an EMT, they maintain the resulting mesenchymal/CSC state by cell-autonomous autocrine loops [111]. An autocrine PDGF/PDGFR loop, which contributes to the maintenance of EMT, is established in breast cancer cells as a consequence of TGF-β signaling [89].

One of the most well-studied paracrine mechanisms involved in the early metastatic step of breast cancer is the urokinase plasminogen activator (uPA) system, composed by the protease uPA and its receptor uPAR, which converts the plasminogen in plasmin. Plasmin in turn degrades—either directly or indirectly through the activation of matrix metalloprotease—several ECM proteins, including fibronectin, laminin, and others. This releases growth factors that stimulate proliferation, migration, invasion, and metastasis upon binding to their cognate receptors [112,113]. Moreover, the uPA/uPA complex cooperates with integrins, G-protein coupled receptors, caveolins, and lipids rafts for signal transduction. Indeed, uPA and its inhibitor PAI-1 are markers of poor prognosis and metastases in primary breast tumors [114,115], and evidence has been reported that the uPA system facilitates breast cancer metastases by several mechanisms [116].

Exosomes and microRNAs

Several cellular components of the tumor microenvironment and cancer cells secrete exosomes that function in an autocrine or paracrine manner to promote many aspects of cancer cells. These aspects include angiogenesis, invasion, proliferation, and contribution to cancer cell plasticity by regulating EMT in the tumor microenvironment [117]. They are small vesicles that originate from the plasma membrane and released from the cell in the extracellular milieu. They contain a wide variety of biological active material that they can exchange with neighboring cells, thus enabling a potent mode of intercellular communication [118,119]. Unlike soluble factors secreted by cells, exosomes carry a concentrated group of functional molecules, provide protection to the transported molecules, and serve as intercellular communicators not only locally, but also systemically [117]. This group of functional molecules may include oncoproteins and oncomiRNAs. The oncogenic message may be transferred by exosomes in different ways: (i) by releasing ligands in the extracellular milieu; (ii) by fusion with the plasma membrane of recipient cells; and (iii) by endocytosis [117]. Luga et al. observed that Wnt containing exosomes derived from CAFs promoted motility and metastasis by activating Wnt signaling in recipient breast cancer cells [119]. Similarly, exosomes derived from mesenchymal stem cell and macrophages promoted the migration and/or invasion of breast cancer cell lines via activation of Wnt signaling [120,121]. Meanwhile, paracrine Wnt10b transported by exosomes released by CAFs can promote cancer progression via EMT induced by the canonical Wnt pathway [122]. On the other hand, exosomes from breast cancer cells can convert adipose tissue-derived mesenchymal stem cells into myofibroblast-like cells [123]. Exosomes are also involved in mediating hypoxia-induced EMT. Specifically in breast cancer cell lines, the induction of hypoxia has been shown to result in the release of an increased number of exosomes, which contain miR-210 [124]. This could play a role in promoting tumor progression in response to hypoxia, as miR-210 can promote endothelial cell tubulogenesis [125].

Several miRNAs have been implicated in the regulation of EMT in cancer [126], and exosome-mediated exchange of miRNAs (exo-miRNAs) between cells has been reported in recent years [127]. MiR-223, a miRNA transported from exosomes released from IL-4-activated macrophages to breast cancer cells, promote breast cancer cell invasion via modulation of the β-catenin pathway [128]. Therefore, tumor and stromal cells can regulate EMT and metastasis through the exosome-mediated delivery of proteins and miRNAs. Other miRNAs related to EMT in breast cancer include either negative regulators (miR-200 family, miR-34 family, miR-497, miR-125b, miR-206, miR-30a, miR-138, miR-195, miR-143, miR-671-5p, miR-153, and miR-300), or positive regulators (miR-10b, miR-21, miR-155, miR-9, miR-29a, miR-103/107, miR-181b-3p, miR-221/222, miR-183/96/182, miR-373, and miR-100). For a recent detailed review of endogenous miRNAs and networks that participate in breast cancer, see elsewhere [129]. Interestingly, different miRNAs may cross-regulate the tumor EMT process. It has been shown that miR-103/107 induces EMT in breast cancer by downregulating miR-200, which targets the E-cadherin negative regulators ZEB1 and ZEB2 [130,131]. Further, a network involving PDGFRβ, miR-9, miR-200, and EMT has been described in mesenchymal TNBC subtypes. Indeed, it has been shown that the induction of miR-9 by PDGFRβ stimulation strongly increases the VM of TNBC cells, whereas ectopic expression of miR-200 causes the reduction of PDGFRβ levels by suppressing ZEB1, and in turn inhibits vasculogenic properties [92].

6. Therapeutic Perspectives

Highly aggressive breast cancer subtypes, such as the claudin-low group, are clinically resistant to chemotherapy due to their enrichment in CSCs. The association between the EMT program and the CSC state represents an attractive opportunity for drug development that is only recently starting to be experimentally proven. A differentiation therapy that is based on the induction of a MET is indeed a possible road to tread: activation of PKA leads to MET and loss of tumor-initiating ability in breast cancer cells [132]. However, a caveat of using such a MET-induced differentiation therapy is the observed requirement of a MET to complete the colonization stage of the metastasis cascade. Consequently, the induction of a MET might inadvertently support the process of metastatic colonization at distant sites [133]. Weinberg's group has recently employed a therapeutic approach that involves the differentiation of CSCs to their non-stem cell counterparts through the induction of a MET. They showed that the induction of a MET as a form of differentiation therapy may improve the response of advanced carcinomas to chemotherapy and prevent their progression to metastasis [134]. A growing list of compounds that reverse EMT in breast cancer has been used in preclinical studies. Through using erbulin, a non-taxane microtubule dynamics inhibitor, for seven days on TNBC cells, Yoshida et al. demonstrated that the treatment induced MET while resulting in decreased in vitro migration and invasiveness, as well as decreased numbers of lung metastasis, when assessed in an in vivo experimental metastasis model [135]. Similar results have been obtained using luteolin, a natural flavonoid compound [136]; diallyl disulfide, an important garlic (Allium sativum) derivative [137]; and mangiferin, a naturally occurring glucosylxanthone [138], which suggests that these compounds could be potential therapeutic candidates for the treatment of advanced or metastatic breast cancer.

As well as being an essential step in tumor metastases, EMT could also be induced under the selective pressure of clinical cytotoxic drugs. To solve this problem, Fan et al. have synthesized multi-functional epigallocatechin gallate/iron nano-complexes (EIN) as a versatile coating material to improve conventional therapies. They showed in vitro that this strategy could eliminate EMT-type cancer cells, and in vivo studies revealed that EIN inhibits the EMT process and enhances the therapeutic effect of conventional chemotherapy, thus preventing drug chemoresistance [139]. Further, a new approach, the ABC7 regimen (Adjuvant for Breast Cancer treatment using seven repurposed drugs), has been recently proposed for metastatic breast cancer. In addition to the current standard treatment with capecitabine, ABC7 uses an ensemble of seven already-marketed noncancer treatment drugs to block different EMT signaling pathways, as a way to make current traditional cytotoxic chemotherapy more effective. However, it has not yet been experimentally tested for its safety and effectiveness [140].

Another therapeutic strategy against EMT may be using monoclonal antibodies or oligonucleotide aptamers that are able to bind to cancer cell surface proteins and disrupt their attachment to the extracellular matrix via integrins. We recently provided evidence that the anti-EGFR CL4 aptamer impairs the integrin αvβ3-EGFR complex on TNBC cells grown on Matrigel or subcutaneously injected in nude mice to form tumors. This causes the inhibition of integrin binding to matrix and, in turn, VM in vitro and in vivo [92]. A similar effect can be obtained by Transtuzumab, a monoclonal antibody against HER2, which causes the loss of integrin αvβ6 and HER2 in breast cancer xenografts [141]. Another interesting approach involving aptamers consists in the selective delivery of therapeutic siRNAs or drugs to breast tumors by using aptamers as delivery agents. In this context, aptamer targeting EpCAM was shown to inhibit CSCs when linked to siRNAs against *PLK1*, a kinase required for mitosis, and cause tumor regression when injected in the TNBC xenograft model [142].

Finally, strategies to interfere with the loading or delivery of tumor-promoting exo-miRNAs or to replenish tumor-suppressive miRNAs via exosomal delivery are under investigation [143], and they can potentially be employed to deliver either miRNAs that negatively regulate EMT, or antagomirs against miRNAs that positively regulate EMT in breast cancer cells. Functional studies showed that the inhibition of miR-23a suppressed the TGF-β1-induced EMT, migration, invasion, and metastasis of breast cancer cells, both in vitro and in vivo [144]. Other studies reported that: miR520c could inhibit breast cancer EMT by targeting STAT3 [145]; miR-10b antagomirs inhibit metastasis in a mouse mammary tumor model [146]; and that miR200c expression significantly enhanced the chemosensitivity and decreased the metastatic potential of a p53(null) claudin-low tumor model [147], and restored trastuzumab sensitivity while suppressing invasion of breast cancer cells [148]. However, all of these studies did not use exosomes to deliver the miRNAs. A recent study showed that the delivery of miR-134 by exosomes in TNBC cells caused the reduction of cellular migration and invasion [149]. This gave a proof of concept of a possible exo-miR therapy. Furthermore, docosahexaenoic acid alters breast cancer exosome secretion and microRNA contents, including EMT-inducing miRNAs, in breast cancer cells [150], which supports its use for a breast cancer therapy aiming to counteract the paracrine effects of exo-miRNAs.

Acknowledgments: Authors' research work is supported by AIRC (IG 12962 to Gennaro Chiappetta and IG 18753 to Laura Cerchia).

Conflicts of Interest: The authors declare no conflicts of interest.

References

1. Ferlay, J.; Soerjomataram, I.; Ervik, M.; Dikshit, R.; Eser, S.; Mathers, C.; Rebelo, M.; Parkin, D.M.; Forman, D.; Bray, F. GLOBOCAN 2012: Estimated cancer incidence, mortality and prevalence worldwide in 2012 v1.0. *IARC CancerBase* **2014**, *11*. ISBN-13 978-92-832-2447-1. IARC. Available online: http://publications.iarc.fr/Databases/Iarc-Cancerbases/Globocan-2012-Estimated-Cancer-Incidence-Mortality-And-Prevalence-Worldwide-In-2012-V1-0-2012 (accessed on 29 September 2017).
2. Kast, K.; Link, T.; Friedrich, K.; Petzold, A.; Niedostatek, A.; Schoffer, O.; Werner, C.; Klug, S.J.; Werner, A.; Gatzweiler, A.; et al. Impact of breast cancer subtypes and patterns of metastasis on outcome. *Breast Cancer Res. Treat.* **2015**, *150*, 621–629. [CrossRef] [PubMed]
3. Lehmann, B.D.; Bauer, J.A.; Chen, X.; Sanders, M.E.; Chakravarthy, A.B.; Shyr, Y.; Pietenpol, J.A. Identification of human triple-negative breast cancer subtypes and preclinical models for selection of targeted therapies. *J. Clin. Investig.* **2011**, *121*, 2750–2767. [CrossRef] [PubMed]
4. Burstein, M.D.; Tsimelzon, A.; Poage, G.M.; Covington, K.R.; Contreras, A.; Fuqua, S.A.; Savage, M.I.; Osborne, C.K.; Hilsenbeck, S.G.; Chang, J.C.; et al. Comprehensive genomic analysis identifies novel subtypes and targets of triple-negative breast cancer. *Clin. Cancer Res.* **2015**, *21*, 1688–1698. [CrossRef] [PubMed]
5. Cancer Genome Atlas Network. Comprehensive molecular portraits of human breast tumours. *Nature* **2012**, *490*, 61–70. [CrossRef]
6. Prat, A.; Parker, J.S.; Karginova, O.; Fan, C.; Livasy, C.; Herschkowitz, J.I.; He, X.; Perou, C.M. Phenotypic and molecular characterization of the claudin-low intrinsic subtype of breast cancer. *Breast Cancer Res.* **2010**, *12*, R68. [CrossRef] [PubMed]
7. Prat, A.; Perou, C.M. Mammary development meets cancer genomics. *Nat. Med.* **2009**, *15*, 842–844. [CrossRef] [PubMed]
8. Felipe Lima, J.; Nofech-Mozes, S.; Bayani, J.; Bartlett, J.M. EMT in breast carcinoma—A review. *J. Clin. Med.* **2016**, *5*, 65. [CrossRef] [PubMed]
9. Mani, S.A.; Guo, W.; Liao, M.; Eaton, E.N.; Zhou, A.Y.; Brooks, M.; Reinhard, F.; Zhang, C.C.; Campbell, L.L.; Polyak, K.; et al. The epithelial-mesenchymal transition generates cells with properties of stem cells. *Cell* **2008**, *133*, 704–715. [CrossRef] [PubMed]
10. Morel, A.P.; Lièvre, M.; Thomas, C.; Hinkal, G.; Ansieau, S.; Puisieux, A. Generation of breast cancer stem cells through epithelial-mesenchymal transition. *PLoS ONE* **2008**, *3*, e2888. [CrossRef] [PubMed]

11. Al-Hajj, M.; Wicha, M.S.; Benito-Hernandez, A.; Morrison, S.J.; Clarke, M.F. Prospective identification of tumorigenic breast cancer cells. *Proc. Natl. Acad. Sci. USA* **2003**, *100*, 3983–3988. [CrossRef] [PubMed]

12. Reya, T.; Morrison, S.J.; Clarke, M.F.; Weissman, I.L. Stem cells, cancer, and cancer stem cells. *Nature* **2001**, *414*, 105–111. [CrossRef] [PubMed]

13. Moreno-Bueno, G.; Portillo, F.; Cano, A. Transcriptional regulation of cell polarity in EMT and cancer. *Oncogene* **2008**, *27*, 6958–6969. [CrossRef] [PubMed]

14. Ye, X.; Brabletz, T.; Kang, Y.; Longmore, G.D.; Nieto, M.A.; Stanger, B.Z.; Yang, J.; Weinberg, R.A. Upholding a role for EMT in breast cancer metastasis. *Nature* **2017**, *547*, E1–E3. [CrossRef] [PubMed]

15. Fisher, K.R.; Altorki, N.K.; Mittal, V.; Gao, D. Fisher et al. replay. *Nature* **2017**, *547*, E5–E6. [CrossRef] [PubMed]

16. Hennessy, B.T.; Giordano, S.; Broglio, K.; Duan, Z.; Trent, J.; Buchholz, T.A.; Babiera, G.; Hortobagyi, G.N.; Valero, V. Biphasic metaplastic sarcomatoid carcinoma of the breast. *Ann. Oncol.* **2006**, *17*, 605–613. [CrossRef] [PubMed]

17. Hennessy, B.T.; Gonzalez-Angulo, A.M.; Stemke-Hale, K.; Gilcrease, M.Z.; Krishnamurthy, S.; Lee, J.S.; Fridlyand, J.; Sahin, A.; Agarwal, R.; Joy, C.; et al. Characterization of a naturally occurring breast cancer subset enriched in epithelial-to-mesenchymal transition and stem cell characteristics. *Cancer Res.* **2009**, *69*, 4116–4124. [CrossRef] [PubMed]

18. Li, R.; Liang, J.; Ni, S.; Zhou, T.; Qing, X.; Li, H.; He, W.; Chen, J.; Li, F.; Zhuang, Q.; et al. A mesenchymal-to-epithelial transition initiates and is required for the nuclear reprogramming of mouse fibroblasts. *Cell Stem Cell* **2010**, *7*, 51–63. [CrossRef] [PubMed]

19. Samavarchi-Tehrani, P.; Golipour, A.; David, L.; Sung, H.K.; Beyer, T.A.; Datti, A.; Woltjen, K.; Nagy, A.; Wrana, J.L. Functional genomics reveals a BMP-driven mesenchymal-to-epithelial transition in the initiation of somatic cell reprogramming. *Cell Stem Cell* **2010**, *7*, 64–77. [CrossRef] [PubMed]

20. Takaishi, M.; Tarutani, M.; Takeda, J.; Sano, S. Mesenchymal to epithelial transition induced by reprogramming factors attenuates the malignancy of cancer cells. *PLoS ONE* **2016**, *11*, e0156904. [CrossRef] [PubMed]

21. Yao, D.; Dai, C.; Peng, S. Mechanism of the mesenchymal-epithelial transition and its relationship with metastatic tumor formation. *Mol. Cancer Res.* **2011**, *9*, 1608–1620. [CrossRef] [PubMed]

22. Micalizzi, D.S.; Christensen, K.L.; Jedlicka, P.; Coletta, R.D.; Barón, A.E.; Harrell, J.C.; Horwitz, K.B.; Billheimer, D.; Heichman, K.A.; Welm, A.L.; et al. The Six1 homeoprotein induces human mammary carcinoma cells to undergo epithelial-mesenchymal transition and metastasis in mice through increasing TGF-beta signaling. *J. Clin. Investig.* **2009**, *119*, 2678–2690. [CrossRef] [PubMed]

23. Scimeca, M.; Antonacci, C.; Colombo, D.; Bonfiglio, R.; Buonomo, O.C.; Bonanno, E. Emerging prognostic markers related to mesenchymal characteristics of poorly differentiated breast cancers. *Tumour Biol.* **2016**, *37*, 5427–5435. [CrossRef] [PubMed]

24. Massagué, J. TGFβ in Cancer. *Cell* **2008**, *134*, 215–230. [CrossRef] [PubMed]

25. Miettinen, P.J.; Ebner, R.; Lopez, A.R.; Derynck, R. TGF-beta induced transdifferentiation of mammary epithelial cells to mesenchymal cells: Involvement of type I receptors. *J. Cell Biol.* **1994**, *127*, 2021–2036. [CrossRef] [PubMed]

26. Massagué, J. How cells read TGF-beta signals. *Nat. Rev. Mol. Cell Biol.* **2000**, *1*, 169–178. [CrossRef] [PubMed]

27. Derynck, R.; Zhang, Y.E. Smad-dependent and Smad-independent pathways in TGF-beta family signalling. *Nature* **2003**, *425*, 577–584. [CrossRef] [PubMed]

28. Ozdamar, B.; Bose, R.; Barrios-Rodiles, M.; Wang, H.R.; Zhang, Y.; Wrana, J.L. Regulation of the polarity protein Par6 by TGFbeta receptors controls epithelial cell plasticity. *Science* **2005**, *307*, 1603–1609. [CrossRef] [PubMed]

29. Perez-Moreno, M.; Jamora, C.; Fuchs, E. Sticky business: Orchestrating cellular signals at adherens junctions. *Cell* **2003**, *112*, 535–548. [CrossRef]

30. Zavadil, J.; Böttinger, E.P. TGF-beta and epithelial-to-mesenchymal transitions. *Oncogene* **2005**, *24*, 5764–5774. [CrossRef] [PubMed]

31. Bhowmick, N.A.; Ghiassi, M.; Bakin, A.; Aakre, M.; Lundquist, C.A.; Engel, M.E.; Arteaga, C.L.; Moses, H.L. Transforming growth factor-beta1 mediates epithelial to mesenchymal transdifferentiation through a RhoA-dependent mechanism. *Mol. Biol. Cell* **2001**, *12*, 27–36. [CrossRef] [PubMed]

32. Shen, X.; Li, J.; Hu, P.P.; Waddell, D.; Zhang, J.; Wang, X.F. The activity of guanine exchange factor NET1 is essential for transforming growth factor-beta-mediated stress fiber formation. *J. Biol. Chem.* **2001**, *276*, 15362–15368. [CrossRef] [PubMed]
33. Ridley, A.J.; Hall, A. The small GTP-binding protein rho regulates the assembly of focal adhesions and actin stress fibers in response to growth factors. *Cell* **1992**, *70*, 389–399. [CrossRef]
34. Fukata, M.; Kaibuchi, K. Rho-family GTPases in cadherin-mediated cell-cell adhesion. *Nat.Rev. Mol. Cell Biol.* **2001**, *2*, 887–897. [CrossRef] [PubMed]
35. Lehmann, K.; Janda, E.; Pierreux, C.E.; Rytömaa, M.; Schulze, A.; McMahon, M.; Hill, C.S.; Beug, H.; Downward, J. Raf induces TGFbeta production while blocking its apoptotic but not invasive responses: A mechanism leading to increased malignancy in epithelial cells. *Genes Dev.* **2000**, *14*, 2610–2622. [CrossRef] [PubMed]
36. Janda, E.; Lehmann, K.; Killisch, I.; Jechlinger, M.; Herzig, M.; Downward, J.; Beug, H.; Grünert, S. Ras and TGF[beta] cooperatively regulate epithelial cell plasticity and metastasis: Dissection of Ras signaling pathways. *J. Cell Biol.* **2002**, *156*, 299–313. [CrossRef] [PubMed]
37. Zavadil, J.; Cermak, L.; Soto-Nieves, N.; Böttinger, E.P. Integration of TGF-beta/Smad and Jagged1/Notch signalling in epithelial-to-mesenchymal transition. *EMBO J.* **2004**, *23*, 1155–1165. [CrossRef] [PubMed]
38. Eger, A.; Stockinger, A.; Park, J.; Langkopf, E.; Mikula, M.; Gotzmann, J.; Mikulits, W.; Beug, H.; Foisner, R. Beta-Catenin and TGFbeta signalling cooperate to maintain a mesenchymal phenotype after FosER-induced epithelial to mesenchymal transition. *Oncogene* **2004**, *23*, 2672–2680. [CrossRef] [PubMed]
39. Lamb, R.; Ablett, M.P.; Spence, K.; Landberg, G.; Sims, A.H.; Clarke, R.B. Wnt pathway activity in breast cancer sub-types and stem-like cells. *PLoS ONE* **2013**, *8*, e67811. [CrossRef] [PubMed]
40. Yook, J.I.; Li, X.Y.; Ota, I.; Hu, C.; Kim, H.S.; Kim, N.H.; Cha, S.Y.; Ryu, J.K.; Choi, Y.J.; Kim, J.; et al. A Wnt-Axin2-3β cascade regulates Snail1 activity in breast cancer cells. *Nat. Cell Biol.* **2006**, *8*, 1398–1406. [CrossRef] [PubMed]
41. Conacci-Sorrell, M.; Simcha, I.; Ben-Yedidia, T.; Blechman, J.; Savagner, P.; Ben-Ze'ev, A. Autoregulation of E-cadherin expression by cadherin-cadherin interactions: The roles of beta-catenin signaling, Slug, and MAPK. *J. Cell Biol.* **2003**, *163*, 847–857. [CrossRef] [PubMed]
42. Howe, L.R.; Watanabe, O.; Leonard, J.; Brown, A.M. Twist is up-regulated in response to Wnt1 and inhibits mouse mammary cell differentiation. *Cancer Res.* **2003**, *63*, 1906–1913. [PubMed]
43. Clevers, H. Wnt/beta-catenin signaling in development and disease. *Cell* **2006**, *127*, 469–480. [CrossRef] [PubMed]
44. Taelman, V.F.; Dobrowolski, R.; Plouhinec, J.L.; Fuentealba, L.C.; Vorwald, P.P.; Gumper, I.; Sabatini, D.D.; De Robertis, E.M. Wnt signaling requires sequestration of glycogen synthase kinase 3 inside multivesicular endosomes. *Cell* **2010**, *143*, 1136–1148. [CrossRef] [PubMed]
45. Bhat, R.A.; Stauffer, B.; Komm, B.S.; Bodine, P.V. Structure-function analysis of secreted frizzled-related protein-1 for its Wnt antagonist function. *J. Cell. Biochem.* **2007**, *102*, 1519–1528. [CrossRef] [PubMed]
46. Niehrs, C. Function and biological roles of the Dickkopf family of Wnt modulators. *Oncogene* **2006**, *25*, 7469–7481. [CrossRef] [PubMed]
47. Suzuki, H.; Toyota, M.; Carraway, H.; Gabrielson, E.; Ohmura, T.; Fujikane, T.; Nishikawa, N.; Sogabe, Y.; Nojima, M.; Sonoda, T.; et al. Frequent epigenetic inactivation of Wnt antagonist genes in breast cancer. *Br. J. Cancer* **2008**, *98*, 1147–1156. [CrossRef] [PubMed]
48. Pohl, S.G.; Brook, N.; Agostino, M.; Arfuso, F.; Kumar, A.P.; Dharmarajan, A. Wnt signaling in triple-negative breast cancer. *Oncogenesis* **2017**, *6*, e310. [CrossRef] [PubMed]
49. Ishitani, T.; Ninomiya-Tsuji, J.; Matsumoto, K. Regulation of lymphoid enhancer factor 1/T-cell factor by mitogen-activated protein kinase-related Nemo-like kinase-dependent phosphorylation in Wnt/beta-catenin signaling. *Mol. Cell. Biol.* **2003**, *23*, 1379–1389. [CrossRef] [PubMed]
50. DiMeo, T.A.; Anderson, K.; Phadke, P.; Fan, C.; Perou, C.M.; Naber, S.; Kuperwasser, C. A novel lung metastasis signature links Wnt signaling with cancer cell self-renewal and epithelial-mesenchymal transition in basal-like breast cancer. *Cancer Res.* **2009**, *69*, 5364–5373. [CrossRef] [PubMed]
51. Xu, J.; Prosperi, J.R.; Choudhury, N.; Olopade, O.I.; Goss, K.H. β-Catenin is required for the tumorigenic behavior of triple-negative breast cancer cells. *PLoS ONE* **2015**, *10*, e0117097. [CrossRef] [PubMed]
52. Borg, J.-P.; Belotti, E.; Daulat, A.; Lembo, F.; Bertucci, F.; Charafe-Jauffret, E.; Birnbaum, D. Deregulation of the non-canonical pathway in triple-negative breast cancer. *FASEB J.* **2013**, *27*. [CrossRef]

53. Miele, L. Notch signaling. *Clin. Cancer Res.* **2006**, *12*, 1074–1079. [CrossRef] [PubMed]
54. Leong, K.G.; Karsan, A. Recent insights into the role of Notch signaling in tumorigenesis. *Blood* **2006**, *107*, 2223–2233. [CrossRef] [PubMed]
55. Weijzen, S.; Rizzo, P.; Braid, M.; Vaishnav, R.; Jonkheer, S.M.; Zlobin, A.; Osborne, B.A.; Gottipati, S.; Aster, J.C.; Hahn, W.C.; et al. Activation of Notch-1 signaling maintains the neoplastic phenotype in human Ras-transformed cells. *Nat. Med.* **2002**, *8*, 979–986. [CrossRef] [PubMed]
56. Ayyanan, A.; Civenni, G.; Ciarloni, L.; Morel, C.; Mueller, N.; Lefort, K.; Mandinova, A.; Raffoul, W.; Fiche, M.; Dotto, G.P.; Brisken, C. Increased Wnt signaling triggers oncogenic conversion of human breast epithelial cells by a Notch-dependent mechanism. *Proc. Natl. Acad. Sci. USA* **2006**, *103*, 3799–3804. [CrossRef] [PubMed]
57. Guo, S.; Liu, M.; Gonzalez-Perez, R.R. Role of Notch and its oncogenic signaling crosstalk in breast cancer. *Biochim. Biophys. Acta* **2011**, *1815*, 197–213. [CrossRef] [PubMed]
58. Mittal, S.; Sharma, A.; Balaji, S.A.; Gowda, M.C.; Dighe, R.R.; Kumar, R.V.; Rangarajan, A. Coordinate hyperactivation of Notch1 and Ras/MAPK pathways correlates with poor patient survival: Novel therapeutic strategy for aggressive breast cancers. *Mol. Cancer Ther.* **2014**, *13*, 3198–3209. [CrossRef] [PubMed]
59. Leong, K.G.; Niessen, K.; Kulic, I.; Raouf, A.; Eaves, C.; Pollet, I.; Karsan, A. Jagged1-mediated Notch activation induces epithelial-to-mesenchymal transition through Slug-induced repression of E-cadherin. *J. Exp. Med.* **2007**, *204*, 2935–2948. [CrossRef] [PubMed]
60. Li, L.; Zhao, F.; Lu, J.; Li, T.; Yang, H.; Wu, C.; Liu, Y. Notch-1 signaling promotes the malignant features of human breast cancer through NF-κB activation. *PLoS ONE* **2014**, *9*, e95912. [CrossRef] [PubMed]
61. Gallahan, D.; Jhappan, C.; Robinson, G.; Hennighausen, L.; Sharp, R.; Kordon, E.; Callahan, R.; Merlino, G.; Smith, G.H. Expression of a truncated Int3 gene in developing secretory mammary epithelium specifically retards lobular differentiation resulting in tumorigenesis. *Cancer Res.* **1996**, *56*, 1775–1785. [PubMed]
62. Hu, C.; Diévart, A.; Lupien, M.; Calvo, E.; Tremblay, G.; Jolicoeur, P. Overexpression of activated murine Notch1 and Notch3 in transgenic mice blocks mammary gland development and induces mammary tumors. *Am. J. Pathol.* **2006**, *168*, 973–990. [CrossRef] [PubMed]
63. Pannuti, A.; Foreman, K.; Rizzo, P.; Osipo, C.; Golde, T.; Osborne, B.; Miele, L. Targeting Notch to target cancer stem cells. *Clin. Cancer Res.* **2010**, *16*, 3141–3152. [CrossRef] [PubMed]
64. Zhang, X.; Liu, X.; Luo, J.; Xiao, W.; Ye, X.; Chen, M.; Li, Y.; Zhang, G.J. Notch3 inhibits epithelial-mesenchymal transition by activating Kibra-mediated Hippo/YAP signaling in breast cancer epithelial cells. *Oncogenesis* **2016**, *5*, e269. [CrossRef] [PubMed]
65. Harvey, K.F.; Zhang, X.; Thomas, D.M. The Hippo pathway and human cancer. *Nat. Rev. Cancer* **2013**, *13*, 246–257. [CrossRef] [PubMed]
66. Pegoraro, S.; Ros, G.; Ciani, Y.; Sgarra, R.; Piazza, S.; Manfioletti, G. A novel HMGA1-CCNE2-YAP axis regulates breast cancer aggressiveness. *Oncotarget* **2015**, *6*, 19087–190101. [CrossRef] [PubMed]
67. Cordenonsi, M.; Zanconato, F.; Azzolin, L.; Forcato, M.; Rosato, A.; Frasson, C.; Inui, M.; Montagner, M.; Parenti, A.R.; Poletti, A.; et al. The Hippo transducer TAZ confers cancer stem cell-related traits on breast cancer cells. *Cell* **2011**, *147*, 759–772. [CrossRef] [PubMed]
68. Overholtzer, M.; Zhang, J.; Smolen, G.A.; Muir, B.; Li, W.; Sgroi, D.C.; Deng, C.X.; Brugge, J.S.; Haber, D.A. Transforming properties of YAP, a candidate oncogene on the chromosome 11q22 amplicon. *Proc. Natl. Acad. Sci. USA* **2006**, *103*, 12405–12410. [CrossRef] [PubMed]
69. Bartucci, M.; Dattilo, R.; Moriconi, C.; Pagliuca, A.; Mottolese, M.; Federici, G.; Benedetto, A.D.; Todaro, M.; Stassi, G.; Sperati, F.; et al. TAZ is required for metastatic activity and chemoresistance of breast cancer stem cells. *Oncogene* **2015**, *34*, 681–690. [CrossRef] [PubMed]
70. Chen, D.; Sun, Y.; Wei, Y.; Zhang, P.; Rezaeian, A.H.; Teruya-Feldstein, J.; Gupta, S.; Liang, H.; Lin, H.K.; Hung, M.C.; et al. LIFR is a breast cancer metastasis suppressor upstream of the Hippo-YAP pathway and a prognostic marker. *Nat. Med.* **2012**, *18*, 1511–1517. [CrossRef] [PubMed]
71. Kim, N.G.; Koh, E.; Chen, X.; Gumbiner, B.M. E-cadherin mediates contact inhibition of proliferation through Hippo signaling-pathway components. *Proc. Natl. Acad. Sci. USA* **2011**, *108*, 11930–11935. [CrossRef] [PubMed]
72. Wang, Y.; Liu, J.; Ying, X.; Lin, P.C.; Zhou, B.P. Twist-mediated Epithelial-mesenchymal Transition Promotes Breast Tumor Cell Invasion via Inhibition of Hippo Pathway. *Sci. Rep.* **2016**, *6*, 24606. [CrossRef] [PubMed]

73. Lehmann, W.; Mossmann, D.; Kleemann, J.; Mock, K.; Meisinger, C.; Brummer, T.; Herr, R.; Brabletz, S.; Stemmler, M.P.; Brabletz, T. ZEB1 turns into a transcriptional activator by interacting with YAP1 in aggressive cancer types. *Nat. Commun.* **2016**, *7*, 10498. [CrossRef] [PubMed]

74. Salah, Z.; Itzhaki, E.; Aqeilan, R.I. The ubiquitin E3 ligase ITCH enhances breast tumor progression by inhibiting the Hippo tumor suppressor pathway. *Oncotarget* **2014**, *5*, 10886–10900. [CrossRef] [PubMed]

75. Habib, J.G.; O'Shaughnessy, J.A. The hedgehog pathway in triple-negative breast cancer. *Cancer Med.* **2016**, *5*, 2989–3006. [CrossRef] [PubMed]

76. Jiang, J.; Hui, C.C. Hedgehog Signaling in Development and Cancer. *Dev. Cell* **2008**, *15*, 801–812. [CrossRef] [PubMed]

77. Sasaki, H.; Nishizaki, Y.; Hui, C.; Nakafuku, M.; Kondoh, H. Regulation of Gli2 and Gli3 activities by an amino-terminal repression domain: Implication of Gli2 and Gli3 as primary mediators of Shh signaling. *Development* **1999**, *126*, 3915–3924. [PubMed]

78. Brennan, D.; Chen, X.; Cheng, L.; Mahoney, M.; Riobo, N.A. Noncanonical Hedgehog signaling. *Vitam. Horm.* **2012**, *88*, 55–72. [CrossRef] [PubMed]

79. Lauth, M.; Toftgård, R. Non-canonical activation of GLI transcription factors: Implications for targeted anti-cancer therapy. *Cell Cycle* **2007**, *6*, 2458–2463. [CrossRef] [PubMed]

80. Colavito, S.A.; Zou, M.R.; Yan, Q.; Nguyen, D.X.; Stern, D.F. Significance of glioma-associated oncogene homolog 1 (GLI1) expression in claudin-low breast cancer and crosstalk with the nuclear factor kappa-light-chain-enhancer of activated B cells (NFκB) pathway. *Breast Cancer Res.* **2014**, *16*, 444. [CrossRef] [PubMed]

81. Das, S.; Samant, R.S.; Shevde, L.A. Nonclassical activation of Hedgehog signaling enhances multidrug resistance and makes cancer cells refractory to Smoothened-targeting Hedgehog inhibition. *J. Biol. Chem.* **2013**, *288*, 11824–11833. [CrossRef] [PubMed]

82. Lei, J.; Fan, L.; Wei, G.; Chen, X.; Duan, W.; Xu, Q.; Sheng, W.; Wang, K.; Li, X. Gli-1 is crucial for hypoxia-induced epithelial-mesenchymal transition and invasion of breast cancer. *Tumour Biol.* **2015**, *36*, 3119–3126. [CrossRef] [PubMed]

83. Fiaschi, M.; Rozell, B.; Bergström, A.; Toftgård, R. Development of mammary tumors by conditional expression of GLI1. *Cancer Res.* **2009**, *69*, 4810–4817. [CrossRef] [PubMed]

84. Gonzalez, D.M.; Medici, D. Signaling mechanisms of the epithelial-mesenchymal transition. *Sci. Signal.* **2014**, *7*, re8. [CrossRef] [PubMed]

85. Gschwind, A.; Fischer, O.M.; Ullrich, A. The discovery of receptor tyrosine kinases: Targets for cancer therapy. *Nat. Rev. Cancer* **2004**, *4*, 361–370. [CrossRef] [PubMed]

86. Gjerdrum, C.; Tiron, C.; Høiby, T.; Stefansson, I.; Haugen, H.; Sandal, T.; Collett, K.; Li, S.; McCormack, E.; Gjertsen, B.T.; Micklem, D.R.; Akslen, L.A.; Glackin, C.; Lorens, J.B. Axl is an essential epithelial-to-mesenchymal transition-induced regulator of breast cancer metastasis and patient survival. *Proc. Natl. Acad. Sci. USA* **2010**, *107*, 1124–1129. [CrossRef] [PubMed]

87. Schoumacher, M.; Burbridge, M. Key Roles of AXL and MER Receptor Tyrosine Kinases in Resistance to Multiple Anticancer Therapies. *Curr. Oncol. Rep.* **2017**, *19*, 19. [CrossRef] [PubMed]

88. Lindsey, S.; Langhans, S.A. Crosstalk of Oncogenic Signaling Pathways during Epithelial-Mesenchymal Transition. *Front. Oncol.* **2014**, *4*, 358. [CrossRef] [PubMed]

89. Jechlinger, M.; Sommer, A.; Moriggl, R.; Seither, P.; Kraut, N.; Capodiecci, P.; Donovan, M.; Cordon-Cardo, C.; Beug, H.; Grünert, S. Autocrine PDGFR signaling promotes mammary cancer metastasis. *J. Clin. Investig.* **2006**, *116*, 1561–1570. [CrossRef] [PubMed]

90. Wagenblast, E.; Soto, M.; Gutiérrez-Ángel, S.; Hartl, C.A.; Gable, A.L.; Maceli, A.R.; Erard, N.; Williams, A.M.; Kim, S.Y.; Dickopf, S.; et al. A model of breast cancer heterogeneity reveals vascular mimicry as a driver of metastasis. *Nature* **2015**, *520*, 358–362. [CrossRef] [PubMed]

91. Desgrosellier, J.S.; Cheresh, D.A. Integrins in cancer: Biological implications and therapeutic opportunities. *Nat. Rev. Cancer* **2010**, *10*, 9–22. [CrossRef] [PubMed]

92. Camorani, S.; Crescenzi, E.; Gramanzini, M.; Fedele, M.; Zannetti, A.; Cerchia, L. Aptamer-mediated impairment of EGFR-integrin αvβ3 complex inhibits vasculogenic mimicry and growth of triple-negative breast cancers. *Sci. Rep.* **2017**, *7*, 46659. [CrossRef] [PubMed]

93. D'Ippolito, E.; Plantamura, I.; Bongiovanni, L.; Casalini, P.; Baroni, S.; Piovan, C.; Orlandi, R.; Gualeni, A.V.; Gloghini, A.; Rossini, A.; et al. miR-9 and miR-200 regulate pdgfrβ-mediated endothelial differentiation of tumor cells in triple-negative breast cancer. *Cancer Res.* **2016**, *76*, 5562–5572. [CrossRef] [PubMed]

94. Seguin, L.; Kato, S.; Franovic, A.; Camargo, M.F.; Lesperance, J.; Elliott, K.C.; Yebra, M.; Mielgo, A.; Lowy, A.M.; Husain, H.; et al. An integrin β3-KRAS-RalB complex drives tumour stemness and resistance to EGFR inhibition. *Nat. Cell Biol.* **2014**, *16*, 457–468. [CrossRef] [PubMed]

95. De Franceschi, N.; Hamidi, H.; Alanko, J.; Sahgal, P.; Ivaska, J. Integrin traffic—The update. *J. Cell Sci.* **2015**, *128*, 839–852. [CrossRef] [PubMed]

96. Borges, E.; Jan, Y.; Ruoslahti, E. Platelet-derived growth factor receptor beta and vascular endothelial growth factor receptor 2 bind to the beta 3 integrin through its extracellular domain. *J. Biol. Chem.* **2000**, *275*, 39867–39873. [CrossRef] [PubMed]

97. Malanchi, I.; Santamaria-Martínez, A.; Susanto, E.; Peng, H.; Lehr, H.A.; Delaloye, J.F.; Huelsken, J. Interactions between cancer stem cells and their niche govern metastatic colonization. *Nature* **2012**, *481*, 85–89. [CrossRef] [PubMed]

98. Radisky, E.S.; Radisky, D.C. Matrix metalloproteinase-induced epithelial-mesenchymal transition in breast cancer. *J. Mammary Gland Biol. Neoplasia* **2010**, *5*, 201–212. [CrossRef] [PubMed]

99. Paltridge, J.L.; Belle, L.; Khew-Goodall, Y. The secretome in cancer progression. *Biochim. Biophys. Acta* **2013**, *1834*, 2233–2241. [CrossRef] [PubMed]

100. Makridakis, M.; Vlahou, A. Secretome proteomics for discovery of cancer biomarkers. *J. Proteom.* **2010**, *73*, 2291–2305. [CrossRef] [PubMed]

101. Hanahan, D.; Coussens, L.M. Accessories to the crime: Functions of cells recruited to the tumor microenvironment. *Cancer Cell* **2012**, *21*, 309–322. [CrossRef] [PubMed]

102. Yu, Y.; Xiao, C.H.; Tan, L.D.; Wang, Q.S.; Li, X.Q.; Feng, Y.M. Cancer-associated fibroblasts induce epithelial-mesenchymal transition of breast cancer cells through paracrine TGF-β signalling. *Br. J. Cancer* **2014**, *110*, 724–732. [CrossRef] [PubMed]

103. Worthley, D.L.; Si, Y.; Quante, M.; Churchill, M.; Mukherjee, S.; Wang, T.C. Bone marrow cells as precursors of the tumor stroma. *Exp. Cell Res.* **2013**, *319*, 1650–1656. [CrossRef] [PubMed]

104. Barcellos-de-Souza, P.; Gori, V.; Bambi, F.; Chiarugi, P. Tumor microenvironment: Bone marrow-mesenchymal stem cells as key players. *Biochim. Biophys. Acta* **2013**, *1836*, 321–335. [CrossRef] [PubMed]

105. Camorani, S.; Hill, B.S.; Fontanella, R.; Greco, A.; Gramanzini, M.; Auletta, L.; Gargiulo, S.; Albanese, S.; Lucarelli, E.; Cerchia, L.; et al. Inhibition of bone marrow-derived mesenchymal stem cells homing towards triple-negative breast cancer microenvironment using an anti-PDGFRβ aptamer. *Theranostic* **2017**, *7*, 3595–3607. [CrossRef] [PubMed]

106. Shipitsin, M.; Campbell, L.L.; Argani, P.; Weremowicz, S.; Bloushtain-Qimron, N.; Yao, J.; Nikolskaya, T.; Serebryiskaya, T.; Beroukhim, R.; Hu, M.; et al. Molecular definition of breast tumor heterogeneity. *Cancer Cell* **2007**, *11*, 259–273. [CrossRef] [PubMed]

107. Valenti, G.; Quinn, H.M.; Heynen, G.J.J.E.; Lan, L.; Holland, J.D.; Vogel, R.; Wulf-Goldenberg, A.; Birchmeier, W. Cancer stem cells regulate cancer-associated fibroblasts via activation of hedgehog signaling in mammary gland tumors. *Cancer Res* **2017**, *77*, 2134–2147. [CrossRef] [PubMed]

108. Wyckoff, J.; Wang, W.; Lin, E.Y.; Wang, Y.; Pixley, F.; Stanley, E.R.; Graf, T.; Pollard, J.W.; Segall, J.; Condeelis, J. A paracrine loop between tumor cells and macrophages is required for tumor cell migration in mammary tumors. *Cancer Res.* **2004**, *64*, 7022–7029. [CrossRef] [PubMed]

109. Lu, H.; Clauser, K.R.; Tam, W.L.; Fröse, J.; Ye, X.; Eaton, E.N.; Reinhardt, F.; Donnenberg, V.S.; Bhargava, R.; Carr, S.A.; et al. A breast cancer stem cell niche supported by juxtacrine signalling from monocytes and macrophages. *Nat. Cell Biol.* **2014**, *16*, 1105–1117. [CrossRef] [PubMed]

110. Friedl, P.; Alexander, S. Cancer invasion and the microenvironment: Plasticity and reciprocity. *Cell* **2011**, *147*, 992–1009. [CrossRef] [PubMed]

111. Scheel, C.; Eaton, E.N.; Li, S.H.; Chaffer, C.L.; Reinhardt, F.; Kah, K.J.; Bell, G.; Guo, W.; Rubin, J.; Richardson, A.L.; et al. Paracrine and autocrine signals induce and maintain mesenchymal and stem cell states in the breast. *Cell* **2011**, *145*, 926–940. [CrossRef] [PubMed]

112. Deryugina, E.I.; Quigley, J.P. Cell surface remodeling by plasmin: A new function for an old enzyme. *J. Biomed. Biotechnol.* **2012**, *2012*, 564259. [CrossRef] [PubMed]

113. Duffy, M.J.; McGowan, P.M.; Harbeck, N.; Thomssen, C.; Schmitt, M. uPA and PAI-1 as biomarkers in breast cancer: Validated for clinical use in level-of-evidence-1 studies. *Breast Cancer Res.* **2014**, *16*, 428. [CrossRef] [PubMed]

114. Harbeck, N.; Kates, R.E.; Look, M.P.; Meijer-Van Gelder, M.E.; Klijn, J.G.; Krüger, A.; Kiechle, M.; Jänicke, F.; Schmitt, M.; Foekens, J.A. Enhanced benefit from adjuvant chemotherapy in breast cancer patients classified high-risk according to urokinase-type plasminogen activator (uPA) and plasminogen activator inhibitor type 1 (n = 3424). *Cancer Res.* **2002**, *62*, 4617–4622. [PubMed]

115. De Cremoux, P.; Grandin, L.; Diéras, V.; Savignoni, A.; Degeorges, A.; Salmon, R.; Bollet, M.A.; Reyal, F.; Sigal-Zafrani, B.; Vincent-Salomon, A.; et al. Breast Cancer Study Group of the Institut Curie. Urokinase-type plasminogen activator and plasminogen-activator-inhibitor type 1 predict metastases in good prognosis breast cancer patients. *Anticancer Res.* **2009**, *29*, 1475–1482. [PubMed]

116. Tang, L.; Han, X. The urokinase plasminogen activator system in breast cancer invasion and metastasis. *Biomed. Pharmacother.* **2012**, *67*, 179–182. [CrossRef] [PubMed]

117. Vella, L.J. The emerging role of exosomes in epithelial-mesenchymal-transition in cancer. *Front. Oncol.* **2014**, *4*, 361. [CrossRef] [PubMed]

118. Simons, M.; Raposo, G. Exosomes—Vesicular carriers for intercellular communication. *Curr. Opin. Cell Biol.* **2009**, *21*, 575–581. [CrossRef] [PubMed]

119. Luga, V.; Zhang, L.; Viloria-Petit, A.M.; Ogunjimi, A.A.; Inanlou, M.R.; Chiu, E.; Buchanan, M.; Hosein, A.N.; Basik, M.; Wrana, J.L. Exosomes mediate stromal mobilization of autocrine Wnt-PCP signaling in breast cancer cell migration. *Cell* **2012**, *151*, 1542–1556. [CrossRef] [PubMed]

120. Lin, R.; Wang, S.; Zhao, R.C. Exosomes from human adipose-derived mesenchymal stem cells promote migration through Wnt signaling pathway in a breast cancer cell model. *Mol. Cell. Biochem.* **2013**, *383*, 13–20. [CrossRef] [PubMed]

121. Menck, K.; Klemm, F.; Gross, J.C.; Pukrop, T.; Wenzel, D.; Binder, C. Induction and transport of Wnt 5a during macrophage-induced malignant invasion is mediated by two types of extracellular vesicles. *Oncotarget* **2013**, *4*, 2057–2066. [CrossRef] [PubMed]

122. Chen, Y.; Zeng, C.; Zhan, Y.; Wang, H.; Jiang, X.; Li, W. Aberrant low expression of p85α in stromal fibroblasts promotes breast cancer cell metastasis through exosome-mediated paracrine Wnt10b. *Oncogene* **2017**, *36*, 4692–4705. [CrossRef] [PubMed]

123. Cho, J.A.; Park, H.; Lim, E.H.; Lee, K.W. Exosomes from breast cancer cells can convert adipose tissue-derived mesenchymal stem cells into myofibroblast-like cells. *Int. J. Oncol.* **2012**, *40*, 130–138. [CrossRef] [PubMed]

124. King, H.W.; Michael, M.Z.; Gleadle, J.M. Hypoxic enhancement of exosome release by breast cancer cells. *BMC Cancer* **2012**, *12*, 421. [CrossRef] [PubMed]

125. Fasanaro, P.; D'Alessandra, Y.; Di Stefano, V.; Melchionna, R.; Romani, S.; Pompilio, G.; Capogrossi, M.C.; Martelli, F. MicroRNA-210 modulates endothelial cell response to hypoxia and inhibits the receptor tyrosine kinase ligand ephrin-A3. *J. Biol. Chem.* **2008**, *283*, 15878–15883. [CrossRef] [PubMed]

126. Zhang, J.; Ma, L. MicroRNA control of epithelial-mesenchymal transition and metastasis. *Cancer Metastasis Rev.* **2012**, *31*, 653–662. [CrossRef] [PubMed]

127. Valadi, H.; Ekström, K.; Bossios, A.; Sjöstrand, M.; Lee, J.J.; Lötvall, J.O. Exosome-mediated transfer of mRNAs and microRNAs is a novel mechanism of genetic exchange between cells. *Nat. Cell Biol.* **2007**, *9*, 654–659. [CrossRef] [PubMed]

128. Yang, M.; Chen, J.; Su, F.; Yu, B.; Su, F.; Lin, L.; Liu, Y.; Huang, J.D.; Song, E. Microvesicles secreted by macrophages shuttle invasion-potentiating microRNAs into breast cancer cells. *Mol. Cancer* **2011**, *10*, 117. [CrossRef] [PubMed]

129. Zhao, M.; Ang, L.; Huang, J.; Wang, J. MicroRNAs regulate the epithelial-mesenchymal transition and influence breast cancer invasion and metastasis. *Tumour Biol.* **2017**, *39*. [CrossRef] [PubMed]

130. Martello, G.; Rosato, A.; Ferrari, F.; Manfrin, A.; Cordenonsi, M.; Dupont, S.; Enzo, E.; Guzzardo, V.; Rondina, M.; Spruce, T.; et al. A microRNA targeting dicer for metastasis control. *Cell* **2010**, *141*, 1195–1207. [CrossRef] [PubMed]

131. Zhang, H.F.; Xu, L.Y.; Li, E.M. A family of pleiotropically acting microRNAs in cancer progression, miR-200: Potential cancer therapeutic targets. *Curr. Pharm. Des.* **2014**, *20*, 1896–1903. [CrossRef] [PubMed]

132. Pattabiraman, D.R.; Bierie, B.; Kober, K.I.; Thiru, P.; Krall, J.A.; Zill, C.; Reinhardt, F.; Tam, W.L.; Weinberg, R.A. Activation of PKA leads to mesenchymal-to-epithelial transition and loss of tumor-initiating ability. *Science* **2016**, *351*, aad3680. [CrossRef] [PubMed]

133. Pattabiraman, D.R.; Weinberg, R.A. Targeting the epithelial-to-mesenchymal transition: The case for differentiation-based therapy. *Cold Spring Harb. Symp. Quant. Biol.* **2016**, *81*, 11–19. [CrossRef] [PubMed]

134. Pattabiraman, D.; Ostendorp, J.; Weinberg, R. Inducing a mesenchymal-to-epithelial transition for the differentiation therapy of aggressive breast carcinomas. In Proceedings of the EACR-AACR-SIC Special Conference 2017: The Challenges of Optimizing Immuno- and Targeted Therapies: From Cancer Biology to the Clinic, Florence, Italy, 24–27 June 2017.

135. Yoshida, T.; Ozawa, Y.; Kimura, T.; Sato, Y.; Kuznetsov, G.; Xu, S.; Uesugi, M.; Agoulnik, S.; Taylor, N.; Funahashi, Y.; et al. Eribulin mesilate suppresses experimental metastasis of breast cancer cells by reversing phenotype from epithelial-mesenchymal transition (EMT) to mesenchymal-epithelial transition (MET) states. *Br. J. Cancer* **2014**, *110*, 1497–1505. [CrossRef] [PubMed]

136. Lin, D.; Kuang, G.; Wan, J.; Zhang, X.; Li, H.; Gong, X.; Li, H. Luteolin suppresses the metastasis of triple-negative breast cancer by reversing epithelial-to-mesenchymal transition via downregulation of β-catenin expression. *Oncol. Rep.* **2017**, *37*, 1148–1158. [CrossRef] [PubMed]

137. Huang, J.; Yang, B.; Xiang, T.; Peng, W.; Qiu, Z.; Wan, J.; Zhang, L.; Li, H.; Li, H.; Ren, G. Diallyl disulfide inhibits growth and metastatic potential of human triple-negative breast cancer cells through inactivation of the β-catenin signaling pathway. *Mol. Nutr. Food Res.* **2015**, *9*, 1063–1075. [CrossRef] [PubMed]

138. Li, H.; Huang, J.; Yang, B.; Xiang, T.; Yin, X.; Peng, W.; Cheng, W.; Wan, J.; Luo, F.; Li, H.; et al. Mangiferin exerts antitumor activity in breast cancer cells by regulating matrix metalloproteinases, epithelial to mesenchymal transition, and β-catenin signaling pathway. *Toxicol. Appl. Pharmacol.* **2013**, *272*, 180–190. [CrossRef] [PubMed]

139. Fan, J.X.; Zheng, D.W.; Rong, L.; Zhu, J.Y.; Hong, S.; Li, C.; Xu, Z.S.; Cheng, S.X.; Zhang, X.Z. Targeting epithelial-mesenchymal transition: Metal organic network nano-complexes for preventing tumor metastasis. *Biomaterials* **2017**, *139*, 116–126. [CrossRef] [PubMed]

140. Kast, R.E.; Skuli, N.; Cos, S.; Karpel-Massler, G.; Shiozawa, Y.; Goshen, R.; Halatsch, M.E. The ABC7 regimen: A new approach to metastatic breast cancer using seven common drugs to inhibit epithelial-to-mesenchymal transition and augment capecitabine efficacy. *Breast Cancer* **2017**, *9*, 495–514. [CrossRef] [PubMed]

141. Moore, K.M.; Thomas, G.J.; Duffy, S.W.; Warwick, J.; Gabe, R.; Chou, P.; Ellis, I.O.; Green, A.R.; Haider, S.; Brouilette, K.; et al. Therapeutic targeting of integrin αvβ6 in breast cancer. *J. Natl. Cancer Inst.* **2014**, *106*, dju169. [CrossRef] [PubMed]

142. Gilboa-Geffen, A.; Hamar, P.; Le, M.T.; Wheeler, L.A.; Trifonova, R.; Petrocca, F.; Wittrup, A.; Lieberman, J. Gene Knockdown by EpCAM Aptamer-siRNA Chimeras Suppresses Epithelial Breast Cancers and Their Tumor-Initiating Cells. *Mol. Cancer Ther.* **2015**, *14*, 2279–2291. [CrossRef] [PubMed]

143. Sempere, L.F.; Keto, J.; Fabbri, M. Exosomal microRNAs in breast cancer towards diagnostic and therapeutic applications. *Cancers* **2017**, *9*, 71. [CrossRef] [PubMed]

144. Ma, F.; Li, W.; Liu, C.; Li, W.; Yu, H.; Lei, B.; Ren, Y.; Li, Z.; Pang, D.; Qian, C. MiR-23a promotes TGF-β1-induced EMT and tumor metastasis in breast cancer cells by directly targeting CDH1 and activating Wnt/β-catenin signaling. *Oncotarget* **2017**, *8*, 69538–69550. [CrossRef]

145. Wang, N.; Wei, L.; Huang, Y.; Wu, Y.; Su, M.; Pang, X.; Wang, N.; Ji, F.; Zhong, C.; Chen, T.; et al. miR520c blocks EMT progression of human breast cancer cells by repressing STAT3. *Oncol. Rep.* **2017**, *37*, 1537–1544. [CrossRef] [PubMed]

146. Ma, L.; Reinhard, F.; Pan, E.; Soutschek, J.; Bhat, B.; Marcusson, E.G.; Teruya-Feldstein, J.; Bell, G.W.; Weinberg, R.A. Therapeutic silencing of miR-10b inhibits metastasis in a mouse mammary tumor model. *Nat. Biotechnol.* **2010**, *28*, 341–347. [CrossRef] [PubMed]

147. Knezevic, J.; Pfefferle, A.D.; Petrovic, I.; Greene, S.B.; Perou, C.M.; Rosen, J.M. Expression of miR-200c in claudin-low breast cancer alters stem cell functionality, enhances chemosensitivity and reduces metastatic potential. *Oncogene* **2015**, *34*, 5997–6006. [CrossRef] [PubMed]

148. Bai, W.D.; Ye, X.M.; Zhang, M.Y.; Zhu, H.Y.; Xi, W.J.; Huang, X.; Zhao, J.; Gu, B.; Zheng, G.X.; Yang, A.G.; et al. MiR-200c suppresses TGF-β signaling and counteracts trastuzumab resistance and metastasis by targeting ZNF217 and ZEB1 in breast cancer. *Int. J. Cancer* **2014**, *135*, 1356–1368. [CrossRef] [PubMed]

149. O'Brien, K.; Lowry, M.C.; Corcoran, C.; Martinez, V.G.; Daly, M.; Rani, S.; Gallagher, W.M.; Radomski, M.W.; MacLeod, R.A.; O'Driscoll, L. miR-134 in extracellular vesicles reduces triple-negative breast cancer aggression and increases drug sensitivity. *Oncotarget* **2015**, *6*, 32774–32789. [CrossRef] [PubMed]
150. Hannafon, B.N.; Carpenter, K.J.; Berry, W.L.; Janknecht, R.; Dooley, W.C.; Ding, W.Q. Exosome-mediated microRNA signaling from breast cancer cells is altered by the anti-angiogenesis agent docosahexaenoic acid (DHA). *Mol. Cancer* **2015**, *14*, 133. [CrossRef] [PubMed]

Review

Complex Determinants of Epithelial: Mesenchymal Phenotypic Plasticity in Ovarian Cancer

Yuliya Klymenko [1,2], Oleg Kim [3,4] and M. Sharon Stack [1,*]

[1] Department of Chemistry and Biochemistry, Harper Cancer Research Institute, University of Notre Dame, Notre Dame, IN 46617, USA; yklymenk@nd.edu
[2] Medical Sciences Program, Indiana University School of Medicine, Bloomington, IN 47405, USA
[3] Department of Applied and Computational Mathematics and Statistics, Harper Cancer Research Institute, University of Notre Dame, Notre Dame, IN 46617, USA; okim@nd.edu
[4] Department of Mathematics, University of California Riverside, Riverside, CA 92521, USA
* Correspondence: sstack@nd.edu; Tel.: +1-574-631-4100

Received: 9 June 2017; Accepted: 6 August 2017; Published: 9 August 2017

Abstract: Unlike most epithelial malignancies which metastasize hematogenously, metastasis of epithelial ovarian cancer (EOC) occurs primarily via transcoelomic dissemination, characterized by exfoliation of cells from the primary tumor, avoidance of detachment-induced cell death (anoikis), movement throughout the peritoneal cavity as individual cells and multi-cellular aggregates (MCAs), adhesion to and disruption of the mesothelial lining of the peritoneum, and submesothelial matrix anchoring and proliferation to generate widely disseminated metastases. This exceptional microenvironment is highly permissive for phenotypic plasticity, enabling mesenchymal-to-epithelial (MET) and epithelial-to-mesenchymal (EMT) transitions. In this review, we summarize current knowledge on EOC heterogeneity in an EMT context, outline major regulators of EMT in ovarian cancer, address controversies in EMT and EOC chemoresistance, and highlight computational modeling approaches toward understanding EMT/MET in EOC.

Keywords: ovarian cancer; intraperitoneal metastasis; cadherins; heterogeneity; epithelial-to-mesenchymal transition (EMT); mesenchymal-to-epithelial transition (MET); intraperitoneal tumor microenvironment; computational modeling of EMT

1. Introduction

Most epithelial carcinomas disseminate via the bloodstream or lymphatic system, utilizing a classical invasion-metastasis cascade mechanism which involves the local invasion of primary tumor epitheliocytes into the surrounding stroma and extracellular matrix (ECM), intravasation and transport through blood/lymph vessels, arrest at distant organ sites, extravasation into the organ parenchyma, and subsequent proliferation to form micro- and macro-metastases [1–5]. The triggering and ultimate success of these events depends on the epithelial-to-mesenchymal transition (EMT) and its key players, E-cadherin (epithelial, Ecad) [6] and N-cadherin (neural, Ncad) [7]–calcium–dependent transmembrane adhesion molecules which are responsible for maintaining cell-cell junctions between adjacent cells, thereby regulating the epithelial integrity and tissue architecture. During EMT, epithelial-type cancer cells undergo a set of molecular, morphological and functional changes with the loss of Ecad and gain of Ncad, which result in impaired epithelial cell-cell junctions and cell polarity, acquisition of a mesenchymal motile cell phenotype, and labile bonding with Ncad-expressing fibroblasts [8,9], endothelial cells and pericytes [10,11]. These changes facilitate cancer cell migration through stromal tissue, intravasation, and dissemination throughout the organism. The opposite process, designated mesenchymal-to-epithelial transition (MET), includes reverse cadherin switching (Ncad inhibition and Ecad re-expression) and often occurs at the secondary metastatic site, allowing for anchored and extravasating cancer cells regain epithelial features and proliferate into larger tumor nodules [2,3,12].

Epithelial ovarian cancer (EOC) is the deadliest gynecological malignancy, which stably ranks fifth highest among cancer deaths for women, and the American Cancer Society predicts that 14,080 women will die from ovarian cancer in 2017 [13,14]. The high mortality is primarily due to detection at late stages of the disease with vast intra-peritoneal dissemination [15] and to development of drug resistance after initial good response to treatment [16]. As opposed to other malignancies which progress through the above described canonical hematogenous invasion-metastasis cascade, EOC undertakes a distinct transcoelomic route of spread (through peritoneum-covered surfaces and organs of the abdominal and pelvic cavity), expanding via direct extension of cancer cells from the primary tumor into the intra-abdominal fluid-filled space, where they survive and travel as single cells and multi-cellular aggregates (MCAs) with the peritoneal fluid flow, subsequently adhering to peritoneal tissues, migrating into sub-mesothelial matrix and forming secondary lesions [17–19]. Recently, metastatic spreading of EOC via lymphatic [20] and blood [21,22] systems in vivo were reported. Nevertheless, the proposed new hematogenous models of EOC metastasis further highlight the involvement of the ovary in this process, as oophorectomy resulted in a complete abruption of peritoneal metastases and ascites development in mice [22]. These data suggest that, even in the case of hematogenous EOC cell circulation, metastatic EOC cells home to the ovary prior to further harnessing a predominantly intraperitoneal dissemination mechanism.

Hematogenously metastasizing solid tumors must first invade the tumor stroma and access the vasculature, necessitating an early EMT in order to adopt a motile, invasive phenotype. In contrast, the unique transcoelomic route of EOC dissemination generates an exceptional microenvironment quite distinct from most solid tumors, as cells are exfoliated directly into the peritoneal cavity. Thus, early events in metastatic dissemination do not require a mesenchymal phenotype. Alternatively, EOC exhibits phenotypic plasticity with regard to cadherin switching and exhibits significant cadherin heterogeneity during metastasis. In this review we focus on the peculiarities of the EMT/MET process in ovarian carcinoma, discuss tumor site-of-origin as a premise for EOC epithelial/mesenchymal heterogeneity, assess the potential clinical relevance of this plasticity, outline established and potential mediators of EMT/MET in EOC, and share our thoughts on possible future directions for EMT research.

2. EOC Cell of Origin: A Current Controversy

2.1. Ovarian Surface Epithelum Origin

It was widely accepted for years that EOC arises from transformation of the normal ovarian surface epithelium (OSE), a mesodermally derived and hormone-dependent single cell layer that covers the ovary (Figure 1). The OSE regularly undergoes cycles of rupture and repair associated with ovulation and thereby flexibly shifts between mesenchymal and epithelial phenotypes in response to the need for migration and proliferation to regenerate the intact epithelial surface [23]. The normal OSE exhibits flat morphology with high expression of mesenchymal markers (Ncad, calretinin, mesothelin) and absence of epithelial markers (Ecad, EpCAM, EMA, OVGP1 and ciliary bodies) [24]. Conversion of OSE towards a cuboidal/columnar phenotype, indicative of tubal metaplasia of the OSE, is accompanied by acquisition of epithelial markers (Ecad, EpCAM, OVGP1, ciliary bodies) and suppression of mesenchymal markers (modestly downregulated Ncad and fully abrogated calretinin); alterations also characteristic of MET [24]. Among pathogenic factors suggested to initiate OSE metaplasia and malignant transformation is presence of inclusion cysts. The uneven ovarian surface contains invaginations and inclusion cysts that lead to overcrowding of OSE in these regions. Adaptation to a cuboid epithelial-like shape in these regions with accompanying metaplastic changes can then occur [25]. OSE cells trapped inside the inclusion cysts are more exposed to growth factors which may provide additional cues for neoplastic progression. OSE may also launch autocrine mechanisms through the release of hormones and cytokines [23]. Alternatively, OSE may undergo metaplasia to acquire Müllerian duct features with subsequent neoplastic progression to tumor formation [23].

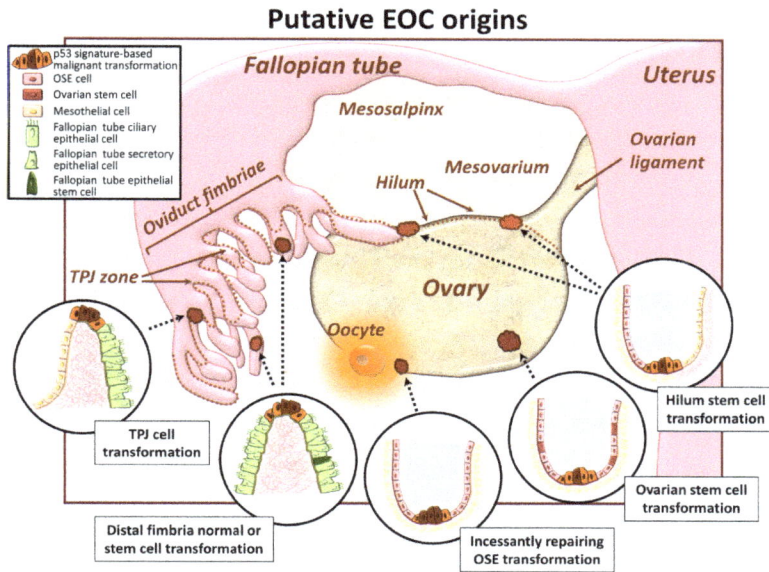

Figure 1. Putative epithelial ovarian cancer (EOC) origins. Represented is a schematic of the female reproductive system and the potential sites for ovarian cancer initiation. In the ovary, EOC may arise from the neoplastic conversion of the normal ovarian surface epithelium (OSE), as a result of frequent ovulatory rupture and post-ovulatory healing accompanied by transitions between mesenchymal and epithelial cell phenotypes; from malignantly transformed ovarian stem cells residing on the ovarian surface; or from the hilum, a translational zone between the ovary and mesothelium (and to some extent oviduct fimbriae), that contains cells with stem cell properties and propensity to tumorigenesis. In the fallopian tube, EOC may arise from the normal or stem-like cells in the distal fimbria actively participating in reproductive function-related fimbria injury; or from the tubal-peritoneal junction (TPJ), a region connecting mesothelium and oviductal fimbriae mucosa, often comprising translational metaplasia changes.

2.2. Oviduct Fimbriae Epithelium Origin

In recent years, the epithelium of the distal fimbriae of the fallopian tube has been recognized as a potential site of origin of ovarian cancer (Figure 1). It is a pseudo-stratified columnar cell monolayer comprised of secretory and ciliated cells and exhibits both epithelial (high levels of Ecad, EpCAM, EMA, OVGP1) and mesenchymal markers (ubiquitous expression of Ncad and vimentin, and varying/absent expression of calretinin and mesothelin) [24,26]. Specifically, serous tubal intra-epithelial carcinoma (STIC), developed via acquisition of a p53 signature in a distal oviduct fimbriae tumor-initiating cell, is considered to be the site of tumor initiation, while tumors on the ovary and peritoneal surfaces are thought to be secondary lesions [27–32]. This hypothesis is supported by the fact that to date no premalignant lesions for ovarian carcinoma has yet been established. Furthermore, STICs can be found in patients without ovarian cancer, while ovarian and peritoneal malignancies are often associated with the presence of STIC [33]. Other aspects supporting this notion include the close physical proximity of the tube distal fimbria to the ovarian surface and the high frequency of identical p53 mutations in STICs and high grade serous adenocarcinomas [33]. Comprehensive analyses specifically focusing on accumulating evidence in support of both OSE and tubal fimbriae as sources of EOC initiation are published elsewhere [26,34].

2.3. Ovarian and/or Tubal Stem Cell Transformation

Recent findings propose stem cell transformation as a potential cause of ovarian cancer initiation (Figure 1) [35]. It has been postulated that ovarian carcinomas are heterogeneous tumors and contain a small number of cells with stem cell-like characteristics that express specific stem cell markers [35]. Retrieved from patient ascites, ovarian cancer cells positive for stem cell-related molecules (such as Oct4, Nestin, and Nanog) exhibit growth in an anchorage-independent manner in vitro and present serial metastatic tumors in vivo [35]. A large gene expression profiling study of normal OSE samples and patient-derived tissues from serous papillary adenocarcinoma patients revealed that a substantial pool of OSE cells are not fully differentiated (multipotent) and retain the capacity to serve as ovarian cancer initiating cells [36]. This side population of cells was identified and characterized from distinct genetically engineered mouse ovarian cancer cell lines, and showed the ability to efflux the DNA-binding dye Hoechst 33342, formed measurable tumors more rapidly and were less sensitive to the chemotherapeutic agent doxorubicin when compared with non-side population cells [18,37]. The detection of ovarian carcinoma stem cells with an ability to self-renew and high epithelial plasticity raises the interesting possibility that these cells have higher metastatic efficiency and may be responsible for the majority of metastasizing ovarian carcinoma cells [18]. Recently, a cancer-prone ovarian cancer stem cell niche was identified [38]. It constitutes the hilum region of the ovary, a translational (or junctional) area between the OSE, mesothelium and tubal (oviductal) epithelium [38]. The hilum region contains a large population of slowly cycling cells that express stem and progenitor cell markers and demonstrate a high propensity to malignant transformation [38]. Another possible "hot spot for carcinogenesis" was identified in a study that characterized the tubal-peritoneal junction (TPJ, a zone where peritoneal mesothelium and oviductal fimbria mucosa meet) with regions of translational metaplasia [39]. Finally, fallopian tube epithelium stem-like cells, lacking markers of ciliated or secretory differentiation, were isolated at the distal (fimbriated) end of the fallopian tube, at the site of frequent, reproductive function-associated fimbria injury, and may also serve as a starting point for EOC carcinogenesis [40]. An elegant report illustrating current knowledge about ovarian and tubal epithelium stem-like cells, their molecular properties and cell niches is published in [41].

3. Phenotypic Heterogeneity and Relevance of Intraperitoneally Residing Cells/MCAs

As described above, tumors classified as ovarian cancer can arise from histologically and molecularly distinct neoplastic progenitors: either from mesenchymal-type OSE that has undergone metaplastic MET changes; from the highly differentiated columnar fallopian tube epithelum which contains both epithelial (predominantly) and mesenchymal (moderately) traits; or from a variety of ovarian surface, oviductal fimbriae or junctional area-localized stem-like precursors which may diversely express epithelial, mesenchymal and stem cell markers. Given the unique intraperitoneal metastasis mode and the diversity of progenitors, it is not surprising that malignant ascites contains a heterogeneous pool of individual and clustered (as MCAs) free-floating metastatic units within the peritoneal cavity [17,42–44]. Furthermore, mechanisms that regulate metastatic MCA formation are currently unknown, however options include exfoliation of single cells from the primary tumor followed by association into aggregates; exfoliation of cell-cell adherent sheets from the primary tumor that subsequently condense into MCAs; or through proliferation of individual cells in suspension into aggregates.

The implications of the clinically observed epithelial, mesenchymal and intermediate cell phenotypes are a subject of continued active investigation. Recent studies show that ovarian cancer cell metastatic properties, aggregation dynamics, MCA surface morphology, inner ultrastructure and behavior differ among epithelial, intermediate and mesenchymal phenotypes and are regulated by E- and Ncad expression both in their free-floating state [43] and at the secondary metastatic site [45].

It has been reported that mesenchymal-type EOC cells tend to aggregate into very compact solid spheroids, whereas epithelial-type EOC form loose and easily disaggregated MCAs [43], and these properties are modulated by shifts in E- and N-cadherin expression. These findings are interesting in the content of MCA survival and chemosensitivity, as such differences in aggregate cohesivity may drive distinct responses of epithelial and mesenchymal MCAs to hypoxic conditions, glucose starvation, and drug penetration to cells in the MCA core.

3.1. Mesenchymal Phenotype

A recent study [45] delineated that within a mixed population of EOC cells and MCAs, only mesenchymal cells and mesenchymal cell-containing aggregates exhibit the ability to invade through 3-dimensional (3D) collagen-rich matrices. Moreover, this process relies on competent Ncad cell-cell junctions and can be successfully blocked by N-cadherin antagonists. Overexpression of Ncad also stimulated mesothelial cell clearance by cancer MCAs [45]. These findings may provide a partial explanation for differential metastatic efficiency among various ovarian cancer cell lines in murine models. Epithelial type ovarian cancer cells, such as OvCa433, do not readily form intraperitoneal tumor xenografts [46], and in vivo growth of this cell line is limited to subcutaneously injected regions [47]. Meanwhile, mesenchymal-type (Ncad+) cell lines easily create large numerous widely-disseminated metastatic lesions in vivo, often accompanied by cancerous cachexia and ascites in mice [48–50]. An immunoblot analysis of two metastatically successful EOC cell lines (SKOV3 and OVCAR8) and their highly metastatic in vivo-selected intraperitoneal (ip) derivatives (SKOV3ip and OVCAR8ip), revealed elevated Ncad in the absence of Ecad in aggressive ip sublines (Figure 2). This is similar to data obtained from analysis of primary tumors and patient-matched metastatic lesions, wherein enhanced Ncad expression is observed in metastases relative to the corresponding primary tumor [17]. Together these data suggest that intraperitoneal delivery of Ncad blocking molecules could be beneficial for suppressing development of metastatic lesions. Moreover, given that EMT is often a reversible process, selective targeting of EMT mediators to induce a MET program in mesenchymal-type EOC cells [12] may become potentially advantageous against peritoneal lesion formation.

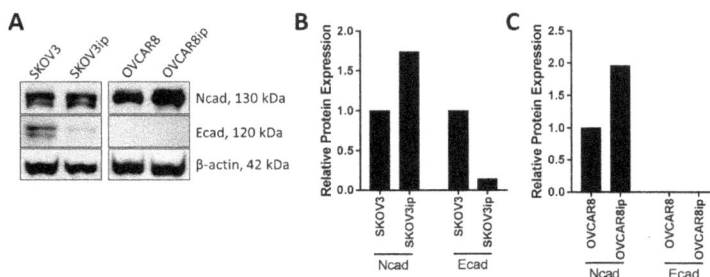

Figure 2. Cadherin expression alterations in human EOC cell lines and their highly metastatic derivatives. (**A**) Ecad and Ncad expression in parental SKOV3 and OVCAR8 cells and their matching highly metastatic derivatives, SKOV3ip and OVCAR8ip (isolated from murine intraperitoneal metastases), was assessed by Western blot analysis with the rabbit monoclonal anti-Ecad (Abcam, 1:1000 dilution) or mouse anti-Ncad (Life Technologies, 1:1000 dilution) primary antibody, respectively, followed by peroxidase-conjugated anti-rabbit or anti-mouse secondary antibody (Sigma-Aldrich, 1:4000 dilution) and enhanced chemiluminescence detection by ImageQuant LAS4000 biomolecular imager. (**B**,**C**) The densitometric analysis of immunoblot bands represented in Figure 2A was conducted in ImageJ, and protein expression (band relative intensity) was normalized against β-actin loading control.

3.2. Epithelial Phenotype

Epithelial phenotype cells may also promote EOC progression. Existing literature suggests that enhanced Ecad expression in early stages of EOC progression is necessary to avoid detachment-induced apoptosis and for resistance to radiation and chemotherapy [51–55]. In agreement with these observations are studies reporting enhanced chemoresistance in EOC cells of epithelial phenotype as compared to mesenchymal phenotype (see Section 5 for further discussion) [56,57]. One the other hand, loss of Ecad is canonically associated with the disruption of adherens junctions, and shedding of the Ecad extracellular domain induced by different bioactive molecules (lysophosphatidic acid, epidermal growth factor, etc.) leads to increased detachment of EOC cells from the tumor surface [23,58–61]. Besides the loss of cell-cell junctions and exfoliation/dissemination of more tumor cells into the peritoneal cavity, the shed Ecad ectodomain retains functional significance and was documented to further disrupt existing cell junctions between EOC cells [62], stimulate cancer cell invasion by triggering signaling pathways [63], and upregulate matrix metalloproteinase (MMP) MMP-2, MMP-9 and membrane type 1 (MT1-MMP) expression [64]. Meanwhile, the freed cytoplasmic domain of Ecad may also potentiate cancer progression via Wnt signaling, as the release of junctional Ecad and β-catenin leads to nuclear accumulation of soluble β-catenin and enhanced transcriptional activity [65]. In some cases (as reported for esophageal squamous cell carcinoma), nuclear translocation of the cytoplasmic Ecad fragment can alone promote transcriptional regulation, independently of β-catenin recruitment [66]. More on the role of Ecad and the functional significance of its cleaved domains is reviewed in [67,68].

3.3. Intermediate (Hybrid) Phenotype

It is currently well recognized for many malignancies that cancer cells do not transiently exhibit hybrid epithelial/mesenchymal (E/M) properties, but rather stably maintain a certain level of intermediate (partial EMT, semi-EMT) phenotype [69–72], which provides benefits over purely epithelial or mesenchymal morphological states. As widely reported in literature, the hybrid E/M phenotype allows cells exist dynamically, adjusting their differentiation and functionality in response to the environmental milieu [69,73–76], garner resistance to cell death, radiation and chemotherapeutic agents [69,77–79]. Moreover, multiple studies indicate that in contrast to cells in a finite epithelial or mesenchymal state, hybrid E/M cells more readily display stemness properties, such as multipotency and self-perpetuation [80,81], (re)-expression of stem cell markers [76,80,81], sphere formation [80,82] and tumor-initiating potential [81,82]. For ovarian cancer in particular, a subset of in vivo tumorigenic cells have been detected in a hybrid E/M state [81], which simultaneously express epithelial, mesenchymal and cancer stem cell markers, and retain the capacity for self-renewal as well as generation of more differentiated progenies. The authors demonstrated that the differentiation fate of the hybrid E/M ovarian cancer cells is associated with EMT pathways and may be regulated by environmental stimuli or the state of adjacent (epithelial or mesenchymal) cells [81]. Similarly, dual cadherin-expressing (intermediate) EOC cell lines OVCAR3 and OvCa432 proliferate to create a heterogeneous mixture containing hybrid Ecad+/Ncad+, predominantly Ecad+ and predominantly Ncad+ progenies within the same subclone, as confirmed by immunostain analysis [43]. Furthermore, the newly generated hybrid Ecad+/Ncad+ subclones exhibit higher proliferation indices in comparison with their parental, purely epithelial or purely mesenchymal EOC cells [43]. Additionally, substratum-free EOC cell aggregation relies on homotypic Ecad:Ecad or Ncad:Ncad interactions; as a consequence, EOC cells in co-culture exhibit the propensity to sort into purely epithelial Ecad-expressing and purely mesenchymal Ncad+ MCAs, whereas hybrid cells presenting both cadherins can form heterogeneous MCAs with all cell phenotypes [43]. This may serve an additional tumor-supporting role through recruitment of these stem-like cells adjacent to fully differentiated cells (e.g., to the metastatic niche together with mesenchymal-type cells or as an apoptosis-resistant component of epithelial-type aggregates). Our current understanding of EOC cell/MCA phenotypic heterogeneity, plasticity and the significance for ovarian cancer progression is summarized in Figure 3.

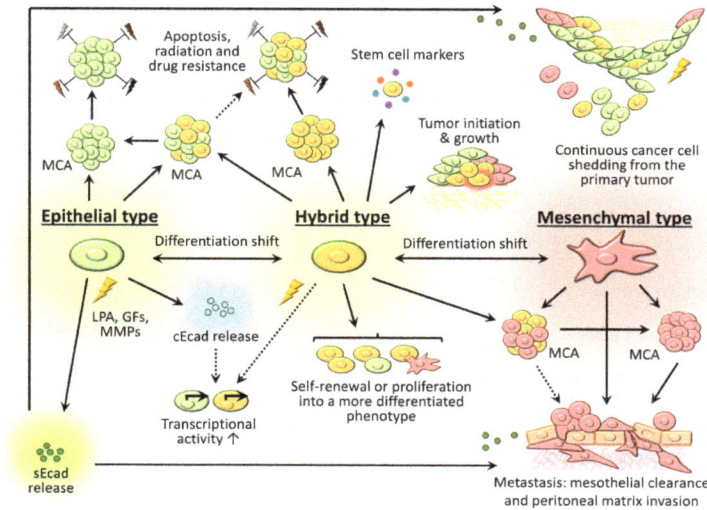

Figure 3. Phenotypic plasticity of EOC cells in the peritoneal cavity. Triggered by the soluble factors abundant in intraperitoneal fluid such as the bioactive lipid lysophosphatidic acid (LPA), growth factors (GFs), and matrix metalloproteinases (MMPs) (all designated as yellow lightening), ovarian cancer cells exist as a heterogeneous mixture of free-floating epithelial, mesenchymal and intermediate (hybrid) cells inside the peritoneal cavity. These cells can dynamically shift between phenotypes along the EMT spectrum adjusting to microenvironmental cues. Multicellular aggregation (MCA) of epithelial-type cells (green) is reported to play a pro-survival role through developing resistance to cell death, radiation and chemotherapeutics (designated grey, brown and black lightening). Bioactive molecules promote cleavage of the Ecad ectodomain to generate a soluble Ecad fragment (sEcad, green filled dots). The sEcad fragment may promote further disruption of cell-cell junctions between the epithelial and hybrid cells and enhance cell shedding from the primary tumor surface. The released Ecad cytoplasmic domain (cEcad, green open dots) may stimulate cell transcriptional activity. Mesenchymal-type cells (pink) readily form MCAs and are predominantly involved in metastasis-associated behaviors, exhibiting mesothelial cell (orange cuboid) clearance activity and peritoneal matrix invasion. Additionally, cell invasion may be amplified by stimuli from sEcad. The hybrid cells (yellow) retain stemness properties: expression of stem cell markers (colored dots), ability for self-renewal or proliferation into a more differentiated phenotype, tumor initiation (tumor-initiating cell designated red-circled) and growth. Hybrid cells may also form mixed MCAs with either epithelial or mesenchymal cells, undergo differentiation consistent with the phenotype of the adjacent cells in co-culture and share their subsequent fate. Black arrows represent interactions reported in ovarian carcinomas; dashed arrows designate patterns that were observed in other cancer types and may potentially be applicable towards ovarian cancer, but require further validation.

4. Factors Contributing to Dynamic EMT Shifts in Ovarian Cancer Cells

4.1. Components of Ascitic Fluid

While MET processes accompanying ovarian carcinoma initiation usually occur due to genetic alterations (neoplastic metamorphosis built on a p53 signature background as a consequence of incessantly damaged OSE or distal oviduct fimbriae, or vulnerabilities of the translational metaplasia regions in the areas of junction between two distinct epithelia), the subsequent shifts along the MET/EMT spectrum in intraperitoneally-residing cells are due not to genetic mutations, but rather occur as a response to a variety of external cues arising from the ascitic microenvironment (Figure 4).

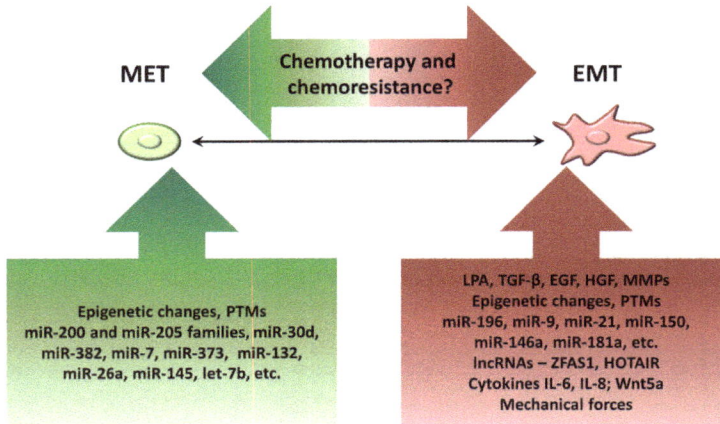

Figure 4. Overview of EMT regulators in ovarian cancer. The diagram summarizes major factors involved in MET (green box) and EMT (red box) programs during EOC progression. The interplay between EMT shifts and ovarian cancer chemotherapy/chemoresistance currently remains controversial.

One of the most well-characterized factors implicated in ovarian carcinogenesis and metastasis is lysophosphatidic acid (LPA), a bioactive lipid molecule which is notably abundant in ascites and plasma of patients with ovarian cancer and activates a subfamily of G-protein coupled cell surface receptors, eliciting a variety of cellular responses such as aberrant proliferation, adhesion, migration, invasion and anoikis-resistance [83–88]. It was shown that LPA is constitutively produced by the mesothelial cells of the peritoneum, increasing EOC cell adhesive, migratory and invasive properties [89]. In the content of EMT, LPA was shown to induce shedding of the 80 kDa extracellular domain of Ecad from EOC cells in a urokinase plasminogen activator (uPA)-dependent manner, disrupting cell-cell junctions and promoting a cellular mesenchymal phenotype with enhanced motility and invasion [63]. Notably, in this work the authors showed that the cleaved soluble Ecad itself also stimulates EOC cell invasion [63]. Additionally, LPA upregulates expression of matrix metalloproteinase-9 (MMP-9) and triggers MMP-9-catalyzed Ecad ectodomain shedding in a concentration- and time-dependent manner, disrupting junctional integrity in ovarian cancer cells and contributing to EMT [60]. Notably, blocking LPA receptors successfully suppressed MMP-9 expression levels and restored EOC cell-cell adhesion [60], suggesting a possible approach for therapeutic intervention. LPA is also involved in reorganization of the actin cytoskeleton, causing rearrangement of stress fibers and cell adhesion together with enhancement of MMP-2 enzymatic activity [90]. MMP-2 activation is enhanced in LPA-treated cells and results in enhanced MMP-dependent migratory and invasive behavior of EOC cells [91]. Following the disruption of cell-cell junctions, LPA also initiates loss of junctional β-catenin, promotes its nuclear translocation and reinforces the EMT program in ovarian cancer cells via activation of Wnt/β-catenin signaling nexus [92]. Recently, LPA was elucidated to stimulate the epithelial-to-mesenchymal switch in ovarian cancer cells via downregulation of SIRT1, an established inactivator of ZEB1 and suppressor of EMT [93]. Additionally, LPA intensifies EMT in ovarian cancer through upregulation of Slug/Snail2 EMT markers via Gαi2, Src, and HIF1α signaling pathway [94].

In addition to LPA, other growth factors, such as epidermal growth factor (EGF), hepatocyte growth factor (HGF), transforming growth factor-β (TGF-β), present in ovarian neoplastic microenvironment, are thought to influence EMT by inducing cleavage of Ecad fragments to impair cell-cell cohesion and via enhancing multiple complex signaling networks and transcriptional activity. These studies are described elsewhere [95–98].

Several MMP family members are known to be involved in EMT processes for different types of epithelial cancers, including ovarian carcinomas. Due to their proteolytic activity, multiple MMPs reportedly cause cleavage of the Ecad ectodomain and reorganization of zonula occludens (repression of tight junction protein ZO-1), hence, abrogating cell-cell adhesion, augmenting β-catenin nuclear translocation and activation of Wnt signaling, boosting acquisition of mesenchymal traits by cancer cells and enhancing cell invasiveness both in vitro and in vivo [60,91,99–103]. The later process is also reinforced due to the well-known role of MMPs in proteolysis and remodeling of diverse components of extracellular matrices [104–106]. Notably, MMP expression and activity may in turn be regulated by EMT-associated factors, such as Snail [107] or β-catenin [108–110]. More on the diverse effects of MMPs on cancer progression and metastasis, their targets, regulation mechanisms and potential therapeutic implications is discussed in [111–113].

Ovarian cancer cells are also subjected to cytokines and immune cells present in EOC-associated malignant ascites, that may potentially contribute to EMT shifts. In particular, interleukins (IL)-1β, IL-6, IL-8 and IL-10 are substantially abundant in ascitic effusions from EOC patients ([114–116] and reviewed in [117]). Notably, the literature reports a pro-EMT role for IL-6 and IL-8 in ovarian cancer. Attenuated Ecad expression, upregulation of β-catenin and enhanced SKOV3 and OVCAR3 cell migration in response to IL8 has been demonstrated [118]. Similarly, exposure of EOC and other gynecologic cancer cells to IL-6 treatment mitigates epithelial marker expression and elevates mesenchymal markers expression, increases MMP-2 and MMP-9 activity and avidly enhances migratory and invasive properties [119]. Additionally, IL-6 is known to be involved in EMT shifts, self-renewal, inducing stemness properties in cancer cells and drug resistance (outlined in [120]). Interestingly, production of both IL-6 and IL-8 is reinforced by LPA through transcriptional activation of IL gene promoters and upregulation of LPA receptors [121]. Evidence for chemokine ligand 5 (CCL-5) and chemokine receptors (CCR)-1/3/5 [122], CCL-19/21 and CCR-7 [123] involvement in EMT events and ovarian carcinoma metastasis have also been reported.

Other bioactive molecules may also participate in the EMT/MET program in EOC. For example, the mucin CA125/MUC16 may be associated with EMT in ovarian cancer due to its binding to Ecad and β-catenin complexes, as its downregulation results in epithelial-mesenchymal switch (as evidenced by loss of Ecad and cytokeratin-18 and gain of Ncad and vimentin), re-activation of EGFR signaling and increases in MMP-2 and MMP-9 expression and functional activity [124]. Contrary to these results are data [125] demonstrating increased EOC tumor growth, colony formation, cell motility, invasive and metastatic properties upon overexpression of MUC16, accompanied by loss of Ecad and enhancement of Ncad and vimentin expression. Non-canonical Wnt5a signaling is another potential EMT-regulator and has recently been the subject of a comprehensive review [126]. Wnt5a protein is rich in ovarian cancer patients' malignant ascites [126] and its expression is shown to correlate with vascular mimicry and metastatic success [127]. Therefore, additional research to elucidate the role of Wnt5a in EOC progression and metastasis is warranted.

4.2. Epigenetic Changes

EMT-related epigenetic changes have been described for many cancers including EOC [128]. Silencing of CDH1 (Ecad) by promoter hypermethylation takes place in breast, prostate, gastric and urinary bladder cancers [129–131]. Zinc finger E-box-binding homeobox 1 (ZEB1), a known transcriptional repressor of Ecad, can also work epigenetically through recruitment of a DNA methyltransferase 1 (DNMT1) to the CDH1 promoter [132]. Among the most common histone modifications involved in EMT are histone demethylation at the SNAI1 promoter (activate transcription of SNAI1), di- or trimethylation at the TWIST promoter (TWIST activation), and histone mono-, di- and trimethylation at the CDH1 promoter by multiple methyltransferases (CDH1 transcription suppression) [128,133–136]. Histone acetylation (recruitment of histone acetyl transferases, HATs, and transcription activation) and deacetylation (recruitment of histone deacetylases, HDACs, and transcription inactivation) are mechanisms often employed for regulation of Ecad expression. SNAI1 and ZEB1 use HDACs for Ecad silencing [137,138]. TWIST recruits a multi-protein complex capable of deacetylation and nucleosome remodeling to the Ecad promoter, thus suppressing Ecad transcription [139]. On the contrary, reactivation of Ecad expression can result from deacetylation of its transcriptional repressor SNAI2 (SLUG) [140]. A comparative analysis of epigenetic events in breast and ovarian cancer related to tumorigenesis [141] identified genetic, microenvironmental, stromal, and epigenetic changes common between breast and ovarian cancer cells, as well as the clinical relevance of these changes. Some of the most striking commonalities included epigenetic alterations in H3K27me3, H3K9me2, H3K9me3, H4K20me3, and H3K4me suggesting shared features of pathogenesis in EOC and breast cancer and pointing out novel directions for managing EOC progression. Epigenetic profiling of ovarian cancer cells following TGF-β treatment discovered significant hypermethylation changes in the loci associated with EMT and cellular movement amongst others [142]. Moreover, the authors showed that TGF-β also stimulated expression and activity of DNMTs, while treatment with DNMT inhibitor SGI-110 abrogated TGF-β-mediated EMT [142]. Epigenetic silencing of secreted frizzled-related protein 5 (SFRP5), important in Wnt signaling regulation, activates the later and thus promotes EOC progression and drug resistance through TWIST-mediated EMT and AKT2 signaling [143]. IQGAP2 was found to be significantly hypermethylated in EOC [144] showing an inverse correlation between IQGAP2 DNA methylation and mRNA expression as IQGAP2 expression was downregulated in EOC. The subsequent survival analysis revealed that decreased IQGAP2 was associated with the low progression-free survival of EOC patients. Moreover, IQGAP2 was shown to suppress Wnt-induced β-catenin nuclear translocation and transcriptional activity, therefore inhibiting EMT, cell invasion and migration. Thus, IQGAP2 was identified as a novel EOC tumor suppressor via repression of invasion through Wnt/β-catenin signaling and was suggested as a new potential therapeutic strategy for EOC treatment. Numerous DNA methylation aberrations have been recognized in EOC. The examples provided above and summarized in Table 1 highlight the potential clinical implications as novel biomarkers for EOC diagnosis and disease progression. More details on candidate genes for epigenetic therapy of EOC can be found in [145–147].

Table 1. Target and candidate genes with EMT-associated epigenetic modifications.

Target or Candidate Gene	Epigenetic Modification	Reference
CDH1 (Ecad), silenced	hypermethylation via 5′ CpG island	[129–131]
	methylation via ZEB1 through recruitment of DNMT1 to CDH1 promotor	[132]
	histone H3K27m3 demethylation at the SNAI1 promotor; mono-, di- and trimethylation of histones H3K36m2, H3K4m2, H3K9m3, H4K20m1, H3K9m1/2 by methyltransferases MMSET, LSD1, Suv39H1, SET8 and G9a at the TWIST, CDH1s and CDH2 promoters	[128,133–136]
CDH1 (Ecad), reactivated	deacetylation of SNAI2	[140]
SFRP5, silenced	hypermethylation in EOC through Wnt signaling pathway	[143]
IQGAP2, silenced	hypermethylated in EOC via Wnt/β-catenin signaling	[144]
CDH1, ERBB3, FGFBP1, IGFBP4, IL1RN, MMP9, SNAI3, SPP1, WNT11, WNT5B (downregulated)	TGF-β induced methylation	[142]
BMP1, COL1A2, COL3A1, COL5A2, FOXC2, GSC, KRT14, KRT7, MMP2, MMP3, RGS2, SNAI1, TCF4, TFPI2, TGFB2, WNT5A, ZEB2 (upregulated)	TGF-β induced methylation	[142]

4.3. Posttranslational Modifications (PTMs)

The most commonly present PTMs mediating EMT include phosphorylation and glycosylation. Phosphorylation of SNAI1 motifs by different protein kinases (such as GSK-3β and protein kinase

D1) subsequently initiates its ubiquitination and degradation, thereby restoring expression of Ecad [148–150]. Inhibition of these protein kinases leads to Ecad repression and EMT. By contrast, glycosylation of SNAIL stabilizes it by preventing its phosphorylation and subsequent degradation, thus supporting EMT activation [151]. Another PTM, SUMOylation (binding of a small ubiquitin-like modifier, SUMO, to the target transcription factor) acts in a suppressive manner and is important for stabilizing and deactivating transcription factors engaged in EMT. In particular, in breast cancer the transcription factor Forkhead box protein M1 (FoxM1) is known to induce EMT via SNAI2 (SLUG) promoter stimulation [152], while in pancreatic cancer, FoxM1 upregulates ZEB1, ZEB2, SNAI2 and vimentin [153]. Yet another transcription factor Smad-interacting protein 1 (SIP1) acts as a downregulator of Ecad [154]. Impaired SUMOylation contributes to constant activation of these factors and, thus, enhances cadherin switching and EMT. One of the important promoters of EOC is the Hippo pathway signaling, which has been shown to affect several key signaling molecules via various types of PTMs [155,156]. In particular, the malfunction of critical Hippo signaling modules such as YAP/TAZ, MAT1/2 and LATS1/2 due to deregulated PTMs has been linked to different types of cancer including EOC. The current knowledge of PTMs with respect to the Hippo signaling pathway and possible therapeutic interventions targeting PTMs and Hippo signaling have been recently reviewed in [155]. Additional information on mediation of EMT at the post-translational level and an overview of various therapeutic approaches currently being investigated to undermine EMT can be found in [128].

4.4. MicroRNAs

Multiple microRNAs have been reported to serve as EMT promoters or inhibitors [157,158]. Members of the miR-205 and miR-200 family suppress EMT by binding the mRNAs encoding ZEB1 and ZEB2 [159,160]. Interestingly, both miR-205 and miR-200 family members negatively correlate with the expression of another EMT driver TWIST1, and are themselves transcriptionally silenced by TWIST1, which directly binds miR-205/miR-200 promoters and further intensifies EMT, complementing its well-known Ecad-suppressive role [161]. MiR-132/212 tandem miRNAs inhibit TGF-β-associated EMT by blocking the SOX4 gene in prostate cancer [162] and alleviate EMT and invasion of cervical cancer via SMAD2 downregulation [163]. Mir-132 directly silences ZEB2, thus attenuating EMT, invasion and metastasis in colorectal [164] and lung [165] cancers, and reduces EMT and migratory/invasive capacity of human non-small cell lung carcinoma (NSCLC) through the EMT-related TGF-β1/Smad2 pathway [166]. MiR-150 triggers EMT and metastatic behavior in NSCLC in vitro and in vivo through downregulation of FOXO4 [167]. In contrast, in esophageal squamous cell carcinoma, miR-150 induces MET-like changes and blocks murine xenograft tumor growth by targeting ZEB1 [168]. MiR-9 stimulates EMT via immediate repression of Ecad-encoding mRNA [169]. In lung cancer-initiating cells, miR-145 downregulates stem-like properties and EMT via blocking Oct4 [170]; alternatively, in NSCLC, miR-145 and -203 are involved in TGF-β-related EMT inhibition through targeting SMAD3 [171]. Among other known pro-metastatic (pro-EMT) microRNAs are miR-27 [172,173], miR-29a [174–176], miR-103/107 [177,178], miR-221/222 [179–181] and miR-661 [182]; alternatively, miR-26a [183,184], miR-26a-5p [185,186], miR-30a [187–189], miR-30a-5p [190], miR-134 [191], miR-194 [192–194], miR-192 and -215 [194], and miR-204 [195–198] exhibit anti-EMT activity.

A large number of microRNAs are implicated in ovarian carcinogenesis, tumor progression and in EMT in particular. A comparative study investigating expression of the miR-200 family, ZEB1 and ZEB2 transcriptional repressors in normal OSE vs. 15 EOC cell lines and 70 ovarian carcinoma tissues revealed that malignant transformation is associated with acquisition of more epithelial traits, such as upregulation of miR-200 family members and attenuation of ZEB1/2 [199], supporting the occurrence of MET, not EMT, during early stages of ovarian cancer progression. The miR-196 family, located in the Hox gene cluster, regulates (usually by inhibition) the HOX genes, in particular HOXA7, which is responsible for controlling the differentiation status in ovarian epithelium. HOXA7 overexpression is associated with the initiation of MET in ovarian epithelium and generation of low-grade Ecad-positive ovarian tumors [200]. Mir-9 levels are often elevated in ovarian cancer tissues in comparison with normal control tissues and are associated with EMT via targeting of Ecad by miR-9 and consequent enhancement of Ncad and vimentin expression [201]. Transcriptional regulator Snail is the direct target of miR-30d, which inhibits TGF-β1- induced EMT in ovarian cancer cells [202]. MiR-424 can suppress cell invasion and EMT via downregulation of DCLK1 in ovarian clear cell carcinoma, a subtype of EOC, associated with drug resistance and low survival rate [203]. MiR-382 serves as an ovarian cancer suppressor due to regulating EMT by targeting ROR1 and negatively impacting EOC cell migration/invasion [204]. A recently discovered miR-506 is one of the crucial down-regulators of EMT and metastasis in EOC through both immediate repression of transcriptional repressor SNAI2 (thus, restoring Ecad expression) and negative regulation of Ncad and vimentin, and is associated with a better survival prognosis [205]. Another EOC EMT- and metastasis-suppressor is miR-7, which acts through EGFR and AKT/ERK1/2 pathway inactivation [206]. More microRNA aberrations observed in EOC are concisely summarized in a review [207].

The concept of microRNA detection in physiological fluids in order to establish new predictive and diagnostic markers is gaining in popularity. A recent analysis of serum-circulating exosomal microRNAs revealed a broad increase in miR-373 and miR-200a in patients with ovarian serous adenocarcinoma across all stages (I-IV), while miR-200b and miR200c were more significantly elevated in later stages (III-IV) and correlated with worse survival outcome, suggesting that these microRNAs may differentially modulate EMT/MET shifts during certain EOC progression steps [208]. Another study [209] evaluating 2222 total microRNAs from ovarian cancer patient serum samples identified the most stably and markedly downregulated microRNAs as miR-132, miR-26a, and miR-145 (which are known to act as EMT-repressors in other tissue types, as discussed above), as well as let-7b, a microRNA that was recently shown to play multiple anti-tumor roles, including anti-EMT (through attenuation of p-AKT, Twist and β-catenin) and pro-apoptosis in malignant mesothelioma cells [210]. A 1,170 patient-based meta-analysis of global transcriptome data delineated let-7b as a stratification factor for molecular and clinical classification, and a predictor of poor survival outcome in high grade serous ovarian carcinoma [211]. Urinary miR-30a-5p was recently reported to be exclusively elevated in ovarian serous adenocarcinoma patients, both at early and metastatic stages of the malignancy and derived specifically from EOC tissue [212]. In contrast, elevation of miR-30a-5p in patients with benign gynecological diseases, gastrointestinal tumors and in the healthy control group was not detected. Moreover, targeted inhibition of miR-30a-5p considerably mitigated ovarian cancer pro-metastatic behavior in vitro [212]. In addition to EMT mediation, the majority of involved microRNAs play multiple roles in ovarian cancer initiation, progression, metastasis, stemness, microenvironmental and chemotherapeutic responses. In a paired analysis, miR-21, miR-150, and miR-146a were shown to be significantly overexpressed in EOC omental metastatic lesions in comparison with their matching primary tumor samples from the same patients [213]; moreover, miR-150 and -146a were also shown to regulate the size of EOC multicellular spheroids, cell survival and chemoresistance to cisplatin [213]. Amongst others, miR-181a is shown to facilitate TGF-β-mediated EMT through downregulation of SMAD7, contributing to cell survival, metastasis-associated behavior and chemoresistance in high grade serous ovarian carcinomas [214]. Further elucidation of signaling pathways and molecular targets of clinically dysregulated microRNAs in EOC is of profound importance, warranting discovery of new predictive markers, diagnostic tools and targeted therapeutic strategies. A comprehensive analysis (2014) based on evaluation of almost 100 research publications provides a structured summary of potentially relevant prognostic, diagnostic and therapeutic microRNAs for ovarian cancer [215]. Yet, another thorough review summarizes the most recent advances in clinical applications (including current clinical trials) of microRNAs for ovarian cancer precision medicine [216].

4.5. Long Noncoding RNAs (lncRNAs)

LncRNAs are a large class of RNAs transcripts having a length of more than 200 nucleotides which do not encode proteins. LncRNAs have gained widespread attention in recent years due to the vast implications in cancer biology, contributing to essential cellular functions including invasion, proliferation, differentiation, apoptosis, cell cycle progression and metastasis [217–220]. Although emerging as a new class of regulators in cancer progression, the role of lncRNAs in EOC and the relationship to EMT/MET in EOC has just begun to be explored.

A comprehensive study using clinical data from 544 ovarian cancer patients from The Cancer Genome Atlas (TCGA) examined lncRNA expression profiles [221], identifying an eight-lncRNA signature (RP4-799P18.3; PTPRD-AS1; RP11-57P19.1; RP11-307C12.11; RP11-254I22.1; RP11-80H5.7; RP1-223E5.4; GACAT3) permitting classification of patients into two groups characterized as high-risk + poor outcome and low-risk + significantly improved outcome. In particular, a superior prognosis performance in BRCA1/2-mutated and BRCA1/2 wild-type tumors was achieved by associating predictions with BRCA1 and BRCA2 mutations. It was shown that the eight-lncRNA signature may serve as a metric to predict chemotherapy response in patients and identify resistance to platinum treatment suggesting other more efficient therapies. These findings support that lncRNAs can be used as diagnostic or prognostic biomarkers in patients with EOC.

Upregulation of the lncRNA ZFAS1 was recently reported in EOC and negatively correlated with overall survival of patients with ovarian carcinomas [218]. It was established that overexpression of ZFAS1 increased proliferation, migration and chemoresistance in EOC cells. miR-150-5p was identified as a potential target of ZFAS1, suppressing the transcription factor Sp1. Meanwhile, inhibition of miR-150-5p partially restored proliferation and migration resulting from depletion of ZFAS1. Thus, the ZFAS1/miR-150-5p/Sp1 pathway was shown to be critical in inducing migration, differentiation and chemoresistance in EOC.

In another study [222], lncRNA HOX transcript antisense RNA (HOTAIR) expression in EOC tissues was evaluated and its correlation with clinico-pathological factors was established. Suppression of HOTAIR in three highly metastatic EOC cell lines (SKOV3.ip1, HO8910-PM, and HEY-A8) significantly reduced cell migration and invasion. Furthermore, the pro-metastatic effects were partially mediated by MMPs along with EMT-related genes. Specifically, siRNA-mediated silencing of HOTAIR increased expression of Ecad and decreased expression of vimentin and Snail.

TGF-β signaling has been shown to serve as a major promoting factor of EMT, facilitating EOC and breast cancer metastasis. Although the relationship between lncRNA and TGF-β in EOC is not known, the lncRNA profile in mouse mammary epithelial NMuMG cells upon TGF-β-mediated induction of EMT has been reported [223], identifying a subset of lncRNAs dysregulated upon TGF-β induced EMT with lncRNA-HIT mediating this process by targeting Ecad. These findings reveal a pivotal role that lncRNAs may play in EMT in breast cancer progression and warrant further studies examining relation between lncRNA profile and TGF-β-induced EMT in EOC. Collectively, these studies suggest a direct or indirect relationship between lncRNA and regulation of EOC invasion and metastasis, as well as new mechanisms involved in EOC EMT, which can potentially result in novel markers and therapeutic targets for epithelial ovarian cancer.

4.6. Biomechanical Forces

The unique transcoelomic route of ovarian cancer metastasis with excessive dissemination within the abdominal cavity leads to obstruction of lymphatic tissues, enriched in peritoneal walls, and early malfunction of the peritoneal lymphatic drainage system [224–226]. Excessive neovascularization of the primary tumor and peritoneal lining [227–231], secretion of factors increasing vascular permeability [232–236] and as a consequence, high protein concentration [237], motivate intraperitoneal transudate aspiration. A disrupted balance between fluid production and fluid clearance generates large volumes of ascites, followed by a dramatic increase in intra-peritoneal pressure (IPP) [238] which subsequently alters intraperitoneal mechanobiology. Meanwhile, biomechanical cues, driven by fluid build-up, ascitic currents, stretching of the peritoneal tissues and organ dislocation currently remain relatively unexplored. One published study has shown that laminar fluid flow drives a more aggressive phenotype in 3-dimensional ovarian micro-nodules through induction of EMT, partially via post-translational upregulation of EGFR, as well as significant downregulation of Ecad and CDC2 expression [239]. These results are in conformity with a study reporting triggering of EMT in laryngeal squamous cell carcinoma by fluid shear stress through integrin-ILK/PI3K-AKT-Snail signaling pathways [240]. Exposure to shear stress caused repression of Ecad with a simultaneous increase in Ncad expression and translocation of β-catenin into the nucleus, alteration of cell morphology to an elongated shape with invadopodia enrichment, and stimulated cell migration; furthermore, these alterations were reversible upon removal of mechanical stress [240]. Conversely, another study addressing the effects of fluid sheer stress on EMT and cancer stem cell (CSC) properties in breast cancer cells documented CSC-like signature promotion with no changes in EMT markers [241]. Peritoneal fluid dwell was shown to drive EMT and hyperplasia of mesothelial cells in an in vitro reconstructed peritoneal cavity model [242]. A transcriptome-wide analysis of microRNA profiles in breast cancer cells and cancer-associated fibroblasts in response to compressive strain revealed alterations in expression levels of microRNAs associated with EMT, migration, invasion, angiogenesis and apoptosis [243]. In 3D breast cancer aggregates, elevated interstitial fluid pressure guided EMT shifting and collective cell invasion through Snail, vimentin and Ecad gene expression alterations [244]. In normal and hypertrophic scar fibroblasts, mechanical compression altered expression of TGF-β signaling target genes SMAD2 and SMAD3 and further upregulated expression of MMP-2 and MMP-9 [245]. Potential alterations in ovarian cancer progression mediated by ascites-induced changes in peritoneal mechanobiology have yet to be uncovered. One recent study has revealed significant enhancement of SNAI1 gene expression in epithelial-type EOC cells and MCAs upon continuous patho-physiologically relevant compression (~22 mmHg), applied to mimic ascites-driven elevations in IPP as observed in the clinic [246]. Additionally, elevated mRNA and protein expression levels of Ecad and Ncad in EOC epithelial- and mesenchymal-type MCAs, respectively, heightened the distinction between these morphological phenotypes [246].

5. Controversy on EMT and Chemoresistance in EOC

Progress made in elucidation of EMT as a continuously evolving process rather than two distinct end stages has heightened controversies related to the divergent behavior of epithelial and mesenchymal cancer cells during metastatic spreading and their response to therapeutic interventions. The link between induction of EMT and the development of drug resistance has been confirmed for multiple cancer types and has gained the attention of researchers worldwide to identify new therapeutic targets that abrogate EMT and re-sensitize cancers to chemotherapeutic drugs. A recent report elegantly illustrates the mechanisms underlying EMT-dependent acquisition of resistance to diverse therapeutic agents observed in different types of cancers, and stratifies approaches for targeting the EMT program as a part of cancer treatment, focusing on prevention of EMT induction, selective targeting of cells that have undergone the EMT shift, and launching the reverse MET program in mesenchymal cancer cells [12].

In ovarian cancer, however, the association between EMT and drug resistance remains disputable, as numerous research data provide contradictory conclusions. For example, activation of Notch3 signaling in ovarian cancer OvCa429 cells (epithelial-type) launched EMT and attenuated carboplatin-induced apoptosis in these cells through inhibited ERK phosphorylation [247]. Similarly, promotion of EMT by hematopoietic PBX interacting protein [248], NANOG [249,250], TWIST1 [251,252], SNAI1 [253] FOXM1 [254] accompanied acquisition of ovarian cancer cell resistance to conventional therapeutic agents, such as cisplatin, carboplatin or paclitaxel. Contrary to these are data from a study that exploited sensitivity to cisplatin among 46 ovarian cancer cell lines and established a higher level of drug resistance in the cells with epithelial status [57]. Subsequent pathway analyses discovered activation of the NF-κB pathway by administered cisplatin exclusively in epithelial cells, resulting in defective apoptosis and cisplatin resistance [57]. Another study showed downregulation of miRNA-200 family members (microRNAs that act as supressors of EMT) in paclitaxel- and carboplatin-resistant ovarian cancer cells with a strong mesenchymal phenotype, but surprisingly, overexpression of miR-200c/miR-141 in these cells and partial restoration of epithelial traits led to a 6 − 8x higher resistance to carboplatin while showing no change in response to paclitaxel. Yet another study characterized tumor cell clusters from ovarian cancer patient ascites and identified that cells/clusters with mesenchymal markers were predominantly derived from chemo-naïve patients, whereas cellular components of the epithelial phenotype were found in ascites of patients with recurrent disease and chemoresistance [56]. Further, comparison of ascites-driven single cells of epithelial and mesenchymal morphology (confirmed by respective marker expression) depicted a significantly higher level of resistance to cisplatin in epithelial-type cells relative to mesenchymal-type cells [56].

Clearly, additional research is needed to fully understand the relationship between EMT/MET and chemoresistance in ovarian cancer, as it will allow recruitment of EMT regulators to combat EOCs significant challenges—multidrug resistance and metastatic aggressiveness. Given the distinct mechanism of ovarian cancer metastatic dissemination and the unique aspects of epithelial vs. mesenchymal EOC cells, nontrivial EMT-targeting therapeutic approaches might be required. In particular, current scientific knowledge highlights ovarian cancer epithelial-state cells as more chemo- and radiation resistant (as outlined in Sections 3 and 5 of this commentary), raising a speculation that new therapeutic interventions may actually lie in EMT *promotion* for re-sensitization of EOC cells to therapeutic agents—a strategy opposite to that suggested for other cancer types [12]. However, pre-clinical data including those of our group (see Section 3 of the current review) indicate that acquisition of the mesenchymal phenotype in EOC is particularly associated with aggressive metastatic invasion. In this case, as our latest report concludes [45], targeting Ncad on the surface of mesenchymal-type EOC cells with Ncad-blocking peptides, such as the HAV-motif harboring drug ADH-1 (Exherin) or monoclonal antibodies may represent a promising anti-metastatic strategy. Future studies designed to resolve the EOC EMT/chemoresistance controversies and target the unique characteristics of EOC cells are warranted.

6. Computational Modeling Approaches to Understanding EMT/MET in EOC

Computational systems biology models have become an indispensable tool in analyzing highly empirical cancer progression data and can greatly contribute to elucidating the underlying principles of EMT/MET in EOC. Regulatory networks underlying these transitions in EOC as well as other cancer types involve multiple signaling pathways including TGF-β, EGF, HGF, FGF, NF-kB, Wnt, Notch, Hedgehog, JAK/STAT, Hippo [255], and hypoxia [256]. In addition, the mechanical properties of the extracellular matrix (ECM) such as density [257] and stiffness [258] also play role in EMT/MET. These signals trigger activation of EMT-inducing transcription factors involving ZEB1/2, SNAIL1/2, TWIST1, and Goosecoid, thereby repressing epithelial genes including Ecad. As mentioned previously, microRNA-mediated control of translation, splicing of mRNAs and epigenetic modifiers can also regulate EMT/MET [259,260]. Various feedback loops discussed can alter plasticity of the cell and enable the existence of intermediate phenotypes. Understanding how these multiple factors govern epithelial-hybrid-mesenchymal states stimulated the development of mathematical models to study the underlying mechanisms, as well as the dynamics, stability and reversibility of EMT. Although EOC-specific EMT/MET computational models are not well-represented in the literature, the existence of similar EMT/MET signaling pathways in different cancer types suggests logical extension of existent models to EOC.

6.1. Regulatory Networks-Based Models of EMT/MET

To delineate the emergent dynamics of EMT/MET regulatory networks, low- and high-dimensional kinetic models have been developed [261–263].

6.1.1. Low-Dimensional Models

The two major low-dimensional models focus on describing individual reactions between a set of micro-RNAs families and comprise miR-34, miR-200 and EMT-TF ZEB and SNAIL players. As was reported recently [261,262] these networks allow for co-existence of epithelial (E) and mesenchymal (M) phenotypes along with a hybrid epithelial-mesenchymal (E-M) phenotype, observed experimentally in many studies revealing subpopulations of E, M, and E-M cells in various cell lines [80]. The fact that E-M clustering can result in a significantly larger amount of EOC secondary tumors as compared to pure E or M phenotype [81], therefore impacting metastatic success, makes the small-scale model a critical component in predicting the outcome of E, M and E-M cell interactions. The modeling approach developed by Lu et al. [261] uses a theoretical framework to account for microRNA- and transcription factor-mediated interactions. The model suggests that miR-200/ZEB feedback loop works as a switch allowing for three stable states and that hybrid E-M cells correspond to intermediate miR-200 and ZEB levels. In contrast, Tian et al. [262] proposed a simplified model applying mathematical forms to consider translational and transcriptional interactions. In their work, it is hypothesized that both miR-200/ZEB and miR-34/SNAIL act as bi-stable switches and the hybrid E-M phenotype is caused by low ZEB and high SNAIL levels.

The impact of other transcription factors modulating EMT/MET in the low-dimensional approach was also considered. In particular, GRHL2 and OVOL2 were shown to act as 'phenotypic stability factors' (PSFs) allowing for the existence of a hybrid E-M phenotype at a wider range of model parameters [72,264]. The regulatory network in the later study [264] coupled OVOL with miR-34/SNAIL and miR-200/ZEB circuits. The core of the EMT regulatory network comprised of self-inhibitory OVOL which formed a mutually inhibitory loop with ZEB and indirectly inhibited miR-200 via STAT3. TGF-β activated SNAIL, and BMP7/Smad4 pathway and C/EBP-β activated OVOL, whereas Wg signaling (Armadillo/dTCF) inhibited OVOL.

In application to ovarian cancer modeling, suppression of GRHL2 was recently shown to inhibit proliferation, invasion, and migration of ovarian cancer cells [265], emphasizing the importance of incorporating this factor into a low-dimensional EOC EMT/MET model. Additionally, extracellular communications such as those mediated by JAG1 were shown to be able to perform the role of PSF via Notch-Jagged signaling [266]. Furthermore, to quantify global stability of the hybrid phenotype in EOC EMT and transition dynamics among different phenotypes, the landscape and kinetic paths approach needs to be used to aid in understanding the mechanisms of EMT processes, and unveil possible roles for the intermediate E-M states [267,268].

6.1.2. High-Dimensional Models

In contrast to low-dimensional kinetic models, high-dimensional, large-scale regulatory network models have been developed based on Boolean formalism focusing on the logical topology of the networks. Following this approach an EOC EMT network of multiple nodes and edges can be constructed to simulate the dynamics of transcription factors and microRNAs. This methodology has been successfully used in a study [263] that considered a network of 70 nodes and 135 edges to follow the dynamics of TGF-β-driven EMT in hepatocellular carcinoma. The nodes of the network represented molecular entities (proteins, small molecules, mRNAs) while the edges depicted activating and inhibitory relationships between nodes. Upstream signals (SHH, Wnt, HGF, PDGF, IGF1, EGF, FGR, Jagged, TGF-β, DELTA, CHD1L, Goosecoid, Hypoxia) regulated transcriptional regulators (SNAI1, SNAI2, FOXC2, TWIST1, ZEB1, ZEB2) through signal transduction pathways, which all converged on the regulation of E-cadherin state, defining the EMT output of the network. A related study [269] simulated multiple signaling pathways, microRNAs and transcription factors to derive mutations and functional changes in network nodes that result in altered metastatic behavior.

One of the advantages of using the logical network formalism is that it provides a larger set of steady states compared to kinetic-mechanism based models due to higher dimensionality of the system considered. The large scale network approach allows for better characterization of EMT by providing a more detailed profile of transient hybrid E-M states. Yet, Boolean modeling framework considers each node to take discrete 'on' and 'off' states which makes it difficult to simulate transitions involving continuous-like switches such as the transcription factor ZEB.

Overall, both low- and high-dimensional models complement each other in identifying regulatory network-based mechanisms of EMT/MET and can perform as an efficient computational tool to recognize stability factors allowing for existence of E-M hybrid phenotypes in EOC.

6.2. Omics-Based Models of EMT

Typical omics-level models of EMT use gene expression data which is then analyzed using statistical methods for identifying transition patterns and characterization of trajectories between phenotypes. By applying methods of chemical reaction dynamics, changes in transcriptional profile in the course of EMT were studied [270]. It was shown that a stable low-energy intermediate state exists in the landscape of the free energy changes during TGF-β1-induced EMT for the lung cancer cells coinciding with metabolic shifts. In another study [271], integration of time-course EMT transcriptomic data with public cistromic data allowed investigators to identify three synergistic master transcription factors, namely ETS2, HNF4A and JUNB, that regulated the transition between partial EMT states. Removal of these factors abrogated TGF-β-induced EMT. Likewise, application of these models to TGF-β-induced EMT in EOC can identify transcription factors responsible for the hybrid E-M phenotype which can be further examined experimentally.

Another useful model approach which was applied to EMT in EOC along with other cancer types was suggested in [272], enabling quantification of the EMT spectrum in the course of cancer progression. In this model, transcriptomics data are used to identify a generic EMT signature involving common molecular signatures of EMT across tumors and cell lines of different origins comprising bladder, breast, colorectal, gastric, lung and ovarian cancers. For a given sample the EMT score is then calculated using a two-sample Kolmogorov-Smirnov test in the range from −1 to +1 with the positive scores indicating mesenchymal, and negative scores corresponding to epithelial phenotype. Interestingly, the generic EMT scores of EOC tumors were slightly negative (around −0.1) pointing out a slight shift towards epithelial phenotype, whereas ovarian cancer cell lines revealed more mesenchymal phenotype with positive scores (+0.25). The results of the model suggest that EMT status does not necessarily translate to chemotherapeutic resistance and correlate with poorer survival.

6.3. Multiscale Models of EMT

The goal of multiscale models is to link the intracellular dynamics of EMT signaling pathways to the adhesion molecules on the cell surface and changes in tumor cell invasion through the extracellular matrix. A multiscale individual-based lattice-free model accounting for the intracellular dynamics of the Ecad-β-catenin interaction and external mechanical forces was developed [273]. Each cell was modeled as an isotropic elastic body permitting migration and division calibrated using cell-kinetic, biophysical and cell-biological experimental data. The results indicate that some of migratory features of EMT can be reproduced in the model when the extracellular gradient of chemoattractant is induced and the system of proteasomes responsible for degrading β-catenin is downregulated.

Monte Carlo-based model simulations of EMT were presented [274] with focus on the formation of cardiac cushions during embryonic development of the heart. Cell rearrangements are provided in terms of Steinberg's differential adhesion hypothesis suggesting type-dependent adhesion along with motility levels sufficient to result in tissue conformations with the largest number of strong bonds. The model couples differential adhesion, EMT, cell proliferation and matrix production by mesenchymal cells, and predicts that increases in cell-ECM interactions might be more important in promoting EMT than decreases in cell-cell adhesion.

Stochastic and deterministic models describing tumor growth based on the cancer stem cell hypothesis in application to EMT were given [275], wherein the possibilities of using quantitative approaches for identification of an increase in stem cell activity following promotion of EMT were discussed. The cancer stem cell hypothesis implies that not all tumor cells are equal in terms of their roles in treatment resistance, therefore suggesting possible therapeutic targets in EOC. If EMT can lead to formation of CSCs from non-stem cancer cells, then EMT must be targeted along with CSCs, as non-stem population may recover the CSC pool.

Various computational and statistical models of EMT can contribute significantly by providing a quantitative assessment of EMT dynamics that incorporate EOC genetic and biophysical parameters, which can then be tested experimentally to yield further insight into mechanisms driving EMT.

7. Conclusions

Successful metastatic dissemination of EOC in the complex peritoneal microenvironment necessitates dynamic and reversible changes in cell-cell interaction as cancer progresses from the primary tumor to free-floating MCAs and matrix-anchored metastases. This unique metastatic niche provides a diversity of biochemical and biomechanical cues that, together with the inherent phenotypic plasticity of EOC cells, promotes EMT and MET at various stages in metastatic progression. Additional research designed to mechanistically integrate key drivers of the EMT/MET program will undoubtedly identify a wealth of potential therapeutic targets to inhibit successful metastatic spread and thereby improve survival of women with EOC.

Acknowledgments: This work was supported in part by Research Grants RO1CA109545 (M.S.S.) and RO1CA086984 (M.S.S.) from the National Institutes of Health/National Cancer Institute; the Leo and Anne Albert Charitable Trust (M.S.S.); the Research Like a Champion grant (Y.K.); the Walther Cancer Foundation Seeding Research in Cancer grant (O.K.); and the Scientist Development Grant SDG33680177 (O.K.) from American Heart Association.

Conflicts of Interest: The authors declare no conflict of interest.

References

1. Chaffer, C.L.; Weinberg, R.A. A Perspective on Cancer Cell Metastasis. *Science* **2011**, *331*, 1559–1564. [CrossRef] [PubMed]
2. Hanahan, D.; Weinberg, R.A. Hallmarks of Cancer: The Next Generation. *Cell* **2011**, *144*, 646–674. [CrossRef] [PubMed]
3. Weinberg, R. *The Biology of Cancer*; Garland Science: New York, NY, USA, 2013.
4. Valastyan, S.; Weinberg, R.A. Tumor Metastasis: Molecular Insights and Evolving Paradigms. *Cell* **2011**, *147*, 275–292. [CrossRef] [PubMed]
5. Fidler, I.J. The Pathogenesis of Cancer Metastasis: The 'Seed and Soil' Hypothesis Revisited. *Nat. Rev. Cancer* **2003**, *3*, 453–458. [CrossRef] [PubMed]
6. Yoshida-Noro, C.; Suzuki, N.; Takeichi, M. Molecular Nature of the Calcium-Dependent Cell-Cell Adhesion System in Mouse Teratocarcinoma and Embryonic Cells Studied with a Monoclonal Antibody. *Dev. Biol.* **1984**, *101*, 19–27. [CrossRef]
7. Takeichi, M. The Cadherins: Cell-Cell Adhesion Molecules Controlling Animal Morphogenesis. *Development* **1988**, *102*, 639–655. [PubMed]
8. Li, G.; Satyamoorthy, K.; Herlyn, M. N-Cadherin-Mediated Intercellular Interactions Promote Survival and Migration of Melanoma Cells. *Cancer Res.* **2001**, *61*, 3819–3825. [PubMed]
9. Mariotti, A.; Perotti, A.; Sessa, C.; Rüegg, C. N-Cadherin as a Therapeutic Target in Cancer. *Expert Opin. Investig. Drugs* **2007**, *16*, 451–465. [CrossRef] [PubMed]
10. Gerhardt, H.; Wolburg, H.; Redies, C. N-cadherin Mediates Pericytic-endothelial Interaction during Brain Angiogenesis in the Chicken. *Dev. Dyn.* **2000**, *218*, 472–479. [CrossRef]
11. Blaschuk, O.W. N-Cadherin Antagonists as Oncology Therapeutics. *Philos. Trans. R. Soc. Lond. E. Biol. Sci.* **2015**, *370*, 20140039. [CrossRef] [PubMed]
12. Shibue, T.; Weinberg, R.A. EMT, CSCs, and Drug Resistance: The Mechanistic Link and Clinical Implications. *Nat. Rev. Clin. Oncol.* **2017**. [CrossRef] [PubMed]
13. American Cancer Society. *Cancer Facts & Figures 2017*; American Cancer Society: Atlanta, GA, USA, 2017.

14. Siegel, R.L.; Miller, K.D.; Jemal, A. Cancer Statistics, 2016. *CA Cancer J. Clin.* **2016**, *66*, 7–30. [CrossRef] [PubMed]
15. Howlader, N.; Noone, A.M.; Krapcho, M.; Garshell, J.; Miller, D.; Altekruse, S.F.; Kosary, C.L.; Yu, M.; Ruhl, J.; Tatalovich, Z.; et al. *SEER Cancer Statistics Review, 1975–2008*; National Cancer Institute: Bethesda, MD, USA, 2011.
16. Marcus, C.S.; Maxwell, G.L.; Darcy, K.M.; Hamilton, C.A.; McGuire, W.P. Current Approaches and Challenges in Managing and Monitoring Treatment Response in Ovarian Cancer. *J. Cancer* **2014**, *5*, 25. [CrossRef] [PubMed]
17. Hudson, L.G.; Zeineldin, R.; Stack, M.S. Phenotypic Plasticity of Neoplastic Ovarian Epithelium: Unique Cadherin Profiles in Tumor Progression. *Clin. Exp. Metastasis* **2008**, *25*, 643–655. [CrossRef] [PubMed]
18. Lengyel, E. Ovarian Cancer Development and Metastasis. *Am. J. Pathol.* **2010**, *177*, 1053–1064. [CrossRef] [PubMed]
19. Shield, K.; Ackland, M.L.; Ahmed, N.; Rice, G.E. Multicellular Spheroids in Ovarian Cancer Metastases: Biology and Pathology. *Gynecol. Oncol.* **2009**, *113*, 143–148. [CrossRef] [PubMed]
20. Feki, A.; Berardi, P.; Bellingan, G.; Major, A.; Krause, K.; Petignat, P.; Zehra, R.; Pervaiz, S.; Irminger-Finger, I. Dissemination of Intraperitoneal Ovarian Cancer: Discussion of Mechanisms and Demonstration of Lymphatic Spreading in Ovarian Cancer Model. *Crit. Rev. Oncol.* **2009**, *72*, 1–9. [CrossRef] [PubMed]
21. Pradeep, S.; Kim, S.W.; Wu, S.Y.; Nishimura, M.; Chaluvally-Raghavan, P.; Miyake, T.; Pecot, C.V.; Kim, S.; Choi, H.J.; Bischoff, F.Z. Hematogenous Metastasis of Ovarian Cancer: Rethinking Mode of Spread. *Cancer Cell* **2014**, *26*, 77–91. [CrossRef] [PubMed]
22. Coffman, L.G.; Burgos-Ojeda, D.; Wu, R.; Cho, K.; Bai, S.; Buckanovich, R.J. New Models of Hematogenous Ovarian Cancer Metastasis Demonstrate Preferential Spread to the Ovary and a Requirement for the Ovary for Abdominal Dissemination. *Trans. Res.* **2016**, *175*, 92–102. [CrossRef] [PubMed]
23. Auersperg, N.; Wong, A.S.T.; Choi, K.; Kang, S.K.; Leung, P.C.K. Ovarian Surface Epithelium: Biology, Endocrinology, and Pathology. *Endocr. Rev.* **2001**, *22*, 255–288. [CrossRef] [PubMed]
24. Auersperg, N. The Origin of Ovarian Carcinomas: A Unifying Hypothesis. *Int. J. Gynecol. Pathol.* **2011**, *30*, 12–21. [CrossRef] [PubMed]
25. Gillett, W.R.; Mitchell, A.; Hurst, P.R. A Scanning Electron Microscopic Study of the Human Ovarian Surface Epithelium: Characterization of Two Cell Types. *Hum. Reprod.* **1991**, *6*, 645–650. [CrossRef] [PubMed]
26. Auersperg, N. The Origin of Ovarian Cancers—Hypotheses and Controversies. *Front. Biosci.* **2013**, *5*, 709–719. [CrossRef]
27. Jarboe, E.; Folkins, A.; Nucci, M.R.; Kindelberger, D.; Drapkin, R.; Miron, A.; Lee, Y.; Crum, C.P. Serous Carcinogenesis in the Fallopian Tube: A Descriptive Classification. *Int. J. Gynecol. Pathol.* **2008**, *27*, 1–9. [CrossRef] [PubMed]
28. Carlson, J.W.; Miron, A.; Jarboe, E.A.; Parast, M.M.; Hirsch, M.S.; Lee, Y.; Muto, M.G.; Kindelberger, D.; Crum, C.P. Serous Tubal Intraepithelial Carcinoma: Its Potential Role in Primary Peritoneal Serous Carcinoma and Serous Cancer Prevention. *J. Clin. Oncol.* **2008**, *26*, 4160–4165. [CrossRef] [PubMed]
29. Kindelberger, D.W.; Lee, Y.; Miron, A.; Hirsch, M.S.; Feltmate, C.; Medeiros, F.; Callahan, M.J.; Garner, E.O.; Gordon, R.W.; Birch, C.; et al. Intraepithelial Carcinoma of the Fimbria and Pelvic Serous Carcinoma: Evidence for a Causal Relationship. *Am. J. Surg. Pathol.* **2007**, *31*, 161–169. [CrossRef] [PubMed]
30. Przybycin, C.G.; Kurman, R.J.; Ronnett, B.M.; Shih, I.; Vang, R. Are all Pelvic (Nonuterine) Serous Carcinomas of Tubal Origin? *Am. J. Surg. Pathol.* **2010**, *34*, 1407–1416. [CrossRef] [PubMed]
31. Roh, M.H.; Kindelberger, D.; Crum, C.P. Serous Tubal Intraepithelial Carcinoma and the Dominant Ovarian Mass: Clues to Serous Tumor Origin? *Am. J. Surg. Pathol.* **2009**, *33*, 376–383. [CrossRef] [PubMed]
32. Salvador, S.; Rempel, A.; Soslow, R.A.; Gilks, B.; Huntsman, D.; Miller, D. Chromosomal Instability in Fallopian Tube Precursor Lesions of Serous Carcinoma and Frequent Monoclonality of Synchronous Ovarian and Fallopian Tube Mucosal Serous Carcinoma. *Gynecol. Oncol.* **2008**, *110*, 408–417. [CrossRef] [PubMed]
33. Vang, R.; Shih, I.; Kurman, R.J. Fallopian Tube Precursors of Ovarian Low-and High-grade Serous Neoplasms. *Histopathology* **2013**, *62*, 44–58. [CrossRef] [PubMed]
34. Chene, G.; Dauplat, J.; Radosevic-Robin, N.; Cayre, A.; Penault-Llorca, F. Tu-be Or Not Tu-be: That is the Question . . . about Serous Ovarian Carcinogenesis. *Crit. Rev. Oncol.* **2013**, *88*, 134–143. [CrossRef] [PubMed]

35. Bapat, S.A.; Mali, A.M.; Koppikar, C.B.; Kurrey, N.K. Stem and Progenitor-Like Cells Contribute to the Aggressive Behavior of Human Epithelial Ovarian Cancer. *Cancer Res.* **2005**, *65*, 3025–3029. [CrossRef] [PubMed]

36. Bowen, N.J.; Walker, L.D.; Matyunina, L.V.; Logani, S.; Totten, K.A.; Benigno, B.B.; McDonald, J.F. Gene Expression Profiling Supports the Hypothesis that Human Ovarian Surface Epithelia are Multipotent and Capable of Serving as Ovarian Cancer Initiating Cells. *BMC Med. Genom.* **2009**, *2*, 71. [CrossRef] [PubMed]

37. Szotek, P.P.; Pieretti-Vanmarcke, R.; Masiakos, P.T.; Dinulescu, D.M.; Connolly, D.; Foster, R.; Dombkowski, D.; Preffer, F.; Maclaughlin, D.T.; Donahoe, P.K. Ovarian Cancer Side Population Defines Cells with Stem Cell-Like Characteristics and Mullerian Inhibiting Substance Responsiveness. *Proc. Natl. Acad. Sci. USA* **2006**, *103*, 11154–11159. [CrossRef] [PubMed]

38. Flesken-Nikitin, A.; Hwang, C.; Cheng, C.; Michurina, T.V.; Enikolopov, G.; Nikitin, A.Y. Ovarian Surface Epithelium at the Junction Area Contains a Cancer-Prone Stem Cell Niche. *Nature* **2013**, *495*, 241–245. [CrossRef] [PubMed]

39. Seidman, J.D.; Yemelyanova, A.; Zaino, R.J.; Kurman, R.J. The Fallopian Tube-Peritoneal Junction: A Potential Site of Carcinogenesis. *Int. J. Gynecol. Pathol.* **2011**, *30*, 4–11. [CrossRef] [PubMed]

40. Paik, D.Y.; Janzen, D.M.; Schafenacker, A.M.; Velasco, V.S.; Shung, M.S.; Cheng, D.; Huang, J.; Witte, O.N.; Memarzadeh, S. Stem-Like Epithelial Cells are Concentrated in the Distal End of the Fallopian Tube: A Site for Injury and Serous Cancer Initiation. *Stem Cells* **2012**, *30*, 2487–2497. [CrossRef] [PubMed]

41. Ng, A.; Barker, N. Ovary and Fimbrial Stem Cells: Biology, Niche and Cancer Origins. *Nat. Rev. Mol. Cell Biol.* **2015**, *16*, 625–638. [CrossRef] [PubMed]

42. Klymenko, Y.; Stack, M.S. Abstract B6: Cadherin Switching, Multicellular Aggregate Dynamics and Metastatic Success. *Clin. Cancer Res.* **2013**, *19*. [CrossRef]

43. Klymenko, Y.; Johnson, J.; Bos, B.; Lombard, R.; Campbell, L.; Loughran, E.; Stack, M. Heterogeneous Cadherin Expression and Multi-Cellular Aggregate Dynamics in Ovarian Cancer Dissemination. *Neoplasia* **2017**, *19*, 549–563. [CrossRef] [PubMed]

44. Sivertsen, S.; Berner, A.; Michael, C.W.; Bedrossian, C.; Davidson, B. Cadherin Expression in Ovarian Carcinoma and Malignant Mesothelioma Cell Effusions. *Acta Cytol.* **2006**, *50*, 603. [CrossRef] [PubMed]

45. Klymenko, Y.; Kim, O.; Loughran, E.A.; Yang, J.; Lombard, R.; Alber, M.; Stack, M. Cadherin Composition and Multicellular Aggregate Dynamics in Organotypic Models of Epithelial Ovarian Cancer Intraperitoneal Metastasis. *Oncogene* **2017**. [CrossRef] [PubMed]

46. Shaw, T.J.; Senterman, M.K.; Dawson, K.; Crane, C.A.; Vanderhyden, B.C. Characterization of Intraperitoneal, Orthotopic, and Metastatic Xenograft Models of Human Ovarian Cancer. *Mol. Ther.* **2004**, *10*, 1032–1042. [CrossRef] [PubMed]

47. Takai, N.; Jain, A.; Kawamata, N.; Popoviciu, L.M.; Said, J.W.; Whittaker, S.; Miyakawa, I.; Agus, D.B.; Koeffler, H.P. 2C4, a Monoclonal Antibody Against HER2, Disrupts the HER Kinase Signaling Pathway and Inhibits Ovarian Carcinoma Cell Growth. *Cancer* **2005**, *104*, 2701–2708. [CrossRef] [PubMed]

48. Afzal, S.; Lalani, E.; Poulsom, R.; Stubbs, A.; Rowlinson, G.; Sato, H.; Seiki, M.; Stamp, G.W.H. MT1-MMP and MMP-2 mRNA Expression in Human Ovarian Tumors: Possible Implications for the Role of Desmoplastic Fibroblasts. *Hum. Pathol.* **1998**, *29*, 155–165. [CrossRef]

49. Liu, Y.; Metzinger, M.N.; Lewellen, K.A.; Cripps, S.N.; Carey, K.D.; Harper, E.I.; Shi, Z.; Tarwater, L.; Grisoli, A.; Lee, E.; et al. Obesity Contributes to Ovarian Cancer Metastatic Success through Increased Lipogenesis, Enhanced Vascularity, and Decreased Infiltration of M1 Macrophages. *Cancer Res.* **2015**, *75*, 5046–5057. [CrossRef] [PubMed]

50. Mitra, A.K.; Davis, D.A.; Tomar, S.; Roy, L.; Gurler, H.; Xie, J.; Lantvit, D.D.; Cardenas, H.; Fang, F.; Liu, Y.; et al. In Vivo Tumor Growth of High-Grade Serous Ovarian Cancer Cell Lines. *Gynecol. Oncol.* **2015**, *138*, 372–377. [CrossRef] [PubMed]

51. Bates, R.C.; Edwards, N.S.; Yates, J.D. Spheroids and Cell Survival. *Crit. Rev. Oncol.* **2000**, *36*, 61–74. [CrossRef]

52. Desoize, B.; Jardillier, J. Multicellular Resistance: A Paradigm for Clinical Resistance? *Crit. Rev. Oncol.* **2000**, *36*, 193–207. [CrossRef]

53. Frankel, A.; Rosen, K.; Filmus, J.; Kerbel, R.S. Induction of Anoikis and Suppression of Human Ovarian Tumor Growth in Vivo by Down-Regulation of Bcl-X(L). *Cancer Res.* **2001**, *61*, 4837–4841. [PubMed]

54. Green, S.K.; Francia, G.; Isidoro, C.; Kerbel, R.S. Antiadhesive Antibodies Targeting E-Cadherin Sensitize Multicellular Tumor Spheroids to Chemotherapy in Vitro. *Mol. Cancer. Ther.* **2004**, *3*, 149–159. [PubMed]

55. Sutherland, R.M. Cell and Environment Interactions in Tumor Microregions: The Multicell Spheroid Model. *Science* **1988**, *240*, 177–184. [CrossRef] [PubMed]

56. Latifi, A.; Luwor, R.B.; Bilandzic, M.; Nazaretian, S.; Stenvers, K.; Pyman, J.; Zhu, H.; Thompson, E.W.; Quinn, M.A.; Findlay, J.K. Isolation and Characterization of Tumor Cells from the Ascites of Ovarian Cancer Patients: Molecular Phenotype of Chemoresistant Ovarian Tumors. *PLoS ONE* **2012**, *7*, e46858. [CrossRef] [PubMed]

57. Miow, Q.; Tan, T.; Ye, J.; Lau, J.; Yokomizo, T.; Thiery, J.; Mori, S. Epithelial–mesenchymal Status Renders Differential Responses to Cisplatin in Ovarian Cancer. *Oncogene* **2015**, *34*, 1899–1907. [CrossRef] [PubMed]

58. Reddy, P.; Liu, L.; Ren, C.; Lindgren, P.; Boman, K.; Shen, Y.; Lundin, E.; Ottander, U.; Rytinki, M.; Liu, K. Formation of E-Cadherin-Mediated Cell-Cell Adhesion Activates AKT and Mitogen Activated Protein Kinase Via Phosphatidylinositol 3 Kinase and Ligand-Independent Activation of Epidermal Growth Factor Receptor in Ovarian Cancer Cells. *Mol. Endocrinol.* **2005**, *19*, 2564–2578. [CrossRef] [PubMed]

59. Burkhalter, R.J.; Symowicz, J.; Hudson, L.G.; Gottardi, C.J.; Stack, M.S. Integrin Regulation of Beta-Catenin Signaling in Ovarian Carcinoma. *J. Biol. Chem.* **2011**, *286*, 23467–23475. [CrossRef] [PubMed]

60. Liu, Y.; Burkhalter, R.; Symowicz, J.; Chaffin, K.; Ellerbroek, S.; Stack, M.S. Lysophosphatidic Acid Disrupts Junctional Integrity and Epithelial Cohesion in Ovarian Cancer Cells. *J. Oncol.* **2012**, *2012*, 501492. [CrossRef] [PubMed]

61. Wu, C.; Cipollone, J.; Maines-Bandiera, S.; Tan, C.; Karsan, A.; Auersperg, N.; Roskelley, C.D. The Morphogenic Function of E-cadherin-mediated Adherens Junctions in Epithelial Ovarian Carcinoma Formation and Progression. *Differentiation* **2008**, *76*, 193–205. [CrossRef] [PubMed]

62. Symowicz, J.; Adley, B.P.; Gleason, K.J.; Johnson, J.J.; Ghosh, S.; Fishman, D.A.; Hudson, L.G.; Stack, M.S. Engagement of Collagen-Binding Integrins Promotes Matrix Metalloproteinase-9-Dependent E-Cadherin Ectodomain Shedding in Ovarian Carcinoma Cells. *Cancer Res.* **2007**, *67*, 2030–2039. [CrossRef] [PubMed]

63. Gil, O.D.; Lee, C.; Ariztia, E.V.; Wang, F.; Smith, P.J.; Hope, J.M.; Fishman, D.A. Lysophosphatidic Acid (LPA) Promotes E-Cadherin Ectodomain Shedding and OVCA429 Cell Invasion in an uPA-Dependent Manner. *Gynecol. Oncol.* **2008**, *108*, 361–369. [CrossRef] [PubMed]

64. Nawrocki-Raby, B.; Gilles, C.; Polette, M.; Bruyneel, E.; Laronze, J.; Bonnet, N.; Foidart, J.; Mareel, M.; Birembaut, P. Upregulation of MMPs by Soluble E-cadherin in Human Lung Tumor Cells. *Int. J. Cancer* **2003**, *105*, 790–795. [CrossRef] [PubMed]

65. Marambaud, P.; Shioi, J.; Serban, G.; Georgakopoulos, A.; Sarner, S.; Nagy, V.; Baki, L.; Wen, P.; Efthimiopoulos, S.; Shao, Z.; et al. A Presenilin-1/Gamma-Secretase Cleavage Releases the E-Cadherin Intracellular Domain and Regulates Disassembly of Adherens Junctions. *EMBO J.* **2002**, *21*, 1948–1956. [CrossRef] [PubMed]

66. Salahshor, S.; Naidoo, R.; Serra, S.; Shih, W.; Tsao, M.; Chetty, R.; Woodgett, J.R. Frequent Accumulation of Nuclear E-Cadherin and Alterations in the Wnt Signaling Pathway in Esophageal Squamous Cell Carcinomas. *Mod. Pathol.* **2008**, *21*, 271–281. [CrossRef] [PubMed]

67. Grabowska, M.M.; Day, M.L. Soluble E-Cadherin: More than a Symptom of Disease. *Front. Biosci.* **2012**, *17*, 1948–1964. [CrossRef]

68. Rodriguez, F.J.; Lewis-Tuffin, L.J.; Anastasiadis, P.Z. E-Cadherin's Dark Side: Possible Role in Tumor Progression. *Biochim. Biophys. Acta Rev. Cancer* **2012**, *1826*, 23–31. [CrossRef] [PubMed]

69. Jolly, M.K.; Boareto, M.; Huang, B.; Jia, D.; Lu, M.; Ben-Jacob, E.; Onuchic, J.N.; Levine, H. Implications of the Hybrid Epithelial/Mesenchymal Phenotype in Metastasis. *Front. Oncol.* **2015**, *5*. [CrossRef] [PubMed]

70. Tam, W.L.; Weinberg, R.A. The Epigenetics of Epithelial-Mesenchymal Plasticity in Cancer. *Nat. Med.* **2013**, *19*, 1438–1449. [CrossRef] [PubMed]

71. Huang, R.Y.; Wong, M.; Tan, T.; Kuay, K.; Ng, A.; Chung, V.; Chu, Y.; Matsumura, N.; Lai, H.; Lee, Y. An EMT Spectrum Defines an Anoikis-Resistant and Spheroidogenic Intermediate Mesenchymal State that is Sensitive to E-Cadherin Restoration by a Src-Kinase Inhibitor, Saracatinib (AZD0530). *Cell Death Dis.* **2013**, *4*, e915. [CrossRef] [PubMed]

72. Jolly, M.K.; Tripathi, S.C.; Jia, D.; Mooney, S.M.; Celiktas, M.; Hanash, S.M.; Mani, S.A.; Pienta, K.J.; Ben-Jacob, E.; Levine, H. Stability of the Hybrid Epithelial/Mesenchymal Phenotype. *Oncotarget* **2016**, *7*, 27067–27084. [CrossRef] [PubMed]

73. Chao, Y.; Wu, Q.; Acquafondata, M.; Dhir, R.; Wells, A. Partial Mesenchymal to Epithelial Reverting Transition in Breast and Prostate Cancer Metastases. *Cancer Microenviron.* **2012**, *5*, 19–28. [CrossRef] [PubMed]

74. Hecht, I.; Bar-El, Y.; Balmer, F.; Natan, S.; Tsarfaty, I.; Schweitzer, F.; Ben-Jacob, E. Tumor Invasion Optimization by Mesenchymal-Amoeboid Heterogeneity. *Sci. Rep.* **2015**, *5*. [CrossRef] [PubMed]

75. Friedl, P.; Wolf, K. Plasticity of Cell Migration: A Multiscale Tuning Model. *J. Cell Biol.* **2010**, *188*, 11–19. [CrossRef] [PubMed]

76. Andriani, F.; Bertolini, G.; Facchinetti, F.; Baldoli, E.; Moro, M.; Casalini, P.; Caserini, R.; Milione, M.; Leone, G.; Pelosi, G. Conversion to Stem-cell State in Response to Microenvironmental Cues is Regulated by Balance between Epithelial and Mesenchymal Features in Lung Cancer Cells. *Mol. Oncol.* **2016**, *10*, 253–271. [CrossRef] [PubMed]

77. Bastos, L.G.D.R.; de Marcondes, P.G.; de-Freitas-Junior, J.C.M.; Leve, F.; Mencalha, A.L.; de Souza, W.F.; de Araujo, W.M.; Tanaka, M.N.; Abdelhay, E.S.F.W.; Morgado-Díaz, J.A. Progeny from Irradiated Colorectal Cancer Cells Acquire an EMT-Like Phenotype and Activate Wnt/β-Catenin Pathway. *J. Cell. Biochem.* **2014**, *115*, 2175–2187. [CrossRef] [PubMed]

78. Hiscox, S.; Jiang, W.G.; Obermeier, K.; Taylor, K.; Morgan, L.; Burmi, R.; Barrow, D.; Nicholson, R.I. Tamoxifen Resistance in MCF7 Cells Promotes EMT-like Behaviour and Involves Modulation of B-catenin Phosphorylation. *Int. J. Cancer* **2006**, *118*, 290–301. [CrossRef] [PubMed]

79. Wu, Y.; Ginther, C.; Kim, J.; Mosher, N.; Chung, S.; Slamon, D.; Vadgama, J.V. Expression of Wnt3 Activates Wnt/Beta-Catenin Pathway and Promotes EMT-Like Phenotype in Trastuzumab-Resistant HER2-Overexpressing Breast Cancer Cells. *Mol. Cancer. Res.* **2012**, *10*, 1597–1606. [CrossRef] [PubMed]

80. Grosse-Wilde, A.; d'Hérouël, A.F.; McIntosh, E.; Ertaylan, G.; Skupin, A.; Kuestner, R.E.; del Sol, A.; Walters, K.; Huang, S. Stemness of the Hybrid Epithelial/Mesenchymal State in Breast Cancer and its Association with Poor Survival. *PLoS ONE* **2015**, *10*, e0126522. [CrossRef] [PubMed]

81. Strauss, R.; Li, Z.; Liu, Y.; Beyer, I.; Persson, J.; Sova, P.; Möller, T.; Pesonen, S.; Hemminki, A.; Hamerlik, P. Analysis of Epithelial and Mesenchymal Markers in Ovarian Cancer Reveals Phenotypic Heterogeneity and Plasticity. *PLoS ONE* **2011**, *6*, e16186. [CrossRef]

82. Ruscetti, M.; Quach, B.; Dadashian, E.L.; Mulholland, D.J.; Wu, H. Tracking and Functional Characterization of Epithelial-Mesenchymal Transition and Mesenchymal Tumor Cells during Prostate Cancer Metastasis. *Cancer Res.* **2015**, *75*, 2749–2759. [CrossRef] [PubMed]

83. Mills, G.B.; Moolenaar, W.H. The Emerging Role of Lysophosphatidic Acid in Cancer. *Nat. Rev. Cancer* **2003**, *3*, 582–591. [CrossRef] [PubMed]

84. Mills, G.B.; Eder, A.; Fang, X.; Hasegawa, Y.; Mao, M.; Lu, Y.; Tanyi, J.; Tabassam, F.H.; Wiener, J.; Lapushin, R. Critical role of lysophospholipids in the pathophysiology, diagnosis, and management of ovarian cancer. In *Ovarian Cancer*; Springer: New York, NY, USA, 2002; pp. 259–283.

85. Baker, D.L.; Morrison, P.; Miller, B.; Riely, C.A.; Tolley, B.; Westermann, A.M.; Bonfrer, J.M.; Bais, E.; Moolenaar, W.H.; Tigyi, G. Plasma Lysophosphatidic Acid Concentration and Ovarian Cancer. *JAMA* **2002**, *287*, 3081–3082. [CrossRef] [PubMed]

86. Xu, Y.; Shen, Z.; Wiper, D.W.; Wu, M.; Morton, R.E.; Elson, P.; Kennedy, A.W.; Belinson, J.; Markman, M.; Casey, G. Lysophosphatidic Acid as a Potential Biomarker for Ovarian and Other Gynecologic Cancers. *JAMA* **1998**, *280*, 719–723. [CrossRef] [PubMed]

87. Westermann, A.M.; Havik, E.; Postma, F.R.; Beijnen, J.H.; Dalesio, O.; Moolenaar, W.H.; Rodenhuis, S. Malignant Effusions Contain Lysophosphatidic Acid (LPA)-Like Activity. *Ann. Oncol.* **1998**, *9*, 437–442. [CrossRef] [PubMed]

88. Sutphen, R.; Xu, Y.; Wilbanks, G.D.; Fiorica, J.; Grendys, E.C., Jr.; LaPolla, J.P.; Arango, H.; Hoffman, M.S.; Martino, M.; Wakeley, K.; et al. Lysophospholipids are Potential Biomarkers of Ovarian Cancer. *Cancer Epidemiol. Prev. Biomark.* **2004**, *13*, 1185–1191.

89. Ren, J.; Xiao, Y.J.; Singh, L.S.; Zhao, X.; Zhao, Z.; Feng, L.; Rose, T.M.; Prestwich, G.D.; Xu, Y. Lysophosphatidic Acid is Constitutively Produced by Human Peritoneal Mesothelial Cells and Enhances Adhesion, Migration, and Invasion of Ovarian Cancer Cells. *Cancer Res.* **2006**, *66*, 3006–3014. [CrossRef] [PubMed]

90. Do, T.V.; Symowicz, J.C.; Berman, D.M.; Liotta, L.A.; Petricoin, E.F.; Stack, M.S.; Fishman, D.A. Lysophosphatidic Acid Down-Regulates Stress Fibers and Up-Regulates Pro-Matrix Metalloproteinase-2 Activation in Ovarian Cancer Cells. *Mol. Cancer Res.* **2007**, *5*, 121–131. [CrossRef] [PubMed]

91. Fishman, D.A.; Liu, Y.; Ellerbroek, S.M.; Stack, M.S. Lysophosphatidic Acid Promotes Matrix Metalloproteinase (MMP) Activation and MMP-Dependent Invasion in Ovarian Cancer Cells. *Cancer Res.* **2001**, *61*, 3194–3199. [PubMed]
92. Burkhalter, R.J.; Westfall, S.D.; Liu, Y.; Stack, M.S. Lysophosphatidic Acid Initiates Epithelial to Mesenchymal Transition and Induces Beta-Catenin-Mediated Transcription in Epithelial Ovarian Carcinoma. *J. Biol. Chem.* **2015**, *290*, 22143–22154. [CrossRef] [PubMed]
93. Ray, U.; Roy, S.S.; Chowdhury, S.R. Lysophosphatidic Acid Promotes Epithelial to Mesenchymal Transition in Ovarian Cancer Cells by Repressing SIRT1. *Cell. Physiol. Biochem.* **2017**, *41*, 795–805. [CrossRef] [PubMed]
94. Ha, J.H.; Ward, J.D.; Radhakrishnan, R.; Jayaraman, M.; Song, Y.S.; Dhanasekaran, D.N. Lysophosphatidic Acid Stimulates Epithelial to Mesenchymal Transition Marker Slug/Snail2 in Ovarian Cancer Cells Via Galphai2, Src, and HIF1alpha Signaling Nexus. *Oncotarget* **2016**, *7*, 37664–37679. [CrossRef] [PubMed]
95. Vergara, D.; Merlot, B.; Lucot, J.; Collinet, P.; Vinatier, D.; Fournier, I.; Salzet, M. Epithelial–mesenchymal Transition in Ovarian Cancer. *Cancer Lett.* **2010**, *291*, 59–66. [CrossRef] [PubMed]
96. Xu, Z.; Jiang, Y.; Steed, H.; Davidge, S.; Fu, Y. TGFβ and EGF Synergistically Induce a More Invasive Phenotype of Epithelial Ovarian Cancer Cells. *Biochem. Biophys. Res. Commun.* **2010**, *401*, 376–381. [CrossRef] [PubMed]
97. Thiery, J.P.; Sleeman, J.P. Complex Networks Orchestrate Epithelial–mesenchymal Transitions. *Nat. Rev. Mol. Cell Boil.* **2006**, *7*, 131–142. [CrossRef] [PubMed]
98. Thiery, J.P.; Acloque, H.; Huang, R.Y.; Nieto, M.A. Epithelial-Mesenchymal Transitions in Development and Disease. *Cell* **2009**, *139*, 871–890. [CrossRef] [PubMed]
99. Liu, Y.; Sun, X.; Feng, J.; Deng, L.L.; Liu, Y.; Li, B.; Zhu, M.; Lu, C.; Zhou, L. MT2-MMP Induces Proteolysis and Leads to EMT in Carcinomas. *Oncotarget* **2016**, *7*, 48193–48205. [CrossRef] [PubMed]
100. Dahl, K.D.C.; Symowicz, J.; Ning, Y.; Gutierrez, E.; Fishman, D.A.; Adley, B.P.; Stack, M.S.; Hudson, L.G. Matrix Metalloproteinase 9 is a Mediator of Epidermal Growth Factor-Dependent E-Cadherin Loss in Ovarian Carcinoma Cells. *Cancer Res.* **2008**, *68*, 4606–4613. [CrossRef] [PubMed]
101. Covington, M.D.; Burghardt, R.C.; Parrish, A.R. Ischemia-Induced Cleavage of Cadherins in NRK Cells Requires MT1-MMP (MMP-14). *Am. J. Physiol. Renal Physiol.* **2006**, *290*, F43–F51. [CrossRef] [PubMed]
102. Lochter, A.; Galosy, S.; Muschler, J.; Freedman, N.; Werb, Z.; Bissell, M.J. Matrix Metalloproteinase Stromelysin-1 Triggers a Cascade of Molecular Alterations that Leads to Stable Epithelial-to-Mesenchymal Conversion and a Premalignant Phenotype in Mammary Epithelial Cells. *J. Cell Biol.* **1997**, *139*, 1861–1872. [CrossRef] [PubMed]
103. Noe, V.; Fingleton, B.; Jacobs, K.; Crawford, H.C.; Vermeulen, S.; Steelant, W.; Bruyneel, E.; Matrisian, L.M.; Mareel, M. Release of an Invasion Promoter E-Cadherin Fragment by Matrilysin and Stromelysin-1. *J. Cell. Sci.* **2001**, *114*, 111–118. [PubMed]
104. Kessenbrock, K.; Plaks, V.; Werb, Z. Matrix Metalloproteinases: Regulators of the Tumor Microenvironment. *Cell* **2010**, *141*, 52–67. [CrossRef] [PubMed]
105. Rowe, R.G.; Weiss, S.J. Navigating ECM Barriers at the Invasive Front: The Cancer Cell–stroma Interface. *Ann. Rev. Cell Dev.* **2009**, *25*, 567–595. [CrossRef] [PubMed]
106. Itoh, Y. Membrane-Type Matrix Metalloproteinases: Their Functions and Regulations. *Matrix Biol.* **2015**, *44*, 207–223. [CrossRef] [PubMed]
107. Yokoyama, K.; Kamata, N.; Fujimoto, R.; Tsutsumi, S.; Tomonari, M.; Taki, M.; Hosokawa, H.; Nagayama, M. Increased Invasion and Matrix Metalloproteinase-2 Expression by Snail-Induced Mesenchymal Transition in Squamous Cell Carcinomas. *Int. J. Oncol.* **2003**, *22*, 891–898. [CrossRef] [PubMed]
108. Brabletz, T.; Jung, A.; Dag, S.; Hlubek, F.; Kirchner, T. B-Catenin Regulates the Expression of the Matrix Metalloproteinase-7 in Human Colorectal Cancer. *Am. J. Pathol.* **1999**, *155*, 1033–1038. [CrossRef]
109. Crawford, H.C.; Fingleton, B.; Gustavson, M.D.; Kurpios, N.; Wagenaar, R.A.; Hassell, J.A.; Matrisian, L.M. The PEA3 Subfamily of Ets Transcription Factors Synergizes with Beta-Catenin-LEF-1 to Activate Matrilysin Transcription in Intestinal Tumors. *Mol. Cell. Biol.* **2001**, *21*, 1370–1383. [CrossRef] [PubMed]
110. Takahashi, M.; Tsunoda, T.; Seiki, M.; Nakamura, Y.; Furukawa, Y. Identification of Membrane-Type Matrix Metalloproteinase-1 as a Target of the [Beta]-Catenin/Tcf4 Complex in Human Colorectal Cancers. *Oncogene* **2002**, *21*, 5861. [CrossRef] [PubMed]
111. Turunen, S.P.; Tatti-Bugaeva, O.; Lehti, K. Membrane-Type Matrix Metalloproteases as Diverse Effectors of Cancer Progression. *Biochim. Biophys. Acta Mol. Cell Res.* **2017**. [CrossRef] [PubMed]

112. Gilles, C.; Newgreen, D.F.; Sato, H.; Thompson, E.W. Matrix Metalloproteases and Epithelial-to-Mesenchymal Transition. In *Rise and Fall of Epithelial Phenotype*; Springer: New York, NY, USA, 2005; pp. 297–315.

113. Roy, R.; Yang, J.; Moses, M.A. Matrix Metalloproteinases as Novel Biomarker s and Potential Therapeutic Targets in Human Cancer. *J. Clin. Oncol.* **2009**, *27*, 5287–5297. [CrossRef] [PubMed]

114. Milliken, D.; Scotton, C.; Raju, S.; Balkwill, F.; Wilson, J. Analysis of Chemokines and Chemokine Receptor Expression in Ovarian Cancer Ascites. *Clin. Cancer Res.* **2002**, *8*, 1108–1114. [PubMed]

115. Matte, I.; Lane, D.; Laplante, C.; Rancourt, C.; Piché, A. Profiling of Cytokines in Human Epithelial Ovarian Cancer Ascites. *Am. J. Cancer Res.* **2012**, *2*, 566–580. [PubMed]

116. Penson, R.T.; Kronish, K.; Duan, Z.; Feller, A.J.; Stark, P.; Cook, S.E.; Duska, L.R.; Fuller, A.F.; Goodman, A.; Nikrui, N. Cytokines IL-1β, IL-2, IL-6, IL-8, MCP-1, GM-CSF and TNFα in Patients with Epithelial Ovarian Cancer and their Relationship to Treatment with Paclitaxel. *Int. J. Gynecol. Cancer* **2000**, *10*, 33–41. [CrossRef] [PubMed]

117. Kipps, E.; Tan, D.S.; Kaye, S.B. Meeting the Challenge of Ascites in Ovarian Cancer: New Avenues for Therapy and Research. *Nat. Rev. Cancer* **2013**, *13*, 273–282. [CrossRef] [PubMed]

118. Yin, J.; Zeng, F.; Wu, N.; Kang, K.; Yang, Z.; Yang, H. Interleukin-8 Promotes Human Ovarian Cancer Cell Migration by Epithelial–mesenchymal Transition Induction in Vitro. *Clin. Transl. Oncol.* **2015**, *17*, 365–370. [CrossRef] [PubMed]

119. So, K.A.; Min, K.J.; Hong, J.H.; Lee, J. Interleukin-6 Expression by Interactions between Gynecologic Cancer Cells and Human Mesenchymal Stem Cells Promotes Epithelial-Mesenchymal Transition. *Int. J. Oncol.* **2015**, *47*, 1451–1459. [CrossRef] [PubMed]

120. Bharti, R.; Dey, G.; Mandal, M. Cancer Development, Chemoresistance, Epithelial to Mesenchymal Transition and Stem Cells: A Snapshot of IL-6 Mediated Involvement. *Cancer Lett.* **2016**, *375*, 51–61. [CrossRef] [PubMed]

121. Fang, X.; Yu, S.; Bast, R.C.; Liu, S.; Xu, H.J.; Hu, S.X.; LaPushin, R.; Claret, F.X.; Aggarwal, B.B.; Lu, Y.; et al. Mechanisms for Lysophosphatidic Acid-Induced Cytokine Production in Ovarian Cancer Cells. *J. Biol. Chem.* **2004**, *279*, 9653–9661. [CrossRef] [PubMed]

122. Long, H.; Xiang, T.; Qi, W.; Huang, J.; Chen, J.; He, L.; Liang, Z.; Guo, B.; Li, Y.; Xie, R.; et al. CD133+ Ovarian Cancer Stem-Like Cells Promote Non-Stem Cancer Cell Metastasis Via CCL5 Induced Epithelial-Mesenchymal Transition. *Oncotarget* **2015**, *6*, 5846–5859. [CrossRef] [PubMed]

123. Cheng, S.; Han, L.; Guo, J.; Yang, Q.; Zhou, J.; Yang, X. The Essential Roles of CCR7 in Epithelial-to-Mesenchymal Transition Induced by Hypoxia in Epithelial Ovarian Carcinomas. *Tumor Biol.* **2014**, *35*, 12293–12298. [CrossRef] [PubMed]

124. Comamala, M.; Pinard, M.; Theriault, C.; Matte, I.; Albert, A.; Boivin, M.; Beaudin, J.; Piche, A.; Rancourt, C. Downregulation of Cell Surface CA125/MUC16 Induces Epithelial-to-Mesenchymal Transition and Restores EGFR Signalling in NIH: OVCAR3 Ovarian Carcinoma Cells. *Br. J. Cancer* **2011**, *104*, 989–999. [CrossRef] [PubMed]

125. Thériault, C.; Pinard, M.; Comamala, M.; Migneault, M.; Beaudin, J.; Matte, I.; Boivin, M.; Piché, A.; Rancourt, C. MUC16 (CA125) Regulates Epithelial Ovarian Cancer Cell Growth, Tumorigenesis and Metastasis. *Gynecol. Oncol.* **2011**, *121*, 434–443. [CrossRef] [PubMed]

126. Asem, M.S.; Buechler, S.; Wates, R.B.; Miller, D.L.; Stack, M.S. Wnt5a Signaling in Cancer. *Cancers* **2016**, *8*, 79. [CrossRef] [PubMed]

127. Qi, H.; Sun, B.; Zhao, X.; Du, J.; Gu, Q.; Liu, Y.; Cheng, R.; Dong, X. Wnt5a Promotes Vasculogenic Mimicry and Epithelial-Mesenchymal Transition Via Protein Kinase Cα in Epithelial Ovarian Cancer. *Oncol. Rep.* **2014**, *32*, 771–779. [CrossRef] [PubMed]

128. Serrano-Gomez, S.J.; Maziveyi, M.; Alahari, S.K. Regulation of Epithelial-Mesenchymal Transition through Epigenetic and Post-Translational Modifications. *Mol. Cancer* **2016**, *15*, 18. [CrossRef] [PubMed]

129. Graff, J.R.; Herman, J.G.; Lapidus, R.G.; Chopra, H.; Xu, R.; Jarrard, D.F.; Isaacs, W.B.; Pitha, P.M.; Davidson, N.E.; Baylin, S.B. E-Cadherin Expression is Silenced by DNA Hypermethylation in Human Breast and Prostate Carcinomas. *Cancer Res.* **1995**, *55*, 5195–5199. [PubMed]

130. Bornman, D.M.; Mathew, S.; Alsruhe, J.; Herman, J.G.; Gabrielson, E. Methylation of the E-Cadherin Gene in Bladder Neoplasia and in Normal Urothelial Epithelium from Elderly Individuals. *Am. J. Pathol.* **2001**, *159*, 831–835. [CrossRef]

131. Tamura, G.; Yin, J.; Wang, S.; Fleisher, A.S.; Zou, T.; Abraham, J.M.; Kong, D.; Smolinski, K.N.; Wilson, K.T.; James, S.P.; et al. E-Cadherin Gene Promoter Hypermethylation in Primary Human Gastric Carcinomas. *J. Natl. Cancer Inst.* **2000**, *92*, 569–573. [CrossRef] [PubMed]

132. Fukagawa, A.; Ishii, H.; Miyazawa, K.; Saitoh, M. δEF1 Associates with DNMT1 and Maintains DNA Methylation of the E-cadherin Promoter in Breast Cancer Cells. *Cancer Med.* **2015**, *4*, 125–135. [CrossRef] [PubMed]

133. Kassambara, A.; Klein, B.; Moreaux, J. MMSET is overexpressed in Cancers: Link with Tumor Aggressiveness. *Biochem. Biophys. Res. Commun.* **2009**, *379*, 840–845. [CrossRef] [PubMed]

134. Dong, C.; Wu, Y.; Yao, J.; Wang, Y.; Yu, Y.; Rychahou, P.G.; Evers, B.M.; Zhou, B.P. G9a Interacts with Snail and is Critical for Snail-Mediated E-Cadherin Repression in Human Breast Cancer. *J. Clin. Invest.* **2012**, *122*, 1469–1486. [CrossRef] [PubMed]

135. Dong, C.; Wu, Y.; Wang, Y.; Wang, C.; Kang, T.; Rychahou, P.G.; Chi, Y.; Evers, B.M.; Zhou, B.P. Interaction with Suv39H1 is Critical for Snail-Mediated E-Cadherin Repression in Breast Cancer. *Oncogene* **2013**, *32*, 1351–1362. [CrossRef] [PubMed]

136. Jorgensen, S.; Elvers, I.; Trelle, M.B.; Menzel, T.; Eskildsen, M.; Jensen, O.N.; Helleday, T.; Helin, K.; Sorensen, C.S. The Histone Methyltransferase SET8 is Required for S-Phase Progression. *J. Cell Biol.* **2007**, *179*, 1337–1345. [CrossRef] [PubMed]

137. Peinado, H.; Ballestar, E.; Esteller, M.; Cano, A. Snail Mediates E-Cadherin Repression by the Recruitment of the Sin3A/Histone Deacetylase 1 (HDAC1)/HDAC2 Complex. *Mol. Cell. Biol.* **2004**, *24*, 306–319. [CrossRef] [PubMed]

138. Aghdassi, A.; Sendler, M.; Guenther, A.; Mayerle, J.; Behn, C.O.; Heidecke, C.D.; Friess, H.; Buchler, M.; Evert, M.; Lerch, M.M.; et al. Recruitment of Histone Deacetylases HDAC1 and HDAC2 by the Transcriptional Repressor ZEB1 Downregulates E-Cadherin Expression in Pancreatic Cancer. *Gut* **2012**, *61*, 439–448. [CrossRef] [PubMed]

139. Fu, J.; Qin, L.; He, T.; Qin, J.; Hong, J.; Wong, J.; Liao, L.; Xu, J. The TWIST/Mi2/NuRD Protein Complex and its Essential Role in Cancer Metastasis. *Cell Res.* **2011**, *21*, 275–289. [CrossRef] [PubMed]

140. Adhikary, A.; Chakraborty, S.; Mazumdar, M.; Ghosh, S.; Mukherjee, S.; Manna, A.; Mohanty, S.; Nakka, K.K.; Joshi, S.; De, A.; et al. Inhibition of Epithelial to Mesenchymal Transition by E-Cadherin Up-Regulation Via Repression of Slug Transcription and Inhibition of E-Cadherin Degradation: Dual Role of Scaffold/Matrix Attachment Region-Binding Protein 1 (SMAR1) in Breast Cancer Cells. *J. Biol. Chem.* **2014**, *289*, 25431–25444. [CrossRef] [PubMed]

141. Longacre, M.; Snyder, N.A.; Housman, G.; Leary, M.; Lapinska, K.; Heerboth, S.; Willbanks, A.; Sarkar, S. A Comparative Analysis of Genetic and Epigenetic Events of Breast and Ovarian Cancer Related to Tumorigenesis. *Int. J. Mol. Sci.* **2016**, *17*, 759. [CrossRef] [PubMed]

142. Cardenas, H.; Vieth, E.; Lee, J.; Segar, M.; Liu, Y.; Nephew, K.P.; Matei, D. TGF-B Induces Global Changes in DNA Methylation during the Epithelial-to-Mesenchymal Transition in Ovarian Cancer Cells. *Epigenetics* **2014**, *9*, 1461–1472. [CrossRef] [PubMed]

143. Su, H.; Lai, H.; Lin, Y.; Liu, C.; Chen, C.; Chou, Y.; Lin, S.; Lin, W.; Lee, H.; Yu, M. Epigenetic Silencing of SFRP5 is Related to Malignant Phenotype and Chemoresistance of Ovarian Cancer through Wnt Signaling Pathway. *Int. J. Cancer* **2010**, *127*, 555–567. [CrossRef] [PubMed]

144. Deng, Z.; Wang, L.; Hou, H.; Zhou, J.; Li, X. Epigenetic Regulation of IQGAP2 Promotes Ovarian Cancer Progression Via Activating Wnt/B-Catenin Signaling. *Int. J. Oncol.* **2016**, *48*, 153–160. [CrossRef] [PubMed]

145. Smith, H.J.; Straughn, J.M.; Buchsbaum, D.J.; Arend, R.C. Epigenetic Therapy for the Treatment of Epithelial Ovarian Cancer: A Clinical Review. *Gynecol. Oncol. Rep.* **2017**, *20*, 81–86. [CrossRef] [PubMed]

146. Gloss, B.S.; Samimi, G. Epigenetic Biomarkers in Epithelial Ovarian Cancer. *Cancer Lett.* **2014**, *342*, 257–263. [CrossRef] [PubMed]

147. Wang, Y.; Cardenas, H.; Fang, F.; Condello, S.; Taverna, P.; Segar, M.; Liu, Y.; Nephew, K.P.; Matei, D. Epigenetic Targeting of Ovarian Cancer Stem Cells. *Cancer Res.* **2014**, *74*, 4922–4936. [CrossRef] [PubMed]

148. Zhou, B.P.; Deng, J.; Xia, W.; Xu, J.; Li, Y.M.; Gunduz, M.; Hung, M. Dual Regulation of Snail by GSK-3β-Mediated Phosphorylation in Control of Epithelial–mesenchymal Transition. *Nat. Cell Biol.* **2004**, *6*, 931–940. [CrossRef] [PubMed]

149. Du, C.; Zhang, C.; Hassan, S.; Biswas, M.H.; Balaji, K.C. Protein Kinase D1 Suppresses Epithelial-to-Mesenchymal Transition through Phosphorylation of Snail. *Cancer Res.* **2010**, *70*, 7810–7819. [CrossRef] [PubMed]

150. Zheng, H.; Shen, M.; Zha, Y.; Li, W.; Wei, Y.; Blanco, M.A.; Ren, G.; Zhou, T.; Storz, P.; Wang, H. PKD1 Phosphorylation-Dependent Degradation of SNAIL by SCF-FBXO11 Regulates Epithelial-Mesenchymal Transition and Metastasis. *Cancer Cell* **2014**, *26*, 358–373. [CrossRef] [PubMed]

151. Park, S.Y.; Kim, H.S.; Kim, N.H.; Ji, S.; Cha, S.Y.; Kang, J.G.; Ota, I.; Shimada, K.; Konishi, N.; Nam, H.W.; et al. Snail1 is Stabilized by O-GlcNAc Modification in Hyperglycaemic Condition. *EMBO J.* **2010**, *29*, 3787–3796. [CrossRef] [PubMed]

152. Yang, C.; Chen, H.; Tan, G.; Gao, W.; Cheng, L.; Jiang, X.; Yu, L.; Tan, Y. FOXM1 Promotes the Epithelial to Mesenchymal Transition by Stimulating the Transcription of Slug in Human Breast Cancer. *Cancer Lett.* **2013**, *340*, 104–112. [CrossRef] [PubMed]

153. Bao, B.; Wang, Z.; Ali, S.; Kong, D.; Banerjee, S.; Ahmad, A.; Li, Y.; Azmi, A.S.; Miele, L.; Sarkar, F.H. Over-expression of FoxM1 Leads to Epithelial–mesenchymal Transition and Cancer Stem Cell Phenotype in Pancreatic Cancer Cells. *J. Cell. Biochem.* **2011**, *112*, 2296–2306. [CrossRef] [PubMed]

154. Long, J.; Zuo, D.; Park, M. Pc2-Mediated Sumoylation of Smad-Interacting Protein 1 Attenuates Transcriptional Repression of E-Cadherin. *J. Biol. Chem.* **2005**, *280*, 35477–35489. [CrossRef] [PubMed]

155. He, M.; Zhou, Z.; Shah, A.A.; Hong, Y.; Chen, Q.; Wan, Y. New Insights into Posttranslational Modifications of Hippo Pathway in Carcinogenesis and Therapeutics. *Cell Div.* **2016**, *11*, 4. [CrossRef] [PubMed]

156. Hall, C.A.; Wang, R.; Miao, J.; Oliva, E.; Shen, X.; Wheeler, T.; Hilsenbeck, S.G.; Orsulic, S.; Goode, S. Hippo Pathway Effector Yap is an Ovarian Cancer Oncogene. *Cancer Res.* **2010**, *70*, 8517–8525. [CrossRef] [PubMed]

157. Zhang, J.; Ma, L. MicroRNA Control of Epithelial–mesenchymal Transition and Metastasis. *Cancer Metastasis Rev.* **2012**, *31*, 653–662. [CrossRef] [PubMed]

158. Abba, M.L.; Patil, N.; Leupold, J.H.; Allgayer, H. MicroRNA Regulation of Epithelial to Mesenchymal Transition. *J. Clin. Med.* **2016**, *5*, 8. [CrossRef] [PubMed]

159. Gregory, P.A.; Bert, A.G.; Paterson, E.L.; Barry, S.C.; Tsykin, A.; Farshid, G.; Vadas, M.A.; Khew-Goodall, Y.; Goodall, G.J. The miR-200 Family and miR-205 Regulate Epithelial to Mesenchymal Transition by Targeting ZEB1 and SIP1. *Nat. Cell Biol.* **2008**, *10*, 593–601. [CrossRef] [PubMed]

160. Park, S.M.; Gaur, A.B.; Lengyel, E.; Peter, M.E. The miR-200 Family Determines the Epithelial Phenotype of Cancer Cells by Targeting the E-Cadherin Repressors ZEB1 and ZEB2. *Genes Dev.* **2008**, *22*, 894–907. [CrossRef] [PubMed]

161. Wiklund, E.D.; Bramsen, J.B.; Hulf, T.; Dyrskjøt, L.; Ramanathan, R.; Hansen, T.B.; Villadsen, S.B.; Gao, S.; Ostenfeld, M.S.; Borre, M. Coordinated Epigenetic Repression of the miR-200 Family and miR-205 in Invasive Bladder Cancer. *Int. J. Cancer* **2011**, *128*, 1327–1334. [CrossRef] [PubMed]

162. Fu, W.; Tao, T.; Qi, M.; Wang, L.; Hu, J.; Li, X.; Xing, N.; Du, R.; Han, B. MicroRNA-132/212 Upregulation Inhibits TGF-β-Mediated Epithelial–Mesenchymal Transition of Prostate Cancer Cells by Targeting SOX4. *Prostate* **2016**, *76*, 1560–1570. [CrossRef] [PubMed]

163. Zhao, J.; Zhang, L.; Guo, X.; Wang, J.; Zhou, W.; Liu, M.; Li, X.; Tang, H. miR-212/132 Downregulates SMAD2 Expression to Suppress the G1/S Phase Transition of the Cell Cycle and the Epithelial to Mesenchymal Transition in Cervical Cancer Cells. *IUBMB Life* **2015**, *67*, 380–394. [CrossRef] [PubMed]

164. Zheng, Y.; Luo, H.; Shi, Q.; Hao, Z.; Ding, Y.; Wang, Q.; Li, S.; Xiao, G.; Tong, S. miR-132 Inhibits Colorectal Cancer Invasion and Metastasis Via Directly Targeting ZEB2. *World J. Gastroenterol.* **2014**, *20*, 6515–6522. [CrossRef] [PubMed]

165. You, J.; Li, Y.; Fang, N.; Liu, B.; Zu, L.; Chang, R.; Li, X.; Zhou, Q. MiR-132 Suppresses the Migration and Invasion of Lung Cancer Cells Via Targeting the EMT Regulator ZEB2. *PLoS ONE* **2014**, *9*, e91827. [CrossRef] [PubMed]

166. Zhang, J.X.; Zhai, J.F.; Yang, X.T.; Wang, J. MicroRNA-132 Inhibits Migration, Invasion and Epithelial-Mesenchymal Transition by Regulating TGFbeta1/Smad2 in Human Non-Small Cell Lung Cancer. *Eur. Rev. Med. Pharmacol. Sci.* **2016**, *20*, 3793–3801. [PubMed]

167. Li, H.; Ouyang, R.; Wang, Z.; Zhou, W.; Chen, H.; Jiang, Y.; Zhang, Y.; Liao, M.; Wang, W.; Ye, M. MiR-150 Promotes Cellular Metastasis in Non-Small Cell Lung Cancer by Targeting FOXO4. *Sci. Rep.* **2016**, *6*, 39001. [CrossRef] [PubMed]

168. Yokobori, T.; Suzuki, S.; Tanaka, N.; Inose, T.; Sohda, M.; Sano, A.; Sakai, M.; Nakajima, M.; Miyazaki, T.; Kato, H. MiR-150 is Associated with Poor Prognosis in Esophageal Squamous Cell Carcinoma Via Targeting the EMT Inducer ZEB1. *Cancer Sci.* **2013**, *104*, 48–54. [CrossRef] [PubMed]

169. Ma, L.; Young, J.; Prabhala, H.; Pan, E.; Mestdagh, P.; Muth, D.; Teruya-Feldstein, J.; Reinhardt, F.; Onder, T.T.; Valastyan, S. miR-9, a MYC/MYCN-Activated microRNA, Regulates E-Cadherin and Cancer Metastasis. *Nat. Cell Biol.* **2010**, *12*, 247–256. [CrossRef] [PubMed]

170. Hu, J.; Qiu, M.; Jiang, F.; Zhang, S.; Yang, X.; Wang, J.; Xu, L.; Yin, R. MiR-145 Regulates Cancer Stem-Like Properties and Epithelial-to-Mesenchymal Transition in Lung Adenocarcinoma-Initiating Cells. *Tumor Biol.* **2014**, *35*, 8953–8961. [CrossRef] [PubMed]

171. Hu, H.; Xu, Z.; Li, C.; Xu, C.; Lei, Z.; Zhang, H.; Zhao, J. MiR-145 and miR-203 Represses TGF-B-Induced Epithelial-Mesenchymal Transition and Invasion by Inhibiting SMAD3 in Non-Small Cell Lung Cancer Cells. *Lung Cancer* **2016**, *97*, 87–94. [CrossRef] [PubMed]

172. Zhang, Z.; Liu, S.; Shi, R.; Zhao, G. miR-27 Promotes Human Gastric Cancer Cell Metastasis by Inducing Epithelial-to-Mesenchymal Transition. *Cancer Genet.* **2011**, *204*, 486–491. [CrossRef] [PubMed]

173. Li, J.; Wang, Y.; Song, Y.; Fu, Z.; Yu, W. miR-27a Regulates Cisplatin Resistance and Metastasis by Targeting RKIP in Human Lung Adenocarcinoma Cells. *Mol. Cancer* **2014**, *13*, 193. [CrossRef] [PubMed]

174. Pei, Y.; Lei, Y.; Liu, X. MiR-29a Promotes Cell Proliferation and EMT in Breast Cancer by Targeting Ten Eleven Translocation 1. *Biochim. Biophys. Acta Mol. Basis Dis.* **2016**, *1862*, 2177–2185. [CrossRef] [PubMed]

175. Gebeshuber, C.A.; Zatloukal, K.; Martinez, J. miR-29a Suppresses Tristetraprolin, which is a Regulator of Epithelial Polarity and Metastasis. *EMBO Rep.* **2009**, *10*, 400–405. [CrossRef] [PubMed]

176. Jiang, H.; Zhang, G.; Wu, J.; Jiang, C. Diverse Roles of miR-29 in Cancer (Review). *Oncol. Rep.* **2014**, *31*, 1509–1516. [CrossRef] [PubMed]

177. Zheng, Y.; Xiao, K.; Xiao, G.; Tong, S.; Ding, Y.; Wang, Q.; Li, S.; Hao, Z. MicroRNA-103 Promotes Tumor Growth and Metastasis in Colorectal Cancer by Directly Targeting LATS2. *Oncol. Lett.* **2016**, *12*, 2194–2200. [CrossRef] [PubMed]

178. Chen, H.Y.; Lin, Y.M.; Chung, H.C.; Lang, Y.D.; Lin, C.J.; Huang, J.; Wang, W.C.; Lin, F.M.; Chen, Z.; Huang, H.D.; et al. miR-103/107 Promote Metastasis of Colorectal Cancer by Targeting the Metastasis Suppressors DAPK and KLF4. *Cancer Res.* **2012**, *72*, 3631–3641. [CrossRef] [PubMed]

179. Shah, M.Y.; Calin, G.A. MicroRNAs miR-221 and miR-222: A New Level of Regulation in Aggressive Breast Cancer. *Genome Med.* **2011**, *3*, 56. [CrossRef] [PubMed]

180. Hwang, M.S.; Yu, N.; Stinson, S.Y.; Yue, P.; Newman, R.J.; Allan, B.B.; Dornan, D. miR-221/222 Targets Adiponectin Receptor 1 to Promote the Epithelial-to-Mesenchymal Transition in Breast Cancer. *PLoS ONE* **2013**, *8*, e66502. [CrossRef] [PubMed]

181. Stinson, S.; Lackner, M.R.; Adai, A.T.; Yu, N.; Kim, H.J.; O'Brien, C.; Spoerke, J.; Jhunjhunwala, S.; Boyd, Z.; Januario, T.; et al. miR-221/222 Targeting of Trichorhinophalangeal 1 (TRPS1) Promotes Epithelial-to-Mesenchymal Transition in Breast Cancer. *Sci. Signal.* **2011**, *4*, pt5. [CrossRef] [PubMed]

182. Vetter, G.; Saumet, A.; Moes, M.; Vallar, L.; Le Béchec, A.; Laurini, C.; Sabbah, M.; Arar, K.; Theillet, C.; Lecellier, C. miR-661 Expression in SNAI1-Induced Epithelial to Mesenchymal Transition Contributes to Breast Cancer Cell Invasion by Targeting Nectin-1 and StarD10 Messengers. *Oncogene* **2010**, *29*, 4436–4448. [CrossRef] [PubMed]

183. Ma, D.; Chai, Z.; Zhu, X.; Zhang, N.; Zhan, D.; Ye, B.; Wang, C.; Qin, C.; Zhao, Y.; Zhu, W. MicroRNA-26a Suppresses Epithelial-Mesenchymal Transition in Human Hepatocellular Carcinoma by Repressing Enhancer of Zeste Homolog 2. *J. Hematol. Oncol.* **2016**, *9*, 1. [CrossRef] [PubMed]

184. Liang, H.; Liu, S.; Chen, Y.; Bai, X.; Liu, L.; Dong, Y.; Hu, M.; Su, X.; Chen, Y.; Huangfu, L. miR-26a Suppresses EMT by Disrupting the Lin28B/Let-7d Axis: Potential Cross-Talks among miRNAs in IPF. *J. Mol. Med.* **2016**, *94*, 655–665. [CrossRef] [PubMed]

185. Chang, L.; Li, K.; Guo, T. miR-26a-5p Suppresses Tumor Metastasis by Regulating EMT and is Associated with Prognosis in HCC. *Clin. Transl. Oncol.* **2016**, *19*, 695–703. [CrossRef] [PubMed]

186. Wang, Y.; Sun, B.; Zhao, X.; Zhao, N.; Sun, R.; Zhu, D.; Zhang, Y.; Li, Y.; Gu, Q.; Dong, X.; et al. Twist1-Related miR-26b-5p Suppresses Epithelial-Mesenchymal Transition, Migration and Invasion by Targeting SMAD1 in Hepatocellular Carcinoma. *Oncotarget* **2016**, *7*, 24383–24401. [CrossRef] [PubMed]

187. Zhang, L.; Wang, Y.; Li, W.; Tsonis, P.A.; Li, Z.; Xie, L.; Huang, Y. MicroRNA-30a Regulation of Epithelial-Mesenchymal Transition in Diabetic Cataracts through Targeting SNAI1. *Sci. Rep.* **2017**, *7*. [CrossRef] [PubMed]

188. Peng, R.; Zhou, L.; Zhou, Y.; Zhao, Y.; Li, Q.; Ni, D.; Hu, Y.; Long, Y.; Liu, J.; Lyu, Z. MiR-30a Inhibits the Epithelial—Mesenchymal Transition of Podocytes through Downregulation of NFATc3. *Int. J. Mol. Sci.* **2015**, *16*, 24032–24047. [CrossRef] [PubMed]

189. Zhou, Q.; Yang, M.; Lan, H.; Yu, X. miR-30a Negatively Regulates TGF-β1–Induced Epithelial-Mesenchymal Transition and Peritoneal Fibrosis by Targeting Snai1. *Am. J. Pathol.* **2013**, *183*, 808–819. [CrossRef] [PubMed]

190. Wei, W.; Yang, Y.; Cai, J.; Cui, K.; Li, R.; Wang, H.; Shang, X.; Wei, D. MiR-30a-5p Suppresses Tumor Metastasis of Human Colorectal Cancer by Targeting ITGB3. *Cell. Physiol. Biochem.* **2016**, *39*, 1165–1176. [CrossRef] [PubMed]

191. Li, J.; Wang, Y.; Luo, J.; Fu, Z.; Ying, J.; Yu, Y.; Yu, W. miR-134 Inhibits Epithelial to Mesenchymal Transition by Targeting FOXM1 in Non-Small Cell Lung Cancer Cells. *FEBS Lett.* **2012**, *586*, 3761–3765. [CrossRef] [PubMed]

192. Dong, P.; Kaneuchi, M.; Watari, H.; Hamada, J.; Sudo, S.; Ju, J.; Sakuragi, N. MicroRNA-194 Inhibits Epithelial to Mesenchymal Transition of Endometrial Cancer Cells by Targeting Oncogene BMI-1. *Mol. Cancer* **2011**, *10*, 99. [CrossRef] [PubMed]

193. Meng, Z.; Fu, X.; Chen, X.; Zeng, S.; Tian, Y.; Jove, R.; Xu, R.; Huang, W. miR-194 is a Marker of Hepatic Epithelial Cells and Suppresses Metastasis of Liver Cancer Cells in Mice. *Hepatology* **2010**, *52*, 2148–2157. [CrossRef] [PubMed]

194. Khella, H.W.; Bakhet, M.; Allo, G.; Jewett, M.A.; Girgis, A.H.; Latif, A.; Girgis, H.; Von Both, I.; Bjarnason, G.A.; Yousef, G.M. miR-192, miR-194 and miR-215: A Convergent microRNA Network Suppressing Tumor Progression in Renal Cell Carcinoma. *Carcinogenesis* **2013**, *34*, 2231–2239. [CrossRef] [PubMed]

195. Zhang, L.; Wang, X.; Chen, P. MiR-204 Down Regulates SIRT1 and Reverts SIRT1-Induced Epithelial-Mesenchymal Transition, Anoikis Resistance and Invasion in Gastric Cancer Cells. *BMC Cancer* **2013**, *13*, 290. [CrossRef] [PubMed]

196. Qiu, Y.H.; Wei, Y.P.; Shen, N.J.; Wang, Z.C.; Kan, T.; Yu, W.L.; Yi, B.; Zhang, Y.J. miR-204 Inhibits Epithelial to Mesenchymal Transition by Targeting Slug in Intrahepatic Cholangiocarcinoma Cells. *Cell. Physiol. Biochem.* **2013**, *32*, 1331–1341. [CrossRef] [PubMed]

197. Sun, Y.; Yu, X.; Bai, Q. miR-204 Inhibits Invasion and Epithelial-Mesenchymal Transition by Targeting FOXM1 in Esophageal Cancer. *Int. J. Clin. Exp. Pathol.* **2015**, *8*, 12775–12783. [PubMed]

198. Liu, Z.; Long, J.; Du, R.; Ge, C.; Guo, K.; Xu, Y. miR-204 Regulates the EMT by Targeting Snail to Suppress the Invasion and Migration of Gastric Cancer. *Tumor Biol.* **2016**, *37*, 8327–8335. [CrossRef] [PubMed]

199. Bendoraite, A.; Knouf, E.C.; Garg, K.S.; Parkin, R.K.; Kroh, E.M.; O'Briant, K.C.; Ventura, A.P.; Godwin, A.K.; Karlan, B.Y.; Drescher, C.W.; et al. Regulation of miR-200 Family microRNAs and ZEB Transcription Factors in Ovarian Cancer: Evidence Supporting a Mesothelial-to-Epithelial Transition. *Gynecol. Oncol.* **2010**, *116*, 117–125. [CrossRef] [PubMed]

200. Cheng, W.; Liu, J.; Yoshida, H.; Rosen, D.; Naora, H. Lineage Infidelity of Epithelial Ovarian Cancers is Controlled by HOX Genes that Specify Regional Identity in the Reproductive Tract. *Nat. Med.* **2005**, *11*, 531–537. [CrossRef] [PubMed]

201. Zhou, B.; Xu, H.; Xia, M.; Sun, C.; Li, N.; Guo, E.; Guo, L.; Shan, W.; Lu, H.; Wu, Y. Overexpressed miR-9 Promotes Tumor Metastasis Via Targeting E-Cadherin in Serous Ovarian Cancer. *Front. Med.* **2017**, *11*, 214–222. [CrossRef] [PubMed]

202. Ye, Z.; Zhao, L.; Li, J.; Chen, W.; Li, X. miR-30d Blocked Transforming Growth Factor Beta1-Induced Epithelial-Mesenchymal Transition by Targeting Snail in Ovarian Cancer Cells. *Int. J. Gynecol. Cancer* **2015**, *25*, 1574–1581. [CrossRef] [PubMed]

203. Wu, X.; Ruan, Y.; Jiang, H.; Xu, C. MicroRNA-424 Inhibits Cell Migration, Invasion, and Epithelial Mesenchymal Transition by Downregulating Doublecortin-Like Kinase 1 in Ovarian Clear Cell Carcinoma. *Int. J. Biochem. Cell Biol.* **2017**, *85*, 66–74. [CrossRef] [PubMed]

204. Tan, H.; He, Q.; Gong, G.; Wang, Y.; Li, J.; Wang, J.; Zhu, D.; Wu, X. miR-382 Inhibits Migration and Invasion by Targeting ROR1 through Regulating EMT in Ovarian Cancer. *Int. J. Oncol.* **2016**, *48*, 181–190. [CrossRef] [PubMed]

205. Sun, Y.; Hu, L.; Zheng, H.; Bagnoli, M.; Guo, Y.; Rupaimoole, R.; Rodriguez-Aguayo, C.; Lopez-Berestein, G.; Ji, P.; Chen, K. MiR-506 Inhibits Multiple Targets in the Epithelial-to-mesenchymal Transition Network and is Associated with Good Prognosis in Epithelial Ovarian Cancer. *J. Pathol.* **2015**, *235*, 25–36. [CrossRef] [PubMed]

206. Zhou, X.; Hu, Y.; Dai, L.; Wang, Y.; Zhou, J.; Wang, W.; Di, W.; Qiu, L. MicroRNA-7 Inhibits Tumor Metastasis and Reverses Epithelial-Mesenchymal Transition through AKT/ERK1/2 Inactivation by Targeting EGFR in Epithelial Ovarian Cancer. *PLoS ONE* **2014**, *9*, e96718. [CrossRef] [PubMed]

207. Mezzanzanica, D.; Bagnoli, M.; De Cecco, L.; Valeri, B.; Canevari, S. Role of microRNAs in Ovarian Cancer Pathogenesis and Potential Clinical Implications. *Int. J. Biochem. Cell Biol.* **2010**, *42*, 1262–1272. [CrossRef] [PubMed]

208. Meng, X.; Muller, V.; Milde-Langosch, K.; Trillsch, F.; Pantel, K.; Schwarzenbach, H. Diagnostic and Prognostic Relevance of Circulating Exosomal miR-373, miR-200a, miR-200b and miR-200c in Patients with Epithelial Ovarian Cancer. *Oncotarget* **2016**, *7*, 16923–16935. [CrossRef] [PubMed]

209. Chung, Y.W.; Bae, H.S.; Song, J.Y.; Lee, J.K.; Lee, N.W.; Kim, T.; Lee, K.W. Detection of microRNA as Novel Biomarkers of Epithelial Ovarian Cancer from the Serum of Ovarian Cancer Patients. *Int. J. Gynecol. Cancer* **2013**, *23*, 673–679. [CrossRef] [PubMed]

210. Sohn, E.J.; Won, G.; Lee, J.; Yoon, S.W.; Lee, I.; Kim, H.J.; Kim, S. Blockage of Epithelial to Mesenchymal Transition and Upregulation of Let 7b are Critically Involved in Ursolic Acid Induced Apoptosis in Malignant Mesothelioma Cell. *Int. J. Biol. Sci.* **2016**, *12*, 1279–1288. [CrossRef] [PubMed]

211. Tang, Z.; Ow, G.S.; Thiery, J.P.; Ivshina, A.V.; Kuznetsov, V.A. Meta-analysis of Transcriptome Reveals Let-7b as an Unfavorable Prognostic Biomarker and Predicts Molecular and Clinical Subclasses in High-grade Serous Ovarian Carcinoma. *Int. J. Cancer* **2014**, *134*, 306–318. [CrossRef] [PubMed]

212. Zhou, J.; Gong, G.; Tan, H.; Dai, F.; Zhu, X.; Chen, Y.; Wang, J.; Liu, Y.; Chen, P.; Wu, X. Urinary microRNA-30a-5p is a Potential Biomarker for Ovarian Serous Adenocarcinoma. *Oncol. Rep.* **2015**, *33*, 2915–2923. [CrossRef] [PubMed]

213. Vang, S.; Wu, H.; Fischer, A.; Miller, D.H.; MacLaughlan, S.; Douglass, E.; Steinhoff, M.; Collins, C.; Smith, P.J.; Brard, L. Identification of Ovarian Cancer Metastatic miRNAs. *PLoS ONE* **2013**, *8*, e58226. [CrossRef] [PubMed]

214. Parikh, A.; Lee, C.; Joseph, P.; Marchini, S.; Baccarini, A.; Kolev, V.; Romualdi, C.; Fruscio, R.; Shah, H.; Wang, F. microRNA-181a has a Critical Role in Ovarian Cancer Progression through the Regulation of the Epithelial–mesenchymal Transition. *Nat. Commun.* **2014**, *5*, 2977. [CrossRef] [PubMed]

215. Kinose, Y.; Sawada, K.; Nakamura, K.; Kimura, T. The Role of microRNAs in Ovarian Cancer. *Biomed. Res. Int.* **2014**, *2014*, 249393. [CrossRef] [PubMed]

216. Smith, B.; Agarwal, P.; Bhowmick, N.A. MicroRNA Applications for Prostate, Ovarian and Breast Cancer in the Era of Precision Medicine. *Endocr. Relat. Cancer* **2017**, *24*, R157–R172. [CrossRef] [PubMed]

217. Venkatraman, A.; He, X.C.; Thorvaldsen, J.L.; Sugimura, R.; Perry, J.M.; Tao, F.; Zhao, M.; Christenson, M.K.; Sanchez, R.; Jaclyn, Y.Y. Maternal Imprinting at the H19-Igf2 Locus Maintains Adult Haematopoietic Stem Cell Quiescence. *Nature* **2013**, *500*, 345–349. [CrossRef] [PubMed]

218. Xia, B.; Hou, Y.; Chen, H.; Yang, S.; Liu, T.; Lin, M.; Lou, G. Long Non-Coding RNA ZFAS1 Interacts with miR-150–5p to Regulate Sp1 Expression and Ovarian Cancer Cell Malignancy. *Oncotarget* **2017**, *8*, 19534–19546. [CrossRef] [PubMed]

219. Tsai, M.C.; Spitale, R.C.; Chang, H.Y. Long Intergenic Noncoding RNAs: New Links in Cancer Progression. *Cancer Res.* **2011**, *71*, 3–7. [CrossRef] [PubMed]

220. Yuan, J.; Yang, F.; Wang, F.; Ma, J.; Guo, Y.; Tao, Q.; Liu, F.; Pan, W.; Wang, T.; Zhou, C. A Long Noncoding RNA Activated by TGF-B Promotes the Invasion-Metastasis Cascade in Hepatocellular Carcinoma. *Cancer Cell* **2014**, *25*, 666–681. [CrossRef] [PubMed]

221. Zhou, M.; Sun, Y.; Sun, Y.; Xu, W.; Zhang, Z.; Zhao, H.; Zhong, Z.; Sun, J. Comprehensive Analysis of lncRNA Expression Profiles Reveals a Novel lncRNA Signature to Discriminate Nonequivalent Outcomes in Patients with Ovarian Cancer. *Oncotarget* **2016**, *7*, 32433–32448. [CrossRef] [PubMed]

222. Qiu, J.; Lin, Y.; Ye, L.; Ding, J.; Feng, W.; Jin, H.; Zhang, Y.; Li, Q.; Hua, K. Overexpression of Long Non-Coding RNA HOTAIR Predicts Poor Patient Prognosis and Promotes Tumor Metastasis in Epithelial Ovarian Cancer. *Gynecol. Oncol.* **2014**, *134*, 121–128. [CrossRef] [PubMed]

223. Richards, E.J.; Zhang, G.; Li, Z.P.; Permuth-Wey, J.; Challa, S.; Li, Y.; Kong, W.; Dan, S.; Bui, M.M.; Coppola, D.; et al. Long Non-Coding RNAs (LncRNA) Regulated by Transforming Growth Factor (TGF) Beta: LncRNA-Hit-Mediated TGFbeta-Induced Epithelial to Mesenchymal Transition in Mammary Epithelia. *J. Biol. Chem.* **2015**, *290*, 6857–6867. [CrossRef] [PubMed]

224. Holm-Nielsen, P. Pathogenesis of Ascites in Peritoneal Carcinomatosis1. *Acta Pathol. Microbiol. Scand.* **1953**, *33*, 10–21. [CrossRef] [PubMed]

225. Feldman, G.B.; Knapp, R.C.; Order, S.E.; Hellman, S. The Role of Lymphatic Obstruction in the Formation of Ascites in a Murine Ovarian Carcinoma. *Cancer Res.* **1972**, *32*, 1663–1666. [PubMed]

226. Garrison, R.N.; Galloway, R.H.; Heuser, L.S. Mechanisms of Malignant Ascites Production. *J. Surg. Res.* **1987**, *42*, 126–132. [CrossRef]

227. Bamberger, E.; Perrett, C. Angiogenesis in Epithelial Ovarian Cancer. *J. Clin. Pathol.* **2002**, *55*, 348. [CrossRef]

228. Ueda, M.; Terai, Y.; Kanda, K.; Kanemura, M.; Takehara, M.; Futakuchi, H.; Yamaguchi, H.; Yasuda, M.; NISHKAMA, K.; Ueki, M. Tumor Angiogenesis and Molecular Target Therapy in Ovarian Carcinomas. *Hum. Cell* **2005**, *18*, 1–16. [CrossRef] [PubMed]

229. Nagy, J.A.; Herzberg, K.T.; Dvorak, J.M.; Dvorak, H.F. Pathogenesis of Malignant Ascites Formation: Initiating Events that Lead to Fluid Accumulation. *Cancer Res.* **1993**, *53*, 2631–2643. [PubMed]

230. Salani, D.; Di Castro, V.; Nicotra, M.R.; Rosano, L.; Tecce, R.; Venuti, A.; Natali, P.G.; Bagnato, A. Role of Endothelin-1 in Neovascularization of Ovarian Carcinoma. *Am. J. Pathol.* **2000**, *157*, 1537–1547. [CrossRef]

231. Chen, Y.; Gou, X.; Ke, X.; Cui, H.; Chen, Z. Human Tumor Cells Induce Angiogenesis through Positive Feedback between CD147 and Insulin-Like Growth Factor-I. *PLoS ONE* **2012**, *7*, e40965. [CrossRef] [PubMed]

232. Senger, D.R.; Galli, S.J.; Dvorak, A.M.; Perruzzi, C.A.; Harvey, V.S.; Dvorak, H.F. Tumor Cells Secrete a Vascular Permeability Factor that Promotes Accumulation of Ascites Fluid. *Science* **1983**, *219*, 983–985. [CrossRef] [PubMed]

233. Neufeld, G.; Cohen, T.; Gengrinovitch, S.; Poltorak, Z. Vascular Endothelial Growth Factor (VEGF) and its Receptors. *FASEB J.* **1999**, *13*, 9–22. [PubMed]

234. Geva, E.; Jaffe, R.B. Role of Vascular Endothelial Growth Factor in Ovarian Physiology and Pathology. *Fertil. Steril.* **2000**, *74*, 429–438. [CrossRef]

235. Kassim, S.K.; El-Salahy, E.M.; Fayed, S.T.; Helal, S.A.; Helal, T.; Azzam, E.E.; Khalifa, A. Vascular Endothelial Growth Factor and Interleukin-8 are Associated with Poor Prognosis in Epithelial Ovarian Cancer Patients. *Clin. Biochem.* **2004**, *37*, 363–369. [CrossRef] [PubMed]

236. Santin, A.D.; Hermonat, P.L.; Ravaggi, A.; Cannon, M.J.; Pecorelli, S.; Parham, G.P. Secretion of Vascular Endothelial Growth Factor in Ovarian Cancer. *Eur. J. Gynaecol. Oncol.* **1999**, *20*, 177–181.

237. Garrison, R.N.; Kaelin, L.D.; Galloway, R.H.; Heuser, L.S. Malignant Ascites. Clinical and Experimental Observations. *Ann. Surg.* **1986**, *203*, 644–651. [CrossRef] [PubMed]

238. Parsons, S.L.; Lang, M.W.; Steele, R.J.C. Malignant Ascites: A 2-Year Review from a Teaching Hospital. *Eur. J. Surg. Oncol.* **1996**, *22*, 237–239. [CrossRef]

239. Rizvi, I.; Gurkan, U.A.; Tasoglu, S.; Alagic, N.; Celli, J.P.; Mensah, L.B.; Mai, Z.; Demirci, U.; Hasan, T. Flow Induces Epithelial-Mesenchymal Transition, Cellular Heterogeneity and Biomarker Modulation in 3D Ovarian Cancer Nodules. *Proc. Natl. Acad. Sci. USA* **2013**, *110*, E1974–E1983. [CrossRef] [PubMed]

240. Liu, S.; Zhou, F.; Shen, Y.; Zhang, Y.; Yin, H.; Zeng, Y.; Liu, J.; Yan, Z.; Liu, X. Fluid Shear Stress Induces Epithelial-Mesenchymal Transition (EMT) in Hep-2 Cells. *Oncotarget* **2016**, *7*, 32876–32892. [CrossRef] [PubMed]

241. Triantafillu, U.L.; Park, S.; Klaassen, N.L.; Raddatz, A.D.; Kim, Y. Fluid Shear Stress Induces Cancer Stem Cell-Like Phenotype in MCF7 Breast Cancer Cell Line without Inducing Epithelial to Mesenchymal Transition. *Int. J. Oncol.* **2017**, *50*, 993–1001. [CrossRef] [PubMed]

242. Aoki, S.; Noguchi, M.; Takezawa, T.; Ikeda, S.; Uchihashi, K.; Kuroyama, H.; Chimuro, T.; Toda, S. Fluid Dwell Impact Induces Peritoneal Fibrosis in the Peritoneal Cavity Reconstructed in Vitro. *J. Artif. Organs* **2016**, *19*, 87–96. [CrossRef] [PubMed]

243. Kim, B.G.; Kang, S.; Han, H.H.; Lee, J.H.; Kim, J.E.; Lee, S.H.; Cho, N.H. Transcriptome-Wide Analysis of Compression-Induced microRNA Expression Alteration in Breast Cancer for Mining Therapeutic Targets. *Oncotarget* **2016**, *7*, 27468–27478. [PubMed]

244. Piotrowski-Daspit, A.S.; Tien, J.; Nelson, C.M. Interstitial Fluid Pressure Regulates Collective Invasion in Engineered Human Breast Tumors via Snail, Vimentin, and E-Cadherin. *Integr. Biol.* **2016**, *8*, 319–331. [CrossRef] [PubMed]

245. Huang, D.; Liu, Y.; Huang, Y.; Xie, Y.; Shen, K.; Zhang, D.; Mou, Y. Mechanical Compression Upregulates MMP9 through SMAD3 but Not SMAD2 Modulation in Hypertrophic Scar Fibroblasts. Connect. *Tissue Res.* **2014**, *55*, 391–396. [CrossRef] [PubMed]

246. Burkhalter, R.J. *Microenvironmental Regulation of Ovarian Cancer Dissemination via Activation of the Wnt Signaling Pathway*; University of Missouri-Columbia: Columbia, MO, USA, 2012.

247. Gupta, N.; Xu, Z.; El-Sehemy, A.; Steed, H.; Fu, Y. Notch3 Induces Epithelial–mesenchymal Transition and Attenuates Carboplatin-Induced Apoptosis in Ovarian Cancer Cells. *Gynecol. Oncol.* **2013**, *130*, 200–206. [CrossRef] [PubMed]

248. Bugide, S.; Gonugunta, V.K.; Penugurti, V.; Malisetty, V.L.; Vadlamudi, R.K.; Manavathi, B. HPIP Promotes Epithelial-Mesenchymal Transition and Cisplatin Resistance in Ovarian Cancer Cells through PI3K/AKT Pathway Activation. *Cell. Oncol.* **2017**, *40*, 133–144. [CrossRef] [PubMed]

249. Qin, S.; Li, Y.; Cao, X.; Du, J.; Huang, X. NANOG Regulates Epithelial-Mesenchymal Transition and Chemoresistance in Ovarian Cancer. *Biosci. Rep.* **2017**, *37*. [CrossRef] [PubMed]

250. Liu, S.; Sun, J.; Cai, B.; Xi, X.; Yang, L.; Zhang, Z.; Feng, Y.; Sun, Y. NANOG Regulates Epithelial-Mesenchymal Transition and Chemoresistance through Activation of the STAT3 Pathway in Epithelial Ovarian Cancer. *Tumor Biol.* **2016**, *37*, 9671–9680. [CrossRef] [PubMed]

251. Roberts, C.M.; Tran, M.A.; Pitruzzello, M.C.; Wen, W.; Loeza, J.; Dellinger, T.H.; Mor, G.; Glackin, C.A. TWIST1 Drives Cisplatin Resistance and Cell Survival in an Ovarian Cancer Model, Via Upregulation of GAS6, L1CAM, and Akt Signalling. *Sci. Rep.* **2016**, *6*, 37652. [CrossRef] [PubMed]

252. Zhu, X.; Shen, H.; Yin, X.; Long, L.; Xie, C.; Liu, Y.; Hui, L.; Lin, X.; Fang, Y.; Cao, Y. miR-186 Regulation of Twist1 and Ovarian Cancer Sensitivity to Cisplatin. *Oncogene* **2016**, *35*, 323–332. [CrossRef] [PubMed]

253. Lim, S.; Becker, A.; Zimmer, A.; Lu, J.; Buettner, R.; Kirfel, J. SNAI1-Mediated Epithelial-Mesenchymal Transition Confers Chemoresistance and Cellular Plasticity by Regulating Genes Involved in Cell Death and Stem Cell Maintenance. *PLoS ONE* **2013**, *8*, e66558. [CrossRef] [PubMed]

254. Chiu, W.; Huang, Y.; Tsai, H.; Chen, C.; Chang, C.; Huang, S.; Hsu, K.; Chou, C. FOXM1 Confers to Epithelial-Mesenchymal Transition, Stemness and Chemoresistance in Epithelial Ovarian Carcinoma Cells. *Oncotarget* **2015**, *6*, 2349–2365. [CrossRef] [PubMed]

255. Zhang, X.; George, J.; Deb, S.; Degoutin, J.; Takano, E.; Fox, S.; Bowtell, D.; Harvey, K. The Hippo Pathway Transcriptional Co-Activator, YAP, is an Ovarian Cancer Oncogene. *Oncogene* **2011**, *30*, 2810–2822. [CrossRef] [PubMed]

256. Giannakakis, A.; Sandaltzopoulos, R.; Greshock, J.; Liang, S.; Huang, J.; Hasegawa, K.; Li, C.; O'Brien-Jenkins, A.; Katsaros, D.; Weber, B.L. miR-210 Links Hypoxia with Cell Cycle Regulation and is Deleted in Human Epithelial Ovarian Cancer. *Cancer Boil. Ther.* **2008**, *7*, 255–264. [CrossRef]

257. Leight, J.L.; Wozniak, M.A.; Chen, S.; Lynch, M.L.; Chen, C.S. Matrix Rigidity Regulates a Switch between TGF-Beta1-Induced Apoptosis and Epithelial-Mesenchymal Transition. *Mol. Biol. Cell* **2012**, *23*, 781–791. [CrossRef] [PubMed]

258. Wei, S.C.; Fattet, L.; Tsai, J.H.; Guo, Y.; Pai, V.H.; Majeski, H.E.; Chen, A.C.; Sah, R.L.; Taylor, S.S.; Engler, A.J. Matrix Stiffness Drives Epithelial-Mesenchymal Transition and Tumour Metastasis through a TWIST1-G3BP2 Mechanotransduction Pathway. *Nat. Cell Biol.* **2015**, *17*, 678–688. [CrossRef] [PubMed]

259. Bedi, U.; Mishra, V.K.; Wasilewski, D.; Scheel, C.; Johnsen, S.A. Epigenetic Plasticity: A Central Regulator of Epithelial-to-Mesenchymal Transition in Cancer. *Oncotarget* **2014**, *5*, 2016–2029. [CrossRef] [PubMed]

260. Wu, C.; Tsai, Y.; Wu, M.; Teng, S.; Wu, K. Epigenetic Reprogramming and Post-Transcriptional Regulation during the Epithelial–mesenchymal Transition. *Trends Genet.* **2012**, *28*, 454–463. [CrossRef] [PubMed]

261. Lu, M.; Jolly, M.K.; Levine, H.; Onuchic, J.N.; Ben-Jacob, E. MicroRNA-Based Regulation of Epithelial-Hybrid-Mesenchymal Fate Determination. *Proc. Natl. Acad. Sci. USA* **2013**, *110*, 18144–18149. [CrossRef] [PubMed]

262. Tian, X.; Zhang, H.; Xing, J. Coupled Reversible and Irreversible Bistable Switches Underlying TGFβ-Induced Epithelial to Mesenchymal Transition. *Biophys. J.* **2013**, *105*, 1079–1089. [CrossRef] [PubMed]

263. Steinway, S.N.; Zanudo, J.G.; Ding, W.; Rountree, C.B.; Feith, D.J.; Loughran, T.P., Jr.; Albert, R. Network Modeling of TGFbeta Signaling in Hepatocellular Carcinoma Epithelial-to-Mesenchymal Transition Reveals Joint Sonic Hedgehog and Wnt Pathway Activation. *Cancer Res.* **2014**, *74*, 5963–5977. [CrossRef] [PubMed]

264. Jia, D.; Jolly, M.K.; Boareto, M.; Parsana, P.; Mooney, S.M.; Pienta, K.J.; Levine, H.; Ben-Jacob, E. OVOL Guides the Epithelial-Hybrid-Mesenchymal Transition. *Oncotarget* **2015**, *6*, 15436–15448. [CrossRef] [PubMed]

265. Faddaoui, A.; Sheta, R.; Bachvarova, M.; Plante, M.; Gregoire, J.; Renaud, M.; Sebastianelli, A.; Gobeil, S.; Morin, C.; Ghani, K. Suppression of the Grainyhead Transcription Factor 2 Gene (GRHL2) Inhibits the Proliferation, Migration, Invasion and Mediates Cell Cycle Arrest of Ovarian Cancer Cells. *Cell Cycle* **2017**, *16*, 693–706. [CrossRef] [PubMed]

266. Boareto, M.; Jolly, M.K.; Goldman, A.; Pietila, M.; Mani, S.A.; Sengupta, S.; Ben-Jacob, E.; Levine, H.; Onuchic, J.N. Notch-Jagged Signalling can Give Rise to Clusters of Cells Exhibiting a Hybrid Epithelial/Mesenchymal Phenotype. *J. R. Soc. Interface* **2016**, *13*. [CrossRef] [PubMed]

267. Li, C.; Hong, T.; Nie, Q. Quantifying the Landscape and Kinetic Paths for Epithelial–Mesenchymal Transition from a Core Circuit. *Phys. Chem. Chem. Phys.* **2016**, *18*, 17949–17956. [CrossRef] [PubMed]

268. Li, C.; Wang, J. Quantifying the Underlying Landscape and Paths of Cancer. *J. R. Soc. Interface* **2014**, *11*, 20140774. [CrossRef] [PubMed]

269. Cohen, D.P.; Martignetti, L.; Robine, S.; Barillot, E.; Zinovyev, A.; Calzone, L. Mathematical Modelling of Molecular Pathways Enabling Tumour Cell Invasion and Migration. *PLoS Comput. Biol.* **2015**, *11*, e1004571. [CrossRef] [PubMed]

270. Zadran, S.; Arumugam, R.; Herschman, H.; Phelps, M.E.; Levine, R.D. Surprisal Analysis Characterizes the Free Energy Time Course of Cancer Cells Undergoing Epithelial-to-Mesenchymal Transition. *Proc. Natl. Acad. Sci. USA* **2014**, *111*, 13235–13240. [CrossRef] [PubMed]

271. Chang, H.; Liu, Y.; Xue, M.; Liu, H.; Du, S.; Zhang, L.; Wang, P. Synergistic Action of Master Transcription Factors Controls Epithelial-to-Mesenchymal Transition. *Nucleic Acids Res.* **2016**, *44*, 2514–2527. [CrossRef] [PubMed]

272. Tan, T.Z.; Miow, Q.H.; Miki, Y.; Noda, T.; Mori, S.; Huang, R.Y.; Thiery, J.P. Epithelial-Mesenchymal Transition Spectrum Quantification and its Efficacy in Deciphering Survival and Drug Responses of Cancer Patients. *EMBO Mol. Med.* **2014**, *6*, 1279–1293. [CrossRef] [PubMed]

273. Ramis-Conde, I.; Drasdo, D.; Anderson, A.R.; Chaplain, M.A. Modeling the Influence of the E-Cadherin-B-Catenin Pathway in Cancer Cell Invasion: A Multiscale Approach. *Biophys. J.* **2008**, *95*, 155–165. [CrossRef] [PubMed]

274. Neagu, A.; Mironov, V.; Kosztin, I.; Barz, B.; Neagu, M.; Moreno-Rodriguez, R.A.; Markwald, R.R.; Forgacs, G. Computational Modeling of Epithelial–mesenchymal Transformations. *BioSystems* **2010**, *100*, 23–30. [CrossRef] [PubMed]

275. Turner, C.; Kohandel, M. Quantitative Approaches to Cancer Stem Cells and Epithelial–Mesenchymal Transition. *Sem. Cancer Biol.* **2012**, 374–378. [CrossRef] [PubMed]

cancers

MDPI

Review

Epithelial-to-Mesenchymal Transition and MicroRNAs in Lung Cancer

Antoine Legras [1,2], Nicolas Pécuchet [1,3], Sandrine Imbeaud [4], Karine Pallier [1], Audrey Didelot [1], Hélène Roussel [5,6], Laure Gibault [5], Elizabeth Fabre [1,3], Françoise Le Pimpec-Barthes [2,4], Pierre Laurent-Puig [1,7] and Hélène Blons [1,7,*]

[1] INSERM UMR-S1147, CNRS SNC 5014, Saints-Pères Research Center, 45 rue des Saints-Pères Paris-Descartes University, Sorbonne Paris Cité University, 75006 Paris, France; antlegras@gmail.com (A.L.); nicolas.pecuchet@me.com (N.P.); pallier.karine@gmail.com (K.P.); audrey.didelot@parisdescartes.fr (A.D.); elizabeth.fabre@aphp.fr (E.F.); pierre.laurent-puig@parisdescartes.fr (P.L.-P.)

[2] Thoracic Surgery and Lung Transplantation Department, Georges Pompidou European Hospital, 20 rue Leblanc, Assistance Publique-Hôpitaux de Paris, 75015 Paris, France; francoise.lepimpec-barthes@aphp.fr

[3] Medical Thoracic Oncology Department, Georges Pompidou European Hospital, 20 rue Leblanc, Assistance Publique-Hôpitaux de Paris, 75015 Paris, France

[4] INSERM UMR-S1162, 27 rue Juliette Dodu, 75010 Paris, France; sandrine.imbeaud@inserm.fr

[5] Pathology Department, Georges Pompidou European Hospital, 20 rue Leblanc, Assistance Publique-Hôpitaux de Paris, 75015 Paris, France; helene.roussel@aphp.fr (H.R.); laure.gibault@aphp.fr (L.G.)

[6] INSERM UMR-S970, Paris Centre de Recherche Cardiovasculaire, Georges Pompidou European Hospital, 20 rue Leblanc, 75015 Paris, France

[7] Molecular Biology Department, Georges Pompidou European Hospital, 20 rue Leblanc, Assistance Publique-Hôpitaux de Paris, 75015 Paris, France

* Correspondence: helene.blons@parisdescartes.fr; Tel.: +33-1-42-86-40-67; Fax: +33-1-42-86-20-72

Academic Editor: Joëlle Roche
Received: 24 June 2017; Accepted: 26 July 2017; Published: 3 August 2017

Abstract: Despite major advances, non-small cell lung cancer (NSCLC) remains the major cause of cancer-related death in developed countries. Metastasis and drug resistance are the main factors contributing to relapse and death. Epithelial-to-mesenchymal transition (EMT) is a complex molecular and cellular process involved in tissue remodelling that was extensively studied as an actor of tumour progression, metastasis and drug resistance in many cancer types and in lung cancers. Here we described with an emphasis on NSCLC how the changes in signalling pathways, transcription factors expression or microRNAs that occur in cancer promote EMT. Understanding the biology of EMT will help to define reversing process and treatment strategies. We will see that this complex mechanism is related to inflammation, cell mobility and stem cell features and that it is a dynamic process. The existence of intermediate phenotypes and tumour heterogeneity may be debated in the literature concerning EMT markers, EMT signatures and clinical consequences in NSCLC. However, given the role of EMT in metastasis and in drug resistance the development of EMT inhibitors is an interesting approach to counteract tumour progression and drug resistance. This review describes EMT involvement in cancer with an emphasis on NSCLC and microRNA regulation.

Keywords: epithelial-mesenchymal transition; microRNAs; lung neoplasms; biomarkers; tumour

1. Introduction

Epithelial-to-mesenchymal transition (EMT) is an evolutionarily conserved but complex molecular and cellular program in which cells undergo conversion from epithelial to mesenchymal state [1]. Epithelial differentiated characteristics are lost, including cell-cell adhesion, planar and apical-basal polarity and lack of mobility. On the contrary, mesenchymal features are briefly acquired, such as cell mobility, invasiveness, gain of stem cell properties and a reinforced resistance to apoptosis. EMT was initially described as a cell culture phenomenon before being recognized in vivo and studied in embryonic development. This process is obviously reversible and is highly conserved from diploblasts (medusa) 800 million years ago to nowadays [2]. It aims at dissociating epithelium and degrading basement membranes so that cells acquire a fibroblastic or glial-cell phenotype and finally migrate. This state was defined as "mesenchyme" (from Greek, μεσο, μεσο, middle and ενκηυμα in fusion) to describe a poorly organized state between two tissues. After migration, such cells are able to reverse their phenotype with the mesenchymal to epithelial transition (MET). EMT and MET are included under the term epithelial-mesenchymal plasticity [3]. Both play an important play in organogenesis (i.e., in renal epithelium or cardiac organogenesis formation) [4] but also in cancer [5,6].

1.1. Cellular Pathways and EMT

Many signalling pathways control EMT according to the different cellular contexts [5,7]. Transforming Growth Factor β TGFβ) [8] and Epidermal Growth Factor (EGF) [9] pathways have been extensively studied but others are also known to drive EMT in specific situations (tumour cells including NSCLC): the Fibroblast Growth Factor (FGF) [10,11], Hepatocyte Growth Factor (HGF) [12], Platelet-derived Growth Factor (PDGF), Insulin-like Growth Factor (IGF) [13], Vascular Endothelial Growth Factor (VEGF), Oestrogens, Hypoxia [14,15], Autocrine Motility Growth Factor (AMF), bile acids, nicotine, ultraviolet light, integrins, Wnt, Notch [16], Interleukin-related Protein (ILEI), Interleukin-6 (IL-6), Sonic hedgehog (Shh), Bone Morphogenetic Protein (BMP), Stem Cell Factor (SCF), cyclooxygenase-2/prostaglandin E2 (COX-2/PGE2) [17] and also extracellular matrix changes, as shown for collagen I and hyaluronan [18,19]. Depending on the experimental model and the stimulation nearly all signalling pathways may eventually promote a mesenchymal transition.

1.2. Transcriptional Regulation of EMT

Several transcription factors act as molecular switches for the EMT program [20]. The loss of E-cadherin is a hallmark of EMT. This cadherin is responsible for cell cell adhesion and cytoskeleton organisation. Its loss leads to the conversion from epithelial cells to motile and invasive mesenchymal cells [21] which is orchestrated by transcriptional repressors [5]. Direct transcriptional repression of E-cadherin is coordinated by several transcription factors interacting together. SNAIL superfamily members (such as zinc finger proteins SNAI1/SNAIL and SNAI2/SLUG) interact with several co-repressors and epigenetic remodelling complexes to repress E-cadherin through N-terminal SNAG domain binding to the E-box promoter sequence [22]. Other direct repressors are ZEB family members ZEB1 (Zinc finger E-box-binding homeobox 1, previously known as TCF8 and δEF1) and ZEB2/SIP1, E47 and the Krüpple-like factor 8 (KLF8) [23,24]. Furthermore, TWIST family members (basic helix-loop-helix transcription factors TWIST1 and TWIST2), Goosecoid, E2.2/TCF4 and FOXC2 indirectly repress E-cadherin transcription [6]. TWIST is a transcription factor fully implicated in EMT [25]. Its proteins belong to the family of basic helix-loop-helix transcription factors which are able to modulate expression of different target genes through E-box responsive elements [26]. TWIST1 and TWIST2, highly conserved, share the Twist-box protein interaction surface in their C-terminal half [27]. TWIST forms functional homo- and heterodimers with TCF3/E2A and Hand2, the balance between homo- and heterodimers regulating limb development and cranial suture fusion. In adulthood, TWIST1 expression was reported in mesoderm-derived tissues including heart, skeletal muscles, placenta [27] and also brown fat with a specific thermoregulation function [28]. TWIST proteins physically interact with NF-κB and specifically prevent its ability to activate pro-inflammatory cytokine-encoding genes [29]. Furthermore, TWIST proteins reduce TNFα induction response to T-cell receptor activation [30].

EMT network is highly controlled. The high-mobility group protein HMGA2 and the homeodomain-containing protein SIX1 act as a coordinator of EMT inducers [31,32]. In development and in carcinogenesis, SNAIL1 appears at the onset of EMT then SNAIL2, ZEB, E47 and TWIST are induced to maintain the migratory mesenchymal state [33] suggesting an orchestration in time of the process. Moreover EMT transcription factors may undergo post-translational regulation. For example SNAIL1 nuclear stabilization is promoted by NF-κB [34] and the zinc transporter LIV1 [35].

1.3. Junctions, Cytoskeleton and Matrix

The cadherin switch through the loss of E-cadherin represents a universal marker of EMT. E-cadherin is the major constituent of epithelial adherens junctions which mediate intercellular adhesion with tight junctions, forming the zonula adherens. Epithelial cells loose cell-cell adhesion and cell polarity. Following E-cadherin loss, the expression of mesenchymal markers such as vimentin, fibronectin, N-cadherin, alpha-smooth muscle actin (αSMA), and the activity of matrix metalloproteinases (MMP-2, MMP-3, MMP-9) increase. SNAIL1 represses the expression of tight junction components such as claudin-3, -4 and -7 [36]. Furthermore, EMT inducers directly repress the protein complexes involved in apico-basal polarity of epithelial cells (Par, Crumbs and Scribble) [37] and enhance expression of metalloproteinases that degrade the basement membrane, thereby favouring cell migration and invasion. Interestingly, MMP3 can trigger EMT through a positive regulatory feedback loop [38]. Finally, EMT transcription factors induce the expression of mesenchymal proteins such as fibronectin and N-cadherin [39] and are involved in the remodelling of the actin cytoskeleton [21]. The complexity of interactive downstream effector pathways and the fact that EMT is not a simple matter of changes in cell adhesive capabilities or in cytoskeletal organization leads to a wide range of different profiles of expression of the markers described [21]. The choice of the markers and the changes in their expression levels will depend upon the diversity of signals inducing EMT.

2. EMT in Cancer

2.1. Carcinogenesis

Recent evidences show that transcription factors linked to EMT may directly be involved in carcinogenesis. TWIST and ZEB proteins can prevent cells from undergoing oncogene-induced senescence and apoptosis by inhibiting both TP53 and RB-dependent pathways, leading to a deregulation of MYC, RAS, ERBB2 and cell cycle inhibitors P14, P16 and P21 expression [40,41]. This may explain how TWIST1 and TWIST2 can cooperate with an activated version of RAS to transform mouse embryonic fibroblasts [41]. Furthermore, TWIST proteins also downregulate PP2A phosphatase activity and efficiently cooperate with an oncogenic version of H-RAS in malignant transformation of human mammary epithelial cells, leading to claudin-low tumours, which are believed to be the most primitive breast malignancies [42]. Thus, by downregulating crucial tumour suppressor functions, EMT inducers make cells particularly prone to malignant conversion. Beside this effect, EMT was linked to progression and metastasis through its effects on matrix, motility and gain of stem cell properties.

Both *TWIST* genes were reported overexpressed in many types of cancers: prostate, breast, cervical, endometrial, ovarian, head and neck cancer, oesophageal squamous cell carcinoma, gastric, hepatocellular carcinoma, pancreas, colon, kidney, glioma, melanoma, neuroblastoma, parathyroid, pheochromocytoma, sarcoma [41,43] and also NSCLC [44].

TWIST proteins are reported to induce EMT and to be one of its main actor: TWIST induces loss of E-cadherin-mediated cell-cell adhesion and an EMT in epithelial cells [45,46]. In NSCLC, the overexpression rate of TWIST proteins was 38% in tissues samples and results correlated with mRNA high level and N-cadherin expression [47]. Such a pattern was associated with worst prognosis. In cell lines, TWIST down expression inhibited cell invasion and increased apoptosis [47]. In *Epidermal Growth Factor Receptor* (*EGFR*)-mutated lung adenocarcinoma, we previously reported on surgical samples that TWIST1 expression was linked to *EGFR* mutations, low E-cadherin expression and low disease-free survival [44]. In cell lines, we demonstrated that EMT and the associated cell mobility were dependent upon TWIST1 expression in cells with *EGFR* mutation. Moreover a decrease of EGFR pathway stimulation through EGF retrieval or an inhibition of TWIST1 expression by small RNA technology reversed the phenomenon [44]. Such results are consistent with with EGF promoting E-cadherin endocytosis to induce EMT [48] but also induction of both SNAIL and TWIST [9,49]. EMT is common in NSCLC [50] and could in some patients be related to tobacco exposure. Indeed, cigarette smoke was proved to induce EMT [51,52]. In lung adenocarcinoma cell lines (H358), cigarette smoke extracts was able to induce EMT with activation of SRC kinase and the SRC kinase inhibitor PP2 inhibited cigarette smoke-stimulated EMT changes, suggesting that SRC is critical in cigarette smoke-stimulated EMT induction [53].

2.2. Lymph Node Metastases

To the best of our knowledge, presence of EMT-MET phenomenon remains poorly studied in the metastatic lymph nodes. In gastric cancer, the expression of N-cadherin in metastatic lymph nodes was associated to a bad prognosis [54]. An other study related the heterogeneity between primary tumours and metastatic lymph nodes in oesophageal cancer, with distinct EMT phenotypes and thus, a novel independent prognostic indicator [55]. Similar results were found in head and neck cancer [56] and breast cancer [57]. Concerning NSCLC, only one study investigated EMT markers in metastatic lymph nodes [58]. Expression of Brachyury in 115 surgically resected primary NSCLC and the corresponding metastatic lymph node samples were evaluated by immunohistochemical staining. Brachyury is a highly conserved cellular protein that belongs to the T-box transcription factor family and was reported to be essential for mesoderm formation in the early embryo [59]. In recent studies Brachyury was also linked to the EMT process during cancer progression [60]. Gene expression was associated with IL-8 expression and inversely associated with E-cadherin expression in NSCLC [14]. In metastatic lymph nodes Brachyury expression was significantly higher than in the primary tumour and lymph node level of expression was inversely associated with survival [58]. In models of NSCLC, lymphangiogenesis leads to proliferation, invasiveness and nodal metastasis [61]. Vascular endothelial growth factors (VEGF) -C and -D and their corresponding receptor (VEGFR3/Flt4) are the main actors in the development of tumour-associated lymphatic vessels [62]. They recruit endothelial cells and others stromal cells to develop and maintain an unrefined lymphatic network within the tumour microenvironment [62,63]. Furthermore, they might be involved in pre-metastatic lymph nodes by preparing the lymphatic vasculature to host the cancer cells [64]. These markers were correlated in NSCLC to nodal metastasis and, thus, patient survival, with heterogeneous results [65]. A meta-analysis found a correlation between VEGF-C, Lymphatic Vessel Density and lymph node metastasis in NSCLC [65]. Links between lymphangiogenesis and EMT remain to be elucidated but EMT markers were already associated with lymphatic vessel density in surgical specimen of NSCLC [66].

2.3. Distant Metastases

Development of metastasis involves distinct steps with specific underlying molecular mechanisms, which are: detachment of tumour cells from the primary tumour, invasion into surrounding tissues, intravasation into blood or lymphatic vessels, dissemination in the blood stream or the lymphatic system and finally, extravasation and outgrowth at a secondary site [21]. Despite solid evidence showing EMT implication in distant metastases [67], the metastatic process remains unclear for many points. Reversion of EMT (MET) seems to play a role in the metastatic process, MET could at some points explain the histologic similarity between metastases and primary tumours. First, different observations reported association of MET with metastatic tumour formation: (i) epithelial phenotype seems to be of importance in the formation of secondary tumours, as epithelial characteristics were highly associated with increased distant colonization after blood stream injections [68]. (ii) E-cadherin positive metastatic foci were found after injections of mesenchymal-like breast cancer cells, presenting a direct demonstration of need for E-cadherin expression [69]. Escape from primary tumour, gain of cell mobility and invasion may at first step be EMT-dependent, then cancer cells should undergo MET in the secondary organ. The regulation of phenotype plasticity remains largely unknown but it can be hypothesized that while the tumour microenvironment promote the induction and continuation of EMT, circulating tumour cells revert to an epithelial state due to the loss of EMT-inducing signal, before entering into metastatic sites [70]. Nevertheless, Aokage and colleagues [71] showed that MET appeared after the tumour cells arrival at the metastatic site, and considered that local microenvironment or local resident cells contribute largely to the MET. Different histopathological analyses suggested very close contact between metastatic carcinoma cells and the neighbouring parenchyma cells, supporting a possible E-cadherin-dependent linkage [72–74]. This ability to create heterotypic E-cadherin adhesions and related survival signals is coherent with the dormancy of the tumour cells and its low metabolism inherent to this micrometastatic stage [75].

Nevertheless, there are some debates about the implication of EMT in cancer progression [76,77] as aggressive tumours may, in some cases, develop metastases in the absence of EMT-MET. A point-counterpoint review, in 2005, proclaimed that the biological machinery of normal and malignant cells is sufficient to account for the events and processes observed without needing to objective radical change in cell phenotype like EMT [78] suggesting that EMT is not always a prerequisite to metastasis and that tumours may undergo partial or incomplete EMT [68]. Indeed, mesenchymal cells derived from epithelial tumour cells are very difficult to distinguish from stromal cells or other tumour-associated fibroblast. However, EMT was described at the invasive front of tumour as small aggregates of tumour cells extending from the tumour mass into adjacent stroma [79]. Cancer cells have a broad repertoire of invasion mechanisms, and cell-type-specific patterns of cell migration can be classified into single cell migration (amoeboid, mesenchymal) and collective migration modes (cell sheets, strands, tubes, clusters) [80]. The collective migrating cells form membrane protrusions, such as ruffles and pseudopods, use cell-matrix adhesion receptors and, in contrast to solitary migration, do not retract their cellular tails but rather exert pulling forces on adjacent adherent cells [81,82]. An other potential mechanism for tumour cell invasion and metastasis is the podoplanin-mediated remodelling of actin cytoskeleton and tumour invasion [83], also described in NSCLC [84].

3. EMT-Related MicroRNAs

MicroRNAs (miRNA) are highly conserved small single-stranded non-coding RNA of 21-23 nucleotides acting as post-transcriptional regulators of gene expression. They function to affect RNA stability and translation in order to negatively regulate gene expression [85]. MicroRNAs are transcribed from DNA by RNA polymerase II or III as a form of long primary transcripts (pri-miRNA) [86] which are then processed by the microprocessor complex containing RNase III enzyme Drosha and DGCR8 (DiGeorge syndrome chromosomal region 8) into shorter stem-loop-structured double-stranded RNA (hairpin precursor miR, pre-miR) [87]. Pre-miR are delocalized from nucleus to cytoplasm and processed into mature-miR by RNase enzyme III Dicer [88]. The RNA interference is finally efficient into the RNA-induced silencing complex (RISC) to target single-stranded complementary messenger-RNA (mRNA) for translation repression or mRNA degradation, by binding to the 3′ untranslated regions of their target genes [89]. The target mRNA could be blocked in case of partial complementarity or degraded in case of perfect complementarity [90]. Thus, the possible imperfect match offers the ability to regulate many genes.

In normal lung, the overall expression profile of miRNAs is 75% similar in mouse and human lung, indicating evolutionary conservation of miRNAs expression [91]. However, miRNAs expression profile varies along development and profiles differ between human foetal, post-natal and adult lung [91]. In adult lung tissues the 30 most highly expressed miRNAs were identified as [91]: miR-103, miR-99a, miR-15b, miR-150, miR-320, miR-23a, miR-200c, miR-195, miR-27b, miR-199s, miR-92, miR-29b, miR-30d, let-7g, miR-223, miR-199a, miR-30c, miR-142-3p, miR-125a, miR-26b, miR-29a, miR-126, miR-29c, miR-16, let-7b, miR-145, miR-21, let-7a, miR-30b and miR-26a. In parallel, tissue-specific miR were found without any correspondence with the previous list [92]: miR-224, miR-137, miR-192, miR-886, miR-31, miR-92b, miR-10a, miR-625, miR-301a, miR-96 and let-7i.

Reports of miRNA implication in diseases began in the 2000s, with a first description that miRNAs were dysregulated in human B-cell chronic lymphocytic leukemia using a microarray containing miRNA probes [93]. MicroRNAs were then studied in many situations, hepatic viral infections [94], Alzheimer disease [95], cardiac hypertrophy [96], diabetes [97] and in various lung diseases such as chronic obstructive pulmonary disease, sarcoidosis, pulmonary fibrosis [98,99]. MicroRNAs have been shown to play a crucial part in cancer development and progression in the past decade, in various solid organ cancers. Half of miRNAs genes are located at fragile sites or genomic regions involved in cancer-related chromosomal abnormalities [100]. MicroRNAs were then characterized as "OncomiR" or "Tumour suppressor miR" depending on the suppressed target genes [101–103]. Finally, miRNAs functions studies showed their capacity to affect pathways regulating EMT [1,76,104]. The miRNAs regulatory network is highly complex and deeply integrated into cell functions. The different studies may be heterogeneous and difficult to compare, different models can be used including cancer tissues and cell lines. Moreover, despite tissue preservation, RNA studies remain shaky and submitted to numerous pitfalls before final interpretation. Two main technologies allow miRNAs analyses qPCR based and specific probes sets or high throughput sequencing. Concerning EMT, the main miRNAs involved are in one hand the miR-200 family and miR-205 that maintain the epithelial cell phenotype [104–108] and in the other hand, miR-21 is up-regulated in many cancers. It facilitates TGF-β-induced EMT [8] and was the first OncomiR to be identified. Based on comparison between tumour and normal tissue levels of expression, approximately 200 miRNAs were shown dysregulated in NSCLC [103]. MicroRNAs from the miR-200 family (miR-200a/200b/200c/141/429) have been shown to inhibit EMT, cell migration and invasion by targeting ZEB1 and ZEB2 mRNA, two repressors of E-cadherin expression [106]. Based on their chromosomal location, the miR-200 family can be split into two different gene clusters: miR-200/200b/429 (chromosome 1) and miR-200c/141 (chromosome 12) [106]. Loss of these genomic loci and subsequently loss of miRNAs expression was reported in mesenchymal cancer cell lines and linked to cancer progression including NSCLC [107]. Down-regulation of miR-200 family was also documented in cases with hypermethylation of the DNA locus [109,110]. Inversely, miR-200c has been shown to inhibit the metastasis in A549 NSCLC cell lines [111,112]. Bracken et al. showed that the promoter for the pri-miR (shared by miR-200a, miR-200b, and miR-429) is located within a 300-bp segment located 4 kb upstream of miR-200b. This promoter region is sufficient to confer expression in epithelial cells and is repressed in mesenchymal cells by ZEB1 and SIP1 through their binding to a conserved pair of ZEB-type E-box elements, located proximal to the transcription start site [113,114]. Therefore depending on extra cellular stimulations the miR-200/ZEB1/2 equilibrium may turn on epithelial or mesenchymal markers. Moreover, several other miRNAs were associated with NSCLC progression as miR-224 [115] or miR-1247 [116].

Several studies have indicated that miRNAs frequently form feedback loops, since they are regulated by transcription factors which they directly or indirectly target [117]. Siemens et al. showed that miR-34a could downregulate SNAIL as well as SLUG and ZEB1. Conversely, SNAIL can repress miR-34a by binding to E-box sequences in the miR-34a promoter thereby forming a double negative feedback loop blocking cell in a mesenchymal state [117,118].

Furthermore, in parallel to EMT regulation, miRNAs play crucial roles in carcinogenesis. Examples are (i) the negative regulatory loop between NF-κB and miR-146. In the highly metastatic human breast cancer cell line MDA-MB-231, lentiviral-mediated expression of miR-146a/b significantly downregulated the IL-1 receptor-associated kinase and the TNF receptor-associated factor 6, two key scaffold proteins in the IL-1 and Toll-like receptor signalling pathway, known to positively regulate NF-κB activity [119]. (ii) The tumour suppressor miR-34a targets CDK4/6, MET, HDAC1, E2F3 and Bcl-2 and induces cell cycle arrest and apoptosis [117].

4. MicroRNAs, EMT and NSCLC

In NSCLC, probably due to tumour heterogeneity and to different technical and analytical issues, it is difficult to find a common signature of miRNAs expression. Wang et al. [103] reviewed 4 studies comparing miRNAs profile in NSCLC tissues versus the corresponding non-cancerous lung tissues and pointed out that miRNAs identified in each study are different from the others [120–122]. However Zadran et al. [123], described a cancer-specific miRNAs signature for different solid organ cancers including lung, and Vosa et al. [124] presented the 30 most differentially expressed miRNAs in NSCLC. Supplementary Table S1 shows different published miRNAs signatures in NSCLC [120–125]. More than 150 miRNAs were identified as markers of NSCLC form 6 large studies. MiR-210, miR-143 and miR-205 were recurrently linked to NSCLC in 3 or 4 out of 6 studies. Three miRNAs (miR-195, miR-224 and miR-124a-1) are either up or downregulated and finally 130/153 miRNAs were associated to NSCLC in only 1 out of 6 studies. This illustrates the difficulties of validating relevant markers in clinics, due to the absence of specific miRNAs signature in NSCLC.

However miRNAs was evaluated as diagnostic tool, in sputum [126–128] or in plasma/serum [129] for the early detection of NSCLC. In plasma, or serum and/or exosome, different tools were elaborated, with various techniques reviewed by Hou et al. [130]. Several authors proposed single miRNA or multiple-miRNA panels useful for NSCLC screening. However, these studies remain to be validated given the heterogeneity of the normalization methods and the starting material used for RNA isolation [129–133].

To summarize published data concerning miRNAs and EMT in NSCLC, a systematic review of literature in Medline identified more than 75 articles based on cell lines studies (Table 1). Another subset of 37 articles was retained, based on studies on human NSCLC specimens leading to identify 35 miRNAs as modulator of EMT (Table 2). Figure 1 summarizes involved miRNAs in EMT in cancer.

Table 1. List of EMT involved miRNAs based on NSCLC cell lines studies, with details on the involved pathways (green, promote EMT; red, suppress EMT; blue, controversial).

MicroRNAs	Pathway	Cell Lines	References
miR-10a	XRN2	H441, A549, HOP62	[134]
miR-15b	PEBP4	A549	[135]
miR-17	TGF-β	A549	[136]
miR-23a	E-cadherin	A549	[137]
miR-26a	EZH2	SPC-A1, H1299	[138]
miR-30a	SNAIL	A549, Calu1/3, H1299, H1395	[139,140]
miR-34a	NOTCH1	H1299, H460	[141]
m.R-129*	MCRS1	801D, SPC-A1, GLC-82, EPLC-32M1, A549, H292, 16HBE, PT67	[142]
miR-134	ITGB1, MAGI2, FOXM1	A549, H1299, A549, LC2/ad, PC3, PC9, RERF-LCKJ, RERF-LCMS, PC14, ABC-1	[143–145]
miR-138	GIT1, SEMA4C	A549, 95D, H23	[146,147]
miR-145a-5p	TWIST1	H1299	[148]
miR-149	FOXM1	A549	[149]
miR-151-5p	TWIST1	H1299	[148]
miR-154	ZEB2	A549	[150]
miR-155	ZEB1	HCC827	[151]
miR-206	PI3K/AKT	A549, 95D	[152]
miR-221	na	H460, H838, H1299, H3255, HCC4006, HCC4011	[153]
miR-222	na	H460, H838, H1299, H3255, HCC4006, HCC4011	[153]
miR-337-3p	TWIST1	H1299	[148]
m:R-374a	Axl	HCC827, Calu1	[154]
m:R-452	BMI1	A549, H460	[155]
miR-483-5p	ALCAM	A549, PC9	[156]
miR-487b	MAGI2	A549	[144]
miR-520f	ADAM9	A549	[157]
miR-543-3p	TWIST1	H1299	[148,158]
miR-544a	Cadherina 1	95C, 95D	[159]
miR-548b	Axl	HCC827, Calu1	[154]
miR-655	MAGI2	A549	[144]
miR-1271	FOXK2	A549, H520, H1299, H358, H460	[160]
miR-1246	Stem cells	A549, HCC1588	[161]
miR-1290	Stem cells	A549, HCC1588	[161]
let-7	HMG2A	H1975, H1299, H1650	[162]
let-7c	Hedgehog	A549, H1299	[163]

Table 2. List of EMT involved miRNAs based on NSCLC tissue samples and patients' series studies, with details on the involved pathways and the clinical impact (green, promote EMT; red, suppress EMT) (TKI, Tyrosine-Kinase Inhibitor).

MicroRNAs	Pathways	Clinical Impact	References
miR-16	HDGF	Cell growth and motility	[164]
miR-21	STAT3, IL-6	Carcinogenesis	[165]
miR-29b	SPARC	Cancer progression	[166]
miR-30	MMP19	Metastases	[167]
miR-30c	E-cadherin, vimentin, SNAIL	Invasion	[168]
miR-31	ERK1/2	Lymph spread, survival	[169]
miR-33a	TWIST1	Metastases	[170]
miR-92b	RECK	Cell growth and motility	[171]
miR-96	Foxf2	Invasion and metastases	[172]
miR-124	CDH2, ZEB1	Cell migration and invasion	[173,174]
miR-127	feed-forward regulatory loop	TKI resistance	[175]
miR-132	ZEB2	Cell migration and invasion	[176]
miR-135a	KLF8	Cell migration and invasion	[177]
miR-135b	Hippo signaling pathway	Metastases	[178]
miR-143	CD44	Cell migration and invasion	[179]
miR-145	SMAD3	Invasion	[180]
miR-146a	IRS2	Cancer progression	[181]
miR-148a	ROCK1	Lymph spread	[182]
miR-150	p53	Cell proliferation	[183]
miR-182	Foxf2	Invasion and metastases	[172]
miR-183	Foxf2	Invasion and metastases	[172]
miR-184	c-Myc	Overall and disease-free survival	[184]
miR-193a	WT1-E-cadherin axis + ERBB4/PIK3R3/mTOR/S6K2 signaling pathway	Metastases	[185,186]
miR-196a	HOXA5	Cell proliferation, invasion	[187]
miR-196b	NF-κB, Homeobox A9	Cell invasion and migration	[188]
miR-200	ZEB1, Foxf2	Invasion and metastases	[172,189]
miR-200c	E-cadherin, ETAR, NFkB	Invasion and metastases	[190–193]
miR-203	SMAD3	Invasion	[180]
miR-205	CRIPTO1	TKI resistance	[134]
miR-214	Sufu	Metastases	[195]
miR-338-3p	Sox4	Metastases	[196]
miR-361	FOXM1	Cell proliferation and invasion	[197]
miR-375	YAP1	Neuroendocrine features	[198]
miR-451	RAB14	pTNM, Lymph spread	[199]
miR-489	SUZ12	Cell invasion	[200]
miR-490	poly r(C)-binding protein 1	Metastases, lymph spread	[201]
miR-589	HDAC5	Cell migration and invasion	[202]
miR-541	TGIF2	Cancer progression	[203]
miR-638	SOX2	Cell proliferation and invasion	[204]

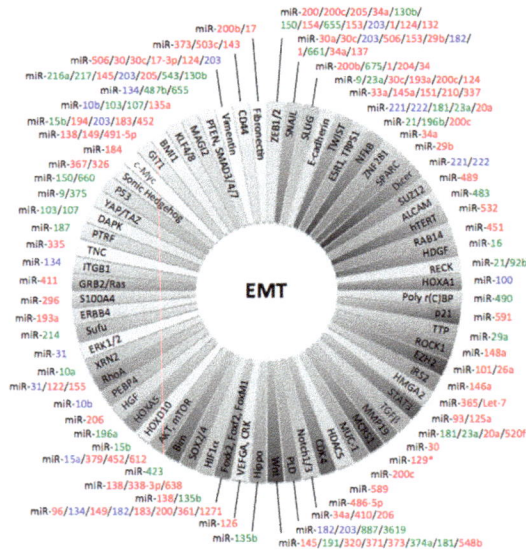

Figure 1. EMT regulation by miR in cancer, a complex network (Colour legend: green, promote EMT; red, suppress EMT; blue, controversial) (adapted from [117,205–210]).

To go further in the interpretation of the data, regulation of TWIST1 by miRNAs was investigated by Nairismägi et al. [148]. They identified 18 miRNAs targeting *TWIST1*'s 3′UTR region but only 3 were able to significantly repress *TWIST1* translation: miR-145a-5p, miR-151-5p and in combination: miR-145a-5p + miR-151-5p and miR-151-5p + miR-337-3p. Among the many roles attributed to miRNAs in cancer, understanding their interaction with oncogenes and signalling pathways is of importance as it could lead to treatment strategies. In NSCLC, activation of the EGFR pathway plays a critical role in tumour development. A subset of miRNAs has been shown to interact with the EGFR pathway either as activators (miR-21, 24, 25 or miR-7) or repressors (miR-133) and may therefore modulate EGFR pathway activation. Concerning RAS activation and miRNAs in NSCLC, miR-31 was shown to target 6 regulators of the RAS/MAPK pathway and regulate lung epithelial cell growth. MiR-31 has been described to cooperate with RAS to drive lung tumorigenesis. Studies have revealed that miRNAs constitute a regulatory network in the post-transcriptional regulation of pathway genes. In parallel, tumour genetic background seems to impact on miRNAs profiles. Indeed, specific miRNAs profiles are associated to *EGFR*, *KRAS* or WT tumours [211]. In a high-throughput screening program in NSCLC, miR-155 was upregulated only in *EGFR/KRAS* negative group, miR-25 was upregulated only in *EGFR* positive group and miR-495 was upregulated only in KRAS positive adenocarcinomas [211].

We have previously reported an oncogenic cooperation between EGFR activation and TWIST1 reactivation in *EGFR* mutated NSCLC. Using NSCLC cell lines, Takeyama et al. showed that *EGFR* mutated cell lines had more epithelial characteristics as compared to non-*EGFR* mutated cell lines and that the mesenchymal state in the second group was related to ZEB1 upregulation [212]. It seems therefore that mutational status could drive specific EMT pathways.

For other pathways involved in NSCLC, experiments suggested that miR-19 induces EMT [213]. MiR-19 and miR-21 modulate PTEN levels [213]. High miR-19 levels were found in NSCLC cells and experiments suggest that PTEN is involved in miR-19-induced EMT, migration and invasion [213].

Inflammatory cytokines and hypoxia have been proven to promote EMT. NF-κB itself is regulated by miRNAs through different regulatory loops of which the mir-146/NF-κB was documented in A549 cell line [214]. Moreover, NF-κB-mediated inflammation was shown to lead to EMT due to decrease in miR-200c [191] and in nuclear stabilization of SNAIL1 [34]. Hypoxia directly triggers EMT through the ubiquitin C-terminal hydrolase-L1 (UCH-L1) and HIF-1α deubiquitinition [215]. *TWIST1* is a direct target of HIF1-α. In cancer cells, hypoxia was shown to reactivate TWIST1 [6], allowing cell to disseminate to a less hostile microenvironment [14,15]. Furthermore, hypoxia induces SNAIL in NSCLC [216]. Relations between ROS and EMT have been established [217,218]. ROS can activate NF-κB signalling and Wnt-β-catenin signalling pathway [217].

As seen, interaction loops between miRNAs and oncogenes or miRNAs and EMT transcription factors were largely described and seem cell and/or model dependent. It is sometimes hard to decipher whether miRNAs dysregulation is at the origin or is a consequence of the EMT process. Approximately half of miRNAs are associated with CpG islands [219] and several studies illustrated that methylation status could be responsible for the dysregulated expression of miRNAs in NSCLC [116,202,220,221]. Such a phenomenon was described for miR-200 inactivation under SNAIL1 control to maintain the mesenchymal phenotype [222]. In addition, histone acetylation may influence miRNAs expression [223]. Histone modifications were identified as the main mechanisms of miR-212 silencing in NSCLC [224]. DNA methylation and histone modifications are also largely associated with EMT regulation [225].

An emergent pathway in cancer is the involvement of Prion protein (PrP) [226]. Several studies reported overexpression of the PrP in cancer and particularly, links with EMT via direct influence of PrP on neural cell adhesion molecules (NCAM) [227], certain matrix metalloproteinases [228] and Fyn activation [229].

As a summary, Figure 2 illustrates the complex regulation of EMT in NSCLC.

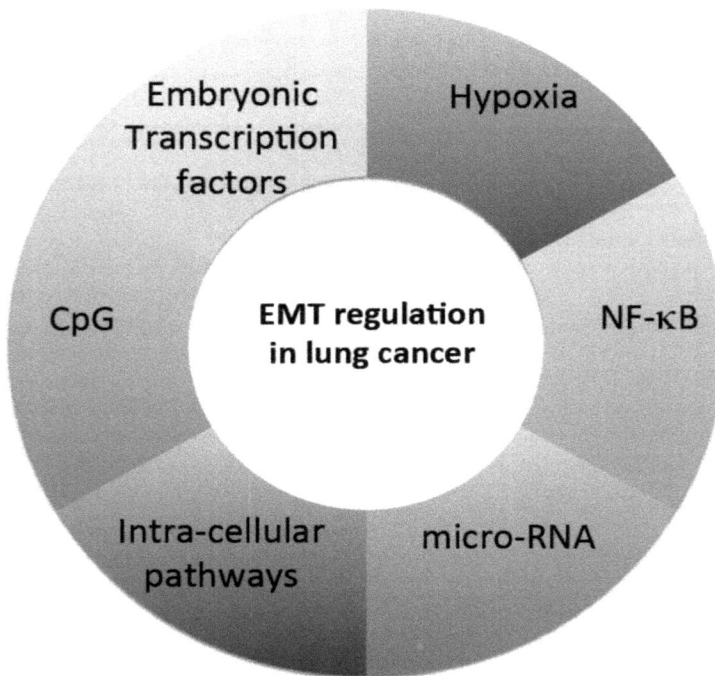

Figure 2. Hallmarks of the complex regulation of EMT in NSCLC. The regulation of EMT in NSCLC is based on several intricate conditions and actors, detailed in the Section 4 (adapted from [24,44,111,116, 148,191,202,206,207,216,220,221]).

5. Clinical Impact of EMT in NSCLC

EMT is largely involved in cancer development and in metastatic progression. For that it was frequently associated with prognosis [14,44,230–232]. The prognostic impact of miRNAs was investigated and some authors proposed prognostic miRNAs signature, in gastric and in NSCLC [233]. Using a cohort of 112 NSCLC patients, the last study identified a 5-miRNA-signature including 3 high-risk miRNAs (miR-137, miR-182*, and miR-372) and 2 protective miRNAs (miR-221 and let-7a). Another study identified a miRNAs profile counting only miR-155 and let-7a-2 [121]. The pooled results of a meta-analyse including 28 articles [234] revealed that high expression of miR-125b, miR-21 and miR-200c were negatively associated with survival. Conversely, high expression of miR-124, miR-365, miR-32, miR-148b, miR-146a and miR-375 were significantly associated with better prognosis [234].

However, the critical point with EMT is the associated chemo-resistance [235–238]. Therefore, understanding the mechanistic of EMT is important to assess treatment strategies for patients with NSCLC.

Upregulation or reactivation of inducing EMT transcription factors activates survival pathways (NF-κB, AKT), proliferation pathways (upregulation of EGFR and MET) and modulates the activity of Bcl-2 family members thereby favouring antiapoptotic signals [25]. Resistance to genotoxic agents such as anthracyclines, platinum-based drugs or spindle poison was associated to EMT. Data suggests that oxaliplatin-resistant colorectal cancer [239], gemcitabine-resistant tumour cells [240] tamoxifen-resistant breast cancer [241], gemcitabine-resistant pancreatic cancer, radiotherapy-resistant endometrial carcinoma [242] and radiotherapy-resistant ovarian cancer cells [243] harbour phenotypes of EMT [25]. Similar results were observed in NSCLC cell lines for cisplatin [244], docetaxel in

A549 and for the pemetrexed-ciplatin combination [245,246]. Moreover resistance to treatment in adjuvant setting was also reported [247]. Increased viability or drug resistance could be due to lower levels of ROS and subsequent better genome protection in cell undergoing EMT [25,248]. In NSCLC with EGFR mutations or ALK fusions, EMT was also related to resistance to tyrosine-kinase inhibitors (TKI) [249,250]. NSCLC lines expressing E-cadherin showed greater sensitivity to EGFR inhibition. In contrast, NSCLC lines expressing vimentin and/or fibronectin were insensitive to EGFR inhibition [39]. Data suggested that, in some cases, TGFβ activation is sufficient to induce TKI resistance and, in smokers population, SRC activation could trigger EMT and subsequent EGFR-TKI resistance [251]. Furthermore, EMT increases membrane transporters expression (ABC family, P-glycoprotein) leading to drug active efflux [246,252,253]. Finally EMT-induced decrease of ceramide was associated to chemo-resistance [254–256].

The association between EMT and stem-like phenotype in NSCLC cells was shown in several in vitro studies [257–259] but the data on this phenomenon in lung cancer patient samples are limited. Koren et al. showed that BMI1 and CD133 (cancer stem-cell markers) were coexpressed in a series of operated NSCLC, giving in vivo evidence of connection between EMT and cancer stem cells [260]. Such tumour subpopulation plays a role in drug resistant tumour cells with abilities of self-renewal, cancer initiation, and further maintenance of tumours [261].

MicroRNAs have been implicated in chemo-resistance as well as in chemo-sensitivity. As seen in the first parts of this review, miRNAs regulate pathways implicated in cell fate: proliferation, apoptosis and differentiation and thereby influence chemotherapy and radiotherapy responses. There are as many models in the literature as different contexts or situations describing a miRNA or a set of miRNAs related to treatment failure. Acquired chemo-resistance, to cisplatin, docetaxel and erlotinib in NSCLC [135,262] was related to up or downregulation of miRNAs linked to EMT: miR-140, miR-628, miR-135b, miR-200b/200c/141, miR-205, miR-197, miR-224, miR-34c, miR301a, miR-636, miR-518f [263]. In *EGFR* mutated tumours CpG island hypermethylation was shown to downregulated miR-200 family members in cells with gefitinib resistance and EMT features. More precisely, miR-200c downregulation was linked to the upregulation of LIN28B a protein favouring stem cell self-renewal [264]. Moreover several afatinib-resistant NSCLC cell lines also displayed EMT features and epigenetic silencing of miR-200c suggesting that miR-200 downregulation could be a common event in TKI acquired resistance [252].

Furthermore, an opposite point-of-view must be presented: some observations lead to consider the epithelial phenotype as drug-resistant. Byers et al. proposed a 76-gene EMT signature to classify NSCLC cell lines into distinct epithelial and mesenchymal groups [265]. Surprisingly, *EGFR*-mutated cell lines H1975 and H820, carrying the acquired-resistance mutation T790M, were classified as epithelial. Furthermore, they found a trend towards greater relative sensitivity in mesenchymal cells as compared to epithelial for cisplatin, gemcitabine and vinorelbine. This observation was also supported by the the work of Miow et al., who reported that ovarian cancer cell lines with an epithelial pattern were more resistant to cisplatin than those with a mesenchymal pattern [266]. This concept remains poorly documented and is largely counterbalanced by abundant literature on mesenchymal status-driven drug resistance. However it may be related with two distinct models of EMT: acquired phenotype of EMT following drug, growth factor or EMT-transcription factor treatment versus inherent phenotype of EMT related to the nature of the cancer cell [1,265,266].

Drug resistance and EMT have been clearly associated in different situations and genetic backgrounds but the mechanisms underlying drug resistance remain under investigation however some explanations could rely on miRNAs expression.

6. Therapies Targeting EMT

Inhibition of EMT could restore senescence and apoptosis capacity [46]. In a recent review, Malek et al. proposed 3 distinct strategies to target EMT through (a) extracellular inducers of EMT, (b) EMT-transcription factors and (c) downstream effectors of EMT inhibition [267].

First (a), inhibition of extracellular EMT-inducer pathways could rely on TGF-β blockade using rapamycin, 17-AAG [268], SB-431542 [269] or TGF-β receptor specific inhibitors such as EW-7195, EW-7203 and EW-7197 [270–273]. Only LY2157299 an oral TGF-βRI tyrosine kinase inhibitor reached clinical trials and is currently tested in several cancers, excluding NSCLC despite previous encouraging preclinical results [274]. Direct blockade of TGFβ using human TGFβ-antibody (fresolimumab) was not yet explored in NSCLC. Besides TGFβ blockade, interesting results were shown using LDN57444, a specific small molecule inhibitor, targeting ubiquitin carboxy-terminal hydrolase L1 (UCH-L1). LDN57444 was shown to downregulate HIF-1α, suppress EMT features and reduce the incidence of distant metastases [275,276]. Concerning extracellular EMT inducers, MMP as metastases-related cellular enzymes could appear as good candidates. However, MMP inhibitors failed in clinical trials likely due to the complexity of the metastatic process [277].

Second (b), inhibition of EMT transcription factors was analysed as a therapeutic option and illustrated by results based on TWIST1 [41,42,45,278–280]; PRMT1 [281] and SNAIL family and ZEB1/2 inhibition [20,282]. EMT-transcription factors are challenging therapeutic targets due to their heterogeneous expression and to the complexity of the EMT regulation network [283,284]. However, several agents were proved efficient in cell lines, including plant extracts: moscatilin [285], fucoidan [286], quercetin [287], thymoquinone [288], imipramine blue [289] and finally, in vitro and in vivo data on NSCLC suggest a potential impact of the harmala alkaloids to downregulate TWIST1 [267]. Targeting glycosylation pathways that regulate EMT transcription factors should be an interesting mean of treatment. Aberrant glycosylation was associated with carcinogenesis and aggressiveness in many cancers including NSCLC [290,291] and inhibitors of one of the main glycosylation enzyme are in preclinical pipelines [292,293]. Moreover chromatin modulators such as drugs targeting histone methyltransferases (G9a and EZH2) and demethylases which contributes to EMT as E-cadherin repressors are promising epigenetic oncotargets [294,295]. Finally, histone deacetylation can be inhibited by HDAC inhibitors (butyrate, trichostatin A and Suberoylanilide hydroxamic acid which is the FDA-approved Vorinostat) and have been shown in preclinical studies to selectively target cancer cells by inducing apoptosis, cell cycle arrest, suppression of tumour angiogenesis, metastasis and invasion at least partially through upregulating E-cadherin [267,296,297].

Third (c), targeting the downstream effectors of EMT such as E-cadherin, N-cadherin, vimentin and HoxA9 may offer some possibilities [267].

MicroRNAs have emerged as a class of therapeutics targets. However there are some limitations due to the fact that miRNAs have typically many targets. This can either be harmful as it limits specificity but can also be interesting to consider if miRNAs blockage leads to the inhibition at different levels of a pathway, using a single agent. Therapeutic strategy could be either direct administration of anti-miRNA (antisens miRNA) to block OncomiR, or restoration of miRNAs expression to reactivate miRNAs with onco-suppressive functions [206,298–300]. Local delivery of miR-200 members into the tumour endothelium showed reduction of metastasis and angiogenesis in several experimental models of ovarian, lung, renal and basal-like breast cancers [301]. In NSCLC, enforced expression of miR-145 inhibited EMT and metastatic ability [208].

Nevertheless, inhibiting or reversing EMT could also lead to a serious adverse event: favouring MET and thus colonisation of metastatic sites by circulating tumour cells [67,302]. Moreover, the intra-tumour and inter-tumour heterogeneity in case of multiple sites, and the dynamic nature of epithelial plasticity in cancer suggests that to be efficient the strategy should be multimodal and not focusing only on a single EMT-related target. As examples, combination of chemotherapy and anti-miRNA strategies and also models of miRNA replacement therapy were tested in vitro and in mouse models [264,303]. Sato et al. showed that the introduction of miR-200c using pre-miR-200c caused LIN28B suppression in cells with acquired EGFR-TKI resistance that harboured EMT features (HCC4006 after chronic exposure to gefinitib) [264]. Van Roosbroeck et al. proposed the use of cisplatin and anti-miR-155 in a mouse model of athymic nude mice (intrapulmonary injections of A549 cells stably infected with lentivirus containing a miR-155-overexpressing lentiviral vector) with significant results in term of primary tumour size and mediastinal lymph nodes [303].

Finally EMT is pointed out as a mechanism of resistance to immunotherapy and is involved in the shaping of the immune microenvironment. It was recently shown in NSCLC that tumours with EMT features expressed PDL1 and other immune checkpoints molecules and suggested that EMT should be further investigated as a predictor of response to immunotherapies [304]. Immunotherapy should then be considered as an anti-EMT therapy [305–307].

7. Conclusions

EMT is a highly regulated multistep process that is implicated in cancer progression through activation of proliferation pathways, loss of response to apoptotic signals, gain of stem cell properties, matrix remodelling and mobility. EMT involves various signalling pathways and crosstalk as well as a network of transcription factors. Upstream non-coding RNAs such as miRNAs have emerged as potent modulators of EMT. The physiopathology of the EMT process is highly dependent upon the cellular model, the environment and the EMT stimulating factors. Therefore cells undergoing EMT may express different markers. Moreover quantifying the degree of EMT in a tumour remains a challenging task due to its transient and reversible nature. However, some features such as loss of E-cadherin, reactivation of TWIST1, ZEB1 and SNAIL and downregulation of the miR-200 cluster could be common features of EMT in NSCLC.

Because of its link with metastasis and resistance to treatment, EMT has emerged as a useful prognosis and predictive marker but there is yet no clinical application in NSCLC. Our understanding of EMT is growing and may enable us to forward EMT characterization to the clinics. The development of methods to investigate molecular profiles or EMT signature using small amounts of tissues and FFPE sample will help validate EMT markers in clinical settings. Finally, EMT is a promising therapeutic target to overcome drug resistance in NSCLC in patients treated with chemotherapy, targeted therapies and immunotherapies. Because of the complexity of the EMT regulation network, the treatment will rely on integrative personalize care and high throughput molecular screenings.

Supplementary Materials: The following are available online at http://www.mdpi.com/2072-6694/9/8/101/s1. Table S1: MicroRNA signatures in NSCLC [120–125].

Acknowledgments: Laura Spurrier (MD student) for English editing. No funding was received for this work.

Author Contributions: Hélène Blons, Pierre Laurent-Puig and Antoine Legras designed the review; Nicolas Pécuchet, Sandrine Imbeaud, Karine Pallier, Audrey Didelot, Hélène Roussel, Laure Gibault and Elizabeth Fabre contributed to literature review; Antoine Legras and Hélène Blons wrote the paper.

Conflicts of Interest: The authors declare no conflict of interest.

References

1. Kalluri, R.; Weinberg, R.A. The basics of epithelial-mesenchymal transition. *J. Clin. Invest.* **2009**, *119*, 1420–1428. [CrossRef] [PubMed]
2. Byrum, C.; Martindale, M. Gastrulation in the cnidaria and ctenophora. In *Gastrulation*; Stern CD: New York, NY, USA, 2004.
3. Thompson, E.W.; Haviv, I. The social aspects of EMT-MET plasticity. *Nat. Med.* **2011**, *17*, 1048–1049. [CrossRef] [PubMed]
4. Lim, J.; Thiery, J.P. Epithelial-mesenchymal transitions: insights from development. *Dev. Camb. Engl.* **2012**, *139*, 3471–3486. [CrossRef] [PubMed]
5. Thiery, J.P.; Acloque, H.; Huang, R.Y.J.; Nieto, M.A. Epithelial-mesenchymal transitions in development and disease. *Cell* **2009**, *139*, 871–890. [CrossRef] [PubMed]
6. Yang, J.; Weinberg, R.A. Epithelial-mesenchymal transition: At the crossroads of development and tumor metastasis. *Dev. Cell* **2008**, *14*, 818–829. [CrossRef] [PubMed]
7. Zaravinos, A. The Regulatory Role of MicroRNAs in EMT and Cancer. *J. Oncol.* **2015**, *2015*, 865816. [CrossRef] [PubMed]

8. Zavadil, J.; Böttinger, E.P. TGF-beta and epithelial-to-mesenchymal transitions. *Oncogene* **2005**, *24*, 5764–5774. [CrossRef] [PubMed]
9. Lo, H.-W.; Hsu, S.-C.; Xia, W.; Cao, X.; Shih, J.-Y.; Wei, Y.; Abbruzzese, J.L.; Hortobagyi, G.N.; Hung, M.-C. Epidermal growth factor receptor cooperates with signal transducer and activator of transcription 3 to induce epithelial-mesenchymal transition in cancer cells via up-regulation of TWIST gene expression. *Cancer Res.* **2007**, *67*, 9066–9076. [CrossRef] [PubMed]
10. Lee, J.M.; Dedhar, S.; Kalluri, R.; Thompson, E.W. The epithelial-mesenchymal transition: New insights in signaling, development, and disease. *J. Cell Biol.* **2006**, *172*, 973–981. [CrossRef] [PubMed]
11. Acevedo, V.D.; Gangula, R.D.; Freeman, K.W.; Li, R.; Zhang, Y.; Wang, F.; Ayala, G.E.; Peterson, L.E.; Ittmann, M.; Spencer, D.M. Inducible FGFR-1 activation leads to irreversible prostate adenocarcinoma and an epithelial-to-mesenchymal transition. *Cancer Cell* **2007**, *12*, 559–571. [CrossRef] [PubMed]
12. Savagner, P.; Yamada, K.M.; Thiery, J.P. The zinc-finger protein slug causes desmosome dissociation, an initial and necessary step for growth factor-induced epithelial-mesenchymal transition. *J. Cell Biol.* **1997**, *137*, 1403–1419. [CrossRef] [PubMed]
13. Graham, T.R.; Zhau, H.E.; Odero-Marah, V.A.; Osunkoya, A.O.; Kimbro, K.S.; Tighiouart, M.; Liu, T.; Simons, J.W.; O'Regan, R.M. Insulin-like growth factor-I-dependent up-regulation of ZEB1 drives epithelial-to-mesenchymal transition in human prostate cancer cells. *Cancer Res.* **2008**, *68*, 2479–2488. [CrossRef] [PubMed]
14. Yang, M.-H.; Wu, M.-Z.; Chiou, S.-H.; Chen, P.-M.; Chang, S.-Y.; Liu, C.-J.; Teng, S.-C.; Wu, K.-J. Direct regulation of TWIST by HIF-1alpha promotes metastasis. *Nat. Cell Biol.* **2008**, *10*, 295–305. [CrossRef] [PubMed]
15. Gort, E.H.; van Haaften, G.; Verlaan, I.; Groot, A.J.; Plasterk, R.H.A.; Shvarts, A.; Suijkerbuijk, K.P.M.; van Laar, T.; van der Wall, E.; Raman, V.; et al. The TWIST1 oncogene is a direct target of hypoxia-inducible factor-2alpha. *Oncogene* **2008**, *27*, 1501–1510. [CrossRef] [PubMed]
16. Leong, K.G.; Niessen, K.; Kulic, I.; Raouf, A.; Eaves, C.; Pollet, I.; Karsan, A. Jagged1-mediated Notch activation induces epithelial-to-mesenchymal transition through Slug-induced repression of E-cadherin. *J. Exp. Med.* **2007**, *204*, 2935–2948. [CrossRef] [PubMed]
17. Takai, E.; Tsukimoto, M.; Kojima, S. TGF-β1 downregulates COX-2 expression leading to decrease of PGE2 production in human lung cancer A549 cells, which is involved in fibrotic response to TGF-β1. *PLoS ONE* **2013**, *8*, e76346. [CrossRef] [PubMed]
18. Shintani, Y.; Maeda, M.; Chaika, N.; Johnson, K.R.; Wheelock, M.J. Collagen I promotes epithelial-to-mesenchymal transition in lung cancer cells via transforming growth factor-beta signaling. *Am. J. Respir. Cell Mol. Biol.* **2008**, *38*, 95–104. [CrossRef] [PubMed]
19. Zoltan-Jones, A.; Huang, L.; Ghatak, S.; Toole, B.P. Elevated hyaluronan production induces mesenchymal and transformed properties in epithelial cells. *J. Biol. Chem.* **2003**, *278*, 45801–45810. [CrossRef] [PubMed]
20. Tania, M.; Khan, M.A.; Fu, J. Epithelial to mesenchymal transition inducing transcription factors and metastatic cancer. *Tumour Biol. J. Int. Soc. Oncodevelopmental Biol. Med.* **2014**, *35*, 7335–7342. [CrossRef] [PubMed]
21. Yilmaz, M.; Christofori, G. EMT, the cytoskeleton, and cancer cell invasion. *Cancer Metastasis Rev.* **2009**, *28*, 15–33. [CrossRef] [PubMed]
22. Wang, Y.; Shi, J.; Chai, K.; Ying, X.; Zhou, B.P. The Role of Snail in EMT and Tumorigenesis. *Curr. Cancer Drug Targets* **2013**, *13*, 963–972. [CrossRef] [PubMed]
23. Zhang, P.; Sun, Y.; Ma, L. ZEB1: At the crossroads of epithelial-mesenchymal transition, metastasis and therapy resistance. *Cell Cycle* **2015**, *14*, 481–487. [CrossRef] [PubMed]
24. Gemmill, R.M.; Roche, J.; Potiron, V.A.; Nasarre, P.; Mitas, M.; Coldren, C.D.; Helfrich, B.A.; Garrett-Mayer, E.; Bunn, P.A.; Drabkin, H.A. ZEB1-responsive genes in non-small cell lung cancer. *Cancer Lett.* **2011**, *300*, 66–78. [CrossRef] [PubMed]
25. Ansieau, S.; Morel, A.-P.; Hinkal, G.; Bastid, J.; Puisieux, A. TWISTing an embryonic transcription factor into an oncoprotein. *Oncogene* **2010**, *29*, 3173–3184. [CrossRef] [PubMed]
26. Yin, Z.; Xu, X.L.; Frasch, M. Regulation of the twist target gene tinman by modular cis-regulatory elements during early mesoderm development. *Dev. Camb. Engl.* **1997**, *124*, 4971–4982.
27. Wang, S.M.; Coljee, V.W.; Pignolo, R.J.; Rotenberg, M.O.; Cristofalo, V.J.; Sierra, F. Cloning of the human twist gene: Its expression is retained in adult mesodermally-derived tissues. *Gene* **1997**, *187*, 83–92. [CrossRef]

28. Pan, D.; Fujimoto, M.; Lopes, A.; Wang, Y.-X. Twist-1 is a PPARdelta-inducible, negative-feedback regulator of PGC-1alpha in brown fat metabolism. *Cell* **2009**, *137*, 73–86. [CrossRef] [PubMed]

29. Šošić, D.; Richardson, J.A.; Yu, K.; Ornitz, D.M.; Olson, E.N. Twist regulates cytokine gene expression through a negative feedback loop that represses NF-kappaB activity. *Cell* **2003**, *112*, 169–180. [CrossRef]

30. Sharif, M.N.; Sosic, D.; Rothlin, C.V.; Kelly, E.; Lemke, G.; Olson, E.N.; Ivashkiv, L.B. Twist mediates suppression of inflammation by type I IFNs and Axl. *J. Exp. Med.* **2006**, *203*, 1891–1901. [CrossRef] [PubMed]

31. Thuault, S.; Tan, E.-J.; Peinado, H.; Cano, A.; Heldin, C.-H.; Moustakas, A. HMGA2 and Smads co-regulate SNAIL1 expression during induction of epithelial-to-mesenchymal transition. *J. Biol. Chem.* **2008**, *283*, 33437–33446. [CrossRef] [PubMed]

32. Micalizzi, D.S.; Christensen, K.L.; Jedlicka, P.; Coletta, R.D.; Barón, A.E.; Harrell, J.C.; Horwitz, K.B.; Billheimer, D.; Heichman, K.A.; Welm, A.L.; et al. The Six1 homeoprotein induces human mammary carcinoma cells to undergo epithelial-mesenchymal transition and metastasis in mice through increasing TGF-beta signaling. *J. Clin. Invest.* **2009**, *119*, 2678–2690. [CrossRef] [PubMed]

33. Peinado, H.; Olmeda, D.; Cano, A. Snail, Zeb and bHLH factors in tumour progression: An alliance against the epithelial phenotype? *Nat. Rev. Cancer* **2007**, *7*, 415–428. [CrossRef] [PubMed]

34. Wu, Y.; Deng, J.; Rychahou, P.G.; Qiu, S.; Evers, B.M.; Zhou, B.P. Stabilization of snail by NF-kappaB is required for inflammation-induced cell migration and invasion. *Cancer Cell* **2009**, *15*, 416–428. [CrossRef] [PubMed]

35. Yamashita, S.; Miyagi, C.; Fukada, T.; Kagara, N.; Che, Y.-S.; Hirano, T. Zinc transporter LIVI controls epithelial-mesenchymal transition in zebrafish gastrula organizer. *Nature* **2004**, *429*, 298–302. [CrossRef] [PubMed]

36. Ikenouchi, J.; Matsuda, M.; Furuse, M.; Tsukita, S. Regulation of tight junctions during the epithelium-mesenchyme transition: Direct repression of the gene expression of claudins/occludin by Snail. *J. Cell Sci.* **2003**, *116*, 1959–1967. [CrossRef] [PubMed]

37. Moreno-Bueno, G.; Portillo, F.; Cano, A. Transcriptional regulation of cell polarity in EMT and cancer. *Oncogene* **2008**, *27*, 6958–6969. [CrossRef] [PubMed]

38. Billottet, C.; Tuefferd, M.; Gentien, D.; Rapinat, A.; Thiery, J.-P.; Broët, P.; Jouanneau, J. Modulation of several waves of gene expression during FGF-1 induced epithelial-mesenchymal transition of carcinoma cells. *J. Cell. Biochem.* **2008**, *104*, 826–839. [CrossRef] [PubMed]

39. Xiao, D.; He, J. Epithelial mesenchymal transition and lung cancer. *J. Thorac. Dis.* **2010**, *2*, 154–159. [CrossRef] [PubMed]

40. Valsesia-Wittmann, S.; Magdeleine, M.; Dupasquier, S.; Garin, E.; Jallas, A.-C.; Combaret, V.; Krause, A.; Leissner, P.; Puisieux, A. Oncogenic cooperation between H-Twist and N-Myc overrides failsafe programs in cancer cells. *Cancer Cell* **2004**, *6*, 625–630. [CrossRef] [PubMed]

41. Ansieau, S.; Bastid, J.; Doreau, A.; Morel, A.-P.; Bouchet, B.P.; Thomas, C.; Fauvet, F.; Puisieux, I.; Doglioni, C.; Piccinin, S.; et al. Induction of EMT by twist proteins as a collateral effect of tumor-promoting inactivation of premature senescence. *Cancer Cell* **2008**, *14*, 79–89. [CrossRef] [PubMed]

42. Morel, A.-P.; Hinkal, G.W.; Thomas, C.; Fauvet, F.; Courtois-Cox, S.; Wierinckx, A.; Devouassoux-Shisheboran, M.; Treilleux, I.; Tissier, A.; Gras, B.; et al. EMT inducers catalyze malignant transformation of mammary epithelial cells and drive tumorigenesis towards claudin-low tumors in transgenic mice. *PLoS Genet.* **2012**, *8*, e1002723. [CrossRef] [PubMed]

43. Puisieux, A.; Valsesia-Wittmann, S.; Ansieau, S. A twist for survival and cancer progression. *Br. J. Cancer* **2006**, *94*, 13–17. [CrossRef] [PubMed]

44. Pallier, K.; Cessot, A.; Côté, J.-F.; Just, P.-A.; Cazes, A.; Fabre, E.; Danel, C.; Riquet, M.; Devouassoux-Shisheboran, M.; Ansieau, S.; Puisieux, A.; et al. TWIST1 a new determinant of epithelial to mesenchymal transition in EGFR mutated lung adenocarcinoma. *PLoS ONE* **2012**, *7*, e29954. [CrossRef] [PubMed]

45. Yang, J.; Mani, S.A.; Donaher, J.L.; Ramaswamy, S.; Itzykson, R.A.; Come, C.; Savagner, P.; Gitelman, I.; Richardson, A.; Weinberg, R.A. Twist, a master regulator of morphogenesis, plays an essential role in tumor metastasis. *Cell* **2004**, *117*, 927–939. [CrossRef] [PubMed]

46. Weinberg, R.A. Twisted epithelial-mesenchymal transition blocks senescence. *Nat. Cell Biol.* **2008**, *10*, 1021–1023. [CrossRef] [PubMed]

47. Hui, L.; Zhang, S.; Dong, X.; Tian, D.; Cui, Z.; Qiu, X. Prognostic significance of twist and N-cadherin expression in NSCLC. *PLoS ONE* **2013**, *8*, e62171. [CrossRef] [PubMed]

48. Lu, Z.; Ghosh, S.; Wang, Z.; Hunter, T. Downregulation of caveolin-1 function by EGF leads to the loss of E-cadherin, increased transcriptional activity of beta-catenin, and enhanced tumor cell invasion. *Cancer Cell* **2003**, *4*, 499–515. [CrossRef]

49. Lee, M.-Y.; Chou, C.-Y.; Tang, M.-J.; Shen, M.-R. Epithelial-mesenchymal transition in cervical cancer: Correlation with tumor progression, epidermal growth factor receptor overexpression, and snail up-regulation. *Clin. Cancer Res.* **2008**, *14*, 4743–4750. [CrossRef] [PubMed]

50. Tsoukalas, N.; Aravantinou-Fatorou, E.; Tolia, M.; Giaginis, C.; Galanopoulos, M.; Kiakou, M.; Kostakis, I.D.; Dana, E.; Vamvakaris, I.; Korogiannos, A.; et al. Epithelial-mesenchymal transition in non small-cell lung cancer. *Anticancer Res.* **2017**, *37*, 1773–1778. [CrossRef] [PubMed]

51. Milara, J.; Peiró, T.; Serrano, A.; Cortijo, J. Epithelial to mesenchymal transition is increased in patients with COPD and induced by cigarette smoke. *Thorax* **2013**, *68*, 410–420. [CrossRef] [PubMed]

52. Wang, Q.; Wang, Y.; Zhang, Y.; Zhang, Y.; Xiao, W. The role of uPAR in epithelial-mesenchymal transition in small airway epithelium of patients with chronic obstructive pulmonary disease. *Respir. Res.* **2013**, *14*, 67. [CrossRef] [PubMed]

53. Zhang, H.; Liu, H.; Borok, Z.; Davies, K.J.A.; Ursini, F.; Forman, H.J. Cigarette smoke extract stimulates epithelial-mesenchymal transition through Src activation. *Free Radic. Biol. Med.* **2012**, *52*, 1437–1442. [CrossRef] [PubMed]

54. Okubo, K.; Uenosono, Y.; Arigami, T.; Yanagita, S.; Matsushita, D.; Kijima, T.; Amatatsu, M.; Uchikado, Y.; Kijima, Y.; Maemura, K.; et al. Clinical significance of altering epithelial-mesenchymal transition in metastatic lymph nodes of gastric cancer. *Gastric Cancer* **2017**. [CrossRef] [PubMed]

55. Wen, J.; Luo, K.-J.; Liu, Q.-W.; Wang, G.; Zhang, M.-F.; Xie, X.-Y.; Yang, H.; Fu, J.-H.; Hu, Y. The epithelial-mesenchymal transition phenotype of metastatic lymph nodes impacts the prognosis of esophageal squamous cell carcinoma patients. *Oncotarget* **2016**, *7*, 37581–37588. [CrossRef] [PubMed]

56. Lee, W.-Y.; Shin, D.-Y.; Kim, H.J.; Ko, Y.-H.; Kim, S.; Jeong, H.-S. Prognostic significance of epithelial-mesenchymal transition of extracapsular spread tumors in lymph node metastases of head and neck cancer. *Ann. Surg. Oncol.* **2014**, *21*, 1904–1911. [CrossRef] [PubMed]

57. Yu, H.; Simons, D.L.; Segall, I.; Carcamo-Cavazos, V.; Schwartz, E.J.; Yan, N.; Zuckerman, N.S.; Dirbas, F.M.; Johnson, D.L.; Holmes, S.P.; et al. PRC2/EED-EZH2 complex is up-regulated in breast cancer lymph node metastasis compared to primary tumor and correlates with tumor proliferation in situ. *PLoS ONE* **2012**, *7*, e51239. [CrossRef] [PubMed]

58. Shimamatsu, S.; Okamoto, T.; Haro, A.; Kitahara, H.; Kohno, M.; Morodomi, Y.; Tagawa, T.; Okano, S.; Oda, Y.; Maehara, Y. Prognostic Significance of Expression of the Epithelial-Mesenchymal Transition-Related Factor Brachyury in Intrathoracic Lymphatic Spread of Non-Small Cell Lung Cancer. *Ann. Surg. Oncol.* **2016**, *23*, 1012–1020. [CrossRef] [PubMed]

59. Kispert, A.; Herrmann, B.G. Immunohistochemical analysis of the Brachyury protein in wild-type and mutant mouse embryos. *Dev. Biol.* **1994**, *161*, 179–193. [CrossRef] [PubMed]

60. Larocca, C.; Cohen, J.R.; Fernando, R.I.; Huang, B.; Hamilton, D.H.; Palena, C. An autocrine loop between TGF-β1 and the transcription factor brachyury controls the transition of human carcinoma cells into a mesenchymal phenotype. *Mol. Cancer Ther.* **2013**, *12*, 1805–1815. [CrossRef] [PubMed]

61. Su, J.-L.; Yang, P.-C.; Shih, J.-Y.; Yang, C.-Y.; Wei, L.-H.; Hsieh, C.-Y.; Chou, C.-H.; Jeng, Y.-M.; Wang, M.-Y.; Chang, K.-J.; et al. The VEGF-C/Flt-4 axis promotes invasion and metastasis of cancer cells. *Cancer Cell* **2006**, *9*, 209–223. [CrossRef] [PubMed]

62. Alitalo, A.; Detmar, M. Interaction of tumor cells and lymphatic vessels in cancer progression. *Oncogene* **2012**, *31*, 4499–4508. [CrossRef] [PubMed]

63. Adams, R.H.; Alitalo, K. Molecular regulation of angiogenesis and lymphangiogenesis. *Nat. Rev. Mol. Cell Biol.* **2007**, *8*, 464–478. [CrossRef] [PubMed]

64. Liersch, R.; Hirakawa, S.; Berdel, W.E.; Mesters, R.M.; Detmar, M. Induced lymphatic sinus hyperplasia in sentinel lymph nodes by VEGF-C as the earliest premetastatic indicator. *Int. J. Oncol.* **2012**, *41*, 2073–2078. [CrossRef] [PubMed]

65. Kilvaer, T.K.; Paulsen, E.-E.; Hald, S.M.; Wilsgaard, T.; Bremnes, R.M.; Busund, L.-T.; Donnem, T. Lymphangiogenic Markers and Their Impact on Nodal Metastasis and Survival in Non-Small Cell Lung Cancer–A Structured Review with Meta-Analysis. *PLoS ONE* **2015**, *10*, e0132481. [CrossRef] [PubMed]

66. Zhou, L.; Yu, L.; Wu, S.; Feng, Z.; Song, W.; Gong, X. Clinicopathological significance of KAI1 expression and epithelial-mesenchymal transition in non-small cell lung cancer. *World J. Surg. Oncol.* **2015**, *13*, 234. [CrossRef] [PubMed]

67. Tsai, J.H.; Donaher, J.L.; Murphy, D.A.; Chau, S.; Yang, J. Spatiotemporal regulation of epithelial-mesenchymal transition is essential for squamous cell carcinoma metastasis. *Cancer Cell* **2012**, *22*, 725–736. [CrossRef] [PubMed]

68. Chaffer, C.L.; Brennan, J.P.; Slavin, J.L.; Blick, T.; Thompson, E.W.; Williams, E.D. Mesenchymal-to-epithelial transition facilitates bladder cancer metastasis: Role of fibroblast growth factor receptor-2. *Cancer Res.* **2006**, *66*, 11271–11278. [CrossRef] [PubMed]

69. Chao, Y.L.; Shepard, C.R.; Wells, A. Breast carcinoma cells re-express E-cadherin during mesenchymal to epithelial reverting transition. *Mol. Cancer* **2010**, *9*, 179. [CrossRef] [PubMed]

70. Frisch, S.M. The epithelial cell default-phenotype hypothesis and its implications for cancer. *BioEssays News Rev. Mol. Cell. Dev. Biol.* **1997**, *19*, 705–709. [CrossRef] [PubMed]

71. Aokage, K.; Ishii, G.; Ohtaki, Y.; Yamaguchi, Y.; Hishida, T.; Yoshida, J.; Nishimura, M.; Nagai, K.; Ochiai, A. Dynamic molecular changes associated with epithelial-mesenchymal transition and subsequent mesenchymal-epithelial transition in the early phase of metastatic tumor formation. *Int. J. Cancer* **2011**, *128*, 1585–1595. [CrossRef] [PubMed]

72. Rubin, M.A.; Mucci, N.R.; Figurski, J.; Fecko, A.; Pienta, K.J.; Day, M.L. E-cadherin expression in prostate cancer: A broad survey using high-density tissue microarray technology. *Hum. Pathol.* **2001**, *32*, 690–697. [CrossRef] [PubMed]

73. Kowalski, P.J.; Rubin, M.A.; Kleer, C.G. E-cadherin expression in primary carcinomas of the breast and its distant metastases. *Breast Cancer Res.* **2003**, *5*, R217–R222. [CrossRef] [PubMed]

74. Imai, T.; Horiuchi, A.; Shiozawa, T.; Osada, R.; Kikuchi, N.; Ohira, S.; Oka, K.; Konishi, I. Elevated expression of E-cadherin and alpha-, beta-, and gamma-catenins in metastatic lesions compared with primary epithelial ovarian carcinomas. *Hum. Pathol.* **2004**, *35*, 1469–1476. [CrossRef] [PubMed]

75. Wells, A.; Yates, C.; Shepard, C.R. E-cadherin as an indicator of mesenchymal to epithelial reverting transitions during the metastatic seeding of disseminated carcinomas. *Clin. Exp. Metastasis* **2008**, *25*, 621–628. [CrossRef] [PubMed]

76. Meng, F.; Wu, G. The rejuvenated scenario of epithelial-mesenchymal transition (EMT) and cancer metastasis. *Cancer Metastasis Rev.* **2012**, *31*, 455–467. [CrossRef] [PubMed]

77. Diepenbruck, M.; Christofori, G. Epithelial-mesenchymal transition (EMT) and metastasis: Yes, no, maybe? *Curr. Opin. Cell Biol.* **2016**, *43*, 7–13. [CrossRef] [PubMed]

78. Tarin, D.; Thompson, E.W.; Newgreen, D.F. The fallacy of epithelial mesenchymal transition in neoplasia. *Cancer Res.* **2005**, *65*, 5996–6001. [CrossRef] [PubMed]

79. Prall, F. Tumour budding in colorectal carcinoma. *Histopathology* **2007**, *50*, 151–162. [CrossRef] [PubMed]

80. Friedl, P. Prespecification and plasticity: Shifting mechanisms of cell migration. *Curr. Opin. Cell Biol.* **2004**, *16*, 14–23. [CrossRef] [PubMed]

81. Yilmaz, M.; Christofori, G. Mechanisms of motility in metastasizing cells. *Mol. Cancer Res.* **2010**, *8*, 629–642. [CrossRef] [PubMed]

82. Friedl, P.; Hegerfeldt, Y.; Tusch, M. Collective cell migration in morphogenesis and cancer. *Int. J. Dev. Biol.* **2004**, *48*, 441–449. [CrossRef] [PubMed]

83. Wicki, A.; Christofori, G. The potential role of podoplanin in tumour invasion. *Br. J. Cancer* **2007**, *96*, 1–5. [CrossRef] [PubMed]

84. Wicki, A.; Lehembre, F.; Wick, N.; Hantusch, B.; Kerjaschki, D.; Christofori, G. Tumor invasion in the absence of epithelial-mesenchymal transition: Podoplanin-mediated remodeling of the actin cytoskeleton. *Cancer Cell* **2006**, *9*, 261–272. [CrossRef] [PubMed]

85. Van Rooij, E. The art of microRNA research. *Circ. Res.* **2011**, *108*, 219–234. [CrossRef] [PubMed]

86. Cullen, B.R. Transcription and processing of human microRNA precursors. *Mol. Cell* **2004**, *16*, 861–865. [CrossRef] [PubMed]

87. Denli, A.M.; Tops, B.B.J.; Plasterk, R.H.A.; Ketting, R.F.; Hannon, G.J. Processing of primary microRNAs by the Microprocessor complex. *Nature* **2004**, *432*, 231–235. [CrossRef] [PubMed]
88. Ketting, R.F.; Fischer, S.E.; Bernstein, E.; Sijen, T.; Hannon, G.J.; Plasterk, R.H. Dicer functions in RNA interference and in synthesis of small RNA involved in developmental timing in C. elegans. *Genes Dev.* **2001**, *15*, 2654–2659. [CrossRef] [PubMed]
89. Sontheimer, E.J. Assembly and function of RNA silencing complexes. *Nat. Rev. Mol. Cell Biol.* **2005**, *6*, 127–138. [CrossRef] [PubMed]
90. Gregory, R.I.; Chendrimada, T.P.; Cooch, N.; Shiekhattar, R. Human RISC couples microRNA biogenesis and posttranscriptional gene silencing. *Cell* **2005**, *123*, 631–640. [CrossRef] [PubMed]
91. Williams, A.E.; Moschos, S.A.; Perry, M.M.; Barnes, P.J.; Lindsay, M.A. Maternally imprinted microRNAs are differentially expressed during mouse and human lung development. *Dev. Dyn. Off. Publ. Am. Assoc. Anat.* **2007**, *236*, 572–580. [CrossRef] [PubMed]
92. Landgraf, P.; Rusu, M.; Sheridan, R.; Sewer, A.; Iovino, N.; Aravin, A.; Pfeffer, S.; Rice, A.; Kamphorst, A.O.; Landthaler, M.; et al. A mammalian microRNA expression atlas based on small RNA library sequencing. *Cell* **2007**, *129*, 1401–1414. [CrossRef] [PubMed]
93. Calin, G.A.; Liu, C.-G.; Sevignani, C.; Ferracin, M.; Felli, N.; Dumitru, C.D.; Shimizu, M.; Cimmino, A.; Zupo, S.; Dono, M.; et al. MicroRNA profiling reveals distinct signatures in B cell chronic lymphocytic leukemias. *Proc. Natl. Acad. Sci. USA* **2004**, *101*, 11755–11760. [CrossRef] [PubMed]
94. Gupta, A.; Swaminathan, G.; Martin-Garcia, J.; Navas-Martin, S. MicroRNAs, hepatitis C virus, and HCV/HIV-1 co-infection: New insights in pathogenesis and therapy. *Viruses* **2012**, *4*, 2485–2513. [CrossRef] [PubMed]
95. Cui, L.; Li, Y.; Ma, G.; Wang, Y.; Cai, Y.; Liu, S.; Chen, Y.; Li, J.; Xie, Y.; Liu, G.; et al. A functional polymorphism in the promoter region of microRNA-146a is associated with the risk of Alzheimer disease and the rate of cognitive decline in patients. *PLoS ONE* **2014**, *9*, e89019. [CrossRef] [PubMed]
96. Zhang, Z.; Li, J.; Liu, B.; Luo, C.; Dong, Q.; Zhao, L.; Zhong, Y.; Chen, W.; Chen, M.; Liu, S. MicroRNA-26 was decreased in rat cardiac hypertrophy model and may be a promising therapeutic target. *J. Cardiovasc. Pharmacol.* **2013**, *62*, 312–319. [CrossRef] [PubMed]
97. Karolina, D.S.; Armugam, A.; Tavintharan, S.; Wong, M.T.K.; Lim, S.C.; Sum, C.F.; Jeyaseelan, K. MicroRNA 144 impairs insulin signaling by inhibiting the expression of insulin receptor substrate 1 in type 2 diabetes mellitus. *PLoS ONE* **2011**, *6*, e22839. [CrossRef]
98. Ebrahimi, A.; Sadroddiny, E. MicroRNAs in lung diseases: Recent findings and their pathophysiological implications. *Pulm. Pharmacol. Ther.* **2015**, *34*, 55–63. [CrossRef] [PubMed]
99. Alipoor, S.D.; Adcock, I.M.; Garssen, J.; Mortaz, E.; Varahram, M.; Mirsaeidi, M.; Velayati, A. The roles of miRNAs as potential biomarkers in lung diseases. *Eur. J. Pharmacol.* **2016**, *791*, 395–404. [CrossRef] [PubMed]
100. Calin, G.A.; Sevignani, C.; Dumitru, C.D.; Hyslop, T.; Noch, E.; Yendamuri, S.; Shimizu, M.; Rattan, S.; Bullrich, F.; Negrini, M.; et al. Human microRNA genes are frequently located at fragile sites and genomic regions involved in cancers. *Proc. Natl. Acad. Sci. USA* **2004**, *101*, 2999–3004. [CrossRef] [PubMed]
101. Esquela-Kerscher, A.; Slack, F.J. Oncomirs-microRNAs with a role in cancer. *Nat. Rev. Cancer* **2006**, *6*, 259–269. [CrossRef] [PubMed]
102. Wang, D.; Qiu, C.; Zhang, H.; Wang, J.; Cui, Q.; Yin, Y. Human microRNA oncogenes and tumor suppressors show significantly different biological patterns: From functions to targets. *PLoS ONE* **2010**, *5*. [CrossRef] [PubMed]
103. Wang, Q.Z.; Xu, W.; Habib, N.; Xu, R. Potential uses of microRNA in lung cancer diagnosis, prognosis, and therapy. *Curr. Cancer Drug Targets* **2009**, *9*, 572–594. [CrossRef] [PubMed]
104. Gregory, P.A.; Bracken, C.P.; Bert, A.G.; Goodall, G.J. MicroRNAs as regulators of epithelial-mesenchymal transition. *Cell Cycle Georget. Tex* **2008**, *7*, 3112–3118. [CrossRef] [PubMed]
105. Korpal, M.; Lee, E.S.; Hu, G.; Kang, Y. The miR-200 family inhibits epithelial-mesenchymal transition and cancer cell migration by direct targeting of E-cadherin transcriptional repressors ZEB1 and ZEB2. *J. Biol. Chem.* **2008**, *283*, 14910–14914. [CrossRef] [PubMed]
106. Park, S.-M.; Gaur, A.B.; Lengyel, E.; Peter, M.E. The miR-200 family determines the epithelial phenotype of cancer cells by targeting the E-cadherin repressors ZEB1 and ZEB2. *Genes Dev.* **2008**, *22*, 894–907. [CrossRef] [PubMed]

107. Gregory, P.A.; Bert, A.G.; Paterson, E.L.; Barry, S.C.; Tsykin, A.; Farshid, G.; Vadas, M.A.; Khew-Goodall, Y.; Goodall, G.J. The miR-200 family and miR-205 regulate epithelial to mesenchymal transition by targeting ZEB1 and SIP1. *Nat. Cell Biol.* **2008**, *10*, 593–601. [CrossRef] [PubMed]

108. Paterson, E.L.; Kolesnikoff, N.; Gregory, P.A.; Bert, A.G.; Khew-Goodall, Y.; Goodall, G.J. The microRNA-200 family regulates epithelial to mesenchymal transition. *Sci. World J.* **2008**, *8*, 901–904. [CrossRef] [PubMed]

109. Díaz-Martín, J.; Díaz-López, A.; Moreno-Bueno, G.; Castilla, M.Á.; Rosa-Rosa, J.M.; Cano, A.; Palacios, J. A core microRNA signature associated with inducers of the epithelial-to-mesenchymal transition. *J. Pathol.* **2014**, *232*, 319–329. [CrossRef] [PubMed]

110. Davalos, V.; Moutinho, C.; Villanueva, A.; Boque, R.; Silva, P.; Carneiro, F.; Esteller, M. Dynamic epigenetic regulation of the microRNA-200 family mediates epithelial and mesenchymal transitions in human tumorigenesis. *Oncogene* **2012**, *31*, 2062–2074. [CrossRef] [PubMed]

111. Jiao, A.; Sui, M.; Zhang, L.; Sun, P.; Geng, D.; Zhang, W.; Wang, X.; Li, J. MicroRNA-200c inhibits the metastasis of non-small cell lung cancer cells by targeting ZEB2, an epithelial-mesenchymal transition regulator. *Mol. Med. Rep.* **2016**, *13*, 3349–3355. [CrossRef] [PubMed]

112. Roybal, J.D.; Zang, Y.; Ahn, Y.-H.; Yang, Y.; Gibbons, D.L.; Baird, B.N.; Alvarez, C.; Thilaganathan, N.; Liu, D.D.; Saintigny, P.; et al. miR-200 Inhibits Lung Adenocarinoma Cell Invasion and Metastasis by Targeting Flt1/VEGFR1. *Mol. Cancer Res.* **2011**, *9*, 25–35. [CrossRef] [PubMed]

113. Bracken, C.P.; Gregory, P.A.; Kolesnikoff, N.; Bert, A.G.; Wang, J.; Shannon, M.F.; Goodall, G.J. A double-negative feedback loop between ZEB1-SIP1 and the microRNA-200 family regulates epithelial-mesenchymal transition. *Cancer Res.* **2008**, *68*, 7846–7854. [CrossRef] [PubMed]

114. Hill, L.; Browne, G.; Tulchinsky, E. ZEB/miR-200 feedback loop: At the crossroads of signal transduction in cancer. *Int. J. Cancer* **2013**, *132*, 745–754. [CrossRef] [PubMed]

115. Cui, R.; Meng, W.; Sun, H.-L.; Kim, T.; Ye, Z.; Fassan, M.; Jeon, Y.-J.; Li, B.; Vicentini, C.; Peng, Y.; et al. MicroRNA-224 promotes tumor progression in nonsmall cell lung cancer. *Proc. Natl. Acad. Sci. USA* **2015**, *112*, E4288–E4297. [CrossRef] [PubMed]

116. Zhang, J.; Fu, J.; Pan, Y.; Zhang, X.; Shen, L. Silencing of miR-1247 by DNA methylation promoted non-small-cell lung cancer cell invasion and migration by effects of STMN1. *OncoTargets Ther.* **2016**, *9*, 7297–7307. [CrossRef] [PubMed]

117. Chan, S.-H.; Wang, L.-H. Regulation of cancer metastasis by microRNAs. *J. Biomed. Sci.* **2015**, *22*, 9. [CrossRef] [PubMed]

118. Siemens, H.; Jackstadt, R.; Hünten, S.; Kaller, M.; Menssen, A.; Götz, U.; Hermeking, H. MiR-34 and SNAIL form a double-negative feedback loop to regulate epithelial-mesenchymal transitions. *Cell Cycle Georget. Tex.* **2011**, *10*, 4256–4271. [CrossRef] [PubMed]

119. Bhaumik, D.; Scott, G.K.; Schokrpur, S.; Patil, C.K.; Campisi, J.; Benz, C.C. Expression of microRNA-146 suppresses NF-kappaB activity with reduction of metastatic potential in breast cancer cells. *Oncogene* **2008**, *27*, 5643–5647. [CrossRef] [PubMed]

120. Volinia, S.; Calin, G.A.; Liu, C.-G.; Ambs, S.; Cimmino, A.; Petrocca, F.; Visone, R.; Iorio, M.; Roldo, C.; Ferracin, M.; et al. A microRNA expression signature of human solid tumors defines cancer gene targets. *Proc. Natl. Acad. Sci. USA* **2006**, *103*, 2257–2261. [CrossRef] [PubMed]

121. Yanaihara, N.; Caplen, N.; Bowman, E.; Seike, M.; Kumamoto, K.; Yi, M.; Stephens, R.M.; Okamoto, A.; Yokota, J.; Tanaka, T.; et al. Unique microRNA molecular profiles in lung cancer diagnosis and prognosis. *Cancer Cell* **2006**, *9*, 189–198. [CrossRef] [PubMed]

122. Peltier, H.J.; Latham, G.J. Normalization of microRNA expression levels in quantitative RT-PCR assays: Identification of suitable reference RNA targets in normal and cancerous human solid tissues. *RNA* **2008**, *14*, 844–852. [CrossRef] [PubMed]

123. Zadran, S.; Remacle, F.; Levine, R.D. MiRNA and mRNA cancer signatures determined by analysis of expression levels in large cohorts of patients. *Proc. Natl. Acad. Sci. USA* **2013**, *110*, 19160–19165. [CrossRef] [PubMed]

124. Võsa, U.; Vooder, T.; Kolde, R.; Fischer, K.; Välk, K.; Tõnisson, N.; Roosipuu, R.; Vilo, J.; Metspalu, A.; Annilo, T. Identification of miR-374a as a prognostic marker for survival in patients with early-stage nonsmall cell lung cancer. *Genes. Chromosomes Cancer* **2011**, *50*, 812–822. [CrossRef] [PubMed]

125. Cazzoli, R.; Buttitta, F.; Di Nicola, M.; Malatesta, S.; Marchetti, A.; Rom, W.N.; Pass, H.I. MicroRNAs derived from circulating exosomes as noninvasive biomarkers for screening and diagnosing lung cancer. *J. Thorac. Oncol. Off. Publ. Int. Assoc. Study Lung Cancer* **2013**, *8*, 1156–1162. [CrossRef] [PubMed]

126. Su, Y.; Fang, H.; Jiang, F. Integrating DNA methylation and microRNA biomarkers in sputum for lung cancer detection. *Clin. Epigenetics* **2016**, *8*, 109. [CrossRef] [PubMed]

127. Razzak, R.; Bédard, E.L.R.; Kim, J.O.; Gazala, S.; Guo, L.; Ghosh, S.; Joy, A.; Nijjar, T.; Wong, E.; Roa, W.H. MicroRNA expression profiling of sputum for the detection of early and locally advanced non-small-cell lung cancer: A prospective case-control study. *Curr. Oncol. Tor. Ont.* **2016**, *23*, e86–e94. [CrossRef] [PubMed]

128. Xie, Y.; Todd, N.W.; Liu, Z.; Zhan, M.; Fang, H.; Peng, H.; Alattar, M.; Deepak, J.; Stass, S.A.; Jiang, F. Altered miRNA expression in sputum for diagnosis of non-small cell lung cancer. *Lung Cancer Amst. Neth.* **2010**, *67*, 170–176. [CrossRef] [PubMed]

129. Zhang, H.; Mao, F.; Shen, T.; Luo, Q.; Ding, Z.; Qian, L.; Huang, J. Plasma miR-145, miR-20a, miR-21 and miR-223 as novel biomarkers for screening early-stage non-small cell lung cancer. *Oncol. Lett.* **2017**, *13*, 669–676. [CrossRef] [PubMed]

130. Hou, J.; Meng, F.; Chan, L.W.C.; Cho, W.C.S.; Wong, S.C.C. Circulating Plasma MicroRNAs As Diagnostic Markers for NSCLC. *Front. Genet.* **2016**, *7*, 193. [CrossRef] [PubMed]

131. Giallombardo, M.; Reclusa, P.; Valentino, A.; Sirera, R.; Pauwels, P.; Rolfo, C. P2.07: Evaluation of different exosomal RNA isolation methods in nsclc liquid biopsies: Track: Biology and pathogenesis. *J. Thorac. Oncol. Off. Publ. Int. Assoc. Study Lung Cancer* **2016**, *11*, S220–S221. [CrossRef]

132. Reclusa, P.; Giallombardo, M.; Castiglia, M.; Sorber, L.; Van Der Steen, N.; Pauwels, P.; Rolfo, C. P2.06: Exosomal miRNA analysis in non-small cell lung cancer: New liquid biomarker?: Track: Biology and pathogenesis. *J. Thorac. Oncol. Off. Publ. Int. Assoc. Study Lung Cancer* **2016**, *11*, S219–S220. [CrossRef]

133. Rolfo, C.; Giallombardo, M.; Reclusa, P.; Sirera, R.; Peeters, M. Exosomes in lung cancer liquid biopsies: Two sides of the same coin? *Lung Cancer Amst. Neth.* **2017**, *104*, 134–135. [CrossRef] [PubMed]

134. Zhang, H.; Lu, Y.; Chen, E.; Li, X.; Lv, B.; Vikis, H.G.; Liu, P. XRN2 promotes EMT and metastasis through regulating maturation of miR-10a. *Oncogene* **2017**. [CrossRef] [PubMed]

135. Zhao, Z.; Zhang, L.; Yao, Q.; Tao, Z. MiR-15b regulates cisplatin resistance and metastasis by targeting PEBP4 in human lung adenocarcinoma cells. *Cancer Gene Ther.* **2015**, *22*, 108–114. [CrossRef] [PubMed]

136. Jiang, Z.; Yin, J.; Fu, W.; Mo, Y.; Pan, Y.; Dai, L.; Huang, H.; Li, S.; Zhao, J. MiRNA 17 family regulates cisplatin-resistant and metastasis by targeting TGFbetaR2 in NSCLC. *PLoS ONE* **2014**, *9*, e94639. [CrossRef] [PubMed]

137. Cao, M.; Seike, M.; Soeno, C.; Mizutani, H.; Kitamura, K.; Minegishi, Y.; Noro, R.; Yoshimura, A.; Cai, L.; Gemma, A. MiR-23a regulates TGF-β-induced epithelial-mesenchymal transition by targeting E-cadherin in lung cancer cells. *Int. J. Oncol.* **2012**, *41*, 869–875. [CrossRef] [PubMed]

138. Chen, J.; Xu, Y.; Tao, L.; Pan, Y.; Zhang, K.; Wang, R.; Chen, L.-B.; Chu, X. MiRNA-26a Contributes to the Acquisition of Malignant Behaviors of Doctaxel-Resistant Lung Adenocarcinoma Cells through Targeting EZH2. *Cell. Physiol. Biochem. Int. J. Exp. Cell. Physiol. Biochem. Pharmacol.* **2017**, *41*, 583–597. [CrossRef] [PubMed]

139. Liu, K.; Guo, L.; Guo, Y.; Zhou, B.; Li, T.; Yang, H.; Yin, R.; Xi, T. AEG-1 3′-untranslated region functions as a ceRNA in inducing epithelial-mesenchymal transition of human non-small cell lung cancer by regulating miR-30a activity. *Eur. J. Cell Biol.* **2015**, *94*, 22–31. [CrossRef] [PubMed]

140. Kumarswamy, R.; Mudduluru, G.; Ceppi, P.; Muppala, S.; Kozlowski, M.; Niklinski, J.; Papotti, M.; Allgayer, H. MicroRNA-30a inhibits epithelial-to-mesenchymal transition by targeting Snail and is downregulated in non-small cell lung cancer. *Int. J. Cancer* **2012**, *130*, 2044–2053. [CrossRef] [PubMed]

141. Kang, J.; Kim, E.; Kim, W.; Seong, K.M.; Youn, H.; Kim, J.W.; Kim, J.; Youn, B. Rhamnetin and cirsiliol induce radiosensitization and inhibition of epithelial-mesenchymal transition (EMT) by miR-34a-mediated suppression of Notch-1 expression in non-small cell lung cancer cell lines. *J. Biol. Chem.* **2013**, *288*, 27343–27357. [CrossRef] [PubMed]

142. Liu, M.-X.; Zhou, K.-C.; Cao, Y. MCRS1 overexpression, which is specifically inhibited by miR-129*, promotes the epithelial-mesenchymal transition and metastasis in non-small cell lung cancer. *Mol. Cancer* **2014**, *13*, 245. [CrossRef] [PubMed]

143. Qin, Q.; Wei, F.; Zhang, J.; Li, B. MiR-134 suppresses the migration and invasion of non-small cell lung cancer by targeting ITGB1. *Oncol. Rep.* **2017**, *37*, 823–830. [CrossRef] [PubMed]

144. Kitamura, K.; Seike, M.; Okano, T.; Matsuda, K.; Miyanaga, A.; Mizutani, H.; Noro, R.; Minegishi, Y.; Kubota, K.; Gemma, A. MiR-134/487b/655 cluster regulates TGF-β-induced epithelial-mesenchymal transition and drug resistance to gefitinib by targeting MAGI2 in lung adenocarcinoma cells. *Mol. Cancer Ther.* **2014**, *13*, 444–453. [CrossRef] [PubMed]

145. Li, J.; Wang, Y.; Luo, J.; Fu, Z.; Ying, J.; Yu, Y.; Yu, W. MiR-134 inhibits epithelial to mesenchymal transition by targeting FOXM1 in non-small cell lung cancer cells. *FEBS Lett.* **2012**, *586*, 3761–3765. [CrossRef] [PubMed]

146. Li, J.; Wang, Q.; Wen, R.; Liang, J.; Zhong, X.; Yang, W.; Su, D.; Tang, J. MiR-138 inhibits cell proliferation and reverses epithelial-mesenchymal transition in non-small cell lung cancer cells by targeting GIT1 and SEMA4C. *J. Cell. Mol. Med.* **2015**, *19*, 2793–2805. [CrossRef] [PubMed]

147. Jin, Z.; Guan, L.; Song, Y.; Xiang, G.-M.; Chen, S.-X.; Gao, B. MicroRNA-138 regulates chemoresistance in human non-small cell lung cancer via epithelial mesenchymal transition. *Eur. Rev. Med. Pharmacol. Sci.* **2016**, *20*, 1080–1086. [PubMed]

148. Nairismägi, M.-L.; Füchtbauer, A.; Labouriau, R.; Bramsen, J.B.; Füchtbauer, E.-M. The proto-oncogene TWIST1 is regulated by microRNAs. *PLoS ONE* **2013**, *8*, e66070. [CrossRef] [PubMed]

149. Ke, Y.; Zhao, W.; Xiong, J.; Cao, R. MiR-149 Inhibits Non-Small-Cell Lung Cancer Cells EMT by Targeting FOXM1. *Biochem. Res. Int.* **2013**, *2013*, 506731. [CrossRef] [PubMed]

150. Lin, X.; Yang, Z.; Zhang, P.; Liu, Y.; Shao, G. MiR-154 inhibits migration and invasion of human non-small cell lung cancer by targeting ZEB2. *Oncol. Lett.* **2016**, *12*, 301–306. [CrossRef] [PubMed]

151. Narita, M.; Shimura, E.; Nagasawa, A.; Aiuchi, T.; Suda, Y.; Hamada, Y.; Ikegami, D.; Iwasawa, C.; Arakawa, K.; Igarashi, K.; et al. Chronic treatment of non-small-cell lung cancer cells with gefitinib leads to an epigenetic loss of epithelial properties associated with reductions in microRNA-155 and -200c. *PLoS ONE* **2017**, *12*, e0172115. [CrossRef] [PubMed]

152. Chen, Q.; Jiao, D.; Wu, Y.; Chen, J.; Wang, J.; Tang, X.; Mou, H.; Hu, H.; Song, J.; Yan, J.; et al. MiR-206 inhibits HGF-induced epithelial-mesenchymal transition and angiogenesis in non-small cell lung cancer via c-Met/PI3k/Akt/mTOR pathway. *Oncotarget* **2016**, *7*, 18247–18261. [CrossRef] [PubMed]

153. Yamashita, R.; Sato, M.; Kakumu, T.; Hase, T.; Yogo, N.; Maruyama, E.; Sekido, Y.; Kondo, M.; Hasegawa, Y. Growth inhibitory effects of miR-221 and miR-222 in non-small cell lung cancer cells. *Cancer Med.* **2015**, *4*, 551–564. [CrossRef] [PubMed]

154. Wang, Y.; Xia, H.; Zhuang, Z.; Miao, L.; Chen, X.; Cai, H. Axl-altered microRNAs regulate tumorigenicity and gefitinib resistance in lung cancer. *Cell Death Dis.* **2014**, *5*, e1227. [CrossRef] [PubMed]

155. Zhang, Y.; Han, L.; Pang, J.; Wang, Y.; Feng, F.; Jiang, Q. Expression of microRNA-452 via adenoviral vector inhibits non-small cell lung cancer cells proliferation and metastasis. *Tumour Biol. J. Int. Soc. Oncodevelopmental Biol. Med.* **2016**, *37*, 8259–8270. [CrossRef] [PubMed]

156. Song, Q.; Xu, Y.; Yang, C.; Chen, Z.; Jia, C.; Chen, J.; Zhang, Y.; Lai, P.; Fan, X.; Zhou, X.; et al. MiR-483-5p promotes invasion and metastasis of lung adenocarcinoma by targeting RhoGDI1 and ALCAM. *Cancer Res.* **2014**, *74*, 3031–3042. [CrossRef] [PubMed]

157. Van Kampen, J.G.M.; van Hooij, O.; Jansen, C.F.; Smit, F.P.; van Noort, P.I.; Schultz, I.; Schaapveld, R.Q.J.; Schalken, J.A.; Verhaegh, G.W. MiRNA-520f Reverses Epithelial-to-Mesenchymal Transition by Targeting ADAM9 and TGFBR2. *Cancer Res.* **2017**, *77*, 2008–2017. [CrossRef] [PubMed]

158. Bi, M.; Chen, W.; Yu, H.; Wang, J.; Ding, F.; Tang, D.J.; Tang, C. MiR-543 is up-regulated in gefitinib-resistant non-small cell lung cancer and promotes cell proliferation and invasion via phosphatase and tensin homolog. *Biochem. Biophys. Res. Commun.* **2016**, *480*, 369–374. [CrossRef] [PubMed]

159. Mo, X.; Zhang, F.; Liang, H.; Liu, M.; Li, H.; Xia, H. MiR-544a promotes the invasion of lung cancer cells by targeting cadherina 1 in vitro. *OncoTargets Ther.* **2014**, *7*, 895–900. [CrossRef] [PubMed]

160. Chen, S.; Jiang, S.; Hu, F.; Xu, Y.; Wang, T.; Mei, Q. Foxk2 inhibits non-small cell lung cancer epithelial-mesenchymal transition and proliferation through the repression of different key target genes. *Oncol. Rep.* **2017**, *37*, 2335–2347. [CrossRef] [PubMed]

161. Kim, G.; An, H.-J.; Lee, M.-J.; Song, J.-Y.; Jeong, J.-Y.; Lee, J.-H.; Jeong, H.-C. Hsa-miR-1246 and hsa-miR-1290 are associated with stemness and invasiveness of non-small cell lung cancer. *Lung Cancer Amst. Neth.* **2016**, *91*, 15–22. [CrossRef] [PubMed]

162. Alam, M.; Ahmad, R.; Rajabi, H.; Kufe, D. MUC1-C Induces the LIN28B→LET-7→HMGA2 Axis to Regulate Self-Renewal in NSCLC. *Mol. Cancer Res.* **2015**, *13*, 449–460. [CrossRef] [PubMed]

163. Ahmad, A.; Maitah, M.Y.; Ginnebaugh, K.R.; Li, Y.; Bao, B.; Gadgeel, S.M.; Sarkar, F.H. Inhibition of Hedgehog signaling sensitizes NSCLC cells to standard therapies through modulation of EMT-regulating miRNAs. *J. Hematol. Oncol.* **2013**, *6*, 77. [CrossRef] [PubMed]

164. Ke, Y.; Zhao, W.; Xiong, J.; Cao, R. Downregulation of miR-16 promotes growth and motility by targeting HDGF in non-small cell lung cancer cells. *FEBS Lett.* **2013**, *587*, 3153–3157. [CrossRef] [PubMed]

165. Luo, F.; Xu, Y.; Ling, M.; Zhao, Y.; Xu, W.; Liang, X.; Jiang, R.; Wang, B.; Bian, Q.; Liu, Q. Arsenite evokes IL-6 secretion, autocrine regulation of STAT3 signaling, and miR-21 expression, processes involved in the EMT and malignant transformation of human bronchial epithelial cells. *Toxicol. Appl. Pharmacol.* **2013**, *273*, 27–34. [CrossRef] [PubMed]

166. Grant, J.L.; Fishbein, M.C.; Hong, L.-S.; Krysan, K.; Minna, J.D.; Shay, J.W.; Walser, T.C.; Dubinett, S.M. A novel molecular pathway for Snail-dependent, SPARC-mediated invasion in non-small cell lung cancer pathogenesis. *Cancer Prev. Res.* **2014**, *7*, 150–160. [CrossRef] [PubMed]

167. Yu, G.; Herazo-Maya, J.D.; Nukui, T.; Romkes, M.; Parwani, A.; Juan-Guardela, B.M.; Robertson, J.; Gauldie, J.; Siegfried, J.M.; Kaminski, N.; et al. Matrix metalloproteinase-19 promotes metastatic behavior in vitro and is associated with increased mortality in non-small cell lung cancer. *Am. J. Respir. Crit. Care Med.* **2014**, *190*, 780–790. [CrossRef] [PubMed]

168. Zhong, Z.; Xia, Y.; Wang, P.; Liu, B.; Chen, Y. Low expression of microRNA-30c promotes invasion by inducing epithelial mesenchymal transition in non-small cell lung cancer. *Mol. Med. Rep.* **2014**, *10*, 2575–2579. [CrossRef] [PubMed]

169. Meng, W.; Ye, Z.; Cui, R.; Perry, J.; Dedousi-Huebner, V.; Huebner, A.; Wang, Y.; Li, B.; Volinia, S.; Nakanishi, H.; et al. MicroRNA-31 predicts the presence of lymph node metastases and survival in patients with lung adenocarcinoma. *Clin. Cancer Res.* **2013**, *19*, 5423–5433. [CrossRef] [PubMed]

170. Yang, L.; Yang, J.; Li, J.; Shen, X.; Le, Y.; Zhou, C.; Wang, S.; Zhang, S.; Xu, D.; Gong, Z. MircoRNA-33a inhibits epithelial-to-mesenchymal transition and metastasis and could be a prognostic marker in non-small cell lung cancer. *Sci. Rep.* **2015**, *5*, 13677. [CrossRef] [PubMed]

171. Lei, L.; Huang, Y.; Gong, W. Inhibition of miR-92b suppresses nonsmall cell lung cancer cells growth and motility by targeting RECK. *Mol. Cell. Biochem.* **2014**, *387*, 171–176. [CrossRef] [PubMed]

172. Kundu, S.T.; Byers, L.A.; Peng, D.H.; Roybal, J.D.; Diao, L.; Wang, J.; Tong, P.; Creighton, C.J.; Gibbons, D.L. The miR-200 family and the miR-183~96~182 cluster target Foxf2 to inhibit invasion and metastasis in lung cancers. *Oncogene* **2016**, *35*, 173–186. [CrossRef] [PubMed]

173. Ma, T.; Zhao, Y.; Wei, K.; Yao, G.; Pan, C.; Liu, B.; Xia, Y.; He, Z.; Qi, X.; Li, Z.; et al. MicroRNA-124 Functions as a Tumor Suppressor by Regulating CDH2 and Epithelial-Mesenchymal Transition in Non-Small Cell Lung Cancer. *Cell. Physiol. Biochem. Int. J. Exp. Cell. Physiol. Biochem. Pharmacol.* **2016**, *38*, 1563–1574. [CrossRef] [PubMed]

174. Li, Z.; Wang, X.; Li, W.; Wu, L.; Chang, L.; Chen, H. MiRNA-124 modulates lung carcinoma cell migration and invasion. *Int. J. Clin. Pharmacol. Ther.* **2016**, *54*, 603–612. [CrossRef] [PubMed]

175. Shi, L.; Wang, Y.; Lu, Z.; Zhang, H.; Zhuang, N.; Wang, B.; Song, Z.; Chen, G.; Huang, C.; Xu, D.; et al. MiR-127 promotes EMT and stem-like traits in lung cancer through a feed-forward regulatory loop. *Oncogene* **2017**, *36*, 1631–1643. [CrossRef] [PubMed]

176. You, J.; Li, Y.; Fang, N.; Liu, B.; Zu, L.; Chang, R.; Li, X.; Zhou, Q. MiR-132 suppresses the migration and invasion of lung cancer cells via targeting the EMT regulator ZEB2. *PLoS ONE* **2014**, *9*, e91827. [CrossRef] [PubMed]

177. Shi, H.; Ji, Y.; Zhang, D.; Liu, Y.; Fang, P. MiR-135a inhibits migration and invasion and regulates EMT-related marker genes by targeting KLF8 in lung cancer cells. *Biochem. Biophys. Res. Commun.* **2015**, *465*, 125–130. [CrossRef] [PubMed]

178. Lin, C.-W.; Chang, Y.-L.; Chang, Y.-C.; Lin, J.-C.; Chen, C.-C.; Pan, S.-H.; Wu, C.-T.; Chen, H.-Y.; Yang, S.-C.; Hong, T.-M.; et al. MicroRNA-135b promotes lung cancer metastasis by regulating multiple targets in the Hippo pathway and LZTS1. *Nat. Commun.* **2013**, *4*, 1877. [CrossRef] [PubMed]

179. Ma, Q.; Jiang, Q.; Pu, Q.; Zhang, X.; Yang, W.; Wang, Y.; Ye, S.; Wu, S.; Zhong, G.; Ren, J.; et al. MicroRNA-143 inhibits migration and invasion of human non-small-cell lung cancer and its relative mechanism. *Int. J. Biol. Sci.* **2013**, *9*, 680–692. [CrossRef] [PubMed]

180. Hu, H.; Xu, Z.; Li, C.; Xu, C.; Lei, Z.; Zhang, H.-T.; Zhao, J. MiR-145 and miR-203 represses TGF-β-induced epithelial-mesenchymal transition and invasion by inhibiting SMAD3 in non-small cell lung cancer cells. *Lung Cancer Amst. Neth.* **2016**, *97*, 87–94. [CrossRef] [PubMed]

181. Park, D.H.; Jeon, H.S.; Lee, S.Y.; Choi, Y.Y.; Lee, H.W.; Yoon, S.; Lee, J.C.; Yoon, Y.S.; Kim, D.S.; Na, M.J.; et al. MicroRNA-146a inhibits epithelial mesenchymal transition in non-small cell lung cancer by targeting insulin receptor substrate 2. *Int. J. Oncol.* **2015**, *47*, 1545–1553. [CrossRef] [PubMed]

182. Li, J.; Song, Y.; Wang, Y.; Luo, J.; Yu, W. MicroRNA-148a suppresses epithelial-to-mesenchymal transition by targeting ROCK1 in non-small cell lung cancer cells. *Mol. Cell. Biochem.* **2013**, *380*, 277–282. [CrossRef] [PubMed]

183. Zhang, N.; Wei, X.; Xu, L. MiR-150 promotes the proliferation of lung cancer cells by targeting P53. *FEBS Lett.* **2013**, *587*, 2346–2351. [CrossRef] [PubMed]

184. Lin, T.-C.; Lin, P.-L.; Cheng, Y.-W.; Wu, T.-C.; Chou, M.-C.; Chen, C.-Y.; Lee, H. MicroRNA-184 Deregulated by the MicroRNA-21 Promotes Tumor Malignancy and Poor Outcomes in Non-small Cell Lung Cancer via Targeting CDC25A and c-Myc. *Ann. Surg. Oncol.* **2015**, *22* (Suppl. 3), S1532–S1539. [CrossRef] [PubMed]

185. Chen, J.; Gao, S.; Wang, C.; Wang, Z.; Zhang, H.; Huang, K.; Zhou, B.; Li, H.; Yu, Z.; Wu, J.; et al. Erratum to: Pathologically decreased expression of miR-193a contributes to metastasis by targeting WT1-E-cadherin axis in non-small cell lung cancers. *J. Exp. Clin. Cancer Res.* **2017**, *36*, 31. [CrossRef] [PubMed]

186. Yu, T.; Li, J.; Yan, M.; Liu, L.; Lin, H.; Zhao, F.; Sun, L.; Zhang, Y.; Cui, Y.; Zhang, F.; et al. MicroRNA-193a-3p and -5p suppress the metastasis of human non-small-cell lung cancer by downregulating the ERBB4/PIK3R3/mTOR/S6K2 signaling pathway. *Oncogene* **2015**, *34*, 413–423. [CrossRef] [PubMed]

187. Liu, X.; Lu, K.; Wang, K.; Sun, M.; Zhang, E.; Yang, J.; Yin, D.; Liu, Z.; Zhou, J.; Liu, Z.; et al. MicroRNA-196a promotes non-small cell lung cancer cell proliferation and invasion through targeting HOXA5. *BMC Cancer* **2012**, *12*, 348. [CrossRef] [PubMed]

188. Yu, S.-L.; Lee, D.C.; Sohn, H.A.; Lee, S.Y.; Jeon, H.S.; Lee, J.H.; Park, C.G.; Lee, H.Y.; Yeom, Y.I.; Son, J.W.; et al. Homeobox A9 directly targeted by miR-196b regulates aggressiveness through nuclear Factor-kappa B activity in non-small cell lung cancer cells. *Mol. Carcinog.* **2016**, *55*, 1915–1926. [CrossRef] [PubMed]

189. Chen, L.; Gibbons, D.L.; Goswami, S.; Cortez, M.A.; Ahn, Y.-H.; Byers, L.A.; Zhang, X.; Yi, X.; Dwyer, D.; Lin, W.; et al. Metastasis is regulated via microRNA-200/ZEB1 axis control of tumour cell PD-L1 expression and intratumoral immunosuppression. *Nat. Commun.* **2014**, *5*, 5241. [CrossRef] [PubMed]

190. Ceppi, P.; Mudduluru, G.; Kumarswamy, R.; Rapa, I.; Scagliotti, G.V.; Papotti, M.; Allgayer, H. Loss of miR-200c expression induces an aggressive, invasive, and chemoresistant phenotype in non-small cell lung cancer. *Mol. Cancer Res.* **2010**, *8*, 1207–1216. [CrossRef] [PubMed]

191. Zhao, Y.; Xu, Y.; Li, Y.; Xu, W.; Luo, F.; Wang, B.; Pang, Y.; Xiang, Q.; Zhou, J.; Wang, X.; et al. NF-κB-mediated inflammation leading to EMT via miR-200c is involved in cell transformation induced by cigarette smoke extract. *Toxicol. Sci. Off. J. Soc. Toxicol.* **2013**, *135*, 265–276. [CrossRef] [PubMed]

192. Zhao, J.; Zhao, Y.; Wang, Z.; Xuan, Y.; Luo, Y.; Jiao, W. Loss expression of micro ribonucleic acid (miRNA)-200c induces adverse post-surgical prognosis of advanced stage non-small cell lung carcinoma and its potential relationship with ETAR messenger RNA. *Thorac. Cancer* **2015**, *6*, 421–426. [CrossRef] [PubMed]

193. Tejero, R.; Navarro, A.; Campayo, M.; Viñolas, N.; Marrades, R.M.; Cordeiro, A.; Ruíz-Martínez, M.; Santasusagna, S.; Molins, L.; Ramirez, J.; et al. MiR-141 and miR-200c as markers of overall survival in early stage non-small cell lung cancer adenocarcinoma. *PLoS ONE* **2014**, *9*, e101899. [CrossRef] [PubMed]

194. Park, K.-S.; Raffeld, M.; Moon, Y.W.; Xi, L.; Bianco, C.; Pham, T.; Lee, L.C.; Mitsudomi, T.; Yatabe, Y.; Okamoto, I.; et al. CRIPTO1 expression in EGFR-mutant NSCLC elicits intrinsic EGFR-inhibitor resistance. *J. Clin. Invest.* **2014**, *124*, 3003–3015. [CrossRef] [PubMed]

195. Long, H.; Wang, Z.; Chen, J.; Xiang, T.; Li, Q.; Diao, X.; Zhu, B. MicroRNA-214 promotes epithelial-mesenchymal transition and metastasis in lung adenocarcinoma by targeting the suppressor-of-fused protein (Sufu). *Oncotarget* **2015**, *6*, 38705–38718. [CrossRef] [PubMed]

196. Li, Y.; Chen, P.; Zu, L.; Liu, B.; Wang, M.; Zhou, Q. Erratum: MicroRNA-338-3p suppresses metastasis of lung cancer cells by targeting the EMT regulator Sox4. *Am. J. Cancer Res.* **2016**, *6*, 1582. [PubMed]

197. Hou, X.W.; Sun, X.; Yu, Y.; Zhao, H.M.; Yang, Z.J.; Wang, X.; Cao, X.C. MiR-361-5p suppresses lung cancer cell lines progression by targeting FOXM1. *Neoplasma* **2017**, *64*. [CrossRef] [PubMed]

198. Nishikawa, E.; Osada, H.; Okazaki, Y.; Arima, C.; Tomida, S.; Tatematsu, Y.; Taguchi, A.; Shimada, Y.; Yanagisawa, K.; Yatabe, Y.; et al. MiR-375 is activated by ASH1 and inhibits YAP1 in a lineage-dependent manner in lung cancer. *Cancer Res.* **2011**, *71*, 6165–6173. [CrossRef] [PubMed]

199. Wang, R.; Wang, Z.-X.; Yang, J.-S.; Pan, X.; De, W.; Chen, L.-B. MicroRNA-451 functions as a tumor suppressor in human non-small cell lung cancer by targeting ras-related protein 14 (RAB14). *Oncogene* **2011**, *30*, 2644–2658. [CrossRef] [PubMed]

200. Xie, Z.; Cai, L.; Li, R.; Zheng, J.; Wu, H.; Yang, X.; Li, H.; Wang, Z. Down-regulation of miR-489 contributes into NSCLC cell invasion through targeting SUZ12. *Tumour Biol. J. Int. Soc. Oncodevelopmental Biol. Med.* **2015**, *36*, 6497–6505. [CrossRef] [PubMed]

201. Li, J.; Feng, Q.; Wei, X.; Yu, Y. MicroRNA-490 regulates lung cancer metastasis by targeting poly r(C)-binding protein 1. *Tumour Biol. J. Int. Soc. Oncodevelopmental Biol. Med.* **2016**, *37*, 15221–15228. [CrossRef] [PubMed]

202. Liu, C.; Lv, D.; Li, M.; Zhang, X.; Sun, G.; Bai, Y.; Chang, D. Hypermethylation of miRNA-589 promoter leads to upregulation of HDAC5 which promotes malignancy in non-small cell lung cancer. *Int. J. Oncol.* **2017**, *50*, 2079–2090. [CrossRef] [PubMed]

203. Lu, Y.-J.; Liu, R.-Y.; Hu, K.; Wang, Y. MiR-541-3p reverses cancer progression by directly targeting TGIF2 in non-small cell lung cancer. *Tumour Biol. J. Int. Soc. Oncodevelopmental Biol. Med.* **2016**, *37*, 12685–12695. [CrossRef] [PubMed]

204. Xia, Y.; Wu, Y.; Liu, B.; Wang, P.; Chen, Y. Downregulation of miR-638 promotes invasion and proliferation by regulating SOX2 and induces EMT in NSCLC. *FEBS Lett.* **2014**, *588*, 2238–2245. [CrossRef] [PubMed]

205. Behbahani, G.D.; Ghahhari, N.M.; Javidi, M.A.; Molan, A.F.; Feizi, N.; Babashah, S. MicroRNA-Mediated Post-Transcriptional Regulation of Epithelial to Mesenchymal Transition in Cancer. *Pathol. Oncol. Res. POR* **2017**, *23*, 1–12. [CrossRef] [PubMed]

206. Guo, F.; Parker Kerrigan, B.C.; Yang, D.; Hu, L.; Shmulevich, I.; Sood, A.K.; Xue, F.; Zhang, W. Post-transcriptional regulatory network of epithelial-to-mesenchymal and mesenchymal-to-epithelial transitions. *J. Hematol. Oncol. J. Hematol. Oncol.* **2014**, *7*, 19. [CrossRef] [PubMed]

207. Saitoh, M. Epithelial-mesenchymal transition is regulated at post-transcriptional levels by transforming growth factor-β signaling during tumor progression. *Cancer Sci.* **2015**, *106*, 481–488. [CrossRef] [PubMed]

208. Garg, M. Targeting microRNAs in epithelial-to-mesenchymal transition-induced cancer stem cells: Therapeutic approaches in cancer. *Expert Opin. Ther. Targets* **2015**, *19*, 285–297. [CrossRef] [PubMed]

209. Lin, C.-W.; Kao, S.-H.; Yang, P.-C. The miRNAs and epithelial-mesenchymal transition in cancers. *Curr. Pharm. Des.* **2014**, *20*, 5309–5318. [CrossRef] [PubMed]

210. Yan, J.; Gumireddy, K.; Li, A.; Huang, Q. Regulation of mesenchymal phenotype by MicroRNAs in cancer. *Curr. Cancer Drug Targets* **2013**, *13*, 930–934. [CrossRef] [PubMed]

211. Dacic, S.; Kelly, L.; Shuai, Y.; Nikiforova, M.N. MiRNA expression profiling of lung adenocarcinomas: Correlation with mutational status. *Mod. Pathol. Off. J. U. S. Can. Acad. Pathol. Inc* **2010**, *23*, 1577–1582. [CrossRef] [PubMed]

212. Takeyama, Y.; Sato, M.; Horio, M.; Hase, T.; Yoshida, K.; Yokoyama, T.; Nakashima, H.; Hashimoto, N.; Sekido, Y.; Gazdar, A.F.; et al. Knockdown of ZEB1, a master epithelial-to-mesenchymal transition (EMT) gene, suppresses anchorage-independent cell growth of lung cancer cells. *Cancer Lett.* **2010**, *296*, 216–224. [CrossRef] [PubMed]

213. Li, J.; Yang, S.; Yan, W.; Yang, J.; Qin, Y.-J.; Lin, X.-L.; Xie, R.-Y.; Wang, S.-C.; Jin, W.; Gao, F.; et al. MicroRNA-19 triggers epithelial-mesenchymal transition of lung cancer cells accompanied by growth inhibition. *Lab. Investig. J. Tech. Methods Pathol.* **2015**, *95*, 1056–1070. [CrossRef] [PubMed]

214. Perry, M.M.; Williams, A.E.; Tsitsiou, E.; Larner-Svensson, H.M.; Lindsay, M.A. Divergent intracellular pathways regulate interleukin-1beta-induced miR-146a and miR-146b expression and chemokine release in human alveolar epithelial cells. *FEBS Lett.* **2009**, *583*, 3349–3355. [CrossRef] [PubMed]

215. Marie-Egyptienne, D.T.; Lohse, I.; Hill, R.P. Cancer stem cells, the epithelial to mesenchymal transition (EMT) and radioresistance: Potential role of hypoxia. *Cancer Lett.* **2013**, *341*, 63–72. [CrossRef] [PubMed]

216. Hung, J.-J.; Yang, M.-H.; Hsu, H.-S.; Hsu, W.-H.; Liu, J.-S.; Wu, K.-J. Prognostic significance of hypoxia-inducible factor-1alpha, TWIST1 and Snail expression in resectable non-small cell lung cancer. *Thorax* **2009**, *64*, 1082–1089. [CrossRef] [PubMed]

217. Cannito, S.; Novo, E.; di Bonzo, L.V.; Busletta, C.; Colombatto, S.; Parola, M. Epithelial-mesenchymal transition: From molecular mechanisms, redox regulation to implications in human health and disease. *Antioxid. Redox Signal.* **2010**, *12*, 1383–1430. [CrossRef] [PubMed]

218. Giannoni, E.; Parri, M.; Chiarugi, P. EMT and oxidative stress: A bidirectional interplay affecting tumor malignancy. *Antioxid. Redox Signal.* **2012**, *16*, 1248–1263. [CrossRef] [PubMed]

219. Weber, B.; Stresemann, C.; Brueckner, B.; Lyko, F. Methylation of human microRNA genes in normal and neoplastic cells. *Cell Cycle Georget. Tex.* **2007**, *6*, 1001–1005. [CrossRef] [PubMed]

220. Xia, W.; Chen, Q.; Wang, J.; Mao, Q.; Dong, G.; Shi, R.; Zheng, Y.; Xu, L.; Jiang, F. DNA methylation mediated silencing of microRNA-145 is a potential prognostic marker in patients with lung adenocarcinoma. *Sci. Rep.* **2015**, *5*, 16901. [CrossRef] [PubMed]

221. Watanabe, K.; Amano, Y.; Ishikawa, R.; Sunohara, M.; Kage, H.; Ichinose, J.; Sano, A.; Nakajima, J.; Fukayama, M.; Yatomi, Y.; et al. Histone methylation-mediated silencing of miR-139 enhances invasion of non-small-cell lung cancer. *Cancer Med.* **2015**, *4*, 1573–1582. [CrossRef] [PubMed]

222. Díaz-López, A.; Díaz-Martín, J.; Moreno-Bueno, G.; Cuevas, E.P.; Santos, V.; Olmeda, D.; Portillo, F.; Palacios, J.; Cano, A. Zeb1 and Snail1 engage miR-200f transcriptional and epigenetic regulation during EMT. *Int. J. Cancer* **2015**, *136*, E62–E73. [CrossRef] [PubMed]

223. Chen, C.-Q.; Chen, C.-S.; Chen, J.-J.; Zhou, L.-P.; Xu, H.-L.; Jin, W.-W.; Wu, J.-B.; Gao, S.-M. Histone deacetylases inhibitor trichostatin A increases the expression of Dleu2/miR-15a/16-1 via HDAC3 in non-small cell lung cancer. *Mol. Cell. Biochem.* **2013**, *383*, 137–148. [CrossRef] [PubMed]

224. Incoronato, M.; Urso, L.; Portela, A.; Laukkanen, M.O.; Soini, Y.; Quintavalle, C.; Keller, S.; Esteller, M.; Condorelli, G. Epigenetic regulation of miR-212 expression in lung cancer. *PLoS ONE* **2011**, *6*, e27722. [CrossRef] [PubMed]

225. Huangyang, P.; Shang, Y. Epigenetic regulation of epithelial to mesenchymal transition. *Curr. Cancer Drug Targets* **2013**, *13*, 973–985. [CrossRef] [PubMed]

226. Mehrabian, M.; Ehsani, S.; Schmitt-Ulms, G. An emerging role of the cellular prion protein as a modulator of a morphogenetic program underlying epithelial-to-mesenchymal transition. *Front. Cell Dev. Biol.* **2014**, *2*, 53. [CrossRef] [PubMed]

227. Evseenko, D.; Zhu, Y.; Schenke-Layland, K.; Kuo, J.; Latour, B.; Ge, S.; Scholes, J.; Dravid, G.; Li, X.; MacLellan, W.R.; et al. Mapping the first stages of mesoderm commitment during differentiation of human embryonic stem cells. *Proc. Natl. Acad. Sci. USA* **2010**, *107*, 13742–13747. [CrossRef] [PubMed]

228. Pan, Y.; Zhao, L.; Liang, J.; Liu, J.; Shi, Y.; Liu, N.; Zhang, G.; Jin, H.; Gao, J.; Xie, H.; et al. Cellular prion protein promotes invasion and metastasis of gastric cancer. *FASEB J.* **2006**, *20*, 1886–1888. [CrossRef] [PubMed]

229. Mouillet-Richard, S.; Ermonval, M.; Chebassier, C.; Laplanche, J.L.; Lehmann, S.; Launay, J.M.; Kellermann, O. Signal transduction through prion protein. *Science* **2000**, *289*, 1925–1928. [CrossRef] [PubMed]

230. Martin, T.A.; Goyal, A.; Watkins, G.; Jiang, W.G. Expression of the transcription factors snail, slug, and twist and their clinical significance in human breast cancer. *Ann. Surg. Oncol.* **2005**, *12*, 488–496. [CrossRef] [PubMed]

231. Kwok, W.K.; Ling, M.-T.; Lee, T.-W.; Lau, T.C.M.; Zhou, C.; Zhang, X.; Chua, C.W.; Chan, K.W.; Chan, F.L.; Glackin, C.; et al. Up-regulation of TWIST in prostate cancer and its implication as a therapeutic target. *Cancer Res.* **2005**, *65*, 5153–5162. [CrossRef] [PubMed]

232. Hosono, S.; Kajiyama, H.; Terauchi, M.; Shibata, K.; Ino, K.; Nawa, A.; Kikkawa, F. Expression of Twist increases the risk for recurrence and for poor survival in epithelial ovarian carcinoma patients. *Br. J. Cancer* **2007**, *96*, 314–320. [CrossRef] [PubMed]

233. Yu, S.-L.; Chen, H.-Y.; Chang, G.-C.; Chen, C.-Y.; Chen, H.-W.; Singh, S.; Cheng, C.-L.; Yu, C.-J.; Lee, Y.-C.; Chen, H.-S.; et al. MicroRNA signature predicts survival and relapse in lung cancer. *Cancer Cell* **2008**, *13*, 48–57. [CrossRef] [PubMed]

234. Zhan, B.; Lu, D.; Luo, P.; Wang, B. Prognostic Value of Expression of MicroRNAs in Non-Small Cell Lung Cancer: A Systematic Review and Meta-Analysis. *Clin. Lab.* **2016**, *62*, 2203–2211. [CrossRef] [PubMed]

235. Shibue, T.; Weinberg, R.A. EMT, CSCs, and drug resistance: The mechanistic link and clinical implications. *Nat. Rev. Clin. Oncol.* **2017**. [CrossRef] [PubMed]

236. Heery, R.; Finn, S.P.; Cuffe, S.; Gray, S.G. Long non-coding RNAs: Key regulators of epithelial-mesenchymal transition, tumour drug resistance and cancer stem cells. *Cancers* **2017**, *9*, 38. [CrossRef] [PubMed]

237. Brozovic, A. The relationship between platinum drug resistance and epithelial-mesenchymal transition. *Arch. Toxicol.* **2017**, *91*, 605–619. [CrossRef] [PubMed]

238. Du, B.; Shim, J.S. Targeting epithelial-mesenchymal transition (EMT) to overcome drug resistance in cancer. *Molecules* **2016**, *21*, 965. [CrossRef] [PubMed]

239. Yang, A.D.; Fan, F.; Camp, E.R.; van Buren, G.; Liu, W.; Somcio, R.; Gray, M.J.; Cheng, H.; Hoff, P.M.; Ellis, L.M. Chronic oxaliplatin resistance induces epithelial-to-mesenchymal transition in colorectal cancer cell lines. *Clin. Cancer Res.* **2006**, *12*, 4147–4153. [CrossRef] [PubMed]

240. Shah, A.N.; Summy, J.M.; Zhang, J.; Park, S.I.; Parikh, N.U.; Gallick, G.E. Development and characterization of gemcitabine-resistant pancreatic tumor cells. *Ann. Surg. Oncol.* **2007**, *14*, 3629–3637. [CrossRef] [PubMed]

241. Hiscox, S.; Jiang, W.G.; Obermeier, K.; Taylor, K.; Morgan, L.; Burmi, R.; Barrow, D.; Nicholson, R.I. Tamoxifen resistance in MCF7 cells promotes EMT-like behaviour and involves modulation of beta-catenin phosphorylation. *Int. J. Cancer* **2006**, *118*, 290–301. [CrossRef] [PubMed]

242. Tsukamoto, H.; Shibata, K.; Kajiyama, H.; Terauchi, M.; Nawa, A.; Kikkawa, F. Irradiation-induced epithelial-mesenchymal transition (EMT) related to invasive potential in endometrial carcinoma cells. *Gynecol. Oncol.* **2007**, *107*, 500–504. [CrossRef] [PubMed]

243. Kurrey, N.K.; Jalgaonkar, S.P.; Joglekar, A.V.; Ghanate, A.D.; Chaskar, P.D.; Doiphode, R.Y.; Bapat, S.A. Snail and slug mediate radioresistance and chemoresistance by antagonizing p53-mediated apoptosis and acquiring a stem-like phenotype in ovarian cancer cells. *Stem Cells Dayt. Ohio* **2009**, *27*, 2059–2068. [CrossRef] [PubMed]

244. Barr, M.P.; Gray, S.G.; Hoffmann, A.C.; Hilger, R.A.; Thomale, J.; O'Flaherty, J.D.; Fennell, D.A.; Richard, D.; O'Leary, J.J.; O'Byrne, K.J. Generation and characterisation of cisplatin-resistant non-small cell lung cancer cell lines displaying a stem-like signature. *PLoS ONE* **2013**, *8*, e54193. [CrossRef] [PubMed]

245. Rho, J.K.; Choi, Y.J.; Lee, J.K.; Ryoo, B.-Y.; Na, I.I.; Yang, S.H.; Kim, C.H.; Lee, J.C. Epithelial to mesenchymal transition derived from repeated exposure to gefitinib determines the sensitivity to EGFR inhibitors in A549, a non-small cell lung cancer cell line. *Lung Cancer Amst. Neth.* **2009**, *63*, 219–226. [CrossRef] [PubMed]

246. Shen, W.; Pang, H.; Liu, J.; Zhou, J.; Zhang, F.; Liu, L.; Ma, N.; Zhang, N.; Zhang, H.; Liu, L. Epithelial-mesenchymal transition contributes to docetaxel resistance in human non-small cell lung cancer. *Oncol. Res.* **2014**, *22*, 47–55. [CrossRef] [PubMed]

247. Gao, W.; Lu, X.; Liu, L.; Xu, J.; Feng, D.; Shu, Y. MiRNA-21: A biomarker predictive for platinum-based adjuvant chemotherapy response in patients with non-small cell lung cancer. *Cancer Biol. Ther.* **2012**, *13*, 330–340. [CrossRef] [PubMed]

248. Diehn, M.; Cho, R.W.; Lobo, N.A.; Kalisky, T.; Dorie, M.J.; Kulp, A.N.; Qian, D.; Lam, J.S.; Ailles, L.E.; Wong, M.; et al. Association of reactive oxygen species levels and radioresistance in cancer stem cells. *Nature* **2009**, *458*, 780–783. [CrossRef] [PubMed]

249. Yauch, R.L.; Januario, T.; Eberhard, D.A.; Cavet, G.; Zhu, W.; Fu, L.; Pham, T.Q.; Soriano, R.; Stinson, J.; Seshagiri, S.; et al. Epithelial versus mesenchymal phenotype determines in vitro sensitivity and predicts clinical activity of erlotinib in lung cancer patients. *Clin. Cancer Res.* **2005**, *11*, 8686–8698. [CrossRef] [PubMed]

250. Thomson, S.; Buck, E.; Petti, F.; Griffin, G.; Brown, E.; Ramnarine, N.; Iwata, K.K.; Gibson, N.; Haley, J.D. Epithelial to mesenchymal transition is a determinant of sensitivity of non-small-cell lung carcinoma cell lines and xenografts to epidermal growth factor receptor inhibition. *Cancer Res.* **2005**, *65*, 9455–9462. [CrossRef] [PubMed]

251. Li, D.; Zhang, L.; Zhou, J.; Chen, H. Cigarette smoke extract exposure induces EGFR-TKI resistance in EGFR-mutated NSCLC via mediating Src activation and EMT. *Lung Cancer Amst. Neth.* **2016**, *93*, 35–42. [CrossRef] [PubMed]

252. Hashida, S.; Yamamoto, H.; Shien, K.; Miyoshi, Y.; Ohtsuka, T.; Suzawa, K.; Watanabe, M.; Maki, Y.; Soh, J.; Asano, H.; et al. Acquisition of cancer stem cell-like properties in non-small cell lung cancer with acquired resistance to afatinib. *Cancer Sci.* **2015**, *106*, 1377–1384. [CrossRef] [PubMed]

253. Sugano, T.; Seike, M.; Noro, R.; Soeno, C.; Chiba, M.; Zou, F.; Nakamichi, S.; Nishijima, N.; Matsumoto, M.; Miyanaga, A.; et al. Inhibition of ABCB1 Overcomes Cancer Stem Cell-like Properties and Acquired Resistance to MET Inhibitors in Non-Small Cell Lung Cancer. *Mol. Cancer Ther.* **2015**, *14*, 2433–2440. [CrossRef] [PubMed]

254. Zhu, X.; Du, X.; Deng, X.; Yi, H.; Cui, S.; Liu, W.; Shen, A.; Cui, Z. C6 ceramide sensitizes pemetrexed-induced apoptosis and cytotoxicity in osteosarcoma cells. *Biochem. Biophys. Res. Commun.* **2014**, *452*, 72–78. [CrossRef] [PubMed]

255. Brizuela, L.; Ader, I.; Mazerolles, C.; Bocquet, M.; Malavaud, B.; Cuvillier, O. First evidence of sphingosine 1-phosphate lyase protein expression and activity downregulation in human neoplasm: Implication for resistance to therapeutics in prostate cancer. *Mol. Cancer Ther.* **2012**, *11*, 1841–1851. [CrossRef] [PubMed]

256. Edmond, V.; Dufour, F.; Poiroux, G.; Shoji, K.; Malleter, M.; Fouqué, A.; Tauzin, S.; Rimokh, R.; Sergent, O.; Penna, A.; et al. Downregulation of ceramide synthase-6 during epithelial-to-mesenchymal transition reduces plasma membrane fluidity and cancer cell motility. *Oncogene* **2015**, *34*, 996–1005. [CrossRef] [PubMed]

257. Gomez-Casal, R.; Bhattacharya, C.; Ganesh, N.; Bailey, L.; Basse, P.; Gibson, M.; Epperly, M.; Levina, V. Non-small cell lung cancer cells survived ionizing radiation treatment display cancer stem cell and epithelial-mesenchymal transition phenotypes. *Mol. Cancer* **2013**, *12*, 94. [CrossRef] [PubMed]

258. Shien, K.; Toyooka, S.; Yamamoto, H.; Soh, J.; Jida, M.; Thu, K.L.; Hashida, S.; Maki, Y.; Ichihara, E.; Asano, H.; et al. Acquired resistance to EGFR inhibitors is associated with a manifestation of stem cell-like properties in cancer cells. *Cancer Res.* **2013**, *73*, 3051–3061. [CrossRef] [PubMed]

259. Akunuru, S.; James Zhai, Q.; Zheng, Y. Non-small cell lung cancer stem/progenitor cells are enriched in multiple distinct phenotypic subpopulations and exhibit plasticity. *Cell Death Dis.* **2012**, *3*, e352. [CrossRef] [PubMed]

260. Koren, A.; Rijavec, M.; Kern, I.; Sodja, E.; Korosec, P.; Cufer, T. BMI1, ALDH1A1, and CD133 Transcripts Connect Epithelial-Mesenchymal Transition to Cancer Stem Cells in Lung Carcinoma. *Stem Cells Int.* **2016**, *2016*, 9714315. [CrossRef] [PubMed]

261. Suresh, R.; Ali, S.; Ahmad, A.; Philip, P.A.; Sarkar, F.H. The Role of Cancer Stem Cells in Recurrent and Drug-Resistant Lung Cancer. *Adv. Exp. Med. Biol.* **2016**, *890*, 57–74. [CrossRef] [PubMed]

262. Cui, S.-Y.; Huang, J.-Y.; Chen, Y.-T.; Song, H.-Z.; Feng, B.; Huang, G.-C.; Wang, R.; Chen, L.-B.; De, W. Let-7c governs the acquisition of chemo- or radioresistance and epithelial-to-mesenchymal transition phenotypes in docetaxel-resistant lung adenocarcinoma. *Mol. Cancer Res.* **2013**, *11*, 699–713. [CrossRef] [PubMed]

263. Bryant, J.L.; Britson, J.; Balko, J.M.; Willian, M.; Timmons, R.; Frolov, A.; Black, E.P. A microRNA gene expression signature predicts response to erlotinib in epithelial cancer cell lines and targets EMT. *Br. J. Cancer* **2012**, *106*, 148–156. [CrossRef] [PubMed]

264. Sato, H.; Shien, K.; Tomida, S.; Okayasu, K.; Suzawa, K.; Hashida, S.; Torigoe, H.; Watanabe, M.; Yamamoto, H.; Soh, J.; et al. Targeting the miR-200c/LIN28B axis in acquired EGFR-TKI resistance non-small cell lung cancer cells harboring EMT features. *Sci. Rep.* **2017**, *7*, 40847. [CrossRef] [PubMed]

265. Byers, L.A.; Diao, L.; Wang, J.; Saintigny, P.; Girard, L.; Peyton, M.; Shen, L.; Fan, Y.; Giri, U.; Tumula, P.K.; et al. An epithelial-mesenchymal transition gene signature predicts resistance to EGFR and PI3K inhibitors and identifies Axl as a therapeutic target for overcoming EGFR inhibitor resistance. *Clin. Cancer Res.* **2013**, *19*, 279–290. [CrossRef] [PubMed]

266. Miow, Q.H.; Tan, T.Z.; Ye, J.; Lau, J.A.; Yokomizo, T.; Thiery, J.-P.; Mori, S. Epithelial-mesenchymal status renders differential responses to cisplatin in ovarian cancer. *Oncogene* **2015**, *34*, 1899–1907. [CrossRef] [PubMed]

267. Malek, R.; Wang, H.; Taparra, K.; Tran, P.T. Therapeutic Targeting of Epithelial Plasticity Programs: Focus on the Epithelial-Mesenchymal Transition. *Cells Tissues Organs* **2017**, *203*, 114–127. [CrossRef] [PubMed]

268. Reka, A.K.; Kuick, R.; Kurapati, H.; Standiford, T.J.; Omenn, G.S.; Keshamouni, V.G. Identifying inhibitors of epithelial-mesenchymal transition by connectivity map-based systems approach. *J. Thorac. Oncol. Off. Publ. Int. Assoc. Study Lung Cancer* **2011**, *6*, 1784–1792. [CrossRef] [PubMed]

269. Halder, S.K.; Beauchamp, R.D.; Datta, P.K. A specific inhibitor of TGF-beta receptor kinase, SB-431542, as a potent antitumor agent for human cancers. *Neoplasia* **2005**, *7*, 509–521. [CrossRef] [PubMed]

270. Nagaraj, N.S.; Datta, P.K. Targeting the transforming growth factor-beta signaling pathway in human cancer. *Expert Opin. Investig. Drugs* **2010**, *19*, 77–91. [CrossRef] [PubMed]

271. Park, C.-Y.; Kim, D.-K.; Sheen, Y.Y. EW-7203, a novel small molecule inhibitor of transforming growth factor-β (TGF-β) type I receptor/activin receptor-like kinase-5, blocks TGF-β1-mediated epithelial-to-mesenchymal transition in mammary epithelial cells. *Cancer Sci.* **2011**, *102*, 1889–1896. [CrossRef] [PubMed]

272. Park, C.-Y.; Son, J.-Y.; Jin, C.H.; Nam, J.-S.; Kim, D.-K.; Sheen, Y.Y. EW-7195, a novel inhibitor of ALK5 kinase inhibits EMT and breast cancer metastasis to lung. *Eur. J. Cancer Oxf. Engl. 1990* **2011**, *47*, 2642–2653. [CrossRef] [PubMed]

273. Son, J.Y.; Park, S.-Y.; Kim, S.-J.; Lee, S.J.; Park, S.-A.; Kim, M.-J.; Kim, S.W.; Kim, D.-K.; Nam, J.-S.; Sheen, Y.Y. EW-7197, a novel ALK-5 kinase inhibitor, potently inhibits breast to lung metastasis. *Mol. Cancer Ther.* **2014**, *13*, 1704–1716. [CrossRef] [PubMed]

274. Bueno, L.; de Alwis, D.P.; Pitou, C.; Yingling, J.; Lahn, M.; Glatt, S.; Trocóniz, I.F. Semi-mechanistic modelling of the tumour growth inhibitory effects of LY2157299, a new type I receptor TGF-beta kinase antagonist, in mice. *Eur. J. Cancer Oxf. Engl. 1990* **2008**, *44*, 142–150. [CrossRef]

275. Jang, M.J.; Baek, S.H.; Kim, J.H. UCH-L1 promotes cancer metastasis in prostate cancer cells through EMT induction. *Cancer Lett.* **2011**, *302*, 128–135. [CrossRef] [PubMed]

276. Goto, Y.; Zeng, L.; Yeom, C.J.; Zhu, Y.; Morinibu, A.; Shinomiya, K.; Kobayashi, M.; Hirota, K.; Itasaka, S.; Yoshimura, M.; et al. UCHL1 provides diagnostic and antimetastatic strategies due to its deubiquitinating effect on HIF-1α. *Nat. Commun.* **2015**, *6*, 6153. [CrossRef] [PubMed]

277. Sparano, J.A.; Bernardo, P.; Stephenson, P.; Gradishar, W.J.; Ingle, J.N.; Zucker, S.; Davidson, N.E. Randomized phase III trial of marimastat versus placebo in patients with metastatic breast cancer who have responding or stable disease after first-line chemotherapy: Eastern Cooperative Oncology Group trial E2196. *J. Clin. Oncol. Off. J. Am. Soc. Clin. Oncol.* **2004**, *22*, 4683–4690. [CrossRef] [PubMed]

278. Tran, P.T.; Shroff, E.H.; Burns, T.F.; Thiyagarajan, S.; Das, S.T.; Zabuawala, T.; Chen, J.; Cho, Y.-J.; Luong, R.; Tamayo, P.; et al. Twist1 suppresses senescence programs and thereby accelerates and maintains mutant Kras-induced lung tumorigenesis. *PLoS Genet.* **2012**, *8*, e1002650. [CrossRef] [PubMed]

279. Burns, T.F.; Dobromilskaya, I.; Murphy, S.C.; Gajula, R.P.; Thiyagarajan, S.; Chatley, S.N.H.; Aziz, K.; Cho, Y.-J.; Tran, P.T.; Rudin, C.M. Inhibition of TWIST1 leads to activation of oncogene-induced senescence in oncogene-driven non-small cell lung cancer. *Mol. Cancer Res.* **2013**, *11*, 329–338. [CrossRef] [PubMed]

280. Da Silva, S.D.; Alaoui-Jamali, M.A.; Soares, F.A.; Carraro, D.M.; Brentani, H.P.; Hier, M.; Rogatto, S.R.; Kowalski, L.P. TWIST1 is a molecular marker for a poor prognosis in oral cancer and represents a potential therapeutic target. *Cancer* **2014**, *120*, 352–362. [CrossRef] [PubMed]

281. Avasarala, S.; Van Scoyk, M.; Karuppusamy Rathinam, M.K.; Zerayesus, S.; Zhao, X.; Zhang, W.; Pergande, M.R.; Borgia, J.A.; DeGregori, J.; Port, J.D.; Winn, R.A.; et al. PRMT1 Is a Novel Regulator of Epithelial-Mesenchymal-Transition in Non-small Cell Lung Cancer. *J. Biol. Chem.* **2015**, *290*, 13479–13489. [CrossRef] [PubMed]

282. Li, H.; Wang, H.; Wang, F.; Gu, Q.; Xu, X. Snail involves in the transforming growth factor β1-mediated epithelial-mesenchymal transition of retinal pigment epithelial cells. *PLoS ONE* **2011**, *6*, e23322. [CrossRef] [PubMed]

283. Voulgari, A.; Pintzas, A. Epithelial-mesenchymal transition in cancer metastasis: Mechanisms, markers and strategies to overcome drug resistance in the clinic. *Biochim. Biophys. Acta* **2009**, *1796*, 75–90. [CrossRef] [PubMed]

284. Iwatsuki, M.; Mimori, K.; Yokobori, T.; Ishi, H.; Beppu, T.; Nakamori, S.; Baba, H.; Mori, M. Epithelial-mesenchymal transition in cancer development and its clinical significance. *Cancer Sci.* **2010**, *101*, 293–299. [CrossRef] [PubMed]

285. Pai, H.-C.; Chang, L.-H.; Peng, C.-Y.; Chang, Y.-L.; Chen, C.-C.; Shen, C.-C.; Teng, C.-M.; Pan, S.-L. Moscatilin inhibits migration and metastasis of human breast cancer MDA-MB-231 cells through inhibition of Akt and Twist signaling pathway. *J. Mol. Med. Berl. Ger.* **2013**, *91*, 347–356. [CrossRef] [PubMed]

286. Hsu, H.-Y.; Lin, T.-Y.; Hwang, P.-A.; Tseng, L.-M.; Chen, R.-H.; Tsao, S.-M.; Hsu, J. Fucoidan induces changes in the epithelial to mesenchymal transition and decreases metastasis by enhancing ubiquitin-dependent TGFβ receptor degradation in breast cancer. *Carcinogenesis* **2013**, *34*, 874–884. [CrossRef] [PubMed]

287. Chang, W.-W.; Hu, F.-W.; Yu, C.-C.; Wang, H.-H.; Feng, H.-P.; Lan, C.; Tsai, L.-L.; Chang, Y.-C. Quercetin in elimination of tumor initiating stem-like and mesenchymal transformation property in head and neck cancer. *Head Neck* **2013**, *35*, 413–419. [CrossRef] [PubMed]

288. Khan, M.A.; Tania, M.; Wei, C.; Mei, Z.; Fu, S.; Cheng, J.; Xu, J.; Fu, J. Thymoquinone inhibits cancer metastasis by downregulating TWIST1 expression to reduce epithelial to mesenchymal transition. *Oncotarget* **2015**, *6*, 19580–19591. [CrossRef] [PubMed]

289. Yang, W.-H.; Su, Y.-H.; Hsu, W.-H.; Wang, C.-C.; Arbiser, J.L.; Yang, M.-H. Imipramine blue halts head and neck cancer invasion through promoting F-box and leucine-rich repeat protein 14-mediated Twist1 degradation. *Oncogene* **2016**, *35*, 2287–2298. [CrossRef] [PubMed]

290. Lemjabbar-Alaoui, H.; McKinney, A.; Yang, Y.-W.; Tran, V.M.; Phillips, J.J. Glycosylation alterations in lung and brain cancer. *Adv. Cancer Res.* **2015**, *126*, 305–344. [CrossRef] [PubMed]

291. Mi, W.; Gu, Y.; Han, C.; Liu, H.; Fan, Q.; Zhang, X.; Cong, Q.; Yu, W. O-GlcNAcylation is a novel regulator of lung and colon cancer malignancy. *Biochim. Biophys. Acta* **2011**, *1812*, 514–519. [CrossRef] [PubMed]

292. Trapannone, R.; Rafie, K.; van Aalten, D.M.F. O-GlcNAc transferase inhibitors: Current tools and future challenges. *Biochem. Soc. Trans.* **2016**, *44*, 88–93. [CrossRef] [PubMed]

293. Ma, Z.; Vocadlo, D.J.; Vosseller, K. Hyper-O-GlcNAcylation is anti-apoptotic and maintains constitutive NF-κB activity in pancreatic cancer cells. *J. Biol. Chem.* **2013**, *288*, 15121–15130. [CrossRef] [PubMed]

294. Wagner, T.; Jung, M. New lysine methyltransferase drug targets in cancer. *Nat. Biotechnol.* **2012**, *30*, 622–623. [CrossRef] [PubMed]

295. Yuan, Y.; Wang, Q.; Paulk, J.; Kubicek, S.; Kemp, M.M.; Adams, D.J.; Shamji, A.F.; Wagner, B.K.; Schreiber, S.L. A small-molecule probe of the histone methyltransferase G9a induces cellular senescence in pancreatic adenocarcinoma. *ACS Chem. Biol.* **2012**, *7*, 1152–1157. [CrossRef] [PubMed]

296. Takai, N.; Desmond, J.C.; Kumagai, T.; Gui, D.; Said, J.W.; Whittaker, S.; Miyakawa, I.; Koeffler, H.P. Histone deacetylase inhibitors have a profound antigrowth activity in endometrial cancer cells. *Clin. Cancer Res.* **2004**, *10*, 1141–1149. [CrossRef] [PubMed]

297. Ma, X.; Ezzeldin, H.H.; Diasio, R.B. Histone deacetylase inhibitors: Current status and overview of recent clinical trials. *Drugs* **2009**, *69*, 1911–1934. [CrossRef] [PubMed]

298. Krützfeldt, J.; Rajewsky, N.; Braich, R.; Rajeev, K.G.; Tuschl, T.; Manoharan, M.; Stoffel, M. Silencing of microRNAs in vivo with "antagomirs". *Nature* **2005**, *438*, 685–689. [CrossRef] [PubMed]

299. Rupaimoole, R.; Han, H.-D.; Lopez-Berestein, G.; Sood, A.K. MicroRNA therapeutics: Principles, expectations, and challenges. *Chin. J. Cancer* **2011**, *30*, 368–370. [CrossRef] [PubMed]

300. Su, C.-M.; Lee, W.-H.; Wu, A.T.H.; Lin, Y.-K.; Wang, L.-S.; Wu, C.-H.; Yeh, C.-T. Pterostilbene inhibits triple-negative breast cancer metastasis via inducing microRNA-205 expression and negatively modulates epithelial-to-mesenchymal transition. *J. Nutr. Biochem.* **2015**, *26*, 675–685. [CrossRef] [PubMed]

301. Pecot, C.V.; Rupaimoole, R.; Yang, D.; Akbani, R.; Ivan, C.; Lu, C.; Wu, S.; Han, H.-D.; Shah, M.Y.; Rodriguez-Aguayo, C.; Bottsford-Miller, J.; et al. Tumour angiogenesis regulation by the miR-200 family. *Nat. Commun.* **2013**, *4*, 2427. [CrossRef] [PubMed]

302. Ocaña, O.H.; Córcoles, R.; Fabra, A.; Moreno-Bueno, G.; Acloque, H.; Vega, S.; Barrallo-Gimeno, A.; Cano, A.; Nieto, M.A. Metastatic colonization requires the repression of the epithelial-mesenchymal transition inducer Prrx1. *Cancer Cell* **2012**, *22*, 709–724. [CrossRef] [PubMed]

303. Van Roosbroeck, K.; Fanini, F.; Setoyama, T.; Ivan, C.; Rodriguez-Aguayo, C.; Fuentes-Mattei, E.; Xiao, L.; Vannini, I.; Redis, R.S.; D'Abundo, L.; et al. Combining Anti-Mir-155 with Chemotherapy for the Treatment of Lung Cancers. *Clin. Cancer Res.* **2017**, *23*, 2891–2904. [CrossRef] [PubMed]

304. Lou, Y.; Diao, L.; Cuentas, E.R.P.; Denning, W.L.; Chen, L.; Fan, Y.H.; Byers, L.A.; Wang, J.; Papadimitrakopoulou, V.A.; Behrens, C.; et al. Epithelial-mesenchymal transition is associated with a distinct tumor microenvironment including elevation of inflammatory signals and multiple immune checkpoints in lung adenocarcinoma. *Clin. Cancer Res.* **2016**, *22*, 3630–3642. [CrossRef] [PubMed]

305. Ardiani, A.; Gameiro, S.R.; Palena, C.; Hamilton, D.H.; Kwilas, A.; King, T.H.; Schlom, J.; Hodge, J.W. Vaccine-mediated immunotherapy directed against a transcription factor driving the metastatic process. *Cancer Res.* **2014**, *74*, 1945–1957. [CrossRef] [PubMed]

306. Hamilton, D.H.; Litzinger, M.T.; Jales, A.; Huang, B.; Fernando, R.I.; Hodge, J.W.; Ardiani, A.; Apelian, D.; Schlom, J.; Palena, C. Immunological targeting of tumor cells undergoing an epithelial-mesenchymal transition via a recombinant brachyury-yeast vaccine. *Oncotarget* **2013**, *4*, 1777–1790. [CrossRef] [PubMed]

307. Palena, C.; Hamilton, D.H. Immune targeting of tumor epithelial-mesenchymal transition via brachyury-based vaccines. *Adv. Cancer Res.* **2015**, *128*, 69–93. [CrossRef] [PubMed]

Review

EMT and Treatment Resistance in Pancreatic Cancer

Nicola Gaianigo [1], Davide Melisi [1,2] and Carmine Carbone [1,*]

[1]   Digestive Molecular Clinical Oncology Research Unit, Section of Medical Oncology, Department of Medicine, University of Verona, Verona 37134, Italy; nicola.gaianigo@studenti.univr.it (N.G.); davide.melisi@univr.it (D.M.)

[2]   Medical Oncology Unit, Azienda Ospedaliera Universitaria Integrata, Verona 37134, Italy

*   Correspondence: carmine.carbone@univr.it; Tel.: +39-045-812-8415

Academic Editor: Joëlle Roche

Received: 27 July 2017; Accepted: 10 September 2017; Published: 12 September 2017

Abstract: Pancreatic cancer (PC) is the third leading cause of adult cancer mortality in the United States. The poor prognosis for patients with PC is mainly due to its aggressive course, the limited efficacy of active systemic treatments, and a metastatic behavior, demonstrated throughout the evolution of the disease. On average, 80% of patients with PC are diagnosed with metastatic disease, and the half of those who undergo surgery and adjuvant therapy develop liver metastasis within two years. Metastatic dissemination is an early event in PC and is mainly attributed to an evolutionary biological process called epithelial-to-mesenchymal transition (EMT). This innate mechanism could have a dual role during embryonic growth and organ differentiation, and in cancer progression, cancer stem cell intravasation, and metastasis settlement. Many of the molecular pathways decisive in EMT progression have been already unraveled, but little is known about the causes behind the induction of this mechanism. EMT is one of the most distinctive and critical features of PC, occurring even in the very first stages of tumor development. This is known as pancreatic intraepithelial neoplasia (PanIN) and leads to early dissemination, drug resistance, and unfavorable prognosis and survival. The intention of this review is to shed new light on the critical role assumed by EMT during PC progression, with a particular focus on its role in PC resistance.

Keywords: pancreatic cancer; EMT; resistance

---

1. Introduction

According to the American Cancer Society, pancreatic cancer (PC) is ahead of breast cancer as the third leading cause of cancer-related death in the United States, and is predicted to become the second leading cause of cancer-related death by 2020 [1,2]. Currently, PC has a distinctive adverse prognosis, with an overall five-year survival rate of <6%. This is primarily due to late diagnosis, which is aggravated by the absence of early recognizable symptoms in patients and by the lack of effective diagnostic and prognostic markers [3]. In the last two decades, development of new therapeutic drugs has been disappointingly stagnant. Indeed, since the late 1990s gemcitabine has represented the standard of care for advanced PC, although it does not show a drastic improvement in median survival rate [4]. This is mainly explained by the unique chemoresistance of PC cells [5].

Although the histology and the genetic of pancreatic carcinogenesis have been well described [6], the molecular mechanisms that promote the metastatic spread of PC are less clear [7]. These mechanisms include the ability of cancer cells to break away from extracellular matrix (ECM) and to overcome apoptosis process. This behavior has been associated with an early epithelial-to-mesenchymal transition (EMT) in premalignant lesions [8].

EMT is a well-coordinated process triggered by many signaling pathways during embryonic development, however it is also a pathological feature in neoplasia and fibrosis [9]. Cells undergoing

EMT progressively lose the expression of components in the epithelial cell junctions. Instead, they produce a mesenchymal vimentin cytoskeleton and acquire both invasive and chemoresistance properties. Recent studies have also proposed that metastasis is an early event in the natural history of PC and could even precede tumor formation [10].

Thus, improving knowledge of molecular mechanisms that impair the response of cancer patients to chemotherapy is essential to designing more effective treatments for this deadly disease. In this review, we summarize the role of EMT in the context of drug resistance and metastasis in PC, with a special focus on inflammation.

2. EMT and Cancer Progression

EMT and the opposite process, mesenchymal-to-epithelial transition (MET), are innate and essential mechanisms implicated in cellular remodeling and tissue repair. The very first function of EMT occurs during gastrulation, when the blastomer differentiates into the three primordial cell lineages. EMT is constantly repeated out over the lifespan, from fetus development to tissue regeneration in adulthood [11]. It essentially relies on the plastic transformation of cells with an epithelial cobblestone phenotype, which are characterized by an apico-basal pattern of polarization and the establishment of tight junctions with the nearby cellular population, into cells showing distinctive mesenchymal features such as loss of three-dimensional organization in space, lack of cell polarity, and secretion of proteins constituting the backbone of extracellular matrix [12]. This same transition can be observed throughout the evolution of many tumors originating from epithelial cells; in fact, the progressive acquisition of gain and loss of mutations increases their chances of evading the solid tumor microenvironment, eventually resulting in metastasis formation and seeding in distant organs [13].

Many studies have determined which genes are regulated during the EMT process, comparing mesenchymal cancer cells and their epithelial counterparts [14,15]. Among these, downregulation of E-cadherin levels and activation of the transforming growth factor (TGF)-β-related signaling pathway are critical steps strictly required for EMT initiation and metastasis in PC [16,17] (Figure 1). This process involves the partial loss of cellular adhesive junctions, and progressive acquisition of typical mesenchymal markers, such as the neural-cadherin (N-cadherin or CH2) membrane protein (an event described as "cadherin switch"), as well as the expression of fibroblast-secreted extracellular proteins like type I collagene and some metalloproteases [18,19]. Notably, diverse kinase-dependent cellular signaling pathways—such as PI3K/AKT, EGFR, platelet-derived growth factor (PDGF), TAK1, and RAS—have been demonstrated to strongly affect E-cadherin subcellular localization and expression patterns [20,21]. Levels of E-cadherin are tightly regulated by the expression of its own suppressor (also called EMT-activating transcription factor, EMT-ATF), which are commonly subdivided into two main categories.

The first group is composed by SNAIL1, SNAIL2, ZEB1, ZEB2, E47, and KLF8 factors, which primarily act as repressors of *CDH1* promoter but also as down-regulators of genes implicated in maintaining cellular polarity, such as *LGL2*, *PATJ*, and *CRB3* [22,23].

The second group includes Twist, E2.2 and FoxC2 factors, which are responsible for *CDH1* transcription repression via indirect approaches and are principally induced under hypoxic conditions [24]. There is evidence to suggest that high levels of SNAIL and ZEB1 proteins are correlated with cancer disease relapse and short-term survival in many different typologies of cancer, highlighting how the EMT process might be one of the key reasons for dismal clinical outcomes in patients [25]. Interestingly, other EMT-driven mechanisms appear to strongly influence cancer progression at a deeper nuclear level. Biamonti et al. for instance, revealed how during the EMT process the CD44 transmembrane protein—a receptor for many proteins residing in the ECM including hyaluronic acid, osteoporin, collagene, and metalloproteases—undergoes selective alternative splicing, generating a mRNA isoform that differs from the "standard" epithelial variant [26]. To the same extent, the impact provided by re-modulation of the epigenetic landscape in cancer cells, both via methylation of target genes and histones modifications, has been intensely explored and classified as a crucial mechanism in EMT [27].

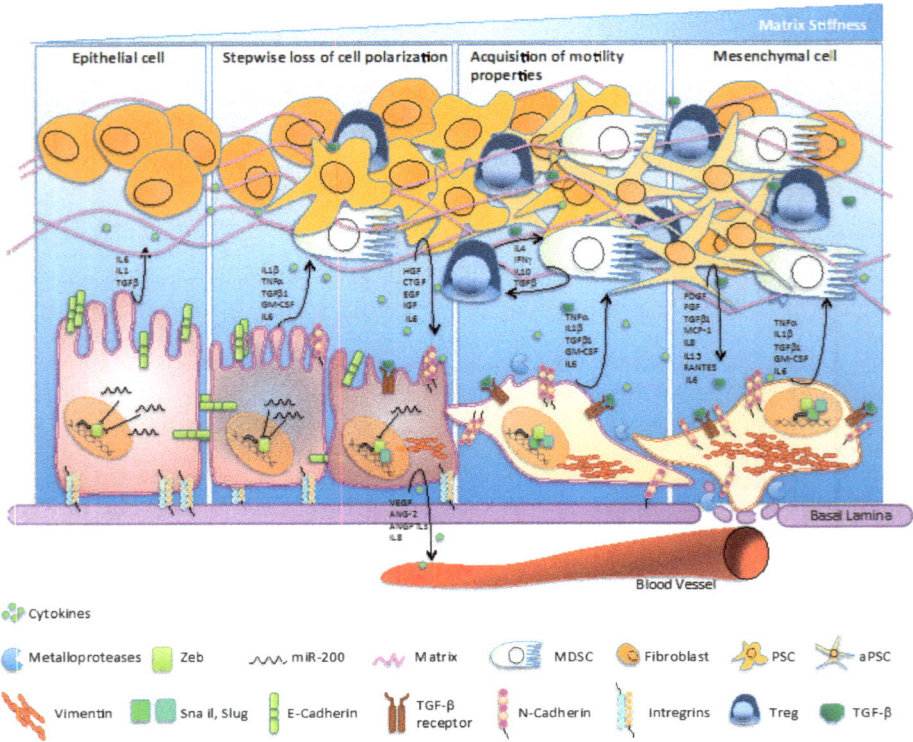

Figure 1. Molecular hallmarks/fluctuations/switching regulating the epithelial-to-mesenchymal transition (EMT) process in pancreatic cancer. The EMT process involves loss of cell polarization, a gain in migratory abilities and progressive acquisition of a mesenchymal phenotype. The EMT mechanism is characterized by the 'cadherin switch', where E-cadherin expression is progressively downregulated and replaced by the expression of N-cadherin. The transition process is associated to a decrease of miR-200 levels and an increase of classical E-cadherin transcriptional suppressors—such as ZEB1, Snail, and Slug—activated upstream by TGF-β. Cells undergoing EMT commonly quit the expression of extracellular matrix (ECM) elements mediating structural rigidity and cell adhesion in favor of proteases, cytokines, growth factors, and ECM components which improve cell migration and intravasation in bloodstream. Pancreatic cancer cell cytokines accelerate transformation of fibroblasts into quiescent pancreatic stellate cells (PSCs) and then into activated pancreatic stellate cells (aPSCs). Furthermore, inflammatory cytokines recruit myeloid progenitor cells and mediate their subsequent differentiation into myeloid-derived suppressive cells (MDSCs), which suppress the immune surveillance function. IL: interleukin; TGF: transforming growth factor; TNF: tumor necrosis factor; GM-CSF: granulocyte-macrophage colony-stimulating factor; HGF: hepatocyte growth factor; CTGF: connective tissue growth factor; EGF: epidermal growth factor; IFN: interferon; PDGF: platelet-derived growth factor; MCP-1: macrophage inflammatory protein 1; RANTES: regulated upon activation normally T-expressed and presumably secreted; VEGF: vascular endothelial growth factor; ANG-2: angiopoietin-2; ANGPTL: angiopoietin-like.

3. KRAS-Addiction of Pancreatic Cancer

Oncogenic KRAS plays a crucial role in the development of PC. Mutation of KRAS occurring in murine pancreas is sufficient to initiate acinar-to-ductal metaplasia (ADM) and pancreatic intraepithelial neoplasia (PanIN), which progress with long latency to invasive metastatic pancreatic ductal adenocarcinoma PDAC, thus recapitulating human disease [28,29].

Recently, a new mice model has been developed, called iKRAS, where pancreatic KRAS is placed under a tetracycline inducible promoter, providing a reliable and accurate model to trace the effects of KRAS contribution at different time points throughout cancer evolution [30]. Interestingly, expression of oncogenic KRAS in adult mice leads to the formation of PanIN lesions under circumstances of long latency and low penetrance, thus proving how a single mutation at KRAS oncogene is not sufficient to affect tissue organization and develop pancreatic neoplasia. At early checkpoints of PDAC evolution, KRAS-deregulated activity is strictly essential for PanIN progression, an effect that is observed also during late stage of tumor evolution, where neoplastic cells seem to undergo apoptosis upon KRAS oncogene inactivation [30].

These data support the idea of a KRAS addiction in which PC onset and evolution is essentially dependent on KRAS mutation. However, specific strategies aimed at KRAS targeting, such as disturbance of its membrane association, developing synthetic lethal interactions, and targeting of its downstream pathway or metabolic processes, showed few benefits in clinical practice [31].

Singh and colleagues, using an RNAi-based assay to deplete KRAS in a panel of KRAS-mutated PC cell lines, identified two classes of cells that do or do not require KRAS to maintain viability. The comparison between these two classes revealed a particular gene expression signature for KRAS-dependent cells, associated with a well-differentiated epithelial phenotype. They established that KRAS dependency is strongly linked to epithelial differentiation status, whereas most KRAS-independent cells appeared to assume a less epithelial phenotype [32].

It is widely accepted that poorly-differentiated tumors are more drug resistant and are associated with poor prognosis, highlighting a crucial role of the KRAS oncogene during the first steps of carcinogenesis of the independent KRAS cell lines [33].

Many fundamental cellular signaling cascades were investigated for their crucial involvement in tumor progression in KRAS-independent pancreatic cancer cell lines. Singh and colleagues hypothesized PI3K as the main driver of neoplastic phenotype in KRAS-independent cell lines [32]. Likewise, many studies supported the role of the nuclear transcription factor yes-associated protein 1 (YAP1) in compensating cancer cell survival and proliferation in KRAS-independent neoplasms, including PDAC [34,35]. The transcription factors YAP and TAZ, the main transducers of the Hippo pathway, have recently emerged for their association with pancreatic cancer ECM increasing stiffness [36]. YAP/TAZ are crucial downstream effectors of physical stimuli originating from the extra cellular environment surrounding tumor cells. ECM stiffness indeed promotes YAP/TAZ nuclear shuttling in cancer cells, leading to transcription of target genes regulating the EMT phenotype and chemoresistance [37].

4. Markers of EMT in Advanced Pancreatic Neoplasia

Roles and functions of EMT driving forces, described in the previous paragraph and normally upregulated in several cancers [38], were also studied in PC. In an immunohistochemistry analysis conducted by Hotz and colleagues in resected PC tissue samples, the expression of Snail and Slug factors accounted for a total of 80% and 50% respectively, whereas Twist showed little or no expression [39]. In a different study performed on 68 PC and 38 normal pancreas tissue samples, Twist nuclear levels were found to be decreased in tumor tissues, while staining of Slug or N-cadherin markers did not show a significant difference among healthy patients [40]. ZEB1 is considered one of the main inducers of EMT conversion in PC tissue in response to stimuli received from TGF-β and NF-κB pathways. Silencing of *ZEB1* in different PC cell lines resulted in a sensitive upregulation of epithelial markers and overall increased sensitivity of chemo-resistant cell lines to different chemotherapeutic agents, measured by an enhanced apoptotic response [41].

Loss of function mutations or downregulation of miRNA-200 family members, principally expressed by epithelial cells and identified as essential negative regulators of EMT and metastatic processes, were determined to indirectly stabilize ZEB1 expression levels and, in the meantime, to reduce E-cadherin expression in pancreatic β-cells, thus enhancing cells progression toward a mesenchymal phenotype [42]. The expression levels of miRNAs assume a tissue-specific pattern both in normal tissue and in PC and could function as potentially predictive diagnostic biomarkers. Downregulation of miR-148a and miR-217 and upregulation of miR-196a, miR-155, miR-203, miR-210, miR-222, and especially miR-21, were associated with a reduction of E-cadherin, an increase of vimentin levels, and with an overall poorer survival rate of patients [43,44]. Interestingly, miR-21 and miR-155 are both widely known as onco-mi-RNAs and their overexpression is linked to enhanced invasiveness, metastasis, and tumorigenesis onset in PC. This effect is thought to take action by compromising anti-inflammatory signaling pathways such as Ship1, or through the suppression of SOCS1 cytokine signaling [45] and the targeted modulation of peculiar tumor suppressor genes in tumor-supporting stromal cells [46].

PC is histologically recognized by an advanced fibrotic response, often referred as pancreatic desmoplasia, which is proved to confer the tumor microenvironment an increased resistance from many external stimuli including chemotherapeutic drugs, hormones, or cytotoxic mediators released by the host immune system [47]. Moreover, the discovery of a small subpopulation of cells with stem cell-like properties, termed cancer stem cells (CSCs), and residing within the tumor microenvironment, opens new possibilities for a more targeted therapy in several types of cancer. Pancreatic CSCs were initially isolated from a niche presenting the cell surface markers CD44 and epithelial-specific antigen (ESA). Many other markers of stemness—including CD133, CD24, c-Met, and CXCR4—were identified to selectively isolated PC stem cell (PCSC) population [48]. These special cells exhibit peculiar stem cell properties of self-renewal, including the ability to produce a differentiated progeny and an enhanced tumorigenic potential compared to control pancreatic cells [49]. Although the direct molecular correlation between EMT and CSC is still largely unknown, it is believed to rely on the similar pattern of activated signaling pathways, like Notch, Wnt/β-catenin, and the sonic hedgehog pathway (SHH), that are commonly shared in both.

5. EMT and Metastasis

Two main models of metastasis progression are hypothesized in the scientific community: a "Darwinian" linear model of evolution, which describes metastasis as a result of stepwise accumulation of genetic and epigenetic mutations that ultimately promote invasive behavior and dissemination during late tumor stage, and a parallel model of progression, where metastatic founder cells are believed to develop into circulating tumor cells (CTCs) and disseminate long before disease is clinically detected [50]. The early dissemination (parallel progression) model is dated back to the early 1950s and states that primary tumor and metastasis develop in parallel and acquire different genetic and epigenetic mutations throughout the progression. A recent mathematical modeling study, using the autopsy and the adjuvant cohort data from PC patient samples, predicted the presence of cells that are able to establish metastasis (not necessarily metastatic disease) even when the size of the primary tumor is still small [51,52]. More recently, in a pioneer study conducted by Rhim et al. [10], a sensitive method to tag and track pancreatic epithelial cells throughout cancer evolution was developed. Surprisingly, those cells were found to enter the bloodstream and seed in the liver even before the original cancer mass could be evidently detected through histologic analysis. In particular, they developed a Cre-Lox based mouse model of PC (called PKCY). In 8–10-week-old PKCY mice, only PanIN lesions were present, and 2.7% and 6.8% of PanIN 2 and 3 lesions, respectively, showed at least one YFP$^+$ ZEB1$^+$ cell on staining. Among these, they identified single YFP$^+$ cells that had crossed the basal lamina and had started to acquire a mesenchymal-like phenotype, making them almost identical to surrounding stromal cells. Surprisingly, cytofluorimetric analysis identified circulating pancreatic cells (CPCs) in the bloodstream of 8–10-week-old PKCY mice; moreover liver seeding of YFP$^+$ cells was detected in 4 out of 11 PanIN mice, although most of them were single cells located near blood vessels but with no evident expression of ZEB1. Overall, these data support a model for PC progression in which metastasis seeding in distant organs occurs before/in parallel to tumor development at the origin site. In another pilot study by Yu et al. [53], RNA sequencing analysis on circulating pancreatic tumor cells (CPCs) originating from genetically engineered mouse models of PC showed a pronounced enrichment for *WNT2* gene, a member of non-canonical WNT pathway, compared to primary tumor cells. Expression of *WNT2* in PC cells suppresses anoikis, enhances anchorage-independent growth of spheres, and increases metastatic propensity in vivo. Moreover, an upregulation for *WNT* genes was also confirmed in the CPCs directly isolated from PC patients. Conversely, Zheng et al. [54] proposed a study where two engineered (KrasLSL.G12D/+; p53R172H/+; PdxCretg/+) (KPC) mouse models were created, one carrying a deletion in Twist1 and the other in Snail1 genes. Unexpectedly, PC evolvement and metastasis formation were not prevented by the absence of the two principal genes involved in EMT progression; indeed, metastasis occurred both in lung and liver. Moreover, engineered mice developed PanIN lesions with the same frequency as normal KPC mice, despite exhibiting a significant loss in the EMT course. Hypothetically, this behavior might be due to a partial balance effect provided by other cellular factors involved in EMT. However, resistance to gemcitabine was significantly lower in the two mouse models. This is the first study claiming that EMT is a side process that is significant for cancer progression and metastasis evolution but not rate-limiting as universally believed [54].

6. Molecular Mechanisms of EMT and PC Treatment Resistance

Emerging evidence suggests a molecular and phenotypic association between increased chemoresistance and gaining of EMT-like phenotype of cancer cells. Different papers demonstrated that PC cells treated with an increasing regimen of chemotherapy drugs developed early resistance. A genome-wide array of the most differential expressed genes among the chemoresistant and chemosensitive PC cell lines indicated a distinctive connection with genes participating in the EMT process. RT-PCR analysis confirmed high levels of ZEB1 and vimentin in the chemo-resistant cells specifically, while target silencing of ZEB1 restored E-cadherin expression and led to an overall increased drug sensitivity [41]. The same effects were achieved when two pancreatic cell lines, L3.6pl and AsPC-1, were exposed to increasing concentration of gemcitabine until they developed an intrinsic resistance. These cells demonstrated a marked loss of cell-cell adhesion, formation of pseudopodia and enhanced mesenchymal-like morphology. Migration and invasion rates were also strongly enhanced in gemcitabine-resistant (GR) cells. Moreover, proteins such as β-catenin, E-cadherin, and vimentin, which are among the principal hallmarks of EMT, were found to radically relocalize into GR cells in a pattern consistent with the gaining of mesenchymal features [55]. Flow cytometry data proved also that the GR cell population carried increased cancer stem cell markers, including CD24, CD44, and ESA. Therefore, it has been hypothesized that the impressive resistance of PC to standard chemotherapy and radiation treatment could be owed to the presence of CSCs, which are documented to express high levels of multidrug-resistant membrane transporters, and abnormally activated signaling cascades promoting cell proliferation, migration and invasion.

Recently, the protein expression of the stem cell marker Nestin has been used to identify progenitor cells in several tissues and was investigated its association with tumor staging and metastasis formation in several cancers comprising PC [56]. Nestin has been linked with pancreas development and PC progression. Moreover, it was demonstrated that the activation of oncogenic KRAS in Nestin$^+$ cell lineage is sufficient to drive the initiation of PanIN lesions in pancreatic tissue [57]. TGF-β1 was proved to induce the expression of Nestin in PDAC cells, predominantly through the SMAD4 pathway. At the same time, overexpression of Nestin itself leads to an autocrine feedback loop triggering the production of TGF-β1 and its receptors both at RNA and protein levels [56]. Efforts were also dedicated in studying the expression pattern of the membrane receptor tyrosine kinase c-KIT (CD117) and its relative ligand, the stem cell factor (SCF), both in normal tissue and in advanced PC. c-KIT expression is normally restricted to the embryonic and fetal life of humans, and to date only few studies suggest its correlation with cancer progression and dissemination, although the data retrieved so far are partially conflicting. Immunohistochemistry tests performed on normal human adult pancreas and on pancreatic ductal adenocarcinoma, revealed that the expression of c-KIT protein is enhanced in the latter, probably due to mutations in its aminoacidic sequence, and seem to be initiated in β-cells of the islets of Langerhans, progressively spreading toward the cancerous ducts. Immunostaining of pancreatic metastasis in liver displayed localization of the c-KIT protein both in the cell membrane and within the cytosol, while its expression was considerably increased compared to the original PC tissue [58]. Wang et al. [59] confirmed that vimentin, α-smooth muscle actin (α-SMA), and ZEB1 are strongly upregulated in gemcitabine-resistant pancreatic cells compared to normal epithelial cells and demonstrated that Notch signaling is one of the leading pathways driving EMT process. RNA and protein levels for Notch pathway were shown to have an increased activation of Notch-2, Notch-4, and Jagged-1 in GR cells [59]. At the same time, expression of NF-κB, one of the principal downstream targets of the activated Notch pathway and the central mediator of EMT in cancer progression, was found to be considerably upregulated. SMAD4 protein is also altered or inactivated in the majority of PC cases, an event that is normally associated with a simultaneous inactivation of INK4A/ARF tumor suppressor gene and activation of KRAS oncogene. In vivo experiments conducted on genetically engineered mice carrying deletion on the *SMAD4* gene showed that the selective knock out of this signal transducer did not influence the physiology of pancreas organ. Interestingly, when combined with *KRAS* activating mutation, SMAD4 deficiency allowed a first rapid evolution of the early stages of pancreatic lesions toward mature PC, but this progression was not combined with further dissemination. *SMAD4*-deficient invasive tumors retained a differentiated ductal histopathology, confirmed by the expression of the E-cadherin epithelial marker [60].

The most recent and advanced targeted therapy showed a slight increment of efficacy in PC, with gemcitabine remaining the most largely used chemotherapeutic drug in PC. This is mainly due to the lack of specific druggable targets and to the substantial resistance that commonly arises along treatments [61].

Constitutive activation of NF-κB in PC represents the main intrinsic mechanisms of resistance due to suppression of apoptosis mechanisms. Experimental pre-clinical evidence supports the potential efficacy of anti-NF-κB strategies, but to date there are no direct NF-κB inhibitors available for patients [62]. However, the possibility of inhibiting NF-κB seems to be achieved indirectly by using inhibitors of specific mediators required for NF-κB activation [63]. In this regard, the use of a TAK1 inhibitor has successfully proven to overcome chemoresistance in PC [64], as well as esophageal cancer models [65].

Pancreatic cells expressing EMT markers also showed an increased resistance to the EGFR inhibitor [66]. The concomitant inhibition of NEH1, a molecular partner of EGFR, with erlotinib results in a decreased three-dimensional colony growth and invasion for both classical and mesenchymal PC cell lines [67].

Another successful approach is the use of monoclonal antibodies against EGFR, cetuximab, and panitumumab, which inhibit receptor dimerization at the extracellular domain and significantly increases radiosensitivity in locally advanced PC. However, the gemcitabine/erlotinib combination also results in an increased toxicities and higher relative cost compared to gemcitabine alone, which are important factors to discuss with patients when reviewing their therapeutic options [68]. In addition, current studies showed that an anti-vascular endothelial growth factor (VEGF) treatment could induce a significantly more aggressive and metastatic phenotype of tumor cells. However, the molecular mechanisms and mediators behind this phenomenon are still unrevealed.

It is increasingly clear that angiogenesis enhances metastatic potential and promotes progression, thus impairing the angiogenic potential of PC, through anti-VEGF therapy, could improve the efficacy of chemotherapy [69]. Recently, in a model of preneoplastic lesions, we showed that autocrine signaling of ANGPTL2 and its receptor, LILRB2, plays key roles in sustaining EMT and the early metastatic behavior of cells in two models of pancreatic preneoplastic lesions [70].

Recently, our group established the tumor cell-initiated mechanisms responsible for the resistance of PC to anti-VEGF treatment. We identified several proinflammatory factors that were expressed at higher levels in cells resistant to anti-VEGF treatment than in treatment-sensitive control cells. These proinflammatory factors acted in a paracrine manner recruiting CD11b$^+$ proangiogenic myeloid cells and thus inducing EMT [71]. More recently we demonstrated that combined inhibition of proinflammatory interleukin (IL) 1, CXCR1/2, and TGF-β signaling pathways might reverse this anti-VEGF resistance, reversing epithelial–mesenchymal transition and inhibiting CD11b$^+$ proangiogenic myeloid cells' tumor infiltration [72]. We integrate the secreted factor responsible of anti-VEGF resistance into the single transcription factor HOXB9. Since the silencing of HOXB9 modulates the anti-VEGF resistance in both pancreatic and colorectal cancer cells, we propose HOXB9 as a candidate predictive biomarker for selecting cancer patients for antiangiogenic therapy [73].

Extrinsic regulators of chemoresistance are represented by cellular components of the stroma, and the extracellular matrix that they produce. Abundant desmoplastic stroma may represent a barrier for chemotherapetic drugs delivering to the tumor. However, approaches to inhibiting the signaling pathways that regulate collagen secretion by fibroblast, such as SHH signaling inhibitors, have not reached the advanced stage of clinical trials so far [74].

7. Role of Desmoplasia in Pancreatic Cancer

PDAC is characterized by a prominent stromal-desmoplasmic reaction [75]. This desmoplasmic/fibrotic state is commonly promoted by the tumor microenvironment itself and is usually characterized by an abundant deposition of structural proteins, an altered organization, increased secretion of growth factors, and enhanced post-translational modifications (PTMs) of ECM proteins [76]. The PDAC stroma is a highly complex structure composed by several extracellular molecules, such as collagen (mainly type I), vitronectin, fibronectin, hyaluronic acid, molecular growth factors, and several specialized cells including endothelial cells (blood vessels), neural cells, cancer stem cells, immune cells, pancreatic stellate cells (PSCs), and cancer-associated fibroblasts (CAFs) [74,77].

Usually, cancer treatment failings are mainly attributed to the stroma enflaming chemoresistance, as well as decreasing microvascularity and, therefore, reduced drug delivery in the PDAC environment [78–80]. Stroma contribution to cancer advancement has been a fundamental subject of study in recent years, an issue that granted tumor-associated extracellular matrix to be listed as one of the main hallmark of cancer disease. This new field of research has shed new expectation for the development of novel therapeutic approaches to overcome tumor–ECM crosstalk and disease progression, however the molecular mechanisms regulating the fine balance between pancreatic carcinoma and desmoplasia are still largely unclear [47].

In normal pancreas, PSCs exist in a very low amount and in quiescent phase, but during carcinogenesis they slowly progress toward an activated phase, when they are referred as activated-PSCs (aPSCs). Similarly, CAFs can arise from PSCs, from residing activated fibroblasts or from close epithelial cells that underwent EMT. CAFs and aPSCs assume a myofibroblast-like phenotype and are mainly responsible for the pancreatic desmoplasmic reaction, through the expression and secretion of α-smooth muscle actin (α-SMA) and other proteins including fibroblast activation protein (FAP) and fibroblast-specific protein (FSP) [81].

Myofibroblasts specifically interact with cancer cells to create a tumor-promoting environment that stimulates resident tumor growth, treatment resistance, and metastasis formation. Their activation is dependent on several extracellular factors secreted by tumor itself, such as PDGF, TGF-β, TNFα, IL-1β and 6, cytokines that are in turn released by aPSCs and CAFs, triggering positive paracrine loops that sustain cancer progression [82,83]. Activation of myofibroblasts and infiltration of immune cells in tumor ECM environment dramatically increase matrix stiffness and stimulate cytoskeletal contractility of transformed epithelial cells, events that further contribute to amplify growth factor and cytokine signaling pathways. This stiffness increment promotes the formation of invadosomes, structures responsible for the production of metalloproteases required to demolish the ECM and to invade the basement membrane by driving focal adhesion assembly [84]. In a study conducted by Laklai et al. [85], using in vivo models they observed that PDAC showing impaired TGF-β activation signaling, increased β-catenin and YAP/TAZ nuclear localization and activity, developed a stiffer, fibrotic matrix associated with a more aggressive tumor and a poorer overall survival. Recapitulating, ECM stiffness directly fosters tumor malignancy and metastasis formation by promoting integrin-dependent matrix adhesion and invasion and regulating tumor plasticity [85].

The ECM is one of the main factors driving tumor resistance onset to many traditional chemotherapeutic treatments. For this reason, recently considerable clinical research has moved toward novel potential therapeutic approaches aiming to selectively targeting the tumor ECM environment rather than focusing solely on neoplastic cells. In particular, such trials have focused on targeting specific signaling molecules or remodeling enzymes, such as sonic hedgehog pathway (SHH), focal adhesion kinase (FAK) protein, and hyaluronidase inhibitors through neoadjuvant therapy [76,86]. The main outcome obtained in PDAC-ECM therapy however was the failure of SHH-targeted treatment, because despite the initial enthusiasm and success achieved by this therapy in preclinical studies, in in vivo SHH-deficient models, tumors showed a reduction in stroma content, but surprisingly, they also showed an increased vascularity, more invasive features, and a more aggressive phenotype [87].

8. Inflammation Sustains EMT in Pancreatic Cancer

PDAC is characterized by a large fibrotic tumor microenvironment, called desmoplasia, that hosts many different cell types including immune cells such as regulatory T cells (Tregs), tumor associated macrophages (TAMs), myeloid-derived suppressive cells (MDSCs), and natural killers (NKs) that support tumor growth, immune suppression and vascular growth (Figure 1). More specifically, PC exhibits a high abundance of regulatory $CD4^+$ $CD25^+$ $FoxP3^+$ [88] and cytotoxic $CD8^+$ T lymphocytes that linger near tumor boundaries. Despite the fact that both innate and adaptive immune responses are active against cancers, PC by itself induces local and systemic immune dysfunction in order to evade the recognition of cancer cells by activated immune effector cells [89].

Many of such infiltrated immune cells sustain mesenchymal transition of pancreatic cancer cells fueled by a local cytokine storm. For instance, TAMs have been described as potent EMT inducers through the secretion of several growth factors (hepatocyte growth factor (HGF), epidermal growth factor (EGF), TGF-β, PDGF, etc.) and inflammatory interleukins and cytokines (IL-1β, IL-6, and tumor necrosis factor (TNF)-α) that promote and support EMT. Interestingly, in vitro experiments demonstrated that the coculture of several PC cell lines with M2-polarized macrophage cells induced the acquisition of EMT properties in cancer cells [90].

MDSCs participate in the cytokine storm, secreting growth factors such as CCL2, TGF-β, and IL28, and inducing EMT in cancer cells [91]. Similarly, Th1 cells have been documented to activate innate immune cells, such as macrophages, and modulate the function of B cells and $CD8^+$ T cells through cytokine secretion and direct cell-cell signaling [88]. Helper T cells are further differentiated in two interchangeable subtypes, Th1 and Th2, with the former principally implicated in cell-mediated immune response by secreting INF-γ and IL2. The latter are predominantly recruited in humoral immune responses, although several studies support their pro tumor-tolerance activity [92]. The cytokine storm in the tumor microenvironment converges mainly on the constitutive activation of NF-κB pathways, implicated in inflammatory-driven EMT intensification, tumor proliferation, and cancer resistance. As a downstream effect, tumor cells express molecules that further supply the inflamed environment such as granulocyte-macrophage colony-stimulating factor (GM-CSF), IDO, IL8, and most importantly TGF-β, IL-17, PDL-1, and FASL, increasing the immunosuppressive state of the tumor environment [93].

In order to evade immuno-surveillance, PC cells express non-functional FAS receptors (members of the TNF receptor family) or increase the secretion of functional FAS ligands, which promote the apoptosis of surrounding cancer infiltrating macrophages, dendritic cells, T lymphocytes and NK cells [94]. Moreover, released TGF-β also exerts an inhibitory function on CD8$^+$ cytotoxic cells by blocking the expression of genes encoding cytolytic proteins, such as granzyme and perforin [95].

Thus, tailored immunotherapy is believed to represent the future of pancreatic cancer treatments and several clinical trials ongoing for pancreatic cancer disease involving the assessment of novel immune checkpoint inhibitors [96]. Immunotherapy efforts have recently focused in particular on three of the most important immune checkpoint participating in PDAC evolution: the cytotoxic T lymphocyte antigen 4 (CTLA-4), programmed cell death 1 (PD-1), and its ligand (PDL-1) [97,98].

CTLA-4 is an antigen expressed on Tregs and on activated CD4$^+$ CD8$^+$ T cells, and its activation modulates the suppression of immune responses by Tregs and effector T cells. Ipilimumab and Tremelimumab are two human therapeutic monoclonal antibodies developed to counteract CTLA-4 effects in cancer [99]. Unfortunately, CTLA-4 inhibition in pancreatic cancer through immunotherapy alone or in combination with standard chemotherapy has shown little or no efficacy so far, an effect probably due to the low number of immune cells residing in the PDAC microenvironment [100,101].

PD-1 is an immune checkpoint antigen expressed on activated dendritic cells (DCs), monocytes, NK cells, and T and B cells. PD-1 selectively binds to antigen-presenting cells (APCs) programmed death ligands PD-L1/2, and to tumor cell ligand PD-L1 [102]. Interaction brings an anergy of effector T cells through the inhibition of many T cells downstream kinases and a reduction of IL-2 and INF-γ secretion. Surprisingly, PDAC and CAFs progressively acquire a potent immune resistance toward effector T cells by the expression of PD-L1, an action that both increases Treg infiltration in tumor microenvironment and induces T-cells inhibition, apoptosis and clearance [103]. Among the drugs designed to hinder the PD-1/PD-L1 interaction currently tested in ongoing clinical trials, there are the humanized monoclonal antibodies nivolumab, pembrolizumab, and pidilizumab, all targeting the PD-1 receptor, and durvalumab, which targets PD-L1 [97,104].

Even in this case, immunotherapy has shown poor results in pancreatic cancer. The only exception is represented by the monoclonal antibody pembrolizumab, which has been recently licensed by the Food and Drug Adminostartion (FDA) in an unprecedented early case of approval, for patients with advanced solid adenocarcinoma, including PDAC, showing mismatch repair defects [105].

In a revolutionary phase II clinical trial, Le and colleagues, evaluated the efficacy of pembrolizumab in a cohort of 41 patients with advanced metastatic carcinoma, with or without mismatch repair defects. Specifically, they hypothesized that the mismatched repair-deficiency mechanisms (MMR) were responsible for the positive responses of patients treated with this mAb. This type of genome instability is known to harbor thousands of somatic mutations, and occurs only on a small fraction of advanced colorectal tumors, characterized also by a prominent lymphocyte infiltrate. The efficacy of pembrolizumab was confirmed in a cohort of both colorectal and non-colorectal cancer patients mutated for MMR, thus representing the first drug based on a predictive cancer marker rather than a tumor type [105].

9. Microbiota, EMT, and Treatment Resistance

In recent years, accumulating evidence indicates that a microbiome alteration influences metabolism, tissue development, inflammation, and immunity of several tumors [106]. The gut microbiota influences both local and systemic inflammation [107], raising the question of whether the microbiota affects inflammatory processes that contribute to cancer progression and its therapy. To date, only few epidemiological studies have examined the association between microbiota or oral microbiota and risk of pancreatic cancer; however, the results from these studies are still controversial [108]. Diverse microbiome alterations exist among several body sites, including pancreatic tissue [109]. However, the role of the microbiome in EMT induction in pancreatic cancer cells is plausible but not yet clarified. The innate immunity is exceeded by microbe infection leading to chronic inflammation and activation of signaling pathways involved in EMT. Thus, growth factors and microbes share common signaling pathways, suggesting that microbes may be considered as EMT inducers [110]. Indeed, the most intriguing results regarding pancreatic cancer treatment resistance and microbiota have been published recently.

Accumulating evidence suggests that specific microbiome and microbial dysbiosis can potentiate both hepatobiliary and pancreatic tumor development by damaging DNA, activating oncogenic signaling pathways, and producing tumor-promoting metabolites [111].

Emerging evidence from mice models suggests that oral administration of microbiota may influence the efficacy of cancer chemotherapies and novel targeted immunotherapies such as anti-CTLA4 and anti-CD274 therapies, improving the function of tumor-specific CD8[+] T cells [112,113].

Recently, Iida et al. demonstrated that optimal responses to cancer therapy require an intact commensal microbiota that mediates its effects by modulating myeloid-derived cell functions in the tumor microenvironment [114].

A recent randomized phase II POC study (IMAGE-1), combining heat-killed *Mycobacterium obuense* (IMM-101) with gemcitabine, suggests a beneficial effect on survival in patients with metastatic PDAC [115]. In a randomized study, Le and colleagues demonstrated that granulocyte-macrophage colony-stimulating factor (GM-CSF)-secreting allogeneic PDA cell lines (GVAX) followed by live-attenuated *Listeria monocytogenes*-expressing mesothelin, (CRS-207) significantly improved OS as compared with GVAX alone in patients with metastatic PDAC [116].

Novel functional analysis of patient-derived microbioma paired with preclinical models will enable the development of new types of anticancer therapy and could improve clinical intervention.

10. Conclusions

Increasing evidences suggest that EMT plays fundamental roles in cancer progression and resistance through several possible mechanisms, leading to a dramatic increase in disease aggressiveness, poorer disease elimination and overall patient survival.

Therefore, a thorough understanding of the underlying molecular features driving PC evolution is of utmost importance in order to develop effective therapies toward the original tumor, but also toward the population of cells responsible for drug resistance and metastasis formation. Moreover, the high inflammatory status and the complex network of immune cells recruited within advanced pancreatic tumor environment positively promote cancer progression and EMT transition of primary cancer cells rather than destroying malignant cells. Thus, new strategies targeting the EMT phenotype could increase sensitivity to both standard and targeted therapies and could improve the outcome of patients with PC.

Acknowledgments: This work was supported by the Basic Research Project 2015 through the University of Verona. Support was also provided by the Associazione Italiana per la Ricerca sul Cancro (AIRC) under Investigator Grant (IG) n°19111 to Davide Melisi. Partial support was also provided by the Nastro Viola Patient Association donations to Davide Melisi. Part of the work was performed at the Laboratorio Universitario di Ricerca Medica (LURM) Research Center, University of Verona.

Conflicts of Interest: The authors declare no conflict of interest.

References

1. Siegel, R.L.; Miller, K.D.; Jemal, A. Cancer statistics, 2016. *CA Cancer J. Clin.* **2016**, *66*, 7–30. [CrossRef] [PubMed]
2. Melisi, D.; Budillon, A. Pancreatic cancer: Between bench and bedside. *Curr. Drug Targets* **2012**, *13*, 729–730. [CrossRef] [PubMed]
3. Tamburrino, A.; Piro, G.; Carbone, C.; Tortora, G.; Melisi, D. Mechanisms of resistance to chemotherapeutic and anti-angiogenic drugs as novel targets for pancreatic cancer therapy. *Front. Pharmacol.* **2013**, *4*, 56. [CrossRef] [PubMed]

4.   Melisi, D.; Calvetti, L.; Frizziero, M.; Tortora, G. Pancreatic cancer: Systemic combination therapies for a heterogeneous disease. *Curr. Pharm. Des.* **2014**, *20*, 6660–6669. [CrossRef] [PubMed]

5.   Vaccaro, V.; Melisi, D.; Bria, E.; Cuppone, F.; Ciuffreda, L.; Pino, M.S.; Gelibter, A.; Tortora, G.; Cognetti, F.; Milella, M. Emerging pathways and future targets for the molecular therapy of pancreatic cancer. *Expert Opin. Ther. Targets* **2011**, *15*, 1183–1196. [CrossRef] [PubMed]

6.   Winter, J.M.; Maitra, A.; Yeo, C.J. Genetics and pathology of pancreatic cancer. *HPB (Oxford)* **2006**, *8*, 324–336. [CrossRef] [PubMed]

7.   Livak, K.J.; Schmittgen, T.D. Analysis of relative gene expression data using real-time quantitative pcr and the 2(-delta delta c(t)) method. *Methods* **2001**, *25*, 402–408. [CrossRef] [PubMed]

8.   Ying, H.; Dey, P.; Yao, W.; Kimmelman, A.C.; Draetta, G.F.; Maitra, A.; DePinho, R.A. Genetics and biology of pancreatic ductal adenocarcinoma. *Genes Dev.* **2016**, *30*, 355–385. [CrossRef] [PubMed]

9.   Nistico, P.; Bissell, M.J.; Radisky, D.C. Epithelial-mesenchymal transition: General principles and pathological relevance with special emphasis on the role of matrix metalloproteinases. *Cold Spring Harb Perspect. Biol.* **2012**, *4*. [CrossRef] [PubMed]

10.   Rhim, A.D.; Mirek, E.T.; Aiello, N.M.; Maitra, A.; Bailey, J.M.; McAllister, F.; Reichert, M.; Beatty, G.L.; Rustgi, A.K.; Vonderheide, R.H.; et al. Emt and dissemination precede pancreatic tumor formation. *Cell* **2012**, *148*, 349–361. [CrossRef] [PubMed]

11.   Kalluri, R.; Weinberg, R.A. The basics of epithelial-mesenchymal transition. *J. Clin. Investig.* **2009**, *119*, 1420–1428. [CrossRef] [PubMed]

12.   Thiery, J.P.; Acloque, H.; Huang, R.Y.; Nieto, M.A. Epithelial-mesenchymal transitions in development and disease. *Cell* **2009**, *139*, 871–890. [CrossRef] [PubMed]

13.   Vogelstein, B.; Papadopoulos, N.; Velculescu, V.E.; Zhou, S.; Diaz, L.A., Jr.; Kinzler, K.W. Cancer genome landscapes. *Science* **2013**, *339*, 1546–1558. [CrossRef] [PubMed]

14.   Groger, C.J.; Grubinger, M.; Waldhor, T.; Vierlinger, K.; Mikulits, W. Meta-analysis of gene expression signatures defining the epithelial to mesenchymal transition during cancer progression. *PLoS ONE* **2012**, *7*, e51136. [CrossRef] [PubMed]

15.   Liang, L.; Sun, H.; Zhang, W.; Zhang, M.; Yang, X.; Kuang, R.; Zheng, H. Meta-analysis of emt datasets reveals different types of emt. *PLoS ONE* **2016**, *11*, e0156839. [CrossRef] [PubMed]

16.   Du, L.; Yamamoto, S.; Burnette, B.L.; Huang, D.; Gao, K.; Jamshidi, N.; Kuo, M.D. Transcriptome profiling reveals novel gene expression signatures and regulating transcription factors of tgfbeta-induced epithelial-to-mesenchymal transition. *Cancer Med.* **2016**, *5*, 1962–1972. [CrossRef] [PubMed]

17.   Melisi, D.; Ishiyama, S.; Sclabas, G.M.; Fleming, J.B.; Xia, Q.; Tortora, G.; Abbruzzese, J.L.; Chiao, P.J. Ly2109761, a novel transforming growth factor beta receptor type I and type II dual inhibitor, as a therapeutic approach to suppressing pancreatic cancer metastasis. *Mol. Cancer Ther.* **2008**, *7*, 829–840. [CrossRef] [PubMed]

18.   Thiery, J.P.; Sleeman, J.P. Complex networks orchestrate epithelial-mesenchymal transitions. *Nat. Rev. Mol. Cell Biol.* **2006**, *7*, 131–142. [CrossRef] [PubMed]

19.   Nakajima, S.; Doi, R.; Toyoda, E.; Tsuji, S.; Wada, M.; Koizumi, M.; Tulachan, S.S.; Ito, D.; Kami, K.; Mori, T.; et al. N-cadherin expression and epithelial-mesenchymal transition in pancreatic carcinoma. *Clin. Cancer Res.* **2004**, *10*, 4125–4133. [CrossRef] [PubMed]

20.   Larue, L.; Bellacosa, A. Epithelial-mesenchymal transition in development and cancer: Role of phosphatidylinositol 3′ kinase/akt pathways. *Oncogene* **2005**, *24*, 7443–7454. [CrossRef] [PubMed]

21.   Lamouille, S.; Xu, J.; Derynck, R. Molecular mechanisms of epithelial-mesenchymal transition. *Nat. Rev. Mol. Cell Biol.* **2014**, *15*, 178–196. [CrossRef] [PubMed]

22.   Aigner, K.; Dampier, B.; Descovich, L.; Mikula, M.; Sultan, A.; Schreiber, M.; Mikulits, W.; Brabletz, T.; Strand, D.; Obrist, P.; et al. The transcription factor zeb1 (deltaef1) promotes tumour cell dedifferentiation by repressing master regulators of epithelial polarity. *Oncogene* **2007**, *26*, 6979–6988. [CrossRef] [PubMed]

23.   Barrallo-Gimeno, A.; Nieto, M.A. The snail genes as inducers of cell movement and survival: Implications in development and cancer. *Development* **2005**, *132*, 3151–3161. [CrossRef] [PubMed]

24.   Xu, J.; Lamouille, S.; Derynck, R. Tgf-beta-induced epithelial to mesenchymal transition. *Cell Res.* **2009**, *19*, 156–172. [CrossRef] [PubMed]

25.   Chiang, S.P.; Cabrera, R.M.; Segall, J.E. Tumor cell intravasation. *Am. J. Physiol. Cell Physiol.* **2016**, *311*, C1–C14. [CrossRef] [PubMed]

26. Biamonti, G.; Bonomi, S.; Gallo, S.; Ghigna, C. Making alternative splicing decisions during epithelial-to-mesenchymal transition (emt). *Cell. Mol. Life Sci.* **2012**, *69*, 2515–2526. [CrossRef] [PubMed]

27. Kiesslich, T.; Pichler, M.; Neureiter, D. Epigenetic control of epithelial-mesenchymal-transition in human cancer. *Mol. Clin. Oncol.* **2013**, *1*, 3–11. [CrossRef] [PubMed]

28. Hingorani, S.R.; Petricoin, E.F.; Maitra, A.; Rajapakse, V.; King, C.; Jacobetz, M.A.; Ross, S.; Conrads, T.P.; Veenstra, T.D.; Hitt, B.A.; et al. Preinvasive and invasive ductal pancreatic cancer and its early detection in the mouse. *Cancer Cell* **2003**, *4*, 437–450. [CrossRef]

29. Guerra, C.; Barbacid, M. Genetically engineered mouse models of pancreatic adenocarcinoma. *Mol. Oncol.* **2013**, *7*, 232–247. [CrossRef] [PubMed]

30. Chin, L.; Tam, A.; Pomerantz, J.; Wong, M.; Holash, J.; Bardeesy, N.; Shen, Q.; O'Hagan, R.; Pantginis, J.; Zhou, H.; et al. Essential role for oncogenic ras in tumour maintenance. *Nature* **1999**, *400*, 468–472. [CrossRef] [PubMed]

31. Eser, S.; Schnieke, A.; Schneider, G.; Saur, D. Oncogenic kras signalling in pancreatic cancer. *Br. J. Cancer* **2014**, *111*, 817–822. [CrossRef] [PubMed]

32. Singh, A.; Greninger, P.; Rhodes, D.; Koopman, L.; Violette, S.; Bardeesy, N.; Settleman, J. A gene expression signature associated with "k-ras addiction" reveals regulators of emt and tumor cell survival. *Cancer Cell* **2009**, *15*, 489–500. [CrossRef] [PubMed]

33. Shah, A.N.; Gallick, G.E. Src, chemoresistance and epithelial to mesenchymal transition: Are they related? *Anticancer Drugs* **2007**, *18*, 371–375. [CrossRef] [PubMed]

34. Corbo, V.; Ponz-Sarvise, M.; Tuveson, D.A. The ras and yap1 dance, who is leading? *EMBO J.* **2014**, *33*, 2437–2438. [CrossRef] [PubMed]

35. Dupont, S.; Morsut, L.; Aragona, M.; Enzo, E.; Giulitti, S.; Cordenonsi, M.; Zanconato, F.; Le Digabel, J.; Forcato, M.; Bicciato, S.; et al. Role of yap/taz in mechanotransduction. *Nature* **2011**, *474*, 179–183. [CrossRef] [PubMed]

36. Dupont, S. Role of yap/taz in cell-matrix adhesion-mediated signalling and mechanotransduction. *Exp. Cell Res.* **2016**, *343*, 42–53. [CrossRef] [PubMed]

37. Rice, A.J.; Cortes, E.; Lachowski, D.; Cheung, B.C.H.; Karim, S.A.; Morton, J.P.; Del Rio Hernandez, A. Matrix stiffness induces epithelial-mesenchymal transition and promotes chemoresistance in pancreatic cancer cells. *Oncogenesis* **2017**, *6*, e352. [CrossRef] [PubMed]

38. Piro, G.; Carbone, C.; Cataldo, I.; Di Nicolantonio, F.; Giacopuzzi, S.; Aprile, G.; Simionato, F.; Boschi, F.; Zanotto, M.; Mina, M.M.; et al. An fgfr3 autocrine loop sustains acquired resistance to trastuzumab in gastric cancer patients. *Clin. Cancer Res.* **2016**, *22*, 6164–6175. [CrossRef] [PubMed]

39. Hotz, B.; Arndt, M.; Dullat, S.; Bhargava, S.; Buhr, H.J.; Hotz, H.G. Epithelial to mesenchymal transition: Expression of the regulators snail, slug, and twist in pancreatic cancer. *Clin. Cancer Res.* **2007**, *13*, 4769–4776. [CrossRef] [PubMed]

40. Cates, J.M.; Byrd, R.H.; Fohn, L.E.; Tatsas, A.D.; Washington, M.K.; Black, C.C. Epithelial-mesenchymal transition markers in pancreatic ductal adenocarcinoma. *Pancreas* **2009**, *38*, e1–e6. [CrossRef] [PubMed]

41. Arumugam, T.; Ramachandran, V.; Fournier, K.F.; Wang, H.; Marquis, L.; Abbruzzese, J.L.; Gallick, G.E.; Logsdon, C.D.; McConkey, D.J.; Choi, W. Epithelial to mesenchymal transition contributes to drug resistance in pancreatic cancer. *Cancer Res.* **2009**, *69*, 5820–5828. [CrossRef] [PubMed]

42. Filios, S.R.; Xu, G.; Chen, J.; Hong, K.; Jing, G.; Shalev, A. MicroRNA-200 is induced by thioredoxin-interacting protein and regulates zeb1 protein signaling and beta cell apoptosis. *J. Biol. Chem.* **2014**, *289*, 36275–36283. [CrossRef] [PubMed]

43. Park, J.Y.; Helm, J.; Coppola, D.; Kim, D.; Malafa, M.; Kim, S.J. MicroRNAs in pancreatic ductal adenocarcinoma. *World J. Gastroenterol.* **2011**, *17*, 817–827. [CrossRef] [PubMed]

44. Xue, Y.; Abou Tayoun, A.N.; Abo, K.M.; Pipas, J.M.; Gordon, S.R.; Gardner, T.B.; Barth, R.J., Jr.; Suriawinata, A.A.; Tsongalis, G.J. MicroRNAs as diagnostic markers for pancreatic ductal adenocarcinoma and its precursor, pancreatic intraepithelial neoplasm. *Cancer Genet.* **2013**, *206*, 217–221. [CrossRef] [PubMed]

45. Calatayud, D.; Dehlendorff, C.; Boisen, M.K.; Hasselby, J.P.; Schultz, N.A.; Werner, J.; Immervoll, H.; Molven, A.; Hansen, C.P.; Johansen, J.S. Tissue microrna profiles as diagnostic and prognostic biomarkers in patients with resectable pancreatic ductal adenocarcinoma and periampullary cancers. *Biomark. Res.* **2017**, *5*, 8. [CrossRef] [PubMed]

46. Kadera, B.E.; Li, L.; Toste, P.A.; Wu, N.; Adams, C.; Dawson, D.W.; Donahue, T.R. MicroRNA-21 in pancreatic ductal adenocarcinoma tumor-associated fibroblasts promotes metastasis. *PLoS ONE* **2013**, *8*, e71978. [CrossRef] [PubMed]

47. Ansari, D.; Carvajo, M.; Bauden, M.; Andersson, R. Pancreatic cancer stroma: Controversies and current insights. *Scand. J. Gastroenterol.* **2017**, *52*, 641–646. [CrossRef] [PubMed]

48. Zhou, P.; Li, B.; Liu, F.; Zhang, M.; Wang, Q.; Liu, Y.; Yao, Y.; Li, D. The epithelial to mesenchymal transition (emt) and cancer stem cells: Implication for treatment resistance in pancreatic cancer. *Mol. Cancer* **2017**, *16*, 52. [CrossRef] [PubMed]

49. Dalla Pozza, E.; Dando, I.; Biondani, G.; Brandi, J.; Costanzo, C.; Zoratti, E.; Fassan, M.; Boschi, F.; Melisi, D.; Cecconi, D.; et al. Pancreatic ductal adenocarcinoma cell lines display a plastic ability to bidirectionally convert into cancer stem cells. *Int. J. Oncol.* **2015**, *46*, 1099–1108. [CrossRef] [PubMed]

50. Klein, C.A. Parallel progression of primary tumours and metastases. *Nat. Rev. Cancer* **2009**, *9*, 302–312. [CrossRef] [PubMed]

51. Haeno, H.; Gonen, M.; Davis, M.B.; Herman, J.M.; Iacobuzio-Donahue, C.A.; Michor, F. Computational modeling of pancreatic cancer reveals kinetics of metastasis suggesting optimum treatment strategies. *Cell* **2012**, *148*, 362–375. [CrossRef] [PubMed]

52. Das, S.; Batra, S.K. Pancreatic cancer metastasis: Are we being pre-emted? *Curr. Pharm. Des.* **2015**, *21*, 1249–1255. [CrossRef] [PubMed]

53. Yu, M.; Ting, D.T.; Stott, S.L.; Wittner, B.S.; Ozsolak, F.; Paul, S.; Ciciliano, J.C.; Smas, M.E.; Winokur, D.; Gilman, A.J.; et al. RNA sequencing of pancreatic circulating tumour cells implicates wnt signalling in metastasis. *Nature* **2012**, *487*, 510–513. [CrossRef] [PubMed]

54. Zheng, X.; Carstens, J.L.; Kim, J.; Scheible, M.; Kaye, J.; Sugimoto, H.; Wu, C.C.; LeBleu, V.S.; Kalluri, R. Epithelial-to-mesenchymal transition is dispensable for metastasis but induces chemoresistance in pancreatic cancer. *Nature* **2015**, *527*, 525–530. [CrossRef] [PubMed]

55. Shah, A.N.; Summy, J.M.; Zhang, J.; Park, S.I.; Parikh, N.U.; Gallick, G.E. Development and characterization of gemcitabine-resistant pancreatic tumor cells. *Ann. Surg. Oncol.* **2007**, *14*, 3629–3637. [CrossRef] [PubMed]

56. Su, H.T.; Weng, C.C.; Hsiao, P.J.; Chen, L.H.; Kuo, T.L.; Chen, Y.W.; Kuo, K.K.; Cheng, K.H. Stem cell marker nestin is critical for tgf-beta1-mediated tumor progression in pancreatic cancer. *Mol. Cancer Res.* **2013**, *11*, 768–779. [CrossRef] [PubMed]

57. Carriere, C.; Seeley, E.S.; Goetze, T.; Longnecker, D.S.; Korc, M. The nestin progenitor lineage is the compartment of origin for pancreatic intraepithelial neoplasia. *Proc. Natl. Acad. Sci. USA* **2007**, *104*, 4437–4442. [CrossRef] [PubMed]

58. Amsterdam, A.; Raanan, C.; Polin, N.; Melzer, E.; Givol, D.; Schreiber, L. Modulation of c-kit expression in pancreatic adenocarcinoma: A novel stem cell marker responsible for the progression of the disease. *Acta Histochem.* **2014**, *116*, 197–203. [CrossRef] [PubMed]

59. Wang, Z.; Li, Y.; Kong, D.; Banerjee, S.; Ahmad, A.; Azmi, A.S.; Ali, S.; Abbruzzese, J.L.; Gallick, G.E.; Sarkar, F.H. Acquisition of epithelial-mesenchymal transition phenotype of gemcitabine-resistant pancreatic cancer cells is linked with activation of the notch signaling pathway. *Cancer Res.* **2009**, *69*, 2400–2407. [CrossRef] [PubMed]

60. Bardeesy, N.; Cheng, K.H.; Berger, J.H.; Chu, G.C.; Pahler, J.; Olson, P.; Hezel, A.F.; Horner, J.; Lauwers, G.Y.; Hanahan, D.; et al. Smad4 is dispensable for normal pancreas development yet critical in progression and tumor biology of pancreas cancer. *Genes Dev.* **2006**, *20*, 3130–3146. [CrossRef] [PubMed]

61. Vaccaro, V.; Sperduti, I.; Vari, S.; Bria, E.; Melisi, D.; Garufi, C.; Nuzzo, C.; Scarpa, A.; Tortora, G.; Cognetti, F.; et al. Metastatic pancreatic cancer: Is there a light at the end of the tunnel? *World J. Gastroenterol.* **2015**, *21*, 4788–4801. [CrossRef] [PubMed]

62. Carbone, C.; Melisi, D. Nf-kappab as a target for pancreatic cancer therapy. *Expert Opin. Ther. Targets* **2012**, *16*, S1–S10. [CrossRef] [PubMed]

63. Melisi, D.; Chiao, P.J. Nf-kappa b as a target for cancer therapy. *Expert Opin. Ther. Targets* **2007**, *11*, 133–144. [CrossRef] [PubMed]

64. Melisi, D.; Xia, Q.; Paradiso, G.; Ling, J.; Moccia, T.; Carbone, C.; Budillon, A.; Abbruzzese, J.L.; Chiao, P.J. Modulation of pancreatic cancer chemoresistance by inhibition of TAK1. *J. Natl. Cancer Inst.* **2011**, *103*, 1190–1204. [CrossRef] [PubMed]

65. Piro, G.; Giacopuzzi, S.; Bencivenga, M.; Carbone, C.; Verlato, G.; Frizziero, M.; Zanotto, M.; Mina, M.M.; Merz, V.; Santoro, R.; et al. Tak1-regulated expression of BIRC3 predicts resistance to preoperative chemoradiotherapy in oesophageal adenocarcinoma patients. *Br. J. Cancer* **2015**, *113*, 878–885. [CrossRef] [PubMed]

66. Byers, L.A.; Diao, L.; Wang, J.; Saintigny, P.; Girard, L.; Peyton, M.; Shen, L.; Fan, Y.; Giri, U.; Tumula, P.K.; et al. An epithelial-mesenchymal transition gene signature predicts resistance to egfr and pi3k inhibitors and identifies axl as a therapeutic target for overcoming egfr inhibitor resistance. *Clin. Cancer Res.* **2013**, *19*, 279–290. [CrossRef] [PubMed]

67. Cardone, R.A.; Greco, M.R.; Zeeberg, K.; Zaccagnino, A.; Saccomano, M.; Bellizzi, A.; Bruns, P.; Menga, M.; Pilarsky, C.; Schwab, A.; et al. A novel nhe1-centered signaling cassette drives epidermal growth factor receptor-dependent pancreatic tumor metastasis and is a target for combination therapy. *Neoplasia* **2015**, *17*, 155–166. [CrossRef] [PubMed]

68. Philip, P.A. Targeted therapies for pancreatic cancer. *Gastrointest. Cancer Res.* **2008**, *2*, S16–S19. [PubMed]

69. Tortora, G.; Melisi, D.; Ciardiello, F. Angiogenesis: A target for cancer therapy. *Curr. Pharm. Des.* **2004**, *10*, 11–26. [CrossRef] [PubMed]

70. Carbone, C.; Piro, G.; Fassan, M.; Tamburrino, A.; Mina, M.M.; Zanotto, M.; Chiao, P.J.; Bassi, C.; Scarpa, A.; Tortora, G.; et al. An angiopoietin-like protein 2 autocrine signaling promotes emt during pancreatic ductal carcinogenesis. *Oncotarget* **2015**, *6*, 13822–13834. [CrossRef] [PubMed]

71. Carbone, C.; Moccia, T.; Zhu, C.; Paradiso, G.; Budillon, A.; Chiao, P.J.; Abbruzzese, J.L.; Melisi, D. Anti-vegf treatment-resistant pancreatic cancers secrete proinflammatory factors that contribute to malignant progression by inducing an emt cell phenotype. *Clin. Cancer Res.* **2011**, *17*, 5822–5832. [CrossRef] [PubMed]

72. Carbone, C.; Tamburrino, A.; Piro, G.; Boschi, F.; Cataldo, I.; Zanotto, M.; Mina, M.M.; Zanini, S.; Sparbati, A.; Scarpa, A.; et al. Combined inhibition of il1, cxcr1/2, and tgfbeta signaling pathways modulates in vivo resistance to anti-vegf treatment. *Anticancer Drugs* **2016**, *27*, 29–40. [CrossRef] [PubMed]

73. Carbone, C.; Piro, G.; Simionato, F.; Ligorio, F.; Cremolini, C.; Loupakis, F.; Ali, G.; Rossini, D.; Merz, V.; Santoro, R.; et al. Homeobox b9 mediates resistance to anti-vegf therapy in colorectal cancer patients. *Clin. Cancer Res.* **2017**, *15*, 4312–4322. [CrossRef] [PubMed]

74. Ohlund, D.; Elyada, E.; Tuveson, D. Fibroblast heterogeneity in the cancer wound. *J. Exp. Med.* **2014**, *211*, 1503–1523. [CrossRef] [PubMed]

75. Kong, X.; Li, L.; Li, Z.; Xie, K. Targeted destruction of the orchestration of the pancreatic stroma and tumor cells in pancreatic cancer cases: Molecular basis for therapeutic implications. *Cytokine Growth Factor Rev.* **2012**, *23*, 343–356. [CrossRef] [PubMed]

76. Pickup, M.W.; Mouw, J.K.; Weaver, V.M. The extracellular matrix modulates the hallmarks of cancer. *EMBO Rep.* **2014**, *15*, 1243–1253. [CrossRef] [PubMed]

77. Nielsen, M.F.; Mortensen, M.B.; Detlefsen, S. Key players in pancreatic cancer-stroma interaction: Cancer-associated fibroblasts, endothelial and inflammatory cells. *World J. Gastroenterol.* **2016**, *22*, 2678–2700. [CrossRef] [PubMed]

78. Jacobetz, M.A.; Chan, D.S.; Neesse, A.; Bapiro, T.E.; Cook, N.; Frese, K.K.; Feig, C.; Nakagawa, T.; Caldwell, M.E.; Zecchini, H.I.; et al. Hyaluronan impairs vascular function and drug delivery in a mouse model of pancreatic cancer. *Gut* **2013**, *62*, 112–120. [CrossRef] [PubMed]

79. Olive, K.P.; Jacobetz, M.A.; Davidson, C.J.; Gopinathan, A.; McIntyre, D.; Honess, D.; Madhu, B.; Goldgraben, M.A.; Caldwell, M.E.; Allard, D.; et al. Inhibition of hedgehog signaling enhances delivery of chemotherapy in a mouse model of pancreatic cancer. *Science* **2009**, *324*, 1457–1461. [CrossRef] [PubMed]

80. Provenzano, P.P.; Cuevas, C.; Chang, A.E.; Goel, V.K.; Von Hoff, D.D.; Hingorani, S.R. Enzymatic targeting of the stroma ablates physical barriers to treatment of pancreatic ductal adenocarcinoma. *Cancer Cell* **2012**, *21*, 418–429. [CrossRef] [PubMed]

81. Shiga, K.; Hara, M.; Nagasaki, T.; Sato, T.; Takahashi, H.; Takeyama, H. Cancer-associated fibroblasts: Their characteristics and their roles in tumor growth. *Cancers* **2015**, *7*, 2443–2458. [CrossRef] [PubMed]

82. Hwang, R.F.; Moore, T.; Arumugam, T.; Ramachandran, V.; Amos, K.D.; Rivera, A.; Ji, B.; Evans, D.B.; Logsdon, C.D. Cancer-associated stromal fibroblasts promote pancreatic tumor progression. *Cancer Res.* **2008**, *68*, 918–926. [CrossRef] [PubMed]

83. Vonlaufen, A.; Phillips, P.A.; Xu, Z.; Goldstein, D.; Pirola, R.C.; Wilson, J.S.; Apte, M.V. Pancreatic stellate cells and pancreatic cancer cells: An unholy alliance. *Cancer Res.* **2008**, *68*, 7707–7710. [CrossRef] [PubMed]

84. Chang, T.T.; Thakar, D.; Weaver, V.M. Force-dependent breaching of the basement membrane. *Matrix Biol.* **2017**, *57–58*, 178–189. [CrossRef] [PubMed]

85. Laklai, H.; Miroshnikova, Y.A.; Pickup, M.W.; Collisson, E.A.; Kim, G.E.; Barrett, A.S.; Hill, R.C.; Lakins, J.N.; Schlaepfer, D.D.; Mouw, J.K.; et al. Genotype tunes pancreatic ductal adenocarcinoma tissue tension to induce matricellular fibrosis and tumor progression. *Nat. Med.* **2016**, *22*, 497–505. [CrossRef] [PubMed]

86. Wang, Z.; Li, J.; Chen, X.; Duan, W.; Ma, Q.; Li, X. Disrupting the balance between tumor epithelia and stroma is a possible therapeutic approach for pancreatic cancer. *Med. Sci. Monit.* **2014**, *20*, 2002–2006. [PubMed]

87. Rhim, A.D.; Oberstein, P.E.; Thomas, D.H.; Mirek, E.T.; Palermo, C.F.; Sastra, S.A.; Dekleva, E.N.; Saunders, T.; Becerra, C.P.; Tattersall, I.W.; et al. Stromal elements act to restrain, rather than support, pancreatic ductal adenocarcinoma. *Cancer Cell* **2014**, *25*, 735–747. [CrossRef] [PubMed]

88. Liyanage, U.K.; Moore, T.T.; Joo, H.G.; Tanaka, Y.; Herrmann, V.; Doherty, G.; Drebin, J.A.; Strasberg, S.M.; Eberlein, T.J.; Goedegebuure, P.S.; et al. Prevalence of regulatory t cells is increased in peripheral blood and tumor microenvironment of patients with pancreas or breast adenocarcinoma. *J. Immunol.* **2002**, *169*, 2756–2761. [CrossRef] [PubMed]

89. Nummer, D.; Suri-Payer, E.; Schmitz-Winnenthal, H.; Bonertz, A.; Galindo, L.; Antolovich, D.; Koch, M.; Buchler, M.; Weitz, J.; Schirrmacher, V.; et al. Role of tumor endothelium in cd4+ cd25+ regulatory t cell infiltration of human pancreatic carcinoma. *J. Natl. Cancer Inst.* **2007**, *99*, 1188–1199. [CrossRef] [PubMed]

90. Liu, C.Y.; Xu, J.Y.; Shi, X.Y.; Huang, W.; Ruan, T.Y.; Xie, P.; Ding, J.L. M2-polarized tumor-associated macrophages promoted epithelial-mesenchymal transition in pancreatic cancer cells, partially through tlr4/il-10 signaling pathway. *Lab. Invest.* **2013**, *93*, 844–854. [CrossRef] [PubMed]

91. Toh, B.; Wang, X.; Keeble, J.; Sim, W.J.; Khoo, K.; Wong, W.C.; Kato, M.; Prevost-Blondel, A.; Thiery, J.P.; Abastado, J.P. Mesenchymal transition and dissemination of cancer cells is driven by myeloid-derived suppressor cells infiltrating the primary tumor. *PLoS Biol.* **2011**, *9*, e1001162. [CrossRef] [PubMed]

92. Chang, J.H.; Jiang, Y.; Pillarisetty, V.G. Role of immune cells in pancreatic cancer from bench to clinical application: An updated review. *Medicine* **2016**, *95*, e5541. [CrossRef] [PubMed]

93. Suarez-Carmona, M.; Lesage, J.; Cataldo, D.; Gilles, C. Emt and inflammation: Inseparable actors of cancer progression. *Mol. Oncol.* **2017**, *11*, 805–823. [CrossRef] [PubMed]

94. Von Bernstorff, W.; Spanjaard, R.A.; Chan, A.K.; Lockhart, D.C.; Sadanaga, N.; Wood, I.; Peiper, M.; Goedegebuure, P.S.; Eberlein, T.J. Pancreatic cancer cells can evade immune surveillance via nonfunctional fas (apo-1/cd95) receptors and aberrant expression of functional fas ligand. *Surgery* **1999**, *125*, 73–84. [CrossRef]

95. Trapani, J.A. The dual adverse effects of tgf-beta secretion on tumor progression. *Cancer Cell* **2005**, *8*, 349–350. [CrossRef] [PubMed]

96. Soares, K.C.; Rucki, A.A.; Wu, A.A.; Olino, K.; Xiao, Q.; Chai, Y.; Wamwea, A.; Bigelow, E.; Lutz, E.; Liu, L.; et al. Pd-1/pd-l1 blockade together with vaccine therapy facilitates effector t-cell infiltration into pancreatic tumors. *J. Immunother.* **2015**, *38*, 1–11. [CrossRef] [PubMed]

97. Johansson, H.; Andersson, R.; Bauden, M.; Hammes, S.; Holdenrieder, S.; Ansari, D. Immune checkpoint therapy for pancreatic cancer. *World J. Gastroenterol.* **2016**, *22*, 9457–9476. [CrossRef] [PubMed]

98. Fokas, E.; O'Neill, E.; Gordon-Weeks, A.; Mukherjee, S.; McKenna, W.G.; Muschel, R.J. Pancreatic ductal adenocarcinoma: From genetics to biology to radiobiology to oncoimmunology and all the way back to the clinic. *Biochim. Biophys. Acta* **2015**, *1855*, 61–82. [CrossRef] [PubMed]

99. Peggs, K.S.; Quezada, S.A.; Chambers, C.A.; Korman, A.J.; Allison, J.P. Blockade of ctla-4 on both effector and regulatory t cell compartments contributes to the antitumor activity of anti-ctla-4 antibodies. *J. Exp. Med.* **2009**, *206*, 1717–1725. [CrossRef] [PubMed]

100. Royal, R.E.; Levy, C.; Turner, K.; Mathur, A.; Hughes, M.; Kammula, U.S.; Sherry, R.M.; Topalian, S.L.; Yang, J.C.; Lowy, I.; et al. Phase 2 trial of single agent ipilimumab (anti-ctla-4) for locally advanced or metastatic pancreatic adenocarcinoma. *J. Immunother.* **2010**, *33*, 828–833. [CrossRef] [PubMed]

101. Myint, Z.W.; Goel, G. Role of modern immunotherapy in gastrointestinal malignancies: A review of current clinical progress. *J. Hematol. Oncol.* **2017**, *10*, 86. [CrossRef] [PubMed]

102. Francisco, L.M.; Salinas, V.H.; Brown, K.E.; Vanguri, V.K.; Freeman, G.J.; Kuchroo, V.K.; Sharpe, A.H. Pd-l1 regulates the development, maintenance, and function of induced regulatory t cells. *J. Exp. Med.* **2009**, *206*, 3015–3029. [CrossRef] [PubMed]

103. Iwai, Y.; Ishida, M.; Tanaka, Y.; Okazaki, T.; Honjo, T.; Minato, N. Involvement of pd-l1 on tumor cells in the escape from host immune system and tumor immunotherapy by pd-l1 blockade. *Proc. Natl. Acad. Sci. USA* **2002**, *99*, 12293–12297. [CrossRef] [PubMed]
104. Iwai, Y.; Hamanishi, J.; Chamoto, K.; Honjo, T. Cancer immunotherapies targeting the pd-1 signaling pathway. *J. Biomed. Sci.* **2017**, *24*, 26. [CrossRef] [PubMed]
105. Le, D.T.; Uram, J.N.; Wang, H.; Bartlett, B.R.; Kemberling, H.; Eyring, A.D.; Skora, A.D.; Luber, B.S.; Azad, N.S.; Laheru, D.; et al. Pd-1 blockade in tumors with mismatch-repair deficiency. *N. Engl. J. Med.* **2015**, *372*, 2509–2520. [CrossRef] [PubMed]
106. Tilg, H.; Kaser, A. Gut microbiome, obesity, and metabolic dysfunction. *J. Clin. Investig.* **2011**, *121*, 2126–2132. [CrossRef] [PubMed]
107. Belkaid, Y.; Hand, T.W. Role of the microbiota in immunity and inflammation. *Cell* **2014**, *157*, 121–141. [CrossRef] [PubMed]
108. Michaud, D.S.; Izard, J. Microbiota, oral microbiome, and pancreatic cancer. *Cancer J.* **2014**, *20*, 203–206. [CrossRef] [PubMed]
109. Ertz-Archambault, N.; Keim, P.; Von Hoff, D. Microbiome and pancreatic cancer: A comprehensive topic review of literature. *World J. Gastroenterol.* **2017**, *23*, 1899–1908. [CrossRef] [PubMed]
110. Said, N.A.; Williams, E.D. Growth factors in induction of epithelial-mesenchymal transition and metastasis. *Cells Tissues Organs* **2011**, *193*, 85–97. [CrossRef] [PubMed]
111. Mima, K.; Nakagawa, S.; Sawayama, H.; Ishimoto, T.; Imai, K.; Iwatsuki, M.; Hashimoto, D ; Baba, Y.; Yamashita, Y.I.; Yoshida, N.; et al. The microbiome and hepatobiliary-pancreatic cancers. *Cancer Lett.* **2017**, *402*, 9–15. [CrossRef] [PubMed]
112. Vetizou, M.; Pitt, J.M.; Daillere, R.; Lepage, P.; Waldschmitt, N.; Flament, C.; Rusakiewicz, S.; Routy, B.; Roberti, M.P.; Duong, C.P.; et al. Anticancer immunotherapy by ctla-4 blockade relies on the gut microbiota. *Science* **2015**, *350*, 1079–1084. [CrossRef] [PubMed]
113. Sivan, A.; Corrales, L.; Hubert, N.; Williams, J.B.; Aquino-Michaels, K.; Earley, Z.M.; Benyamin, F.W.; Lei, Y.M.; Jabri, B.; Alegre, M.L.; et al. Commensal bifidobacterium promotes antitumor immunity and facilitates anti-pd-l1 efficacy. *Science* **2015**, *350*, 1084–1089. [CrossRef] [PubMed]
114. Iida, N.; Dzutsev, A.; Stewart, C.A.; Smith, L.; Bouladoux, N.; Weingarten, R.A.; Molina, D.A.; Salcedo, R.; Back, T.; Cramer, S.; et al. Commensal bacteria control cancer response to therapy by modulating the tumor microenvironment. *Science* **2013**, *342*, 967–970. [CrossRef] [PubMed]
115. Dalgleish, A.G.; Stebbing, J.; Adamson, D.J.; Arif, S.S.; Bidoli, P.; Chang, D.; Cheeseman, S.; Diaz-Beveridge, R.; Fernandez-Martos, C.; Glynne-Jones, R.; et al. Randomised, open-label, phase ii study of gemcitabine with and without imm-101 for advanced pancreatic cancer. *Br. J. Cancer* **2016**, *115*, 789–796. [CrossRef] [PubMed]
116. Le, D.T.; Wang-Gillam, A.; Picozzi, V.; Greten, T.F.; Crocenzi, T.; Springett, G.; Morse, M.; Zeh, H.; Cohen, D.; Fine, R.L.; et al. Safety and survival with gvax pancreas prime and listeria monocytogenes-expressing mesothelin (crs-207) boost vaccines for metastatic pancreatic cancer. *J. Clin. Oncol.* **2015**, *33*, 1325–1333. [CrossRef] [PubMed]

cancers

MDPI

Review

Regulation of EMT in Colorectal Cancer: A Culprit in Metastasis

Trung Vu [1] and Pran K. Datta [1,2,*]

[1]   Division of Hematology and Oncology, Department of Medicine, Comprehensive Cancer Center, University of Alabama at Birmingham, Birmingham, AL 35233, USA; ttvu@uab.edu
[2]   Birmingham Veterans Affairs Medical Center, Birmingham, AL 35233, USA
*   Correspondence: prandatta@uabmc.edu; Tel.: +1-205-975-6039; Fax: +1-205-934-9573

Received: 9 November 2017; Accepted: 5 December 2017; Published: 16 December 2017

Abstract: Epithelial to mesenchymal transition (EMT) is a process during which cells lose their epithelial characteristics, for instance cell polarity and cell–cell contact, and gain mesenchymal properties, such as increased motility. In colorectal cancer (CRC), EMT is associated with an invasive or metastatic phenotype. In this review, we discuss recent studies exploring novel regulation mechanisms of EMT in CRC, including the identification of new CRC EMT regulators. Upregulation of inducers can promote EMT, leading to increased invasiveness and metastasis in CRC. These inducers can downregulate E-cadherin and upregulate N-cadherin and vimentin (VIM) through modulating EMT-related signaling pathways, for instance WNT/β-catenin and TGF-β, and EMT transcription factors, such as zinc finger E-box binding homeobox 1 (ZEB1) and ZEB2. In addition, several microRNAs (miRNAs), including members of the miR-34 and miR-200 families, are found to target mRNAs of EMT-transcription factors, for example ZEB1, ZEB2, or SNAIL. Downregulation of these miRNAs is associated with distant metastasis and advanced stage tumors. Furthermore, the role of EMT in circulating tumor cells (CTCs) is also discussed. Mesenchymal markers on the surface of EMT CTCs were found to be associated with metastasis and could serve as potential biomarkers for metastasis. Altogether, these studies indicate that EMT is orchestrated by a complicated network, involving regulators of different signaling pathways. Further studies are required to understand the mechanisms underlying EMT in CRC.

Keywords: EMT; colorectal cancer; metastasis

1. Introduction

Colorectal cancer (CRC) remains the second leading cause of cancer death in the United States. More than 50% of CRC patients will develop liver metastases during their lifespan [1]. Despite the development of treatment regimens, there is no effective therapy for advanced CRC with metastasis. Although inherited genetic susceptibility has an important role in a subset of CRC cases, the vast majority of CRC cases are sporadic and non-inherited [2]. Sporadic pre-neoplastic lesions gradually accumulate genetic and epigenetic modifications that allow uncontrolled proliferation and cell survival, followed by invasive and metastatic properties typical of colorectal carcinoma [3]. Understanding the molecular mechanisms underlying these transitions, especially how colorectal cancer cells acquire invasive and metastatic properties, is important for the development of optimized strategies to treat CRC.

Epithelial to mesenchymal transition (EMT) is a complicated cellular process, during which epithelial cells acquire a mesenchymal phenotype. EMT has been categorized into three types based on the physiological tissue context: (1) embryonic development and organ formation [4]; (2) wound healing and organ fibrosis [5]; and (3) cancer progression [6]. Type 3 EMT is associated with invasive or metastatic phenotype [7]. In the past decade, increasing numbers of studies have provided strong evidence for the vital role that EMT plays in cancer progression and metastasis in many types of malignancies including CRC [8]. During EMT, tumor cells undergo tight junction dissolution, disruption of apical–basal polarity, and reorganization of the cytoskeletal architecture, which enable cells to develop an invasive phenotype. In cancer cells, EMT is abnormally regulated by extracellular stimuli derived from the tumor microenvironment, including growth factors and inflammatory cytokines, along with intra-tumoral physical stresses such as hypoxia [9]. Therefore, EMT programming allows tumor cells to adapt to the constant changes of the human tumor microenvironment, and thus to successfully metastasize.

In this review, we will describe current knowledge about the mechanisms and regulators of EMT in CRC. These regulators are aberrantly expressed in human CRC and their expression correlates with induction of EMT and cancer metastasis. We will also discuss the importance of studying the role of EMT in metastasis, and how EMT can be considered as an applicable target for therapeutic strategies, especially for patients with CRC metastasis.

2. Regulation of EMT in Colorectal Cancer

The process of EMT requires the cooperation, in a timely manner, of a complex network, consisting of molecular signaling pathways and regulators. These factors are categorized into three groups: the effector molecules which execute the EMT program (EMT effectors), the transcription factors which orchestrate the EMT program (EMT core regulators), and the extracellular cues which activate the EMT program (EMT inducers) [10].

2.1. EMT Effectors

A majority of EMT effectors are subcellular structure proteins that define the epithelial or mesenchymal phenotype of a cell. EMT is characterized by the downregulation of genes encoding for epithelial cell junction proteins (E-cadherin, claudins, and occludins) and the activation of genes, of which the protein products (vimentin, fibronectin and N-cadherin) promote mesenchymal adhesion. Among them, a key feature of EMT is the downregulation of E-cadherin, leading to the destabilization of adherens junctions. In response to a variety of induction signals, E-cadherin is subjected to various levels of regulation, including transcriptional repression [11], promoter methylation [12], and protein phosphorylation and degradation [13]. Loss of E-cadherin expression is associated with a poor prognosis in stage III CRC [14].

Additional key EMT effector molecules are proteins that promote cell migration and invasion during EMT. Fibronectin, an extracellular protein required for mesenchymal cell migration, is frequently induced upon EMT activation. High levels of fibronectin expression are associated with poor prognosis in CRC [15]. Furthermore, knockdown of fibronectin suppresses CRC cell proliferation via the NF-κB/p53-apoptosis signaling pathway and arrested cells in the S phase [16]. Overexpression of N-cadherin, another mesenchymal marker, is correlated with metastasis and worse survival of CRC patients [17].

2.2. EMT Core Regulators-Transcription Factors Driving EMT

The regulation of EMT requires robust transcriptional machinery, consisting largely of developmental transcription factors which regulate epithelial and mesenchymal markers in a coordinated manner. There are three major groups of EMT-activating transcription factors: the SNAIL family of zinc-finger transcription factors SNAIL/SLUG, the zinc finger E-box binding homeobox (ZEB) family of transcription factors ZEB1/ZEB2, and the TWIST family of basic helix-loop-helix (bHLH) transcription factors TWIST1/TWIST2. The expression of these three groups is activated early in EMT, and thus they have central roles in development and cancer. The roles of these transcription factors in EMT have been well established in a variety of cancers including colorectal cancer. In addition to the three major families of EMT transcription factors, other transcription factors were recently shown to induce EMT in colorectal cancer. One of them is Prospero Homeobox 1 (PROX1) which belongs to the family of homeobox transcription factors. Others are forkhead box (FOX) transcription factors, which are defined by a DNA-binding forkhead domain. These transcription factors contribute to EMT by functionally regulating target genes or controlling the expression of each other. The activities of these EMT transcription factors vary by tissue type and depend on regulatory signaling pathways. Aberrant regulation of EMT-related transcription factors and mesenchymal markers has been identified in CRC and is associated with increased rate of cancer recurrence and decreased survival of CRC patients (Table 1). Previous studies reported that overexpression of EMT-related transcription factors, such as SNAIL [18], SLUG [19], TWIST1 [20], TWIST2 [21], ZEB1 [22], ZEB2 [23] PROX1 [24], FOXC2 [25], FOXQ1 [26], FOXC1 [27], and FOXM1 [28] are associated with invasiveness, metastasis, and poor prognosis of CRC.

2.2.1. SNAIL Transcription Factors

The SNAIL family of zinc-finger transcription factors, consisting of SNAIL1, SNAIL2, and SNAIL3 (also known as SNAIL, SLUG, and SMUC, respectively), activate the EMT program during development, fibrosis, and cancer [29]. SNAIL promotes EMT primarily through its suppression of E-cadherin. SNAIL binds to E-box DNA sequences of the E-cadherin promoter and recruits histone deacetylases (HDACs) [30], the Polycomb repressive complex 2 (PRC2) [31,32], Lys-specific demethylase 1 (LSD1) [33], G9a [34], and the suppressor of variegation 3–9 homologue 1 (SUV39H1) [35]. This assembly leads to various histone modifications, including methylation and acetylation at histone H3 Lys 4 (H3K4), H3K9, and H3K27, resulting in the suppression of E-cadherin promoter activity. Additionally, SNAIL can induce expression of genes related to the mesenchymal phenotype, such as fibronectin, N-cadherin, and collagen. In addition to regulating epithelial and mesenchymal genes, SNAIL also upregulates other EMT transcription factors, such as SLUG [36], TWIST [37], and ZEB1 [38]. Furthermore, SNAIL can interact with β-catenin and promote the WNT/β-catenin signaling pathway which is also involved in EMT [39]. Therefore, upregulation of SNAIL is considered a major step in EMT. Diverse signaling pathways—NOTCH [40], TGF-β [41], and WNT/β-catenin [42] signaling pathways, cooperate in the initiation and progression of EMT by activating higher levels of SNAIL expression. The critical roles of SNAIL in CRC have been previously determined. In a recent study, SNAIL contributed to the down-regulation of E-cadherin and the vitamin D receptor in colon cancer, leading to the failure of vitamin D analogue treatment [43]. SNAIL is also a critical player in the link between EMT and stem cell properties of CRC, as SNAIL activates the expression of IL-8 by direct binding to its E3/E4 E-boxes, inducing cancer stem cell activities [44].

2.2.2. bHLH Transcription Factors

Basic helix-loop-helix (bHLH) transcription factors are key players in a wide array of developmental processes, including lineage commitment and differentiation. In this family, TWIST1 and TWIST2 are important regulators of EMT. In cancer, TWIST1 was found to repress E-cadherin and induce the expression of mesenchymal markers, such as fibronectin and N-cadherin, during EMT [45]. TWIST1 recruits the methyltransferase SET8, which represses E-cadherin and activates N-cadherin promoter via its H4K20 mono-methylation activity [46]. In CRC, the expression of TWIST1 and TWIST2 is generally restricted to the tumor stroma. Human fibroblast cell lines stably-transfected with TWIST1 acquire characteristics of activated cancer-associated fibroblasts (CAFs) and an increased ability to migrate [47]. TWIST1-positive stromal cells of human CRC have a fully mesenchymal phenotype and are associated with tumor progression [48]. Additionally, a link between epigenetic regulation of TWIST proteins and the tumor-budding phenotype has been reported. The results of laser capture micro-dissection of a high-grade budding case show positive TWIST1 and TWIST2 stroma and no methylation, while a low-grade budding case was characterized by TWIST1 and TWIST2-negative stroma and strong promoter hypermethylation. TWIST1 stromal cell staining was associated with adverse features—more advanced lymph node metastasis, lymphatic vessel invasion, perineural invasion, and worse overall survival. Stromal cells may influence tumor budding in CRCs through expression of TWIST1. Hypermethylation of TWIST1 and TWIST2 promoters in the tumor stroma may represent an alternative mechanism for regulation of TWIST1 [49].

2.2.3. ZEB Transcription Factors

The ZEB family of transcription factors contains two members (ZEB1 and ZEB2), which bind regulatory gene sequences at E-boxes. These factors function as transcriptional repressors and activators, thereby repressing epithelial genes and activating mesenchymal genes, respectively. ZEBs primarily mediate transcriptional repression by recruiting the co-repressor C-terminal-binding protein (CTBP) to E-boxes [50]. In addition, ZEB1 recruits SWitch/sucrose non-fermentable (SWI/SNF) chromatin remodeling protein BRG1 and represses E-cadherin expression [51]. ZEB1 interacts with the transcriptional coactivator p300/CBP-associated factor (PCAF), which switches it from a transcriptional repressor to a transcriptional activator, promoting SMAD signaling [52]. ZEB1 can also recruit LSD1, which likely associates it to histone demethylation in EMT [53]. Interestingly, the upregulation of ZEB2 at the invasive front significantly correlates with tumor stage in primary CRC. Silencing ZEB2 by siRNA decreases the migration and invasion of CRC cells [23]. A recent study showed that that ZEB1 was detected in invasive regions of colorectal tumors and at the tumor–stroma interface: regions that display reduction in polarity components. ZEB1 induces the loss of basement membrane (BM) by suppressing expression of the epithelial BM component Laminin subunit alpha-3 (LAMA3). It binds specifically to Z1, Z2, and E2 on the LAMA3 promoter and represses transcriptional activity [8]. ZEB1 silencing upregulates the expression of several polarity genes in colorectal carcinoma cells, including crumbs homolog 3 (CRB3) and lethal giant larvae 2 (LGL2) [54]. The LGL2 promoter is a direct target of ZEB1 and loss of LGL2 is associated with increased EMT and metastasis in CRC [55]. Moreover, ZEB1 can also promote tumor invasiveness via the regulation of players involved in stroma remodeling: the urokinase plasminogen activator (uPA), and its inhibitor, plasminogen activator inhibitor-1 (PAI-1). ZEB1 binds to the uPA promoter and activates its transcription through a mechanism implicating histone acetyltransferase p300. On the other hand, ZEB1 inhibits plasminogen activator inhibitor-1 PAI-1 expression by reducing the stability of its mRNA [56].

2.2.4. Other Transcription Factors

Prospero Homeobox 1 (PROX1)

PROX1 promotes dysplasia in colonic adenomas and CRC progression. PROX1 is upregulated in response to abnormally elevated oncogenic signaling of TCF/β-catenin in intestinal epithelium, important for tumor progression via disruption of cell polarity and adhesion [57]. PROX1 overexpression correlates with increased mesenchymal phenotype, advanced tumor stage, and lymph node metastasis. PROX1 binds to the promoter of pre-miR-9-2 and triggers its expression, resulting in the suppression of E-cadherin 3′-UTR reporter activity and protein expression [58].

Forkhead Box Q1 (FOXQ1)

FOXQ1 is a member of the Forkhead box transcription factor family. FOXQ1 is overexpressed in epithelial and stromal tumor compartments along with other EMT genes. A recent report demonstrated the upregulation of FOXQ1 in CRC. Ectopic expression of FOXQ1 promoted an anti-apoptotic effect and enhanced tumor growth [59]. FOXQ1 can repress E-cadherin expression by targeting the E-box in its promoter region [60]. FOXQ1 also mediates EMT through the modulation of other EMT-related transcription factors. FOXQ1 interacts with TWIST1 to reinforce the suppression of E-cadherin transcription in CRC [61].

Forkhead Box Protein C2 (FOXC2)

FOXC2 is upregulated in human CRC cells and tissues, and correlates with colon cancer progression and patient survival. A functional study demonstrated that FOXC2 promoted cell growth, cell migration, and tumor formation in nude mice, whereas knockdown of FOXC2 significantly reversed these effects. FOXC2 enhances AKT activity with subsequent GSK-3β phosphorylation and SNAIL stabilization. This leads to an induction of EMT and subsequent tumor invasion and metastasis [25].

Forkhead Box M1 (FOXM1)

FOXM1 has been found to be aberrantly expressed in nearly all carcinomas. FOXM1 stimulates cell proliferation and cell cycle progression by promoting entry into the S-phase and M-phase. FOXM1 is required for proper execution of mitosis. In accordance with its role in stimulating cell proliferation, its expression is regulated by proliferation and anti-proliferation signals, as well as by proto-oncoproteins and tumor suppressors [62]. FOXM1 upregulates the expression of ZEB1/2, and SLUG, consequently leading to a reduction in the expression of E-cadherin [63]. FOXM1D, a novel isoform of FOXM1, has been found to promote CRC EMT and metastasis through activating Rho-associated kinases (ROCKs). ROCKs are known for their pivotal roles in orchestrating actin cytoskeleton, leading to EMT and cancer invasion. The interaction between FOXM1D and ROCK2 is important for the ability of ROCKs to regulate actin arrangement and EMT in a Rho-dependent manner. Furthermore, FOXM1D-induced ROCK activation could be abolished by the ROCK inhibitors Y-27632 and fasudil [64].

Table 1. Factors/proteins involved in EMT and their relevance to colorectal cancer.

Factors	Description	Relevance to CRC Cancer	Refs
Transcription factors			
SNAIL	Zinc-finger protein, E-box transcriptional repressor	Expression of SNAIL in the tumor stroma correlated with lower survival of cancer patients and with presence of distant metastasis	[18]
SLUG	Zinc-finger protein, E-box transcriptional repressor	Positive expression of SLUG was significantly associated with Dukes stage and distant metastasis	[19]
TWIST1	bHLH factor	Overexpression in primary CRC was associated with shorter overall survival	[20]
TWIST2	bHLH factor	Upregulation in CRC led to poor prognosis, particularly for patients with stage III and IV	[21]
ZEB1	Zinc-finger protein, E-box transcriptional repressor	High expression of ZEB1 correlated with liver metastasis and poor prognosis in CRC	[22]
ZEB2	Zinc-finger protein, E-box transcriptional repressor	High expression of ZEB2 at the tumor invasion front correlated significantly with worsening prognosis	[23]
PROX1	Prospero homeobox protein	High PROX1 expression was associated with poor grade of tumor differentiation, and in colon cancer patients with less favorable patient outcome	[24]
FOXQ1	Forkhead box transcription factor	High expression of FOXM1 and FOXQ1 emerged as independent prognostic factors in CRC patients	[26]
FOXC2	Forkhead box transcription factor	Overexpression of FOXC2 was associated with poor clinical outcome in human colon cancer	[25]
FOXM1	Forkhead box transcription factor	FOXM1 overexpression was a molecular marker predicting increased metastatic potential of CRC and poorer prognosis	[28]
Factors directly associated with EMT			
E-cadherin	Adhesion glycoprotein	Downregulation of E-cadherin expression indicated worse prognosis in CRC patients	[14,65,66]
N-cadherin	Adhesion glycoprotein	Overexpression was correlated with metastasis and worse survival in CRC patients	[17]
Fibronectin	Adhesion glycoprotein	Results indicated that fibronectin levels increased with the progression of CRC	[15]

Abbreviations: EMT, epithelial to mesenchymal transition; CRC, colorectal cancer; FOXC2, forkhead box C2; PROX1, prospero homeobox 1; ZEB, zinc finger E-box binding homeobox; bHLH: basic helix-loop-helix.

2.3. EMT Inducers

2.3.1. Signaling Pathways

TGF-β

TGF-β signaling plays a dual role in tumorigenesis. It is known to mediate tumor-suppressive effects in the early stages of tumor development. However, in later stages, TGF-β signaling may enhance tumor progression, due to its ability to promote cell proliferation and EMT and to suppress immune function in several cancer types, including those of the breast, prostate, and colon [67]. Among SMADs, SMAD4 has been studied with regard to its involvement in the effects of TGF-β in the later stages of CRC carcinogenesis and in the EMT process. SMAD4 suppresses invasion and restores the epithelial phenotype of SW480 CRC cells. Knockdown of SMAD4-induced levels of endogenous TGF-β cytokines, which leads to increased TGF-β signaling and induction of EMT [68]. SMAD4 is also a negative regulator of STAT3 activation [69]. The loss of SMAD4 leads to aberrant activation of STAT3, which may directly contribute to the EMT process and ZEB1 expression in CRC progression. The loss of SMAD4 causes BMP signaling to switch from tumor-suppressive to pro-metastatic, thereby inducing EMT through activation of RHO and ROCK [70]. Furthermore, activation of TGF-β is important for the EMT-inducing effects of platelets on CRC cells. Upon being secreted from platelets, TGF-β stimulates the expression of EMT markers such as SNAIL, vimentin, and fibronectin in tumor cells [71].

WNT/β-Catenin

Over-activation of the WNT/β-catenin pathway promotes EMT-associated dedifferentiation located at the invasive front of colorectal tumors [72]. Recent studies show that both canonical and non-canonical WNT signaling can enhance EMT, depending on the tissue type. Enhanced canonical WNT signaling in CRC cells increases the level of SNAIL, which represses E-cadherin and regulates EMT, thus promoting local invasion [73]. Cytoplasmic SLUG concentration is controlled by GSK3-β phosphorylation and subsequent ubiquitination by β-TrCP. Activation of canonical WNT signaling stabilizes SLUG by inhibiting GSK3β kinase activity and initiates EMT transcriptional programs in cancer cells [74]. WNT3a protein overexpression in CRC patients is concomitant with EMT features, such as reduced expression of the epithelial marker E-cadherin, increased expression of the mesenchymal marker vimentin, and localization of nuclear β-catenin. In vitro and in vivo experiments showed that WNT3a overexpression promotes invasion and induces Snail expression. Dkk1 (an antagonist of WNT/β-catenin signaling) partially reverses the expression of EMT-associated proteins in WNT3a-overexpressing cells [75]. Additionally, the non-canonical WNT pathway is also involved in EMT. There is a high correlation between Fzd2, its ligands WNT5a/b, and EMT markers. It has been shown that Fzd2 expression enhances EMT and cell migration via interaction with STAT3 Stat3. Targeting of Fzd2 by a specific antibody reduces metastasis in a xenograft mouse model of colon cancer [76]. A novel WNT/β-catenin inhibitor IWR-1 suppresses CRC tumor metastasis in relation to EMT [77].

2.3.2. Novel EMT Inducers in Colorectal Cancer

In line with the studies on signaling pathways which contribute to EMT, a number of studies have elucidated several novel EMT inducers that also support EMT. These inducers can regulate expression of EMT-TF or activate other signaling pathways which are involved in EMT (Figure 1).

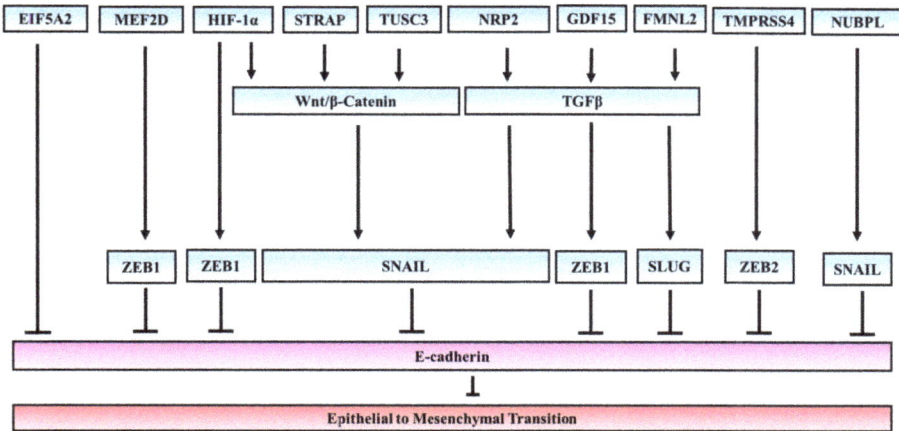

Figure 1. Crosstalk network of novel EMT inducers. Several inducers of EMT in CRC have been identified. These inducers were found to directly regulate expression of EMT-transcriptional factors or do so through other oncogenic signaling pathways. Myocyte enhancer factor 2D (MEF2D), hypoxia-inducible factor 1 alpha (HIF-1α), and transmembrane protease/serine 4 (TMPRSS4) can promote EMT by inducing expression of ZEB transcriptional factors, whereas nucleotide binding protein-like (NUBPL) can upregulate SNAIL. Other proteins such as serine–threonine kinase receptor-associated protein (STRAP) and TUCS3 can induce the WNT/β-Catenin signaling pathway, which has been found to be involved in EMT in CRC. Additionally, neuropilin-2 (NRP2), growth differentiation factor 15 (GDF15), and formin-like2 (FMNL2) can promote TGF-β signaling and induce EMT. Overexpression of eukaryotic initiation factor 5A2 (EIF5A2) can promote EMT through an unknown mechanism which may involve C-Myc and metastasis-associated protein 1 (MTA-1).

Transmembrane Protease/Serine 4 (TMPRSS4)

TMPRSS4 is a member of the type II transmembrane serine protease (TTSP) family and it is highly expressed in CRC tissues and correlates with pathological stages. TMPRSS4 overexpression increases the proliferation and self-renewal ability of CRC cells [78]. Overexpression of TMPRSS4 leads to a significant increase in both in vitro invasion and in vivo metastasis of colon cancer cells. TMPRSS4 induces loss of cell–cell adhesion through E-cadherin downregulation concomitant with the induction in SIP1/ZEB2. TMPRSS4 enhances expression of the integrin subunit α5 [79], which has been centrally implicated in EMT induction and cell motility [80]. Functional blockade of integrin α5β1 demonstrated that this integrin has an important role in TMPRSS4-mediated effects [81]. Blockage of ITG-α5 antibodies (volociximab) is being evaluated in clinical trials for cancer treatment. The possibility of co-targeting both proteins would be an interesting approach to assess synergistic anti-tumor efficacy [82]. In addition, TMPRSS4 overexpression leads to an intracellular signaling cascade that involves FAK, ERK1/2, Akt, Src, and Rac1 activation. Inhibition of PI3K or Src reduces invasiveness and actin rearrangement mediated by TMPRSS4 without restoring E-cadherin expression [83]. A recent study has reported the evaluation of a novel series of 2-hydroxydiarylamide derivatives for the inhibition of TMPRSS4 serine protease activity and the suppression of cancer cell invasion. These derivatives show promising anti-invasive activity of colon cancer cells overexpressing TMPRSS4 [84].

Formin-Like2 (FMNL2)

FMNL2 is a member of the diaphanous-related formins which act as effectors of Rho family GTPases and control actin-dependent processes such as cell motility/invasion. Overexpression of FMNL2 in metastatic cell lines and tissues of colorectal carcinoma is associated with more aggressive tumor behavior [85]. FMNL2 is involved in mesenchymal phenotype maintenance in human CRC cells. Knockdown of FMNL2 leads to a mesenchymal–epithelial transition confirmed by the upregulation of E-cadherin, α-catenin, and γ-catenin; and downregulation of vimentin, SNAIL, and SLUG. Loss of FMNL2 expression lowers the ability of TGF-β to induce EMT, which suggests that FMNL2 contributes to the acquisition of a mesenchymal and highly migrating phenotype in CRC cells induced by TGF-β. The Ras–MAPK pathway is also involved in FMNL2-induced EMT [86]. Furthermore, Zhu et al. revealed that cytoskeletal regulation by the Rho GTPase pathway, the WNT pathway, the G-protein pathway, and the P53 pathway are affected by FMNL2 [85]. FMNL2 is identified as a target of a number of microRNA in CRC, including mir-206 [87], mir-613 [88], mir-34a [89], and mir-137 [90].

EIF5A2

The gene *EIF5A2* encodes eukaryotic initiation factor 5A2 (EIF5A2) and is located on chromosome 3q26, a region frequently amplified in CRC [91]. Ectopic expression of EIF5A2 in CRC cells promotes EMT, cell motility, and invasion in vitro. Overexpression of EIF5A2 is associated with tumor metastasis, determined to be an independent predictor of shortened survival in CRC patients [92]. Overexpression of EIF5A2 in CRC cells enhances the enrichment of c-Myc on the promoter of metastasis-associated protein 1 (MTA1). MTA-1 expression is associated with EMT and metastasis in CRC cells [93].

Growth Differentiation Factor 15 (GDF15)

GDF15 is a divergent member of the BMP-subfamily of the TGF-β superfamily. GDF15 is also referred to as macrophage inhibitory cytokine-1 (MIC-1), prostate-derived factor (PDF), placental bone morphogenetic protein (PLAB), placental transforming growth factor (PTGF), and nonsteroidal anti-inflammatory drug-activated gene-1 (NAG-1) [94]. GDF15 serves as a negative CRC prognostic marker, and high levels of GDF15, both in tumor tissues and plasma, correlate with an increased risk of recurrence and reduced overall survival [95,96]. It has been considered as a target for CRC therapy [97]. GDF15 promotes CRC cell metastasis both in vitro and in vivo through activating EMT. It binds to TGF-β receptor to activate SMAD2 and SMAD3 pathways. Clinical data shows increased GDF15 levels in tumor tissues and serum, which correlate with reduced CRC overall survival [98].

Hypoxia-Inducible Factor 1 Alpha (HIF-1α)

It is well recognized that HIF-1α is involved in cancer metastasis, chemotherapy resistance, and poor prognosis. It induces EMT in a variety of cancer types, including those of the colon, breast, lung, head and neck, thyroid, and prostate. HIF-1α expression is independently associated with poor prognosis in CRC by regulating the expression of EMT-related transcription factors [99]. It directly influences ZEB1 expression through the hypoxia response element 3 (HRE-3), located in the ZEB1 proximal promoter. Inhibition of ZEB1 abrogates HIF-1α-induced EMT and cell invasion [100]. HIFs are also involved in EMT through regulating canonical WNT signaling. HIF-1α silencing decreased the stability and transcriptional activity of β-catenin in CRC cell lines [101]. In a recent study, deferoxamine, a hypoxia-mimetic agent, was found to reduce EMT in CRC. Under hypoxia, dexamethasone treatment inhibits HIF-1α protein levels and decreases mRNA levels of hypoxia-induced SNAIL, SLUG, and TWIST1, and transcriptional factors involved in EMT, as well as the hypoxia-induced integrin αVβ6 protein, a well-known EMT marker for CRC cells [102].

Myocyte Enhancer Factor 2D (MEF2D)

MEF2D is a transcription factor of the MEF2 family, which was first identified as a muscle gene expression regulator [103]. MEF2 plays central roles in the transmission of extracellular signals to the genome and in the activation of genetic programs that control cell differentiation, proliferation, morphogenesis, survival, and apoptosis of a wide range of cell types [104,105]. MEF2D acts as a developmental transcription factor in embryogenic processes in which EMT is active, such as gastrulation and cardiogenesis. MEF2D is abnormally expressed in human CRC and its upregulation correlates with cancer metastasis. It responds to various tumor microenvironment signals, including not only cytokines such as EGF, IL-6, bFGF, and IGF2, but also the physical stimulus of hypoxia. These microenvironmental factors are often over-activated and contribute to tumor initiation, progression, metastasis, and therapy resistance in cancers including CRC. Upon activation, MEF2D transcriptionally regulates ZEB1 expression. Therefore, MEF2D can function as a central integrator, transducing multiple signals to activate EMT-relevant genes and inducing the metastatic capacity of CRC cells [106].

Nucleotide Binding Protein-Like (NUBPL)

NUBPL, also known as IND1 or huInd1, is an assembly factor for human mitochondrial complex I, the largest member of the mitochondrial respiratory chain. NUBPL is significantly overexpressed in CRC tissues compared to normal tissues, and its expression level is positively associated with lymph node metastasis and advanced stage. Overexpression of NUBPL notably promotes the migration and invasive ability of CRC cell lines SW480 and SW620, whereas knockdown of NUBPL leads to the opposite effect. It induces EMT, characterized by downregulation of epithelial marker (E-cadherin) and upregulation of mesenchymal markers (N-cadherin and vimentin). Moreover, it activates ERK signaling, believed to promote EMT and tumor metastasis, as inhibition of ERK suppresses the NUBPL-induced changes in EMT and cell motility [107].

Neuropilin-2 (NRP2)

Neuropilins (NRPs), initially characterized as cell guidance molecule receptors for semaphorins, act as co-receptors for cancer related growth factors and are involved in several signaling pathways leading to cytoskeletal organization, angiogenesis, and cancer progression. NRP2 (neuropilin-2) confers a fibroblastic shape to cancer cells, suggesting an involvement of NRP2 in EMT. Presence of NRP2 in CRC cell lines is correlated with loss of epithelial markers, such as cytokeratin-20 and E-cadherin, and with acquisition of mesenchymal molecules, such as vimentin. NRP2 is identified as a receptor for TGF-β1. NRP2 expression on CRC cell lines has been shown to promote TGF-β1 signaling, leading to a constitutive phosphorylation of the SMAD2/3 complex. Treatment with specific TGF-β type1 receptor kinase inhibitors restores E-cadherin levels and partially inhibits NRP2-induced vimentin expression, suggesting NRP2 cooperates with TGF-β1 receptor to promote EMT in CRC [108]. Interestingly, NRP2b, a spliced isoform of NRP2, was up-regulated during TGFβ-mediated EMT and knockdown of NRP2b significantly inhibited TGFβ-stimulated EMT and migration [109].

Tumor Suppressor Candidate 3 (TUSC3)

TUSC3 has been identified as a putative tumor suppressor in a variety of malignancies, including those of the prostate [110], glioblastoma [111], ovaries [112], and pancreas [113]. The TUSC3 protein is localized to the endoplasmic reticulum and is a subunit of the endoplasmic reticulum-bound OST complex, responsible for the N-glycosylation of nascent proteins. The *TUSC3* gene is located in chromosomal region 8p22, a region in which allelic losses are frequent in cancer [114]. In CRC, an association of 8p allelic loss with poor outcome in CRC has been reported. However, recent studies support the oncogenic function of TUSC3 in CRC [115]. TUSC3-overexpressing CRC cells have increased activities of proliferation, invasiveness, and tumorigenesis. TUSC3 is found to be associated with the MAPK and PI3K/Akt signaling pathways. TUSC3 overexpression in CRC cells increases EMT, accompanied by downregulation of E-cadherin and upregulation of vimentin. Furthermore, TUSC3 promotes EMT through induction of WNT/β-catenin signaling pathways. The TUSC3 protein is co-localized with β-catenin, a key component of the WNT signaling pathway through protein–protein interaction [116].

Serine–Threonine Kinase Receptor-Associated Protein (STRAP)

The multiple roles of STRAP in human cancers have been previously described. Overexpression of STRAP has been reported in lung [117], colon [118], and breast carcinomas [119]. STRAP consists of seven WD40 domains, allowing it to function as a scaffold protein [120]. STRAP was initially identified as a putative inhibitor of the canonical TGF-β signaling pathway [121]. Because the SMAD-dependent TGF-β pathway negatively regulates cellular growth, early studies suggested that STRAP can function as an oncogene. Kashikar et al. showed that loss of STRAP expression induces a mesenchymal-to-epithelial transition through upregulation of E-cadherin, indicating the role of STRAP in EMT [122]. Knockdown of STRAP reduces CRC cell invasion and metastasis in vitro and in vivo. Furthermore, we have observed that STRAP can stabilize β-catenin by inhibiting its ubiquitin-dependent degradation, thus resulting in the inhibition of the expression of its downstream target gene [123].

2.3.3. Non-Coding RNA-Mediated Control of EMT

A number of small non-coding RNAs or microRNAs (miRNAs) regulate the epithelial phenotype and EMT by inhibiting the expression of EMT regulators (Table 2). The expressions of these miRNAs are also regulatory targets of other regulators involved in EMT [124].

Members of the miR-200 family (miR-200a, miR-200b, miR-200c, miR-141, and miR-429) are recognized as regulators of the epithelial phenotype through repression of ZEB1 and ZEB2 mRNA translation [125–127]. The DNA methylation associated with inactivation of various miR-200 members has been described as a major contributing factor of EMT in cancer. miR-200b and miR-200c transcripts undergo a dynamic epigenetic regulation linked to the EMT or mesenchymal-epithelial-transition (MET) phenotype in tumor progression. The 5′-CpG island hypermethylation-associated silencing of both miR-200 loci is observed in transformed cells with mesenchymal characteristics, including low levels of ZEB1/ZEB2 and high E-cadherin expression [128]. Among miR-200 family members, miR-200c plays a pivotal role in the metastatic behavior of CRC cells. MiR-200c decreases migration and invasion in various CRC cell lines via directly targeting ZEB1 [129]. Methylation-induced downregulation of miR-200c allows upregulation of several of its direct target genes—*ZEB1*, *ETS1*, and *FLT1*. In contrast, hypomethylation of miR-200c mediates the MET process, which results in the settlement of metastasized cells at secondary sites. During MET, hypomethylation-induced re-expression of miR-200c suppresses the EMT-driving genes, accompanied by high E-cadherin and low Vimentin expression [125]. miR-429 reverses TGF-β-induced EMT by interfering with Onecut2 in CRC cells [130]. In addition, ZEB2 is also identified as a direct target of miR-132 [131], miR-192 [132], and miR-335 [133]. Downregulation of these miRNAs is associated with distant metastasis and advanced-stage tumors.

Members of the miR-34 family are induced by the tumor suppressor p53 and are known to inhibit EMT, and therefore, presumably suppress the early phases of metastasis. MiR-34 members inhibit metastasis formation in CRC via the EMT-regulating network in SNAIL/ZNF281 [134] and the IL-6 receptor (IL-6R)/STAT3 [135]. MiR-34a/b/c targets a conserved seed-matching sequence in the SNAIL 3′-UTR. Overexpression of miR-34a induces mesenchymal-epithelial-transition (MET) and down-regulation of SNAIL. However, suppression of miR-34a/b/c causes up-regulation of SNAIL with the display of EMT markers and related features and enhanced migration/invasion. MiR-34a also suppresses SLUG and ZEB1. Conversely, the transcription factors SNAIL and ZEB1 bind to E-boxes of the miR-34a/b/c promoters, thereby repressing their expression as a part of the EMT program [136]. Ectopic miR-34a prevents TGF-β-induced EMT. The suppression of miR-34a was required for IL-6-induced EMT and invasion. Exposure of human CRC cells to the cytokine IL-6 activates oncogenic STAT3 transcription factor, which directly represses the *MIR34A* gene via a conserved STAT3-binding site in the first intron. Interestingly, the IL-6 receptor (IL-6R), which mediates IL-6-dependent STAT3 activation, is identified as a direct miR-34a target. This IL-6R/STAT3/miR-34a feedback loop is found in primary colorectal tumors [135].

Additionally, miR-138 [137], miR-212 [138], miR-30b [139], miR-320a [140], miR-598 [141], miR-4775 [142] , miR-675-5p [143], miR-29b [144], miR-363-3p [145], miR-17 [146], miR-139-5p [147], miR-375 [148], and miR-497 [149] also play pivotal roles in regulating the EMT and CRC metastatic processes.

In addition to controlling the expression of EMT transcription factors and inducers, miRNAs also target genes that help to define the epithelial or mesenchymal phenotype; for example, genes encoding adhesion junction and polarity complex proteins and signaling mediators. MiR-9, transcriptionally induced by PROX1, directly represses E-cadherin expression [27]. Integrin-β4 (ITGβ4), exclusively expressed in polarized epithelial cells, is a novel miR-21 target gene and plays a role in EMT regulation. It is remarkably de-repressed after transient miR-21 silencing and downregulated after miR-21 overexpression [150].

Recent studies have also demonstrated major roles of several long non-coding RNAs (lncRNAs) in the regulation of EMT in colorectal cancer. The lncRNA H19 functions as a competing endogenous RNA (ceRNA) for miR-138 and miR-200a to abolish their suppressive effects on mesenchymal marker genes *ZEB1*, *ZEB2*, and *VIM* [151]. BRAF-activated lncRNA (BANCR) induces EMT through a MEK/extracellular signal-regulated kinase-dependent mechanism, as treatment with the MEK inhibitor U0126 can restore the epithelial phenotype in BANCR-overexpressed CRC cells [152]. HOX transcript antisense intergenic RNA (HOTAIR) promotes EMT by downregulating E-cadherin and upregulating vimentin and MMP9 [153]. Other lncRNAs such as long non-coding RNA-activated by TGF-β (lncRNA-ATB) [154], actin filament associated protein 1 antisense RNA1 (AFAP1-AS1) [155], taurine-upregulated gene 1 (TUG1) [156], SPRY4 intronic transcript 1 (SPRY4-IT1) [157], and promoter of CDKN 1A antisense DNA damage-activated RNA (PANDAR) [158] are found to promote EMT and metastasis in CRC through unknown mechanisms.

Table 2. Micro-RNAs (miRNAs) involved in EMT in colorectal cancer.

Factors	Upstream Regulators	Target Genes	Refs
miRNAs involved in inhibiting EMT			
miR-200 family	P53/ZEB1/SIX1	*ZEB1, ZEB2*	[125,126]
miR-429	P53/ZEB1/SIX1	*HOXA5*	[159]
miR-34a	SNAIL, ZEB1, IL-6/STAT3	*SLUG, ZEB1, IL-6R*	[134,135]
miR-9	N/A	*E-Cadherin*	[58]
miR-21	AP-1/ETS1	*ITGβ4*	[150]
miR-138	N/A	*TWIST2*	[137]
miR-132	N/A	*ZEB2*	[131]
miR-30b	N/A	*SIX1*	[139]
miR-598	N/A	*JAG1/Notch2*	[141]
miR-4775	N/A	*TGFβ*	[142]
miR-363-3p	N/A	*SOX4*	[145]
miR-375	N/A	*SP1*	[148]
miRNAs involved in promoting EMT			
miR-17	N/A	*CYP7B1*	[146]
miR-194	N/A	*MMP-2*	[160]
miR-675-5p	N/A	*DDB2*	[143]
miR-150	WNT/β-catenin	*CREB*	[161]
miR-29a	N/A	*KLF4*	[162]

Abbreviations: SIX1, sineoculis homeobox homolog 1; IL-6R, interleukin 6 receptor; ITGβ4, integrin β4; SOX4, SRY-Box 4; CYP7B1, cytochrome P450 family 7 subfamily B member 1; MMP-2, matrix metallopeptidase 2; DDB2, DNA damage-binding protein 2; CREB, cAMP response element binding; KLF4, kruppel like factor 4.

3. EMT–Metastasis

Metastasis is a multistep process by which tumor cells undergo a sequential series of events to disseminate from their primary site and form secondary tumors in distant tissue [163]. In CRC, loss of the epithelial and gain of the mesenchyme-like phenotype of tumor cells at the invasive front results in increased invasiveness [164]. These changes enable tumor cells to migrate through the extracellular matrix and colonize in the lymph/blood vessels, thereby initiating the first step of the metastatic cascade [165].

Recent studies suggest the importance of EMT for circulating tumor cells (CTCs) during the metastatic process [165]. CTCs in blood have been considered recently to be both potential seeds for metastasis and biomarkers for early detection of metastasis [166]. The CTCs of CRC acquire mutations in key genes, such as *KRAS* or *TP53*, that are not identical to those in the corresponding tumor tissue. Gene expression analyses reveal a pronounced upregulation of *CD47* in CTCs as a potential immune-escape mechanism [167]. CTCs which display a mesenchymal phenotype are believed to have increased metastatic potential, [164] with loss of E-cadherin and over-expression of vimentin. Vimentin-positive CTCs displayed higher expression of EMT-related regulators, such as ZEB2, SNAIL, and SLUG. Experimental and clinical data suggest that EMT has an important role in the generation of CTCs. TWIST1 induction dramatically increases CTC numbers in a mouse squamous cell carcinoma tumor model. Successful metastasis may depend on CTCs mesenchymal state maintenance [8].

Studies with metastatic CRC patients show that EMT-induced CTCs can be used as a prognostic marker and as an indicator of therapeutic response [168,169]. In a recent study, Satelli et al. used the 84-1 antibody to detect cell-surface vimentin (CSV) on EMT CTCs from blood of patients with metastatic colon cancer. Overexpression of CSV was found in metastatic tumors compared with primary tumors, suggesting that CSV expression is mainly associated with metastasis and could serve as a potential metastatic biomarker [170]. In another study, plastin3 (PLS3) was discovered as an EMT-induced CTC marker that was expressed in peripheral blood from patients with CRC with distant metastasis. PLS3 codes for an actin-bundling protein known to inhibit cofilin-mediated depolymerization of actin fibers. The association between PLS3-positive CTCs and prognosis was particularly strong in patients with Dukes B and Dukes C [171].

4. Conclusions and Future Perspectives

The process of EMT results from a spectrum of changes and transitions in response to environmental stimuli, dependent on tissue and signaling context. EMT initiation and progression involves a complicated crosstalk among networks of signaling pathways and transcriptional regulators. The hallmark of EMT is the downregulation of E-cadherin, a major epithelial marker, which is regulated by a group of EMT transcription factors. These factors, including the SNAIL, TWIST, and ZEB families, play important roles in both embryogenesis and tumorous settings. Importantly, regulation of these EMT-related transcription factors is associated with invasiveness, metastasis, and poor prognosis of CRC, emphasizing the importance of EMT in tumor progression and metastasis. EMT inducers, including oncogenic signaling pathways, mediate EMT through regulating expression of EMT-TF (transcription factor). In this review, we focused on the current understanding of EMT regulation in CRC with emphasis on the following points:

(1) In addition to well-known signaling pathways which contribute to EMT, there are several novel EMT inducers that also support EMT. These inducers promote EMT by regulating expression of EMT-TF or activate other signaling pathways. Among these inducers, MEF2D and HIF-1α are transcription factors that play central roles in extracellular signal transmission, resulting in the activation of genetic programs that control EMT. The potential of using inhibitors which target these EMT inducers should be considered. Deferoxamine can be used to impair HIF1α function, resulting in the suppression of hypoxia-induced EMT. Additionally, 2-hydroxydiarylamide derivatives that inhibit TMPRSS4 serine protease activity suppress EMT and cell invasion.

(2) A number of microRNAs were found to promote EMT in CRC. These microRNAs have targets which belong to all three groups of EMT-related factors: effectors, transcription factors, and inducers. Among those, miR-200 and miR-34 families target ZEB1, ZEB2, SNAIL, and SLUG transcription factors. Other miRNAs target E-cadherin or integrin-β4. These miRNAs are also

targets of other signaling pathways, suggesting the important roles they play in the crosstalk between oncogenic signaling pathways and EMT.

(3) Current studies provide evidence for the important roles CTCs play in the metastatic process [166]. CTCs are considered as both seeds for metastasis and markers for early detection of metastasis. CTCs constitute a heterogeneous population of tumor-derived cells with different phenotypes. One of the most common approaches for isolating CTCs is the epithelial cell adhesion molecule (EpCAM)-based enrichment technique. However, recent studies demonstrate that this technique failed to detect CTC subpopulations that had undergone EMT [172]. Aberrant activation of the EMT program has been implicated in the dispersion of CTCs from primary tumors. EMT endows CTCs with mesenchymal phenotypes, and is an early event in the metastatic process. Thus it is conceivable that EMT marker detection in CTCs may facilitate the early detection of metastases, as well as the assessment of new drug targets in clinical trials. Overexpression of several EMT-induced CTC markers, such as cell-surface vimentin and PLS3 are correlated with metastasis, and can be used as potential metastatic biomarkers.

In conclusion, the loss of epithelial and gain of mesenchyme-like phenotype of the CRC cells is associated with increased invasiveness and metastasis. The observations discussed in this review indicate that EMT is orchestrated by a complex and multifactorial network, involving regulators of different signaling pathways. Further studies are imperative to identifying novel regulators and to understand the mechanisms underlying EMT in CRC.

Acknowledgments: This study was supported by National Cancer Institute R01 CA95195, the Veterans Affairs Merit Review Award, and a Faculty Development Award from the UAB Comprehensive Cancer Center, P30 CA013148 (to Pran K. Datta). The authors thank Tasha Smith for critically reading the manuscript.

Conflicts of Interest: The authors declare no conflict of interest.

References

1. Misiakos, E.P.; Nikolaos, P.; Kouraklis, G. Current treatment for colorectal liver metastases. *World J. Gastroenterol.* **2011**, *17*, 4067–4075. [CrossRef] [PubMed]
2. Drewes, J.L.; Housseau, F.; Sears, C.L. Sporadic colorectal cancer: Microbial contributors to disease prevention, development and therapy. *Br. J. Cancer* **2016**, *115*, 273–280. [CrossRef] [PubMed]
3. Kinzler, K.W.; Vogelstein, B. Lessons from hereditary colorectal cancer. *Cell* **1996**, *87*, 159–170. [CrossRef]
4. Shook, D.; Keller, R. Mechanisms, mechanics and function of epithelial to mesenchymal transitions in early development. *Mech. Dev.* **2003**, *120*, 1351–1383. [CrossRef] [PubMed]
5. Kalluri, R. EMT: When epithelial cells decide to become mesenchymal-like cells. *J. Clin. Investig.* **2009**, *119*, 1417–1419. [CrossRef] [PubMed]
6. Kang, Y.; Massague, J. Epithelial to mesenchymal transitions: Twist in development and metastasis. *Cell* **2004**, *118*, 277–279. [CrossRef] [PubMed]
7. Singh, A.; Settleman, J. EMT, cancer stem cells and drug resistance: An emerging axis of evil in the war on cancer. *Oncogene* **2010**, *29*, 4741–4751. [CrossRef] [PubMed]
8. Spaderna, S.; Schmalhofer, O.; Hlubek, F.; Berx, G.; Eger, A.; Merkel, S.; Jung, A.; Kirchner, T.; Brabletz, T. A transient, EMT-linked loss of basement membranes indicates metastasis and poor survival in colorectal cancer. *Gastroenterology* **2006**, *131*, 830–840. [CrossRef] [PubMed]
9. Thiery, J.P.; Acloque, H.; Huang, R.Y.; Nieto, M.A. Epithelial-mesenchymal transitions in development and disease. *Cell* **2009**, *139*, 871–890. [CrossRef] [PubMed]
10. Tsai, J.H.; Yang, J. Epithelial to mesenchymal plasticity in carcinoma metastasis. *Genes Dev.* **2013**, *27*, 2192–2200. [CrossRef] [PubMed]
11. Cano, A.; Pérez-Moreno, M.A.; Rodrigo, I.; Locascio, A.; Blanco, M.J.; del Barrio, M.G.; Portillo, F.; Nieto, M.A. The transcription factor snail controls epithelial-mesenchymal transitions by repressing E-cadherin expression. *Nat. Cell Biol.* **2000**, *2*, 76–83. [CrossRef] [PubMed]

12. Graff, J.R.; Herman, J.G.; Lapidus, R.G.; Chopra, H.; Xu, R.; Jarrard, D.F.; Isaacs, W.B.; Pitha, P.M.; Davidson, N.E.; Baylin, S.B. E-cadherin expression is silenced by DNA hypermethylation in human breast and prostate carcinomas. *Cancer Res.* **1995**, *55*, 5195–5199. [PubMed]

13. Palacios, F.; Tushir, J.S.; Fujita, Y.; D'Souza-Schorey, C. Lysosomal targeting of E-cadherin: A unique mechanism for the down-regulation of cell-cell adhesion during epithelial to mesenchymal transitions. *Mol. Cell Biol.* **2005**, *25*, 389–402. [CrossRef] [PubMed]

14. Yun, J.A.; Kim, S.H.; Hong, H.K.; Yun, S.H.; Kim, H.C.; Chun, H.K.; Cho, Y.B.; Lee, W.Y. Loss of E-Cadherin expression is associated with a poor prognosis in stage III colorectal cancer. *Oncology* **2014**, *86*, 318–328. [CrossRef] [PubMed]

15. Saito, N.; Nishimura, H.; Kameoka, S. Clinical significance of fibronectin expression in colorectal cancer. *Mol. Med. Rep.* **2008**, *1*, 77–81. [PubMed]

16. Yi, W.; Xiao, E.; Ding, R.; Luo, P.; Yang, Y. High expression of fibronectin is associated with poor prognosis, cell proliferation and malignancy via the NF-kappaB/p53-apoptosis signaling pathway in colorectal cancer. *Oncol. Rep.* **2016**, *36*, 3145–3153. [CrossRef] [PubMed]

17. Yan, X.; Yan, L.; Liu, S.; Shan, Z.; Tian, Y.; Jin, Z. N-cadherin, a novel prognostic biomarker, drives malignant progression of colorectal cancer. *Mol. Med. Rep.* **2015**, *12*, 2999–3006. [CrossRef] [PubMed]

18. Francí, C.; Gallén, M.; Alameda, F.; Baró, T.; Iglesias, M.; Virtanen, I.; de Herreros, A.G. Snail1 protein in the stroma as a new putative prognosis marker for colon tumours. *PLoS ONE* **2009**, *4*, e5595. [CrossRef] [PubMed]

19. Shioiri, M.; Shida, T.; Koda, K.; Oda, K.; Seike, K.; Nishimura, M.; Takano, S.; Miyazaki, M. Slug expression is an independent prognostic parameter for poor survival in colorectal carcinoma patients. *Br. J. Cancer* **2006**, *94*, 1816–1822. [CrossRef] [PubMed]

20. Gomez, I.; Peña, C.; Herrera, M.; Muñoz, C.; Larriba, M.J.; Garcia, V.; Dominguez, G.; Silva, J.; Rodriguez, R.; de Herreros, A.; et al. TWIST1 is expressed in colorectal carcinomas and predicts patient survival. *PLoS ONE* **2011**, *6*, e18023. [CrossRef] [PubMed]

21. Yu, H.; Jin, G.Z.; Liu, K.; Dong, H.; Yu, H.; Duan, J.C.; Li, Z.; Dong, W.; Cong, W.M.; Yang, J.H. Twist2 is a valuable prognostic biomarker for colorectal cancer. *World J. Gastroenterol.* **2013**, *19*, 2404–2411. [CrossRef] [PubMed]

22. Zhang, G.J.; Zhou, T.; Tian, H.P.; Liu, Z.L.; Xia, S.S. High expression of ZEB1 correlates with liver metastasis and poor prognosis in colorectal cancer. *Oncol. Lett.* **2013**, *5*, 564–568. [PubMed]

23. Kahlert, C.; Lahes, S.; Radhakrishnan, P.; Dutta, S.; Mogler, C.; Herpel, E.; Brand, K.; Steinert, G.; Schneider, M.; Mollenhauer, M.; et al. Overexpression of ZEB2 at the invasion front of colorectal cancer is an independent prognostic marker and regulates tumor invasion in vitro. *Clin. Cancer Res.* **2011**, *17*, 7654–7663. [CrossRef] [PubMed]

24. Skog, M.; Bono, P.; Lundin, M.; Lundin, J.; Louhimo, J.; Linder, N.; Petrova, T.V.; Andersson, L.C.; Joensuu, H.; Alitalo, K.; et al. Expression and prognostic value of transcription factor PROX1 in colorectal cancer. *Br. J. Cancer* **2011**, *105*, 1346–1351. [CrossRef] [PubMed]

25. Li, Q.; Wu, J.; Wei, P.; Xu, Y.; Zhuo, C.; Wang, Y.; Li, D.; Cai, S. Overexpression of forkhead Box C2 promotes tumor metastasis and indicates poor prognosis in colon cancer via regulating epithelial-mesenchymal transition. *Am. J. Cancer Res.* **2015**, *5*, 2022–2034. [PubMed]

26. Weng, W.; Okugawa, Y.; Toden, S.; Toiyama, Y.; Kusunoki, M.; Goel, A. FOXM1 and FOXQ1 are promising prognostic biomarkers and novel targets of tumor-suppressive miR-342 in human colorectal cancer. *Clin. Cancer Res.* **2016**, *22*, 4947–4957. [CrossRef] [PubMed]

27. Li, D.; Li, Q.; Cai, S.; Xie, K. Contribution of FOXC1 to the progression and metastasis and prognosis of human colon cancer. *J. Clin. Oncol.* **2015**, *33*, 636. [CrossRef]

28. Chu, X.Y.; Zhu, Z.M.; Chen, L.B.; Wang, J.H.; Su, Q.S.; Yang, J.R.; Lin, Y.; Xue, L.J.; Liu, X.B.; Mo, X.B. FOXM1 expression correlates with tumor invasion and a poor prognosis of colorectal cancer. *Acta Histochem.* **2012**, *114*, 755–762. [CrossRef] [PubMed]

29. Barrallo-Gimeno, A.; Nieto, M.A. The Snail genes as inducers of cell movement and survival: Implications in development and cancer. *Development* **2005**, *132*, 3151–3161. [CrossRef] [PubMed]

30. Peinado, H.; Ballestar, E.; Esteller, M.; Cano, A. Snail mediates E-cadherin repression by the recruitment of the sin3A/histone deacetylase 1 (HDAC1)/HDAC2 complex. *Mol. Cell. Biol.* **2004**, *24*, 306–319. [CrossRef] [PubMed]

31. Herranz, N.; Pasini, D.; Díaz, V.M.; Francí, C.; Gutierrez, A.; Dave, N.; Escrivà, M.; Hernandez-Muñoz, I.; Di Croce, L.; Helin, K.; et al. Polycomb complex 2 is required for E-cadherin repression by the Snail1 transcription factor. *Mol. Cell. Biol.* **2008**, *28*, 4772–4781. [CrossRef] [PubMed]

32. Tong, Z.T.; Cai, M.Y.; Wang, X.G.; Kong, L.L.; Mai, S.J.; Liu, Y.H.; Zhang, H.B.; Liao, Y.J.; Zheng, F.; Zhu, W.; et al. EZH2 supports nasopharyngeal carcinoma cell aggressiveness by forming a co-repressor complex with HDAC1/HDAC2 and Snail to inhibit E-cadherin. *Oncogene* **2012**, *31*, 583–594. [CrossRef] [PubMed]

33. Lin, T.; Ponn, A.; Hu, X.; Law, B.K.; Lu, J. Requirement of the histone demethylase LSD1 in Snai1-mediated transcriptional repression during epithelial to mesenchymal transition. *Oncogene* **2010**, *29*, 4896–4904. [CrossRef] [PubMed]

34. Dong, C.; Wu, Y.; Yao, J.; Wang, Y.; Yu, Y.; Rychahou, P.G.; Evers, B.M.; Zhou, B.P. G9a interacts with SNAIL and is critical for SNAIL-mediated E-cadherin repression in human breast cancer. *J. Clin. Investig.* **2012**, *122*, 1469–1486. [CrossRef] [PubMed]

35. Dong, C.; Wu, Y.; Wang, Y.; Wang, C.; Kang, T.; Rychahou, P.G.; Chi, Y.I.; Evers, B.M.; Zhou, B.P. Interaction with Suv39H1 is critical for SNAIL-mediated E-cadherin repression in breast cancer. *Oncogene* **2013**, *32*, 1351–1362. [CrossRef] [PubMed]

36. Thuault, S.; Tan, E.J.; Peinado, H.; Cano, A.; Heldin, C.H.; Moustakas, A. HMGA2 and Smads coregulate SNAIL1 expression during induction of epithelial toto-mesenchymal transition. *J. Biol. Chem.* **2008**, *283*, 33437–33446. [CrossRef] [PubMed]

37. Smit, M.A.; Geiger, T.R.; Song, J.Y.; Gitelman, I.; Peeper, D.S. A Twist-SNAIL axis critical for TrkB-induced epithelial to mesenchymal transition-like transformation, anoikis resistance, and metastasis. *Mol. Cell. Biol.* **2009**, *29*, 3722–3737. [CrossRef] [PubMed]

38. Guaita, S.; Puig, I.; Franci, C.; Garrido, M.; Dominguez, D.; Batlle, E.; Sancho, E.; Dedhar, S.; De Herreros, A.G.; Baulida, J. SNAIL induction of epithelial to mesenchymal transition in tumor cells is accompanied by MUC1 repression and ZEB1 expression. *J. Biol. Chem.* **2002**, *277*, 39209–39220. [CrossRef] [PubMed]

39. Stemmer, V.; de Craene, B.; Berx, G.; Behrens, J. Snail promotes WNT target gene expression and interacts with β-catenin. *Oncogene* **2008**, *27*, 5075–5080. [CrossRef] [PubMed]

40. Saad, S.; Stanners, S.R.; Yong, R.; Tang, O.; Pollock, C.A. Notch mediated epithelial to mesenchymal transformation is associated with increased expression of the SNAIL transcription factor. *Int. J. Biochem. Cell Biol.* **2010**, *42*, 1115–1122. [CrossRef] [PubMed]

41. Liu, C.W.; Li, C.H.; Peng, Y.J.; Cheng, Y.W.; Chen, H.W.; Liao, P.L.; Kang, J.J.; Yeng, M.H. SNAIL regulates Nanog status during the epithelial to mesenchymal transition via the Smad1/AKT/GSK3beta signaling pathway in non-small-cell lung cancer. *Oncotarget* **2014**, *5*, 3880–3894. [CrossRef] [PubMed]

42. Yang, X.; Li, L.; Huang, Q.; Xu, W.; Cai, X.; Zhang, J.; Yan, W.; Song, D.; Liu, T.; Zhou, W.; et al. WNT signaling through SNAIL1 and Zeb1 regulates bone metastasis in lung cancer. *Am. J. Cancer Res.* **2015**, *5*, 748–755. [PubMed]

43. Pálmer, H.G.; Larriba, M.J.; García, J.M.; Ordóñez-Morán, P.; Peña, C.; Peiró, S.; Puig, I.; Rodríguez, R.; de la Fuente, R.; Bernad, A. The transcription factor SNAIL represses vitamin D receptor expression and responsiveness in human colon cancer. *Nat. Med.* **2004**, *10*, 917–919. [CrossRef] [PubMed]

44. Hwang, W.L.; Yang, M.H.; Tsai, M.L.; Lan, H.Y.; Su, S.H.; Chang, S.C.; Teng, H.W.; Yang, S.H.; Lan, Y.T.; Chiou, S.H.; et al. SNAIL regulates interleukin-8 expression, stem cell-like activity, and tumorigenicity of human colorectal carcinoma cells. *Gastroenterology* **2011**, *141*, 279–291. [CrossRef] [PubMed]

45. Yang, J.; Mani, S.A.; Donaher, J.L.; Ramaswamy, S.; Itzykson, R.A.; Come, C.; Savagner, P.; Gitelman, I.; Richardson, A.; Weinberg, R.A. Twist, a master regulator of morphogenesis, plays an essential role in tumor metastasis. *Cell* **2004**, *117*, 927–939. [CrossRef] [PubMed]

46. Yang, F.; Sun, L.; Li, Q.; Han, X.; Lei, L.; Zhang, H.; Shang, Y. SET8 promotes epithelial-mesenchymal transition and confers TWIST dual transcriptional activities. *EMBO J.* **2012**, *31*, 110–123. [CrossRef] [PubMed]

47. Garcia-Palmero, I.; Torres, S.; Bartolome, R.A.; Pelaez-Garcia, A.; Larriba, M.J.; Lopez-Lucendo, M.; Pena, C.; Escudero-Paniagua, B.; Munoz, A.; Casal, J.I. Twist1-induced activation of human fibroblasts promotes matrix stiffness by upregulating palladin and collagen alpha1(VI). *Oncogene* **2016**, *35*, 5224–5236. [CrossRef] [PubMed]

48. Celesti, G.; Di Caro, G.; Bianchi, P.; Grizzi, F.; Basso, G.; Marchesi, F.; Doni, A.; Marra, G.; Roncalli, M.; Mantovani, A.; et al. Presence of Twist1-positive neoplastic cells in the stroma of chromosome-unstable colorectal tumors. *Gastroenterology* **2013**, *145*, 647–657. [CrossRef] [PubMed]

49. Galván, J.A.; Helbling, M.; Koelzer, V.H.; Tschan, M.P.; Berger, M.D.; Hädrich, M.; Schnüriger, B.; Karamitopoulou, E.; Dawson, H.; Inderbitzin, D.; et al. TWIST1 and TWIST2 promoter methylation and protein expression in tumor stroma influence the epithelial-mesenchymal transition-like tumor budding phenotype in colorectal cancer. *Oncotarget* **2015**, *6*, 874–885. [CrossRef] [PubMed]

50. Postigo, A.A.; Dean, D.C. ZEB represses transcription through interaction with the corepressor CtBP. *Proc. Natl. Acad. Sci. USA* **1999**, *96*, 6683–6688. [CrossRef] [PubMed]

51. Sánchez-Tilló, E.; Lázaro, A.; Torrent, R.; Cuatrecasas, M.; Vaquero, E.C.; Castells, A.; Engel, P.; Postigo, A. ZEB1 represses E-cadherin and induces an EMT by recruiting the SWI/SNF chromatin-remodeling protein BRG1. *Oncogene* **2010**, *29*, 3490–3500. [CrossRef] [PubMed]

52. Postigo, A.A.; Depp, J.L.; Taylor, J.J.; Kroll, K.L. Regulation of Smad signaling through a differential recruitment of coactivators and corepressors by ZEB proteins. *EMBO J.* **2003**, *22*, 2453–2462. [CrossRef] [PubMed]

53. Wang, J.; Scully, K.; Zhu, X.; Cai, L.; Zhang, J.; Prefontaine, G.G.; Krones, A.; Ohgi, K.A.; Zhu, P.; Garcia-Bassets, I.; et al. Opposing LSD1 complexes function in developmental gene activation and repression programmes. *Nature* **2007**, *446*, 882–887. [CrossRef] [PubMed]

54. Aigner, K.; Dampier, B.; Descovich, L.M.; Mikula, M.; Sultan, A.; Schreiber, M.; Mikulits, W.; Brabletz, T.; Strand, D.; Obrist, P.; et al. The transcription factor ZEB1 (deltaEF1) promotes tumour cell dedifferentiation by repressing master regulators of epithelial polarity. *Oncogene* **2007**, *26*, 6979–6988. [CrossRef] [PubMed]

55. Spaderna, S.; Schmalhofer, O.; Wahlbuhl, M.; Dimmler, A.; Bauer, K.; Sultan, A.; Hlubek, F.; Jung, A.; Strand, D.; Eger, A.; et al. The transcriptional repressor ZEB1 promotes metastasis and loss of cell polarity in cancer. *Cancer Res.* **2008**, *68*, 537–544. [CrossRef] [PubMed]

56. Sanchez-Tillo, E.; de Barrios, O.; Siles, L.; Amendola, P.G.; Darling, D.S.; Cuatrecasas, M.; Castells, A.; Postigo, A. ZEB1 promotes invasiveness of colorectal carcinoma cells through the opposing regulation of uPA and PAI-1. *Clin. Cancer Res.* **2013**, *19*, 1071–1081. [CrossRef] [PubMed]

57. Petrova, T.V.; Nykänen, A.; Norrmén, C.; Ivanov, K.I.; Andersson, L.C.; Haglund, C.; Puolakkainen, P.; Wempe, F.; von Melchner, H.; Gradwohl, G.; et al. Transcription factor PROX1 induces colon cancer progression by promoting the transition from benign to highly dysplastic phenotype. *Cancer Cell* **2008**, *13*, 407–419. [CrossRef] [PubMed]

58. Lu, M.H.; Huang, C.C.; Pan, M.R.; Chen, H.H.; Hung, W.C. Prospero homeobox 1 promotes epithelial-mesenchymal transition in colon cancer cells by inhibiting E-cadherin via miR-9. *Clin. Cancer Res.* **2012**, *18*, 6416–6425. [CrossRef] [PubMed]

59. Kaneda, H.; Arao, T.; Tanaka, K.; Tamura, D.; Aomatsu, K.; Kudo, K.; Sakai, K.; De Velasco, M.A.; Matsumoto, K.; Fujita, Y.; et al. FOXQ1 is overexpressed in colorectal cancer and enhances tumorigenicity and tumor growth. *Cancer Res.* **2010**, *70*, 2053–2063. [CrossRef] [PubMed]

60. Zhang, H.; Meng, F.; Liu, G.; Zhang, B.; Zhu, J.; Wu, F.; Ethier, S.P.; Miller, F.; Wu, G. Forkhead transcription factor foxq1 promotes epithelial-mesenchymal transition and breast cancer metastasis. *Cancer Res.* **2011**, *71*, 1292–1301. [CrossRef] [PubMed]

61. Abba, M.; Patil, N.; Rasheed, K.; Nelson, L.D.; Mudduluru, G.; Leupold, J.H.; Allgayer, H. Unraveling the role of FOXQ1 in colorectal cancer metastasis. *Mol. Cancer Res.* **2013**, *11*, 1017–1028. [CrossRef] [PubMed]

62. Wierstra, I. FOXM1 (Forkhead box M1) in tumorigenesis: Overexpression in human cancer, implication in tumorigenesis, oncogenic functions, tumor-suppressive properties, and target of anticancer therapy. *Adv. Cancer Res.* **2013**, *119*, 191–419. [PubMed]

63. Yang, C.; Chen, H.; Tan, G.; Gao, W.; Cheng, L.; Jiang, X.; Yu, L.; Tan, Y. FOXM1 promotes the epithelial to mesenchymal transition by stimulating the transcription of Slug in human breast cancer. *Cancer Lett.* **2013**, *340*, 104–112. [CrossRef] [PubMed]

64. Zhang, X.; Zhang, L.; Du, Y.; Zheng, H.; Zhang, P.; Sun, Y.; Wang, Y.; Chen, J.; Ding, P.; Wang, N.; et al. A novel FOXM1 isoform, FOXM1D, promotes epithelial-mesenchymal transition and metastasis through ROCKs activation in colorectal cancer. *Oncogene* **2016**, *36*, 807–819. [CrossRef] [PubMed]

65. Mohri, Y. Prognostic significance of E-cadherin expression in human colorectal cancer tissue. *Surg. Today* **1997**, *27*, 606–612. [CrossRef] [PubMed]

66. He, X.; Chen, Z.; Jia, M.; Zhao, X. Downregulated E-cadherin expression indicates worse prognosis in Asian patients with colorectal cancer: Evidence from meta-analysis. *PLoS ONE* **2013**, *8*, e70858. [CrossRef] [PubMed]

67. Roberts, A.B.; Wakefield, L.M. The two faces of transforming growth factor beta in carcinogenesis. *Proc. Natl. Acad. Sci. USA* **2003**, *100*, 8621–8623. [CrossRef] [PubMed]

68. Pohl, M.; Radacz, Y.; Pawlik, N.; Schoeneck, A.; Baldus, S.E.; Munding, J.; Schmiegel, W.; Schwarte-Waldhoff, I.; Reinacher-Schick, A. SMAD4 mediates mesenchymal-epithelial reversion in SW480 colon carcinoma cells. *Anticancer Res.* **2010**, *30*, 2603–2613. [PubMed]

69. Zhao, S.; Venkatasubbarao, K.; Lazor, J.W.; Sperry, J.; Jin, C.; Cao, L.; Freeman, J.W. Inhibition of STAT3 Tyr705 phosphorylation by SMAD4 suppresses transforming growth factor beta-mediated invasion and metastasis in pancreatic cancer cells. *Cancer Res.* **2008**, *68*, 4221–4228. [CrossRef] [PubMed]

70. Voorneveld, P.W.; Kodach, L.L.; Jacobs, R.J.; Liv, N.; Zonnevylle, A.C.; Hoogenboom, J.P.; Biemond, I.; Verspaget, H.W.; Hommes, D.W.; de Rooij, K.; et al. Loss of SMAD4 alters BMP signaling to promote colorectal cancer cell metastasis via activation of Rho and ROCK. *Gastroenterology* **2014**, *147*, 196–208. [CrossRef] [PubMed]

71. Labelle, M.; Begum, S.; Hynes, R.O. Direct signaling between platelets and cancer cells induces an epithelial-mesenchymal-like transition and promotes metastasis. *Cancer Cell* **2011**, *20*, 576–590. [CrossRef] [PubMed]

72. Brabletz, T.; Jung, A.; Reu, S.; Porzner, M.; Hlubek, F.; Kunz-Schughart, L.A.; Knuechel, R.; Kirchner, T. Variable beta-catenin expression in colorectal cancers indicates tumor progression driven by the tumor environment. *Proc. Natl. Acad. Sci. USA* **2001**, *98*, 10356–10361. [CrossRef] [PubMed]

73. Yook, J.I.; Li, X.Y.; Ota, I.; Fearon, E.R.; Weiss, S.J. WNT-dependent regulation of the E-cadherin repressor snail. *J. Biol. Chem.* **2005**, *280*, 11740–11748. [CrossRef] [PubMed]

74. Wu, Z.Q.; Li, X.Y.; Hu, C.Y.; Ford, M.; Kleer, C.G.; Weiss, S.J. Canonical WNT signaling regulates Slug activity and links epithelial-mesenchymal transition with epigenetic Breast Cancer 1, Early Onset (BRCA1) repression. *Proc. Natl. Acad. Sci. USA* **2012**, *109*, 16654–16659. [CrossRef] [PubMed]

75. Qi, L.; Sun, B.; Liu, Z.; Cheng, R.; Li, Y.; Zhao, X. WNT3a expression is associated with epithelial-mesenchymal transition and promotes colon cancer progression. *J. Exp. Clin. Cancer Res.* **2014**, *33*, 107. [CrossRef] [PubMed]

76. Gujral, T.S.; Chan, M.; Peshkin, L.; Sorger, P.K.; Kirschner, M.W.; MacBeath, G. A noncanonical Frizzled2 pathway regulates epithelial-mesenchymal transition and metastasis. *Cell* **2014**, *159*, 844–856. [CrossRef] [PubMed]

77. Lee, S.C.; Kim, O.H.; Lee, S.K.; Kim, S.J. IWR-1 inhibits epithelial-mesenchymal transition of colorectal cancer cells through suppressing WNT/beta-catenin signaling as well as survivin expression. *Oncotarget* **2015**, *6*, 27146–27159. [CrossRef] [PubMed]

78. Huang, A.; Zhou, H.; Zhao, H.; Quan, Y.; Feng, B.; Zheng, M. TMPRSS4 correlates with colorectal cancer pathological stage and regulates cell proliferation and self-renewal ability. *Cancer Biol. Ther.* **2014**, *15*, 297–304. [CrossRef] [PubMed]

79. Jung, H.; Lee, K.P.; Park, S.J.; Park, J.H.; Jang, Y.S.; Choi, S.Y.; Jung, J.G.; Jo, K.; Park, D.Y.; Yoon, J.H.; et al. TMPRSS4 promotes invasion, migration and metastasis of human tumor cells by facilitating an epithelial-mesenchymal transition. *Oncogene* **2008**, *27*, 2635–2647. [CrossRef] [PubMed]

80. Maschler, S.; Wirl, G.; Spring, H.; Bredow, D.V.; Sordat, I.; Beug, H.; Reichmann, E. Tumor cell invasiveness correlates with changes in integrin expression and localization. *Oncogene* **2005**, *24*, 2032–2041. [CrossRef] [PubMed]

81. Larzabal, L.; de Aberasturi, A.L.; Redrado, M.; Rueda, P.; Rodriguez, M.J.; Bodegas, M.E.; Montuenga, L.M.; Calvo, A. TMPRSS4 regulates levels of integrin α5 in NSCLC through miR-205 activity to promote metastasis. *Br. J. Cancer* **2014**, *4*, 764–774. [CrossRef] [PubMed]

82. De Aberasturi, A.L.; Calvo, A. TMPRSS4: An emerging potential therapeutic target in cancer. *Br. J. Cancer* **2015**, *112*, 4–8. [CrossRef] [PubMed]

83. Kim, S.; Kang, H.Y.; Nam, E.H.; Choi, M.S.; Zhao, X.F.; Hong, C.S.; Lee, J.W.; Lee, J.H.; Park, Y.K. TMPRSS4 induces invasion and epithelial-mesenchymal transition through upregulation of integrin alpha5 and its signaling pathways. *Carcinogenesis* **2010**, *31*, 597–606. [CrossRef] [PubMed]

84. Kang, S.; Min, H.J.; Kang, M.S.; Jung, M.G.; Kim, S. Discovery of novel 2-hydroxydiarylamide derivatives as TMPRSS4 inhibitors. *Bioorg. Med. Chem. Lett.* **2013**, *23*, 1748–1751. [CrossRef] [PubMed]

85. Zhu, X.L.; Zeng, Y.F.; Guan, J.; Li, Y.F.; Deng, Y.J.; Bian, X.W.; Ding, Y.Q.; Liang, L. FMNL2 is a positive regulator of cell motility and metastasis in colorectal carcinoma. *J. Pathol.* **2011**, *224*, 377–388. [CrossRef] [PubMed]

86. Li, Y.; Zhu, X.; Zeng, Y.; Wang, J.; Zhang, X.; Ding, Y.Q.; Liang, L. FMNL2 enhances invasion of colorectal carcinoma by inducing epithelial-mesenchymal transition. *Mol. Cancer Res.* **2010**, *8*, 1579–1590. [CrossRef] [PubMed]

87. Ren, X.L.; He, G.Y.; Li, X.M.; Men, H.; Yi, L.Z.; Lu, G.F.; Xin, S.N.; Wu, P.X.; Li, Y.L.; Liao, W.T.; et al. MicroRNA-206 functions as a tumor suppressor in colorectal cancer by targeting FMNL2. *J. Cancer Res. Clin. Oncol.* **2016**, *142*, 581–592. [CrossRef] [PubMed]

88. Li, B.; Xie, Z.; Li, Z.; Chen, S.; Li, B. MicroRNA-613 targets FMNL2 and suppresses progression of colorectal cancer. *Am. J. Transl. Res.* **2016**, *8*, 5475–5484. [PubMed]

89. Lu, G.; Sun, Y.; An, S.; Xin, S.; Ren, X.; Zhang, D.; Wu, P.; Liao, W.; Ding, Y.; Liang, L. MicroRNA-34a targets FMNL2 and E2F5 and suppresses the progression of colorectal cancer. *Exp. Mol. Pathol.* **2015**, *99*, 173–179. [CrossRef] [PubMed]

90. Liang, L.; Li, X.; Zhang, X.; Lv, Z.; He, G.; Zhao, W.; Ren, X.; Li, Y.; Bian, X.; Liao, W.; et al. MicroRNA-137, an HMGA1 target, suppresses colorectal cancer cell invasion and metastasis in mice by directly targeting FMNL2. *Gastroenterology* **2013**, *144*, 624–635. [CrossRef] [PubMed]

91. Jenkins, Z.A.; Haag, P.G.; Johansson, H.E. Human eIF5A2 on chromosome 3q25–q27 is a phylogenetically conserved vertebrate variant of eukaryotic translation initiation factor 5A with tissue-specific expression. *Genomics* **2001**, *71*, 101–109. [CrossRef] [PubMed]

92. Bao, Y.; Lu, Y.; Wang, X.; Feng, W.; Sun, X.; Guo, H.; Tang, C.; Zhang, X.; Shi, Q.; Yu, H. Eukaryotic translation initiation factor 5A2 (eIF5A2) regulates chemoresistance in colorectal cancer through epithelial mesenchymal transition. *Cancer Cell Int.* **2015**, *15*, 109. [CrossRef] [PubMed]

93. Zhu, W.; Cai, M.Y.; Tong, Z.T.; Dong, S.S.; Mai, S.J.; Liao, Y.J.; Bian, X.W.; Lin, M.C.; Kung, H.F.; Zeng, Y.X.; et al. Overexpression of EIF5A2 promotes colorectal carcinoma cell aggressiveness by upregulating MTA1 through C-myc to induce epithelial-mesenchymaltransition. *Gut* **2012**, *61*, 562–575. [CrossRef] [PubMed]

94. Uchiyama, T.; Kawabata, H.; Miura, Y.; Yoshioka, S.; Iwasa, M.; Yao, H.; Sakamoto, S.; Fujimoto, M.; Haga, H.; Kadowaki, N.; et al. The role of growth differentiation factor 15 in the pathogenesis of primary myelofibrosis. *Cancer Med.* **2015**, *4*, 1558–1572. [CrossRef] [PubMed]

95. Mehta, R.S.; Song, M.; Bezawada, N.; Wu, K.; Garcia-Albeniz, X.; Morikawa, T.; Fuchs, C.S.; Ogino, S.; Giovannucci, E.L.; Chan, A.T. A prospective study of macrophage inhibitory cytokine-1 (MIC-1/GDF15) and risk of colorectal cancer. *J. Natl. Cancer Inst.* **2014**, *106*, dju016. [CrossRef] [PubMed]

96. Wallin, U.; Glimelius, B.; Jirstrom, K.; Darmanis, S.; Nong, R.Y.; Ponten, F.; Johansson, C.; Pahlman, L.; Birgisson, H. Growth differentiation factor 15: A prognostic marker for recurrence in colorectal cancer. *Br. J. Cancer* **2011**, *104*, 1619–1627. [CrossRef] [PubMed]

97. Mehta, R.S.; Chong, D.Q.; Song, M.; Meyerhardt, J.A.; Ng, K.; Nishihara, R.; Qian, Z.; Morikawa, T.; Wu, K.; Giovannucci, E.L.; et al. Association between plasma levels of macrophage inhibitory Cytokine-1 before diagnosis of colorectal cancer and mortality. *Gastroenterology* **2015**, *149*, 614–622. [CrossRef] [PubMed]

98. Li, C.; Wang, J.; Kong, J.; Tang, J.; Wu, Y.; Xu, E.; Zhang, H.; Lai, M. GDF15 promotes EMT and metastasis in colorectal cancer. *Oncotarget* **2016**, *7*, 860–872. [CrossRef] [PubMed]

99. Baba, Y.; Nosho, K.; Shima, K.; Irahara, N.; Chan, A.T.; Meyerhardt, J.A.; Chung, D.C.; Giovannucci, E.L.; Fuchs, C.S.; Ogino, S. HIF1A overexpression is associated with poor prognosis in a cohort of 731 colorectal cancers. *Am. J. Pathol.* **2010**, *176*, 2292–2301. [CrossRef] [PubMed]

100. Zhang, W.; Shi, X.; Peng, Y.; Wu, M.; Zhang, P.; Xie, R.; Wu, Y.; Yan, Q.; Liu, S.; Wang, J. HIF-1alpha promotes epithelial-mesenchymal transition and metastasis through direct regulation of ZEB1 in colorectal cancer. *PLoS ONE* **2015**, *10*, e0129603. [CrossRef]

101. Santoyo-Ramos, P.; Likhatcheva, M.; Garcia-Zepeda, E.A.; Castaneda-Patlan, M.C.; Robles-Flores, M. Hypoxia-inducible factors modulate the stemness and malignancy of colon cancer cells by playing opposite roles in canonical WNT signaling. *PLoS ONE* **2014**, *9*, e112580. [CrossRef] [PubMed]

102. Kim, J.H.; Hwang, Y.J.; Han, S.H.; Lee, Y.E.; Kim, S.; Kim, Y.J.; Cho, J.H.; Kwon, K.A.; Kim, J.H.; Kim, S.H. Dexamethasone inhibits hypoxia-induced epithelial-mesenchymal transition in colon cancer. *World J. Gastroenterol.* **2015**, *21*, 9887–9899. [CrossRef] [PubMed]

103. Gossett, L.A.; Kelvin, D.J.; Sternberg, E.A.; Olson, E.N. A new myocyte-specific enhancer-binding factor that recognizes a conserved element associated with multiple muscle-specific genes. *Mol. Cell. Biol.* **1989**, *9*, 5022–5033. [CrossRef] [PubMed]

104. Potthoff, M.J.; Olson, E.N. MEF2: A central regulator of diverse developmental programs. *Development* **2007**, *134*, 4131–4140. [CrossRef] [PubMed]

105. Kinsey, T.A.; Zhang, C.L.; Olson, E.N. MEF2: A calcium-dependent regulator of cell division, differentiation and death. *Trends Biochem. Sci.* **2002**, *27*, 40–47.

106. Su, L.; Luo, Y.; Yang, Z.; Yang, J.; Yao. C.; Cheng, F.; Shan, J.; Chen, J.; Li, F.; Liu, L.; et al. MEF2D transduces microenvironment stimuli to ZEB1 to promote epithelial-mesenchymal transition and metastasis in colorectal cancer. *Cancer Res.* **2016**, *76*, 5054–5067. [CrossRef] [PubMed]

107. Wang, Y.; Wu, N.; Sun, D.; Sun, H.; Tong, D.; Liu, D.; Pang, B.; Li, S.; Wei, J.; Dai, J.; et al. NUBPL, a novel metastasis-related gene, promotes colorectal carcinoma cell motility by inducing epithelial-mesenchymal transition. *Cancer Sci.* **2017**, *108*, 1169–1176. [CrossRef] [PubMed]

108. Grandclement, C.; Pallandre, J.R.; Valmary Degano, S.; Viel, E.; Bouard, A.; Balland, J.; Remy-Martin, J.P.; Simon, B.; Rouleau, A.; Boireau, W.; et al. Neuropilin-2 expression promotes TGF-beta1-mediated epithelial to mesenchymal transition in colorectal cancer cells. *PLoS ONE* **2011**, *6*, e20444. [CrossRef] [PubMed]

109. Gemmill, R.M.; Nasarre, P.; Nair-Menon, J.; Cappuzzo, F.; Landi, L.; D'Incecco, A.; Uramoto, H.; Yoshida, T.; Haura, E.B.; Armeson, K.; et al. The neuropilin 2 isoform NRP2b uniquely supports TGFbeta-mediated progression in lung cancer. *Sci. Signal.* **2017**, *10*, eaag0528. [CrossRef] [PubMed]

110. Horak, P.; Tomasich, E.; Vanhara, P.; Kratochvilova, K.; Anees, M.; Marhold, M.; Lemberger, C.E.; Gerschpacher, M.; Horvat, R.; Sibilia, M.; et al. TUSC3 loss alters the ER stress response and accelerates prostate cancer growth in vivo. *Sci. Rep.* **2014**, *4*. [CrossRef] [PubMed]

111. Jiang, Z.; Guo, M.; Zhang, X.; Yao, L.; Shen, J.; Ma, G.; Liu, L.; Zhao, L.; Xie, C.; Liang, H.; et al. TUSC3 suppresses glioblastoma development by inhibiting Akt signaling. *Tumour Biol.* **2016**, *37*, 12039–12047. [CrossRef] [PubMed]

112. Kratochvilova, K.; Horak, P.; Esner, M.; Soucek, K.; Pils, D.; Anees, M.; Tomasich, E.; Drafi, F.; Jurtikova, V.; Hampl, A.; et al. Tumor suppressor candidate 3 (TUSC3) prevents the epithelial-to-mesenchymal transition and inhibits tumor growth by modulating the endoplasmic reticulum stress response in ovarian cancer cells. *Int. J. Cancer* **2015**, *137*, 1330–1340. [CrossRef] [PubMed]

113. Yachida, S.; Jones, S.; Bozic, I.; Antal, T.; Leary, R.; Fu, B.; Kamiyama, M.; Hruban, R.H.; Eshleman, J.R.; Nowak, M.A.; et al. Distant metastasis occurs late during the genetic evolution of pancreatic cancer. *Nature* **2010**, *467*, 1114–1117. [CrossRef] [PubMed]

114. Levy, A.; Dang, U.C.; Bookstein, R. High-density screen of human tumor cell lines for homozygous deletions of loci on chromosome arm 8p. *Genes Chromosomes Cancer* **1999**, *24*, 42–47. [CrossRef]

115. Takanishi, D.M.; Kim, S.Y.; Kelemen, P.R.; Yaremko, M.L.; Kim, A.H.; Ramesar, J.E.; Horrigan, S.K.; Montag, A.; Michelassi, F.; Westbrook, C.A. Chromosome 8 Losses in Colorectal Carcinoma: Localization and Correlation With Invasive Disease. *J. Mol. Diagn.* **1997**, *2*, 3–10. [CrossRef]

116. Gu, Y.; Wang, Q.; Guo, K.; Qin, W.; Liao, W.; Wang, S.; Ding, Y.; Lin, J. TUSC3 promotes colorectal cancer progression and epithelial-mesenchymal transition (EMT) through WNT/β-catenin and MAPK signalling. *J. Pathol.* **2016**, *239*, 60–71. [CrossRef] [PubMed]

117. Halder, S.K.; Anumanthan, G.; Maddula, R.; Mann, J.; Chytil, A.; Gonzalez, A.L.; Washington, M.K.; Moses, H.L.; Beauchamp, R.D.; Datta, P.K. Oncogenic function of a novel WD-domain protein, STRAP, in human carcinogenesis. *Cancer Res.* **2006**, *66*, 6156–6166. [CrossRef] [PubMed]

118. Anumanthan, G.; Halder, S.K.; Friedman, D.B.; Datta, P.K. Oncogenic serine-threonine kinase receptor-associated protein modulates the function of Ewing sarcoma protein through a novel mechanism. *Cancer Res.* **2006**, *66*, 10824–10832. [CrossRef] [PubMed]

119. Matsuda, S.; Katsumata, R.; Okuda, T.; Yamamoto, T.; Miyazaki, K.; Senga, T.; Machida, K.; Thant, A.A.; Nakatsugawa, S.; Hamaguchi, M. Molecular cloning and characterization of human MAWD, a novel protein containing WD-40 repeats frequently overexpressed in breast cancer. *Cancer Res.* **2000**, *60*, 13–17. [PubMed]

120. Li, D.; Roberts, R. WD-repeat proteins: Structure characteristics, biological function, and their involvement in human diseases. *Cell. Mol. Life Sci.* **2001**, *58*, 2085–2097. [CrossRef] [PubMed]

121. Datta, P.K.; Chytil, A.; Gorska, A.E.; Moses, H.L. Identification of STRAP, a novel WD domain protein in transforming growth factor-beta signaling. *J. Biol. Chem.* **1998**, *273*, 34671–34674. [CrossRef] [PubMed]

122. Kashikar, N.D.; Reiner, J.; Datta, A.; Datta, P.K. Serine threonine receptor-associated protein (STRAP) plays a role in the maintenance of mesenchymal morphology. *Cell Signal.* **2010**, *22*, 138–149. [CrossRef] [PubMed]

123. Yuan, G.; Zhang, B.; Yang, S.; Jin, L.; Datta, A.; Bae, S.; Chen, X.; Datta, P.K. Novel role of STRAP in progression and metastasis of colorectal cancer through WNT/beta-catenin signaling. *Oncotarget* **2016**, *7*, 16023–16037. [CrossRef] [PubMed]

124. Chi, Y.; Zhou, D. MicroRNAs in colorectal carcinoma—From pathogenesis to therapy. *J. Exp. Clin. Cancer Res.* **2016**, *35*, 43. [CrossRef] [PubMed]

125. Hur, K.; Toiyama, Y.; Takahashi, M.; Balaguer, F.; Nagasaka, T.; Koike, J.; Hemmi, H.; Koi, M.; Boland, C.R.; Goel, A. MicroRNA-200c modulates epithelial-to-mesenchymal transition (EMT) in human colorectal cancer metastasis. *Gut* **2013**, *62*, 1315–1326. [CrossRef] [PubMed]

126. Korpal, M.; Lee, E.S.; Hu, G.; Kang, Y. The miR-200 family inhibits epithelial-mesenchymal transition and cancer cell migration by direct targeting of E-cadherin transcriptional repressors ZEB1 and ZEB2. *J. Biol. Chem.* **2008**, *283*, 14910–14914. [CrossRef] [PubMed]

127. Park, S.M.; Gaur, A.B.; Lengyel, E.; Peter, M.E. The miR-200 family determines the epithelial phenotype of cancer cells by targeting the E-cadherin repressors ZEB1 and ZEB2. *Genes Dev.* **2008**, *22*, 894–907. [CrossRef] [PubMed]

128. Davalos, V.; Moutinho, C.; Villanueva, A.; Boque, R.; Silva, P.; Carneiro, F.; Esteller, M. Dynamic epigenetic regulation of the MicroRNA-200 family mediates epithelial and mesenchymal transitions in human tumorigenesis. *Oncogene* **2012**, *31*, 2062–2074. [CrossRef] [PubMed]

129. Chen, M.L.; Liang, L.S.; Wang, X.K. miR-200c inhibits invasion and migration in human colon cancer cells SW480/620 by targeting ZEB1. *Clin. Exp. Metastasis* **2012**, *29*, 457–469. [CrossRef] [PubMed]

130. Sun, Y.; Shen, S.; Liu, X.; Tang, H.; Wang, Z.; Yu, Z.; Li, X.; Wu, M. MiR-429 inhibits cells growth and invasion and regulates EMT-related marker genes by targeting Onecut2 in colorectal carcinoma. *Mol. Cell. Biochem.* **2014**, *390*, 19–30. [CrossRef] [PubMed]

131. Zheng, Y.; Luo, H.; Shi, Q.; Hao, Z.; Ding, Y.; Wang, Q.; Li, S.; Xiao, G.; Tong, S. MiR-132 inhibits colorectal cancer invasion and metastasis via directly targeting ZEB2. *World J. Gastroenterol.* **2014**, *20*, 6515–6522. [CrossRef] [PubMed]

132. Geng, L.; Chaudhuri, A.; Talmon, G.; Wisecarver, J.L.; Are, C.; Brattain, M.; Wang, J. MicroRNA-192 suppresses liver metastasis of colon cancer. *Oncogene* **2014**, *33*, 5332–5340. [CrossRef] [PubMed]

133. Sun, Z.; Zhang, Z.; Liu, Z.; Qiu, B.; Liu, K.; Dong, G. MicroRNA-335 inhibits invasion and metastasis of colorectal cancer by targeting ZEB2. *Med. Oncol.* **2014**, *31*, 982. [CrossRef] [PubMed]

134. Hahn, S.; Jackstadt, R.; Siemens, H.; Hunten, S.; Hermeking, H. SNAIL and miR-34a feed-forward regulation of ZNF281/ZBP99 promotes epithelial-mesenchymal transition. *EMBO J.* **2013**, *32*, 3079–3095. [CrossRef] [PubMed]

135. Rokavec, M.; Oner, M.G.; Li, H.; Jackstadt, R.; Jiang, L.; Lodygin, D.; Kaller, M.; Horst, D.; Ziegler, P.K.; Schwitalla, S.; et al. IL-6R/STAT3/miR-34a feedback loop promotes EMT-mediated colorectal cancer invasion and metastasis. *J. Clin. Investig.* **2014**, *124*, 1853–1867. [CrossRef] [PubMed]

136. Siemens, H.; Jackstadt, R.; Hünten, S.; Kaller, M.; Menssen, A.; Götz, U.; Hermeking, H. MiR-34 and SNAIL form a double-negative feedback loop to regulate epithelial-mesenchymal transitions. *Cell Cycle* **2011**, *10*, 4256–4271. [CrossRef] [PubMed]

137. Long, L.; Huang, G.; Zhu, H.; Guo, Y.; Liu, Y.; Huo, J. Down-regulation of miR-138 promotes colorectal cancer metastasis via directly targeting TWIST2. *J. Transl. Med.* **2013**, *11*, 275. [CrossRef] [PubMed]

138. Meng, X.; Wu, J.; Pan, C.; Wang, H.; Ying, X.; Zhou, Y.; Yu, H.; Zuo, Y.; Pan, Z.; Liu, R.Y.; et al. Genetic and epigenetic down-regulation of microRNA-212 promotes colorectal tumor metastasis via dysregulation of MnSOD. *Gastroenterology* **2013**, *145*, 426–436. [CrossRef] [PubMed]

139. Zhao, H.; Xu, Z.; Qin, H.; Gao, Z.; Gao, L. MiR-30b regulates migration and invasion of human colorectal cancer via SIX1. *Biochem. J.* **2014**, *460*, 117–125. [CrossRef] [PubMed]

140. Zhao, H.; Dong, T.; Zhou, H.; Wang, L.; Huang, A.; Feng, B.; Quan, Y.; Jin, R.; Zhang, W.; Sun, J.; et al. MiR-320a suppresses colorectal cancer progression by targeting Rac1. *Carcinogenesis* **2014**, *35*, 886–895. [CrossRef] [PubMed]

141. Chen, J.; Zhang, H.; Chen, Y.; Qiao, G.; Jiang, W.; Ni, P.; Liu, X.; Ma, L. MiR-598 inhibits metastasis in colorectal cancer by suppressing JAG1/Notch2 pathway stimulating EMT. *Exp. Cell Res.* **2017**, *352*, 104–112. [CrossRef] [PubMed]

142. Zhao, S.; Sun, H.; Jiang, W.; Mi, Y.; Zhang, D.; Wen, Y.; Cheng, D.; Tang, H.; Wu, S.; Yu, Y.; et al. MiR-4775 promotes colorectal cancer invasion and metastasis via the Smad7/TGFβ-mediated epithelial to mesenchymal transition. *Mol. Cancer* **2017**, *16*, 12–20. [CrossRef] [PubMed]

143. Costa, V.; Lo Dico, A.; Rizzo, A.; Rajata, F.; Tripodi, M.; Alessandro, R.; Conigliaro, A. MiR-675-5p supports hypoxia induced epithelial to mesenchymal transition in colon cancer cells. *Oncotarget* **2017**, *8*, 24292–24302. [CrossRef] [PubMed]

144. Wang, B.; Li, W.; Liu, H.; Yang, L.; Liao, Q.; Cui, S.; Wang, H.; Zhao, L. MiR-29b suppresses tumor growth and metastasis in colorectal cancer via downregulating Tiam1 expression and inhibiting epithelial-mesenchymal transition. *Cell Death Dis.* **2014**, *5*, e1335. [CrossRef] [PubMed]

145. Hu, F.; Jiang, M.; Cao, X.; Liu, L.; Ge, Z.; Hu, J.; Li, X. MiR-363-3p inhibits the epithelial-to-mesenchymal transition and suppresses metastasis in colorectal cancer by targeting Sox4. *Biochem. Biophys. Res. Commun.* **2016**, *474*, 35–42. [CrossRef] [PubMed]

146. Xi, X.P.; Zhuang, J.; Teng, M.J.; Xia, L.J.; Yang, M.Y.; Liu, Q.G.; Chen, J.B. MicroRNA-17 induces epithelial-mesenchymal transition consistent with the cancer stem cell phenotype by regulating CYP7B1 expression in colon cancer. *Int. J. Mol. Med.* **2016**, *38*, 499–506. [CrossRef] [PubMed]

147. Li, Q.; Liang, X.; Wang, Y.; Meng, X.; Xu, Y.; Cai, S.; Wang, Z.; Liu, J.; Cai, G. MiR-139-5p inhibits the epithelial-mesenchymal transition and enhances the chemotherapeutic sensitivity of colorectal cancer cells by downregulating BCL2. *Sci. Rep.* **2016**, *6*, 27157. [CrossRef] [PubMed]

148. Cui, F.; Wang, S.; Lao, I.; Zhou, C.; Kong, H.; Bayaxi, N.; Li, J.; Chen, Q.; Zhu, T.; Zhu, H. MiR-375 inhibits the invasion and metastasis of colorectal cancer via targeting SP1 and regulating EMT-associated genes. *Oncol. Rep.* **2016**, *36*, 487–493. [CrossRef] [PubMed]

149. Zhang, N.; Shen, Q.; Zhang, P. MiR-497 suppresses epithelial-mesenchymal transition and metastasis in colorectal cancer cells by targeting fos-related antigen-1. *Onco Targets Ther.* **2016**, *9*, 6597–6604. [CrossRef] [PubMed]

150. Ferraro, A.; Kontos, C.K.; Boni, T.; Bantounas, I.; Siakouli, D.; Kosmidou, V.; Vlassi, M.; Spyridakis, Y.; Tsipras, I.; Zografos, G.; et al. Epigenetic regulation of mir-21 in colorectal cancer: Itgb4 as a novel mir-21 target and a three-gene network (mir-21-itgbeta4-pdcd4) as predictor of metastatic tumor potential. *Epigenetics* **2014**, *9*, 129–141. [CrossRef] [PubMed]

151. Liang, W.C.; Fu, W.M.; Wong, C.W.; Wang, Y.; Wang, W.M.; Hu, G.X.; Zhang, L.; Xiao, L.J.; Wan, D.C.; Zhang, J.F.; et al. The lncRNA H19 promotes epithelial to mesenchymal transition by functioning as miRNA sponges in colorectal cancer. *Oncotarget* **2015**, *6*, 22513–22525. [CrossRef] [PubMed]

152. Guo, Q.; Zhao, Y.; Chen, J.; Hu, J.; Wang, S.; Zhang, D.; Sun, Y. BRAF-activated long non-coding RNA contributes to colorectal cancer migration by inducing epithelial-mesenchymal transition. *Oncol. Lett.* **2014**, *8*, 869–875. [CrossRef] [PubMed]

153. Wu, Z.H.; Wang, X.L.; Tang, H.M.; Jiang, T.; Chen, J.; Lu, S.; Qiu, G.Q.; Peng, Z.H.; Yan, D.W. Long non-coding RNA HOTAIR is a powerful predictor of metastasis and poor prognosis and is associated with epithelial-mesenchymal transition in colon cancer. *Oncol. Rep.* **2014**, *32*, 395–402. [CrossRef] [PubMed]

154. Yue, B.; Qiu, S.; Zhao, S.; Liu, C.; Zhang, D.; Yu, F.; Peng, Z.; Yan, D. LncRNA-ATB mediated E-cadherin repression promotes the progression of colon cancer and predicts poor prognosis. *J. Gastroenterol. Hepatol.* **2016**, *31*, 595–603. [CrossRef] [PubMed]

155. Han, X.; Wang, L.; Ning, Y.; Li, S.; Wang, Z. Long non-coding RNA AFAP1-AS1 facilitates tumor growth and promotes metastasis in colorectal cancer. *Biol. Res.* **2016**, *49*, 36–46. [CrossRef] [PubMed]

156. Wang, L.; Zhao, Z.; Feng, W.; Ye, Z.; Dai, W.; Zhang, C.; Peng, J.; Wu, K. Long non-coding RNA TUG1 promotes colorectal cancer metastasis via EMT pathway. *Oncotarget* **2016**, *7*, 51713–51719. [CrossRef] [PubMed]

157. Shen, F.; Cai, W.S.; Feng, Z.; Chen, J.W.; Feng, J.H.; Liu, Q.C.; Fang, Y.P.; Li, K.P.; Xiao, H.Q.; Cao, J.; et al. Long non-coding RNA SPRY4-IT1 pormotes colorectal cancer metastasis by regulate epithelial-mesenchymal transition. *Oncotarget* **2017**, *8*, 14479–14486. [CrossRef] [PubMed]

158. Lu, M.; Liu, Z.; Li, B.; Wang, G.; Li, D.; Zhu, Y. The high expression of long non-coding RNA PANDAR indicates a poor prognosis for colorectal cancer and promotes metastasis by EMT pathway. *J. Cancer Res. Clin. Oncol.* **2017**, *143*, 71–81. [CrossRef] [PubMed]

159. Han, Y.; Zhao, Q.; Zhou, J.; Shi, R. MiR-429 mediates tumor growth and metastasis in colorectal cancer. *Am. J. Cancer Res.* **2017**, *7*, 218–233. [PubMed]

160. Cai, H.K.; Chen, X.; Tang, Y.H.; Deng, Y.C. MicroRNA-194 modulates epithelial-mesenchymal transition in human colorectal cancer metastasis. *Onco Targets Ther.* **2017**, *10*, 1269–1278. [CrossRef] [PubMed]

161. Guo, Y.H.; Wang, L.Q.; Li, B.; Xu, H.; Yang, J.H.; Zheng, L.S.; Yu, P.; Zhou, A.D.; Zhang, Y.; Xie, S.J.; et al. WNT/β-catenin pathway transactivates microRNA-150 that promotes EMT of colorectal cancer cells by suppressing CREB signaling. *Oncotarget* **2016**, *7*, 42513–42526. [CrossRef] [PubMed]

162. Tang, W.; Zhu, Y.; Gao, J.; Fu, J.; Liu, C.; Liu, Y.; Song, C.; Zhu, S.; Leng, Y.; Wang, G.; et al. MicroRNA-29a promotes colorectal cancer metastasis by regulating matrix metalloproteinase 2 and E-cadherin via KLF4al. *Br. J. Cancer* **2014**, *110*, 450–458. [CrossRef] [PubMed]

163. Gupta, G.P.; Massague, J. Cancer metastasis: Building a framework. *Cell* **2006**, *127*, 679–695. [CrossRef] [PubMed]

164. Lim, S.H.; Becker, T.M.; Chua, W.; Ng, W.L.; de Souza, P.; Spring, K.J. Circulating tumour cells and the epithelial mesenchymal transition in colorectal cancer. *J. Clin. Pathol.* **2014**, *67*, 848–853. [CrossRef] [PubMed]

165. Acloque, H.; Adams, M.S.; Fishwick, K.; Bronner-Fraser, M.; Nieto, M.A. Epithelial-mesenchymal transitions: The importance of changing cell state in development and disease. *J. Clin. Investig.* **2009**, *119*, 1438–1449. [CrossRef] [PubMed]

166. Pantel, K.; Brakenhoff, R.H. Dissecting the metastatic cascade. *Nat. Rev. Cancer* **2004**, *4*, 448–456. [CrossRef] [PubMed]

167. Steinert, G.; Schölch, S.; Niemietz, T.; Iwata, N.; García, S.A.; Behrens, B.; Voigt, A.; Kloor, M.; Benner, A.; Bork, U.; et al. Immune escape and survival mechanisms in circulating tumor cells of colorectal cancer. *Cancer Res.* **2014**, *74*, 1694–1704. [CrossRef] [PubMed]

168. Cohen, S.J.; Punt, C.J.; Iannotti, N.; Saidman, B.H.; Sabbath, K.D.; Gabrail, N.Y.; Picus, J.; Morse, M.; Mitchell, E.; Miller, M.C.; et al. Relationship of circulating tumor cells to tumor response, progression-free survival, and overall survival in patients with metastatic colorectal cancer. *J. Clin. Oncol.* **2008**, *26*, 3213–3221. [CrossRef] [PubMed]

169. Cohen, S.J.; Punt, C.J.; Iannotti, N.; Saidman, B.H.; Sabbath, K.D.; Gabrail, N.Y.; Picus, J.; Morse, M.A.; Mitchell, E.; Miller, M.C.; et al. Prognostic significance of circulating tumor cells in patients with metastatic colorectal cancer. *Ann. Oncol.* **2009**, *20*, 1223–1229. [CrossRef] [PubMed]

170. Satelli, A.; Mitra, A.; Brownlee, Z.; Xia, X.; Bellister, S.; Overman, M.J.; Kopetz, S.; Ellis, L.M.; Meng, Q.H.; Li, S. Epithelial-mesenchymal transitioned circulating tumor cells capture for detecting tumor progression. *Clin. Cancer Res.* **2015**, *21*, 899–906. [CrossRef] [PubMed]

171. Yokobori, T.; Iinuma, H.; Shimamura, T.; Imoto, S.; Sugimachi, K.; Ishii, H.; Iwatsuki, M.; Ota, D.; Ohkuma, M.; Iwaya, T.; et al. Plastin3 is a novel marker for circulating tumor cells undergoing the epithelial-mesenchymal transition and is associated with colorectal cancer prognosis. *Cancer Res.* **2013**, *73*, 2059–2069. [CrossRef] [PubMed]

172. Gorges, T.M.; Tinhofer, I.; Drosch, M.; Rose, L.; Zollner, T.M.; Krahn, T.; von Ahsen, O. Circulating tumour cells escape from EpCAM-based detection due to epithelial-to-mesenchymal transition. *BMC Cancer* **2012**, *12*, 2407–2412. [CrossRef] [PubMed]

Review

Pleiotropic Roles of Non-Coding RNAs in TGF-*β*-Mediated Epithelial-Mesenchymal Transition and Their Functions in Tumor Progression

Simon Grelet [1], Ariel McShane [2], Renaud Geslain [2,*] and Philip H. Howe [1,*]

[1] Department of Biochemistry and Molecular Biology, MUSC, Charleston, SC 29425, USA; grelet@musc.edu
[2] Laboratory of tRNA Biology, Department of Biology, College of Charleston, Charleston, SC 29424, USA; mcshaneab@g.cofc.edu
* Correspondence: geslainr@cofc.edu (R.G.); howep@musc.edu (P.H.H.);
Tel.: +1-843-953-8080 (R.G.); +1-843-792-9318 (P.H.H.)

Academic Editor: Joëlle Roche
Received: 22 May 2017; Accepted: 30 June 2017; Published: 1 July 2017

Abstract: Epithelial-mesenchymal transition (EMT) is a spatially- and temporally-regulated process involved in physiological and pathological transformations, such as embryonic development and tumor progression. While the role of TGF-β as an EMT-inducer has been extensively documented, the molecular mechanisms regulating this transition and their implications in tumor metastasis are still subjects of intensive debates and investigations. TGF-β regulates EMT through both transcriptional and post-transcriptional mechanisms, and recent advances underline the critical roles of non-coding RNAs in these processes. Although microRNAs and lncRNAs have been clearly identified as effectors of TGF-β-mediated EMT, the contributions of other atypical non-coding RNA species, such as piRNAs, snRNAs, snoRNAs, circRNAs, and even housekeeping tRNAs, have only been suggested and remain largely elusive. This review discusses the current literature including the most recent reports emphasizing the regulatory functions of non-coding RNA in TGF-β-mediated EMT, provides original experimental evidence, and advocates in general for a broader approach in the quest of new regulatory RNAs.

Keywords: epithelial-mesenchymal transition; tumor progression; metastasis; TGF-β; non-coding RNA; tRNA; post-transcriptional regulation

1. Introduction

Metastasis represents a critical step in tumor progression that is responsible for more than 90% of cancer-induced mortality. Despite tremendous efforts from the scientific community, the cellular and molecular events that specifically control metastatic colonization are still poorly understood.

Epithelial-mesenchymal transition (EMT) is key in both embryonic development and tumor metastasis. EMT consists of a fine-tuned phenotypic switch characterized by the loss of apical-basal polarity and cellular adhesion in epithelial cells [1,2]. Cells undergoing transition gradually express mesenchymal features, such as enhanced cytoskeletal rearrangement and extracellular matrix (ECM) degradation, both essential for cell motility (Figure 1). While the role of EMT in metastasis progression is still debated, its implication in the increased resistance seen in both conventional and targeted antitumor therapies is well established [1,3–8].

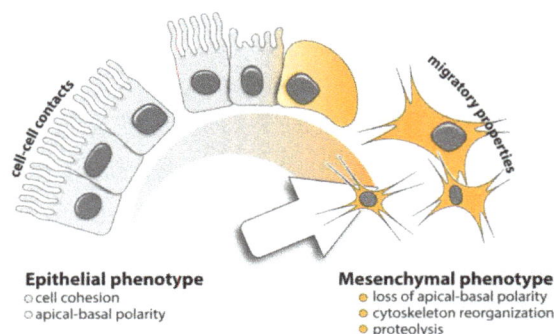

Epithelial phenotype
○ cell cohesion
○ apical-basal polarity

Mesenchymal phenotype
◉ loss of apical-basal polarity
◉ cytoskeleton reorganization
◉ proteolysis

Figure 1. Cell plasticity in EMT. Epithelial-mesenchymal transition is a multistep process allowing epithelial cells to acquire mesenchymal phenotype. Upon TGF-β exposure, epithelial cells lose their apical-basal polarity and cellular junctions leading to a loss of cell-cell cohesion. Through a complex and regimented cellular and molecular program, these cells progressively gain mesenchymal features, including cytoskeleton reorganization and proteolytic capacity favoring efficient cell motility.

The transcriptional mechanisms controlling EMT are particularly well documented, however, evidences of post-transcriptional regulation are now emerging in the literature, urging the scientific community to consider and investigate the synergistic combinations of these two levels of controls. At the cellular level, the tumor microenvironment (TME), including cancer associated fibroblasts (CAF), immune and endothelial cells, as well as the extracellular matrix (ECM) composition, are contributing factors modulating EMT and metastasis [7,9,10]. In addition to its structuring role, the ECM contains numerous cytokines, such as TGF-β. TGF-β signaling has a predominant function in suppressing the growth of normal epithelial cells. It also drives the metastatic process in malignantly-transformed tumor cells. Other growth factors, such as epidermal growth factor (EGF), fibroblast growth factor (FGF), and vascular endothelial growth factor (VEGF) were also clearly identified to be involved in EMT [11].

More recently, the development and improvement of transcriptomics boosted the discovery of new non-coding RNAs harboring regulatory functions. These species include PIWI-interacting RNA (piRNAs), small nuclear RNA (snRNAs), small nucleolar RNA (snoRNAs), circular RNAs (circRNAs), transfer RNAs (tRNAs), microRNAs (miRNAs), and long non-coding RNAs (lncRNAs). These RNA species operate through various molecular mechanisms including transcriptional and post-transcriptional controls.

In this article, we discuss the recent advances on the role of non-coding RNA in the regulation of TGF-β-induced EMT during tumor progression. We also present and comment on unpublished data collected in our laboratory regarding the regulation of tRNA expression in an in vitro model of TGF-β-induced-EMT of human tumor cells.

2. Cellular Basis of TGF-β-Induced EMT

The TGF-β signaling pathway was initially described for its critical role in cell proliferation and EMT during embryonic development of the neural crest, the somites, the heart, and various craniofacial structures [12]. TGF-β-induced EMT also manifests in pathological contexts in adults, specifically, during cancer progression and fibrosis. Due to its transient and reversible nature, EMT is technically challenging to observe throughout tumor progression in vivo. Nevertheless, it was proposed that during cancer progression of epithelial tumors, cells become significantly more invasive after completing EMT. In the current model, EMT-positive tumor cells, displaying newly-acquired mesenchymal features, are capable of invading their surrounding environment and complete extravasation in the circulatory system, resulting in ECM degradation and increased motility. It was proposed that the changes in cell plasticity induced by EMT enhance the ability of the circulating tumor cells (CTCs) to survive in the blood stream [6,13]. CTCs then re-invade distant organs through the extravasation process. Finally, metastatic colonization is initiated by the re-epithelialization of the cells through the reverse mechanism, called MET (mesenchymal-epithelial transition), followed by either a proliferative or dormancy step, which are responsible for secondary tumor growth and drug resistance or later tumor relapse, respectively (Figure 2).

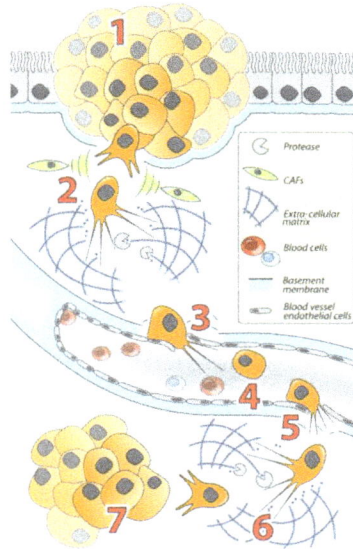

Figure 2. The Role of EMT and MET in carcinomas progression. (**1**) Following carcinogenesis, epithelial tumor cells proliferate to develop primary tumors called carcinoma in situ. In response to acquired mutations and/or exogenous stimuli, tumor cells gain invasive properties allowing them to break the basement membrane. Tumor cells then (**2**) invade and spread to surrounding tissues and structures and interact with numerous TME factors including cytokine-secreting CAFs, which reinforce EMT and invasion processes (cytoskeleton reorganization and increased proteolytic activity allow cells to degrade and invade the extracellular matrix (ECM)); (**3**) penetrate the vascular system (intravasation); (**4**) circulate throughout the body; (**5**) leave the vascular system to invade distant tissues (extravasation); (**6**) colonize distant sites through ECM degradation and invasion; and (**7**) reacquire epithelial phenotypes through MET and proliferate to ultimately form a metastasis.

3. Molecular Mechanisms of TGF-β-Induced EMT

3.1. Transcriptional Regulation of TGF-β-Induced EMT in Tumor Cells

EMT-inducing signals are cell- or tissue-specific and require the cooperation of multiple signaling pathways involving numerous regulators. TGF-β is arguably the most powerful EMT inducer in tumor cells, as it coordinates EMT at several levels and its impact on transcription has now been well documented. The TGF-β pathway is initiated by a superfamily of TGF-β ligands, including three forms of TGF-β ligands (TGF-β1 to -β3) and BMP isoforms (BMP2 to -7), whose secretion depends on tumor context and TME. TGF-β-mediated EMT integrates both Smad and non-Smad signaling pathways and is usually characterized by a loss of epithelial cell markers, such as E-Cadherin, and tight junction proteins in addition to the expression of mesenchymal cell markers, such as N-cadherin and vimentin. The epithelial-to-mesenchymal switch is triggered by a tightly-regulated transcription program that involves EMT-inducing transcription factors (EMT-TFs). These specific transcription factors are themselves activated by Smad signaling or other pathways, such as ErK/MAPK, RhoGTPases and PI3K/Akt, that also contribute to tumor progression through regulation of cytoskeleton organization, cell growth, survival, migration, and invasion [14,15]. Effectors of TGF-β signaling include the Krüppel-like factor 8 (KLF8) [16], the Brachyury factor [17], Goosecoid, TCF4, PRRX1 [18], basic helix-loop-helix factors (Twist and E12/E47), the Snail family of zinc-finger transcription factors (Snail, Slug, and Smuc) [19,20], and the δEF1 family of two-handed zinc-finger factors (δEF1/ZEB1 and Sip1/ZEB2) [2]. With the exception of Twist, Goosecoid and PRRX1, all these effectors directly repress the transcription of the epithelial marker E-Cadherin (CDH1) by binding to the corresponding gene promoter. Conversely, transcriptional activators of E-Cadherin such as Grhl2 or Elf5 inhibit the TGF-β induced EMT in tumor cells [21,22].

Finally, a growing body of evidence suggests that epigenetic modifications support an additional level of transcriptional control. Dynamic changes in the DNA methylome [23], histone modifications, as well as the differential expression of numerous histone modification factors [24,25], were identified as modulators of TGF-β-induced EMT and metastasis. During metastatic colonization, these epigenetic alterations established during EMT are reversed to promote MET [26,27].

3.2. Post-Transcriptional Regulation of TGF-β-Induced EMT in Tumor Cells

Recent data suggest that post-transcriptional regulation of gene expression complements transcriptional regulation during EMT [28–31]. Several RNA binding proteins (RBPs) directly control the translation of EMT-related genes. For instance, the Nanos homolog 3 RBP (Nanos3, Nos3) translationally controls EMT and metastasis in non-small cell lung cancer (NSCLC) cells by binding to the vimentin mRNA and increasing the expression of mesenchymal marker [31]. In addition, overexpression of YB-1 in breast epithelial cells triggers EMT and enhances metastatic potential by directly activating the cap-independent translation of mRNA encoding transcription factors implicated in EMT, such as Snail1 [28]. Finally, the protein hnRNP E1 (PCBP1) silences the translation of a cohort of mesenchymal mRNAs by directly binding to their targets and inhibiting translation at the elongation step [29,30,32].

4. Role of Non-Coding RNAs in TGF-β-Induced EMT

4.1. miRNAs

MicroRNAs are small noncoding RNA that modulate a wide range of biological processes including cell differentiation, proliferation, migration, invasion, and cell death. They silence mRNA translation by binding to the 3′-UTRs of target mRNAs resulting in translation inhibition or mRNA degradation. miRNAs are generated as pri-miRNA precursors and then processed into one or multiple mature miRNAs. miRNAs hybridize their mRNA targets using a complementary 7-nucleotides seed-sequence located at their 5′-end [33]. More than 2000 microRNAs have been characterized in humans. This list was considerably extended by the recent identification of more than 3000 additional miRNAs [34].

Many miRNAs, such as the miR-200 family (miR-200f), miR-205, miR-1, and miR-203, are key effectors of the TGF-β-induced EMT (Table 1). In general, TGF-β downregulates the expression of miRNAs in charge of inhibiting the expression of mesenchymal markers in epithelial cells. Conversely, emerging evidence indicates that several miRNAs regulate TGF-β signaling by targeting various members of its canonical or non-canonical pathways.

The miR-200 family is a group of highly influential miRNAs implicated in the regulation of TGF-β-induced EMT. It consists of five members, organized into two chromosomal clusters, which are expressed as polycistronic transcripts (MiR-200b, miR-200a, and miR-429 on human chromosome 1 and miR-200c and miR-141 on chromosome 12). Although miR-205 does not belong to the miR-200 family, it recognizes common targets and, therefore, complements the role of the miR-200f. Numerous studies demonstrated that TGF-β signaling downregulates the expression or the bioavailability of miR-200f and miR-205 [35]. These miRNAs, expressed by epithelial cells, are known to directly repress both ZEB1 and ZEB2 [35,36] therefore inhibiting the progression of EMT by establishing and maintaining an epithelial phenotype [35]. Interestingly, ZEB proteins and miR200f are reciprocally linked in a feedback loop, each controlling the expression of the other [37]. ZEB factors transcriptionally repress miR-200f by binding to highly-conserved recognition sequences on their promoters, while miR-200f inhibits the expression of ZEBs at the post-transcriptional level by binding to target sequences embedded in their 3'-UTRs. It was, therefore, proposed that the ZEB/miR-200 feedback loop acts as the molecular trigger controlling cellular plasticity in development and disease, and constitutes a significant driving force for cancer progression [5,37].

Although precise alterations of the ZEB/miR-200 balance are able to switch breast cancer cells back and forth between epithelial and mesenchymal states, the induction and maintenance of a stable mesenchymal phenotype requires the establishment of autocrine TGF-β signaling that supports sustained ZEB expression [38]. In addition, the recruitment of histone-modifying complexes by ZEB proteins negatively modulates miR-200f expression through epigenetic modification of the miR-200 loci and therefore amplifies the response to TGF-β [38].

Other miRNAs are known for their contribution in TGF-β-induced EMT. For instance, miR-203 and SLUG (SNAI2) operate as a double negative feedback loop mutually inhibiting their expression and thereby controlling EMT and metastasis [39,40]. Finally, miR-1 and miR-200 were identified as being directly repressed by SLUG during EMT [41].

4.2. Long Non-Coding RNAs

LncRNAs are transcripts greater than 200 nucleotides deprived of any protein coding sequences. They are divided into five broad categories, including sense, antisense, bidirectional, intronic, and intergenic, with respect to the nearest protein-coding transcripts [42]. LncRNAs modulate gene expression trough different cis- or trans-acting mechanisms. For instance, competing endogenous RNAs (ceRNAs) serve as molecular decoys for specific miRNAs, decreasing their bioavailability and, therefore, protecting the corresponding target mRNAs [43]. In synergy with miRNAs, they regulate gene expression by modulating transcription, RNA processing, chromatin remodeling, genomic imprinting, and association with proteins. Interestingly, dysregulation of lncRNA expression is frequently observed in human cancers. Since LncRNAs control mechanisms of many cellular processes, they represent appealing targets for the development of anti-tumor therapies.

4.2.1. LncRNA-ATB

TGF-β-induced lncRNA-ATB competitively binds to miR-200f, favoring the expression of ZEB1 and ZEB2 proteins, therefore promoting EMT and invasion in hepatocellular carcinoma cells [44]. In addition, lncRNA-ATB promotes organ colonization of disseminated hepatocellular carcinoma cells by binding to IL-11 mRNA and subsequently activating IL-11/STAT3 signaling. Aberrant lncRNA-ATB expression is observed in various cancer types, such as prostate carcinoma [45], colorectal cancer [46], NSCLC [47], and breast cancer [48] (Table 1).

4.2.2. MALAT1

Induction of the metastasis-associated lung adenocarcinoma transcript 1 (MALAT1) results in the decrease of E-cadherin expression and the increased expression of mesenchymal markers leading to enhanced EMT [49]. MALAT1 is a prognostic marker in several cancers: including lung, breast, pancreas, liver, colon, uterus, cervix, and prostate [50,51]. In renal cancer, reciprocal crosstalk among MALAT1, miR205, and EZH2 suppresses the expression of E-Cadherin and enhances Wnt signaling to promote cancer metastasis [52] (Table 1).

4.2.3. lncRNA-ZEB2NAT

ZEB2 Natural antisense transcript regulates Zeb2/Sip1 gene expression during Snail1-induced EMT. While Snail1 does not affect the synthesis of the Zeb2 mRNA, it prevents the processing of a specific intron in the 5′- UTR. This intron contains an internal ribosome entry site (IRES) essential for the expression of Zeb2. The maintenance of this intron is dependent on the expression of ZEB2NAT which overlaps with the 5′ splice site [53]. TGFβ1, secreted by cancer-associated fibroblasts, induces an epithelial-mesenchymal transition of bladder cancer cells in a mechanism dependent on ZEB2NAT expression [54] (Table 1).

4.2.4. HOTAIR

HOX antisense intergenic RNA (HOTAIR), a lncRNA encoded by a gene located in the mammalian HOXC locus, binds to the polycomb repressive complex 2 (PRC2), a histone methyltransferase required for epigenetic silencing during development and cancer [55–57]. HOTAIR is overexpressed in both NSCLC and breast cancer and its expression level correlates with poor disease outcome [55,58] (Table 1). HOTAIR upregulation was also observed in several other cancer types where numerous regulatory factors control its expression, including TGF-β. Overall, HOTAIR is involved in multiple processes such as mobility, proliferation, apoptosis and invasion of tumor cells, and is therefore particularly relevant in EMT context. The silencing of HOTAIR in colon cancer cells is followed by the concomitant increase in E-cadherin expression and decrease in vimentin expression, demonstrating the direct link between HOTAIR and EMT [59]. Importantly, it was demonstrated that the control of EMT through HOTAIR contributes to the emergence and maintenance of cancer stem cells (CSCs) [60].

4.2.5. lncRNA-HIT

The HOXA transcript induced by TGF-β (lncRNA-HIT) contributes to EMT in breast carcinoma cells, and its elevated expression is associated with invasion-prone human primary breast carcinoma cells [61] (Table 1).

4.2.6. MEG3

Maternally-expressed 3 (MEG3) forms RNA-DNA triplex structures with genes involved in the TGF-β pathway and ultimately modulates tissue invasion in breast and lung cancer cells [62,63] (Table 1). MEG3 is expressed at significantly lower levels in invasive ductal carcinoma, as well as in the aggressive and difficult-to-treat basal molecular subtype, as compared to normal breast tissue. Finally, tumors maintaining low MEG3 expression feature higher levels of TGF-β-associated genes such as TGFB2, TGFBR1, and SMAD2 [62].

4.3. Other Non-Coding RNA Species

4.3.1. Circular RNAs

Circular RNAs (circRNAs) are expressed in mammalian cells and form a covalently-closed continuous loop. CirRNAs function mainly as sponges for miRNAs and RNA-binding proteins (RBP) [64]. Reports on differential expressions of circular RNAs between cancerous and non-cancerous samples are emerging in the literature suggesting their implication in tumor progression [65–69]. In breast epithelial HMLE cells experiencing TGF-β-induced EMT, the expression of hundreds of circRNAs fluctuates. Under these circumstances, the production of over one-third of the most abundant circRNAs are regulated by the Quaking alternative splicing factor [70].

4.3.2. PIWI-interacting RNAs

Originally observed in the germline, PIWI-interacting RNAs (piRNAs) are small non-coding RNAs which recently emerged as potential markers in tumor progression [71,72]. With over 20,000 genes, piRNA-encoding genes are highly represented in the human genome. piRNAs function primarily in the nucleus and interact with members of the Argonaute family, such as PIWI proteins [71]. Aberrant patterns of piRNA expression, and the dysregulation of key enzymes involved in their biogenesis, are frequently observed in diverse tumor types, such as breast and lung cancers [72,73]. In particular, the PIWI-like RNA-mediated gene silencing 2 (PIWIL2), a member of the Argonaute family involved in piRNA processing, is significantly overexpressed in both breast cancer stem cells and TGF-β-induced EMT [74].

4.3.3. Small Nucleolar and Small Nuclear RNAs

Small nucleolar RNAs (snoRNAs) and small nuclear RNAs (snRNAs) are 60 to 300 nucleotides long. SnoRNAs are involved in the post-transcriptional modification of ribosomal RNA and play integral roles in the formation of small nucleolar ribonucleoprotein particles (snoRNP). SnRNAs support RNA-RNA interactions and spliceosome assembly [75]. Dysregulation of snoRNA and/or snRNA expression is commonly observed in cancers. Several snoRNAs are upregulated in murine and human breast cancer as well as in prostate cancers; interfering with their biogenesis suppresses cell growth and colony formation in MCF-7, U-20S, and A549 cells in vitro, and ultimately compromises tumorigenicity in vivo [76]. Although the exact contribution of small nuclear RNAs (snRNAs) in tumor progression and metastasis is not yet clearly established, the 7SK snRNA indirectly participates in the control of EMT (Table 1). 7SK snRNAs form dynamic complexes with the La-related protein LARP7. These complexes often involve additional molecular partners, such as positive transcription elongation factor b (P-TEFb), whose activity is relevant to EMT-related tumor progression in breast tissue. Decreased levels of LARP7 and 7SK snRNA redistribute P-TEFb toward the transcriptionally active super elongation complex (SEC), resulting in P-TEFb activation and increased transcription of EMT transcription factors including Slug, FOXC2, ZEB2, and Twist1, ultimately promoting EMT, invasion, and metastasis of breast tumor cells [77].

4.3.4. Transfer RNAs

tRNA are abundant molecules which represent 30% of the total RNA pool in eukaryotic cells [78]. The human genome contains over 600 tRNA genes, which are scattered throughout the genome and are present on all but the Y chromosome [79]. Until recently, tRNAs have been considered to be housekeeping molecules dedicated to protein translation. However, a growing body of evidence indicates that differential tRNA expression deeply influences the whole dynamic of translation, supporting or impairing the expression of particular proteins [80–83].

The phenotypic changes associated with EMT and tumor progression imply a radical reprogramming of the proteome and a significant redistribution of the overall codon usage. Considering tRNA molecules in translation as the 'supply' and codon usage as the 'demand', it has been proposed that the codon demand associated with EMT and tumorigenesis is mediated by a transcriptional regulation of the tRNA supply [84,85]. Moreover, the role of tRNAs in the regulation of gene expression was recently supported in a recent study based on the analysis of tRNA content in various human cell lines and tissue samples [85]. It was observed that distinct tRNA signatures correlate with either proliferation or differentiation, two distinct cellular program often described for their role during tumor progression [85–88]. In cancer-derived breast cell lines overall tRNA levels are upregulated by up to three-fold compared to non-cancer control cells [81]. A selective upregulation of tRNA$^{Glu}_{UUC}$ and tRNA$^{Arg}_{CCG}$ was also observed during breast tumor progression, and has been proposed to contribute to metastatic progression of breast tumor cells through the enhanced translation of pro-metastatic transcripts enriched in the corresponding codons [80] (Table 1).

Transfer RNA-derived RNA fragments (tRFs) were also documented for their role in tumor progression. tRFs belong to a family of short non-coding RNAs that are constitutively expressed or activated upon specific growth conditions. This tRNA subspecies are presumably generated through the processing by Dicer and RNAse Z [89,90] or through the action of specific ribonucleases for the stress-induced tRFs fragments (tiRNAs) [91,92]. These small RNAs have distinct functions in various biological processes, including tumor suppression and protein regulation [93]. tRNA fragments have also been reported to promote cell migration as illustrated by miR-720 in cervical cancer cells in vitro [94]. Additionally, in a study on the epithelialization of mouse ovarian tumor cells in Dicer-Pten double-knock-out mice, tRNA overexpression was identified in tumors in the context of a miRNA profiling experiment, and was associated to oncogenic transformation [95].

Table 1. Non-Coding RNA involved in EMT.

Non-Coding RNA	Relevant Examples	Specific Function	Most Described Targets	Related Cancers	References
miRNAs	miR-1 * miR-200 family * miR-205 * miR-203 *	Epithelial maintenance	ZEB1/2↓ Slug↓ Bmi1↓	Breast Lung Prostate	[5,35,37,39,40]
LncRNAs	LncRNA-ATB [†] MALAT1 [†] lncRNA-ZEB2NAT [†] HOTAIR [†] lncRNA-HIT	Tumor cell invasion; Organ colonization; Proliferation; Cancer Stem Cells	ZEB1/2↑ IL-11↑ miR-200↓ miR-205↓ E-cadherin↓	Prostate Lung Breast Kidney Pancreas Liver Colon Uterus	[44–47,49,51–55,57–59,61]
	MEG3 *	TGF-β pathway regulation	TGFBR1↑ TGFB2↑ SMAD2↑	Breast	[62,63]
circRNAs	CDR1as/ciRS-7 *	miRNA sponge	miRNA-7↓	Colon	[66,67,69]
piRNAs	Pir-932 [†]	Stemness properties	Latexin↓	Breast	[74]
snoRNAs snRNAs	7SK snRNA *	Tumor cell invasion	Slug↓ FOXC2↓ ZEB2↓ Twist1↓	Breast	[77]
tRNAs	tRNA$^{Glu}_{UUC}$ [†]	Tumor progression	EXOSC2↓ GRIPAP1↓	Breast	[80]
	MicroRNA-720 [†]	Tumor cell motility	Rab35↓	Uterus	[94]

* Epithelial non-coding RNAs; [†] Mesenchymal non-coding RNAs.

While tRNA expression and enhanced translation capacity appear relevant in tumor progression of carcinomas, their regulation and contribution during TGF-β-induced EMT have not been documented yet. With the intention to further the discussion on the involvement of tRNAs in EMT, our laboratories developed a strategy, based on metabolic RNA labeling followed by microarray analysis, to evaluate global tRNA changes upon TGF-β treatment in tumor cells in vitro. The corresponding original experimental evidence is described and examined in the following section.

5. Evidence of Selective Regulation of tRNA Expression during TGF-β-Induced EMT

The effects of TGFβ-mediated EMT on the expression of tRNAs was examined in three experimental conditions: no treatment (NT), three days of TGFβ exposure (3d), and five days (5d) of TGFβ exposure. This experiment was carried out on the human lung cancer cell line A549. These cells are a common model used in TGFβ mediated EMT, as they respond remarkably to TGFβ treatment. Cells were grown in media containing radioactive orthophosphate (^{32}P), which non-specifically labeled RNA, DNA, phosphoproteins, and other phosphate-containing metabolites. Total RNAs were extracted and hybridized to tRNA microarrays for analysis according to published protocols. Extracted RNAs were also analyzed by gel electrophoresis, to ensure that tRNAs were effectively radiolabeled (Figure 3).

EMT progression was confirmed visually by immunofluorescent microscopy and Western blot. The epithelial cell marker, E-cadherin, and the mesenchymal cell marker, vimentin, were targeted to track the transition. In both experiments, levels of E-cadherin were greatest in the NT culture, and decreased rapidly upon addition of TGFβ. The opposite was true for the expression of vimentin, which increased with prolonged exposure to TGFβ. These data offered positive identification of EMT progression in our cell model.

tRNA expression within each of the experimental cultures was measured using microarray analysis according to published protocols [96]. Array data were collected in triplicate and averaged to generate a heat map of overall tRNA expression throughout EMT (Figure 3). Seleno-cysteine (sel-Cys) represents the least expressed tRNA and valine (val) represents the tRNA with the greatest overall expression in A549 cells. Four tRNAs, including cysteine (GCA), glutamine (yTG), glycine (TCC), and lysine (TTT), displayed statistically significant changes in expression throughout the transition. The relative tRNA expression throughout the three experimental cell cultures for these four tRNA was plotted and standard error bars, as well as statistical significance, was shown (Figure 3).

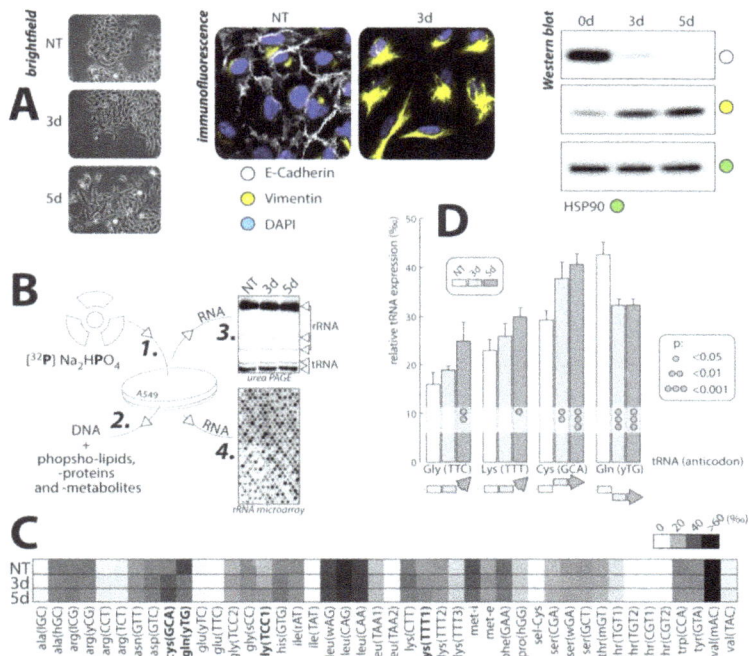

Figure 3. (**A**) Cell morphology and expression of EMT markers: A549 cells pictured after no-treatment (NT), three days of TGFβ treatment (3d), and 5d of TGFβ treatment. Cells begin with a slightly mesenchymal phenotype in the NT, but display a completely mesenchymal phenotype after prolonged exposure to TGFβ. E-cadherin (white) is the epithelial cell marker and vimentin (yellow) is the mesenchymal cell marker. DAPI (blue) is the counterstain used to stain nuclei. HSP90 was used as a loading control. (**B**) Overview of the experimental Procedure: 1. Radioactive orthophosphate was added to cell cultures at onset of experiment. 2. Total RNAs were Trizol extracted and all other labeled molecules were removed from sample. 3. Labeling of tRNA molecules was confirmed via gel electrophoresis. 4. Samples were hybridized to tRNA microarrays and analyzed. (**C**) Average tRNA expression: The heat map shows the average number of each tRNA (per thousand) that is present across the three conditions. tRNA abundances range from close to 0 to over 60‰. (**D**) Statistically significant results: The relative tRNA expression, per thousand, of the four tRNA that displayed significant changes in tRNA expression throughout EMT are shown along with standard error bars. Both 3d and 5d were compared to the NT and the statistical significance of those changes can be seen from the dot on the bars. The overall trend of expression for each tRNA is shown below the graph.

Between the four significant tRNAs, the trends of expression change were not consistent, with only lysine and cysteine showing a similar trend of increasing between NT and 3d, and showing no significant change between the 3d and 5d. Glycine also showed an increasing trend, only changing significantly after five days of treatment with TGFβ. Glutamine, however, displayed a decreasing trend, with both the 3d and 5d cells showed a significantly lower amount of this tRNA than in the NT cells.

There are no commonalities between our results and the type of amino acid carried by the tRNA. Cysteine and glutamine have polar, uncharged side-chains, whereas glycine has a non-polar, aliphatic side-chain, and lysine's side-chain is positively charged. The significance of these tRNA expression patterns with respect to codon usage within proteins in EMT has yet to be explored. Another possible explanation for the change in the expression of these four tRNAs could be an increase in non-translational tRNA activity, such as the production of tRFs as described in the above section.

Though the role of the change in expression of these tRNAs has not been specifically identified at this point, there is the possibility that tRNA may play a role in the progression of EMT, in its regulation, or perhaps could be utilized as EMT markers in the future. Overall non-coding RNAs whose expression fluctuates during EMT deserve special attention as they represent a reservoir of targets and offer potential therapeutic opportunities to prevent EMT-associated metastasis. Fluctuations in abundant and supposedly housekeeping species, such as tRNAs, emphasize the necessity to cast a broad net in the current quest of new regulatory RNAs.

Acknowledgments: This research was supported by a SC INBRE grant from the National Institute of General Medical Science—NIH P20GM103499 to Renaud Geslain and grants CA555536 and CA154664 from National Cancer Institute to Philip H. Howe. We thank Jacob Goldmintz, Morgan Troiano, and Sophia Emetu for helpful discussion.

Conflicts of Interest: The authors declare no conflict of interest.

References

1. Nieto, M.A.; Huang, R.Y.-J.; Jackson, R.A.; Thiery, J.P. EMT: 2016. *Cell* **2016**, *166*, 21–45. [CrossRef] [PubMed]
2. Thiery, J.P.; Acloque, H.; Huang, R.Y.J.; Nieto, M.A. Epithelial-mesenchymal transitions in development and disease. *Cell* **2009**, *139*, 871–890. [CrossRef] [PubMed]
3. Zheng, X.; Carstens, J.L.; Kim, J.; Scheible, M.; Kaye, J.; Sugimoto, H.; Wu, C.-C.; LeBleu, V.S.; Kalluri, R. Epithelial-to-mesenchymal transition is dispensable for metastasis but induces chemoresistance in pancreatic cancer. *Nature* **2015**, *527*, 525–530. [CrossRef] [PubMed]
4. Fischer, K.R.; Durrans, A.; Lee, S.; Sheng, J.; Li, F.; Wong, S.T.C.; Choi, H.; El Rayes, T.; Ryu, S.; Troeger, J.; et al. Epithelial-to-mesenchymal transition is not required for lung metastasis but contributes to chemoresistance. *Nature* **2015**, *527*, 472–476. [CrossRef] [PubMed]
5. Krebs, A.M.; Mitschke, J.; Lasierra Losada, M.; Schmalhofer, O.; Boerries, M.; Busch, H.; Boettcher, M.; Mougiakakos, D.; Reichardt, W.; Bronsert, P.; et al. The EMT-activator Zeb1 is a key factor for cell plasticity and promotes metastasis in pancreatic cancer. *Nat. Cell Biol.* **2017**, *19*, 518–529. [CrossRef] [PubMed]
6. Francart, M.-E.; Lambert, J.; Vanwynsberghe, A.M.; Thompson, E.W.; Bourcy, M.; Polette, M.; Gilles, C. Epithelial-Mesenchymal Plasticity and Circulating Tumor Cells: Travel Companions to Metastases. *Dev. Dyn. Off. Publ. Am. Assoc. Anat.* **2017**. [CrossRef] [PubMed]
7. Kalluri, R.; Weinberg, R.A. The basics of epithelial-mesenchymal transition. *J. Clin. Investig.* **2009**, *119*, 1420–1428. [CrossRef] [PubMed]
8. Smith, B.N.; Bhowmick, N.A. Role of EMT in Metastasis and Therapy Resistance. *J. Clin. Med.* **2016**, *5*, 17. [CrossRef] [PubMed]
9. Schedin, P.; Borges, V. Breaking down barriers: the importance of the stromal microenvironment in acquiring invasiveness in young women's breast cancer. *Breast Cancer Res. BCR* **2009**, *11*, 102. [CrossRef] [PubMed]
10. Kalluri, R.; Zeisberg, M. Fibroblasts in cancer. *Nat. Rev. Cancer* **2006**, *6*, 392–401. [CrossRef] [PubMed]
11. Lamouille, S.; Xu, J.; Derynck, R. Molecular mechanisms of epithelial–mesenchymal transition. *Nat. Rev. Mol. Cell Biol.* **2014**, *15*, 178–196. [CrossRef] [PubMed]
12. Massagué, J. TGFβ in Cancer. *Cell* **2008**, *134*, 215–230. [CrossRef] [PubMed]
13. Bourcy, M.; Suarez-Carmona, M.; Lambert, J.; Francart, M.-E.; Schroeder, H.; Delierneux, C.; Skrypek, N.; Thompson, E.W.; Jérusalem, G.; Berx, G.; et al. Tissue Factor Induced by Epithelial-Mesenchymal Transition Triggers a Procoagulant State That Drives Metastasis of Circulating Tumor Cells. *Cancer Res.* **2016**, *76*, 4270–4282. [CrossRef] [PubMed]

14. Derynck, R.; Zhang, Y.E. Smad-dependent and Smad-independent pathways in TGF-β family signalling. *Nature* **2003**, *425*, 577–584. [CrossRef] [PubMed]

15. Valcourt, U.; Kowanetz, M.; Niimi, H.; Heldin, C.-H.; Moustakas, A. TGF-beta and the Smad signaling pathway support transcriptomic reprogramming during epithelial-mesenchymal cell transition. *Mol. Biol. Cell* **2005**, *16*, 1987–2002. [CrossRef] [PubMed]

16. Zhang, H.; Liu, L.; Wang, Y.; Zhao, G.; Xie, R.; Liu, C.; Xiao, X.; Wu, K.; Nie, Y.; Zhang, H.; et al. KLF8 involves in TGF-beta-induced EMT and promotes invasion and migration in gastric cancer cells. *J. Cancer Res. Clin. Oncol.* **2013**, *139*, 1033–1042. [CrossRef] [PubMed]

17. Larocca, C.; Cohen, J.R.; Fernando, R.I.; Huang, B.; Hamilton, D.H.; Palena, C. An autocrine loop between TGF-β1 and the transcription factor Brachyury controls the transition of human carcinoma cells into a mesenchymal phenotype. *Mol. Cancer Ther.* **2013**, *12*. [CrossRef] [PubMed]

18. Hardin, H.; Guo, Z.; Shan, W.; Montemayor-Garcia, C.; Asioli, S.; Yu, X.-M.; Harrison, A.D.; Chen, H.; Lloyd, R.V. The Roles of the Epithelial-Mesenchymal Transition Marker PRRX1 and miR-146b-5p in Papillary Thyroid Carcinoma Progression. *Am. J. Pathol.* **2014**, *184*, 2342–2354. [CrossRef] [PubMed]

19. Batlle, E.; Sancho, E.; Francí, C.; Domínguez, D.; Monfar, M.; Baulida, J.; García De Herreros, A. The transcription factor snail is a repressor of E-cadherin gene expression in epithelial tumour cells. *Nat. Cell Biol.* **2000**, *2*, 84–89. [CrossRef] [PubMed]

20. Cano, A.; Pérez-Moreno, M.A.; Rodrigo, I.; Locascio, A.; Blanco, M.J.; del Barrio, M.G.; Portillo, F.; Nieto, M.A. The transcription factor snail controls epithelial-mesenchymal transitions by repressing E-cadherin expression. *Nat. Cell Biol.* **2000**, *2*, 76–83. [CrossRef] [PubMed]

21. Xiang, J.; Fu, X.; Ran, W.; Wang, Z. Grhl2 reduces invasion and migration through inhibition of TGFβ-induced EMT in gastric cancer. *Oncogenesis* **2017**, *6*, e284. [CrossRef] [PubMed]

22. Yao, B.; Zhao, J.; Li, Y.; Li, H.; Hu, Z.; Pan, P.; Zhang, Y.; Du, E.; Liu, R.; Xu, Y. Elf5 inhibits TGF-β-driven epithelial-mesenchymal transition in prostate cancer by repressing SMAD3 activation. *Prostate* **2015**, *75*, 872–882. [CrossRef] [PubMed]

23. Cardenas, H.; Vieth, E.; Lee, J.; Segar, M.; Liu, Y.; Nephew, K.P.; Matei, D. TGF-β induces global changes in DNA methylation during the epithelial-to-mesenchymal transition in ovarian cancer cells. *Epigenetics* **2014**, *9*, 1461–1472. [CrossRef] [PubMed]

24. Serrano-Gomez, S.J.; Maziveyi, M.; Alahari, S.K. Regulation of epithelial-mesenchymal transition through epigenetic and post-translational modifications. *Mol. Cancer* **2016**, *15*, 18. [CrossRef] [PubMed]

25. Roche, J.; Nasarre, P.; Gemmill, R.; Baldys, A.; Pontis, J.; Korch, C.; Guilhot, J.; Ait-Si-Ali, S.; Drabkin, H. Global Decrease of Histone H3K27 Acetylation in ZEB1-Induced Epithelial to Mesenchymal Transition in Lung Cancer Cells. *Cancers* **2013**, *5*, 334–356. [CrossRef] [PubMed]

26. Bedi, U.; Mishra, V.K.; Wasilewski, D.; Scheel, C.; Johnsen, S.A. Epigenetic plasticity: A central regulator of epithelial-to-mesenchymal transition in cancer. *Oncotarget* **2014**, *5*, 2016–2029. [CrossRef] [PubMed]

27. De Craene, B.; Berx, G. Regulatory networks defining EMT during cancer initiation and progression. *Nat. Rev. Cancer* **2013**, *13*, 97–110. [CrossRef] [PubMed]

28. Evdokimova, V.; Tognon, C.; Ng, T.; Ruzanov, P.; Melnyk, N.; Fink, D.; Sorokin, A.; Ovchinnikov, L.P.; Davicioni, E.; Triche, T.J.; et al. Translational activation of snail1 and other developmentally regulated transcription factors by YB-1 promotes an epithelial-mesenchymal transition. *Cancer Cell* **2009**, *15*, 402–415. [CrossRef] [PubMed]

29. Chaudhury, A.; Hussey, G.S.; Ray, P.S.; Jin, G.; Fox, P.L.; Howe, P.H. TGF-beta-mediated phosphorylation of hnRNP E1 induces EMT via transcript-selective translational induction of Dab2 and ILEI. *Nat. Cell Biol.* **2010**, *12*, 286–293. [CrossRef] [PubMed]

30. Hussey, G.S.; Chaudhury, A.; Dawson, A.E.; Lindner, D.J.; Knudsen, C.R.; Wilce, M.C.J.; Merrick, W.C.; Howe, P.H. Identification of an mRNP Complex Regulating Tumorigenesis at the Translational Elongation Step. *Mol. Cell* **2011**, *41*, 419–431. [CrossRef] [PubMed]

31. Grelet, S.; Andries, V.; Polette, M.; Gilles, C.; Staes, K.; Martin, A.-P.; Kileztky, C.; Terryn, C.; Dalstein, V.; Cheng, C.-W.; et al. The human NANOS3 gene contributes to lung tumour invasion by inducing epithelial-mesenchymal transition. *J. Pathol.* **2015**, *237*, 25–37. [CrossRef] [PubMed]

32. Hussey, G.S.; Link, L.A.; Brown, A.S.; Howley, B.V.; Chaudhury, A.; Howe, P.H. Establishment of a TGFβ-Induced Post-Transcriptional EMT Gene Signature. *PLOS ONE* **2012**, *7*, e52624. [CrossRef] [PubMed]

33. Bartel, D.P. MicroRNAs: Target recognition and regulatory functions. *Cell* **2009**, *136*, 215–233. [CrossRef] [PubMed]

34. Londin, E.; Loher, P.; Telonis, A.G.; Quann, K.; Clark, P.; Jing, Y.; Hatzimichael, E.; Kirino, Y.; Honda, S.; Lally, M.; et al. Analysis of 13 cell types reveals evidence for the expression of numerous novel primate- and tissue-specific microRNAs. *Proc. Natl. Acad. Sci. USA* **2015**, *112*, E1106–E1115. [CrossRef] [PubMed]

35. Gregory, P.A.; Bert, A.G.; Paterson, E.L.; Barry, S.C.; Tsykin, A.; Farshid, G.; Vadas, M.A.; Khew-Goodall, Y.; Goodall, G.J. The miR-200 family and miR-205 regulate epithelial to mesenchymal transition by targeting ZEB1 and SIP1. *Nat. Cell Biol.* **2008**, *10*, 593–601. [CrossRef] [PubMed]

36. Korpal, M.; Lee, E.S.; Hu, G.; Kang, Y. The miR-200 Family Inhibits Epithelial-Mesenchymal Transition and Cancer Cell Migration by Direct Targeting of E-cadherin Transcriptional Repressors ZEB1 and ZEB2. *J. Biol. Chem.* **2008**, *283*, 14910–14914. [CrossRef] [PubMed]

37. Brabletz, S.; Brabletz, T. The ZEB/miR-200 feedback loop—A motor of cellular plasticity in development and cancer? *EMBO Rep.* **2010**, *11*, 670–677. [CrossRef] [PubMed]

38. Gregory, P.A.; Bracken, C.P.; Smith, E.; Bert, A.G.; Wright, J.A.; Roslan, S.; Morris, M.; Wyatt, L.; Farshid, G.; Lim, Y.-Y.; et al. An autocrine TGF-β/ZEB/miR-200 signaling network regulates establishment and maintenance of epithelial-mesenchymal transition. *Mol. Biol. Cell* **2011**, *22*, 1686–1698. [CrossRef] [PubMed]

39. Ding, X.; Park, S.I.; McCauley, L.K.; Wang, C.-Y. Signaling between Transforming Growth Factor β (TGF-β) and Transcription Factor SNAI2 Represses Expression of MicroRNA miR-203 to Promote Epithelial-Mesenchymal Transition and Tumor Metastasis. *J. Biol. Chem.* **2013**, *288*, 10241–10253. [CrossRef] [PubMed]

40. Wellner, U.; Schubert, J.; Burk, U.C.; Schmalhofer, O.; Zhu, F.; Sonntag, A.; Waldvogel, B.; Vannier, C.; Darling, D.; zur Hausen, A.; et al. The EMT-activator ZEB1 promotes tumorigenicity by repressing stemness-inhibiting microRNAs. *Nat. Cell Biol.* **2009**, *11*, 1487–1495. [CrossRef] [PubMed]

41. Liu, Y.-N.; Yin, J.J.; Abou-Kheir, W.; Hynes, P.G.; Casey, O.M.; Fang, L.; Yi, M.; Stephens, R.M.; Seng, V.; Sheppard-Tillman, H.; et al. MiR-1 and miR-200 inhibit EMT via Slug-dependent and tumorigenesis via Slug-independent mechanisms. *Oncogene* **2013**, *32*, 296–306. [CrossRef] [PubMed]

42. Meseure, D.; Drak Alsibai, K.; Nicolas, A.; Bieche, I.; Morillon, A. Long Noncoding RNAs as New Architects in Cancer Epigenetics, Prognostic Biomarkers, and Potential Therapeutic Targets. *BioMed Res. Int.* **2015**, *2015*, e320214. [CrossRef] [PubMed]

43. Tay, Y.; Rinn, J.; Pandolfi, P.P. The multilayered complexity of ceRNA crosstalk and competition. *Nature* **2014**, *505*, 344–352. [CrossRef] [PubMed]

44. Yuan, J.; Yang, F.; Wang, F.; Ma, J.; Guo, Y.; Tao, Q.; Liu, F.; Pan, W.; Wang, T.; Zhou, C.; et al. A Long Noncoding RNA Activated by TGF-β Promotes the Invasion-Metastasis Cascade in Hepatocellular Carcinoma. *Cancer Cell* **2014**, *25*, 666–681. [CrossRef] [PubMed]

45. Xu, S.; Yi, X.-M.; Tang, C.-P.; Ge, J.-P.; Zhang, Z.-Y.; Zhou, W.-Q. Long non-coding RNA ATB promotes growth and epithelial-mesenchymal transition and predicts poor prognosis in human prostate carcinoma. *Oncol. Rep.* **2016**, *36*, 10–22. [CrossRef] [PubMed]

46. Iguchi, T.; Uchi, R.; Nambara, S.; Saito, T.; Komatsu, H.; Hirata, H.; Ueda, M.; Sakimura, S.; Takano, Y.; Kurashige, J.; et al. A long noncoding RNA, lncRNA-ATB, is involved in the progression and prognosis of colorectal cancer. *Anticancer Res.* **2015**, *35*, 1385–1388. [PubMed]

47. Ke, L.; Xu, S.-B.; Wang, J.; Jiang, X.-L.; Xu, M.-Q. High expression of long non-coding RNA ATB indicates a poor prognosis and regulates cell proliferation and metastasis in non-small cell lung cancer. *Clin. Transl. Oncol.* **2017**, *19*, 599–605. [CrossRef] [PubMed]

48. Shi, S.-J.; Wang, L.-J.; Yu, B.; Li, Y.-H.; Jin, Y.; Bai, X.-Z. LncRNA-ATB promotes trastuzumab resistance and invasion-metastasis cascade in breast cancer. *Oncotarget* **2015**, *6*, 11652–11663. [CrossRef] [PubMed]

49. Fan, Y.; Shen, B.; Tan, M.; Mu, X.; Qin, Y.; Zhang, F.; Liu, Y. TGF-β-induced upregulation of malat1 promotes bladder cancer metastasis by associating with suz12. *Clin. Cancer Res.* **2014**, *20*, 1531–1541. [CrossRef] [PubMed]

50. Shi, X.; Sun, M.; Liu, H.; Yao, Y.; Song, Y. Long non-coding RNAs: A new frontier in the study of human diseases. *Cancer Lett.* **2013**, *339*, 159–166. [CrossRef] [PubMed]

51. Ji, P.; Diederichs, S.; Wang, W.; Böing, S.; Metzger, R.; Schneider, P.M.; Tidow, N.; Brandt, B.; Buerger, H.; Bulk, E.; et al. MALAT-1, a novel noncoding RNA, and thymosin beta4 predict metastasis and survival in early-stage non-small cell lung cancer. *Oncogene* **2003**, *22*, 8031–8041. [CrossRef] [PubMed]

52. Hirata, H.; Hinoda, Y.; Shahryari, V.; Deng, G.; Nakajima, K.; Tabatabai, Z.L.; Ishii, N.; Dahiya, R. Long Noncoding RNA MALAT1 Promotes Aggressive Renal Cell Carcinoma through Ezh2 and Interacts with miR-205. *Cancer Res.* **2015**, *75*, 1322–1331. [CrossRef] [PubMed]

53. Beltran, M.; Puig, I.; Peña, C.; García, J.M.; Álvarez, A.B.; Peña, R.; Bonilla, F.; de Herreros, A.G. A natural antisense transcript regulates Zeb2/Sip1 gene expression during Snail1-induced epithelial–mesenchymal transition. *Genes Dev.* **2008**, *22*, 756–769. [CrossRef] [PubMed]

54. Zhuang, J.; Lu, Q.; Shen, B.; Huang, X.; Shen, L.; Zheng, X.; Huang, R.; Yan, J.; Guo, H. TGFβ1 secreted by cancer-associated fibroblasts induces epithelial-mesenchymal transition of bladder cancer cells through lncRNA-ZEB2NAT. *Sci. Rep.* **2015**, *5*, 11924. [CrossRef] [PubMed]

55. Gupta, R.A.; Shah, N.; Wang, K.C.; Kim, J.; Horlings, H.M.; Wong, D.J.; Tsai, M.-C.; Hung, T.; Argani, P.; Rinn, J.L.; et al. Long non-coding RNA HOTAIR reprograms chromatin state to promote cancer metastasis. *Nature* **2010**, *464*, 1071–1076. [CrossRef] [PubMed]

56. Davidovich, C.; Zheng, L.; Goodrich, K.J.; Cech, T.R. Promiscuous RNA binding by Polycomb Repressive Complex 2. *Nat. Struct. Mol. Biol.* **2013**, *20*, 1250–1257. [CrossRef] [PubMed]

57. Rinn, J.L.; Kertesz, M.; Wang, J.K.; Squazzo, S.L.; Xu, X.; Brugmann, S.A.; Goodnough, L.H.; Helms, J.A.; Farnham, P.J.; Segal, E.; et al. Functional demarcation of active and silent chromatin domains in human HOX loci by noncoding RNAs. *Cell* **2007**, *129*, 1311–1323. [CrossRef] [PubMed]

58. Nakagawa, T.; Endo, H.; Yokoyama, M.; Abe, J.; Tamai, K.; Tanaka, N.; Sato, I.; Takahashi, S.; Kondo, T.; Satoh, K. Large noncoding RNA HOTAIR enhances aggressive biological behavior and is associated with short disease-free survival in human non-small cell lung cancer. *Biochem. Biophys. Res. Commun.* **2013**, *436*, 319–324. [CrossRef] [PubMed]

59. Wu, Z.-H.; Wang, X.-L.; Tang, H.-M.; Jiang, T.; Chen, J.; Lu, S.; Qiu, G.-Q.; Peng, Z.-H.; Yan, D.-W. Long non-coding RNA HOTAIR is a powerful predictor of metastasis and poor prognosis and is associated with epithelial-mesenchymal transition in colon cancer. *Oncol. Rep.* **2014**, *32*, 395–402. [CrossRef] [PubMed]

60. Hajjari, M.; Khoshnevisan, A.; Shin, Y.K. Molecular function and regulation of long non-coding RNAs: paradigms with potential roles in cancer. *Tumour Biol.* **2014**, *35*, 10645–10663. [CrossRef] [PubMed]

61. Richards, E.J.; Zhang, G.; Li, Z.-P.; Permuth-Wey, J.; Challa, S.; Li, Y.; Kong, W.; Dan, S.; Bui, M.M.; Coppola, D.; Mao, W.-M.; et al. Long non-coding RNAs (LncRNA) regulated by transforming growth factor (TGF) β: LncRNA-hit-mediated TGFβ-induced epithelial to mesenchymal transition in mammary epithelia. *J. Biol. Chem.* **2015**, *290*, 6857–6867. [CrossRef] [PubMed]

62. Mondal, T.; Subhash, S.; Vaid, R.; Enroth, S.; Uday, S.; Reinius, B.; Mitra, S.; Mohammed, A.; James, A.R.; Hoberg, E.; et al. MEG3 long noncoding RNA regulates the TGF-β pathway genes through formation of RNA–DNA triplex structures. *Nat. Commun.* **2015**, *6*, 7743. [CrossRef] [PubMed]

63. Terashima, M.; Tange, S.; Ishimura, A.; Suzuki, T. MEG3 long noncoding RNA contributes to the epigenetic regulation of epithelial-mesenchymal transition in lung cancer cell lines. *J. Biol. Chem.* **2016**, *292*, 82–99. [CrossRef] [PubMed]

64. Dong, Y.; He, D.; Peng, Z.; Peng, W.; Shi, W.; Wang, J.; Li, B.; Zhang, C.; Duan, C. Circular RNAs in cancer: an emerging key player. *J. Hematol. Oncol.* **2017**, *10*, 2. [CrossRef] [PubMed]

65. Bachmayr-Heyda, A.; Reiner, A.T.; Auer, K.; Sukhbaatar, N.; Aust, S.; Bachleitner-Hofmann, T.; Mesteri, I.; Grunt, T.W.; Zeillinger, R.; Pils, D. Correlation of circular RNA abundance with proliferation—Exemplified with colorectal and ovarian cancer, idiopathic lung fibrosis, and normal human tissues. *Sci. Rep.* **2015**, *5*, 8057. [CrossRef] [PubMed]

66. Hansen, T.B.; Kjems, J.; Damgaard, C.K. Circular RNA and miR-7 in cancer. *Cancer Res.* **2013**, *73*, 5609–5612. [CrossRef] [PubMed]

67. Li, J.; Yang, J.; Zhou, P.; Le, Y.; Zhou, C.; Wang, S.; Xu, D.; Lin, H.-K.; Gong, Z. Circular RNAs in cancer: novel insights into origins, properties, functions and implications. *Am. J. Cancer Res.* **2015**, *5*, 472–480. [PubMed]

68. Hansen, T.B.; Jensen, T.I.; Clausen, B.H.; Bramsen, J.B.; Finsen, B.; Damgaard, C.K.; Kjems, J. Natural RNA circles function as efficient microRNA sponges. *Nature* **2013**, *495*, 384–388. [CrossRef] [PubMed]

69. Weng, W.; Wei, Q.; Toden, S.; Yoshida, K.; Nagasaka, T.; Fujiwara, T.; Cai, S.; Qin, H.; Ma, Y.; Goel, A. Circular RNA ciRS-7—A promising prognostic biomarker and a potential therapeutic target in colorectal cancer. *Clin. Cancer Res.* **2017**. [CrossRef] [PubMed]

70. Conn, S.J.; Pillman, K.A.; Toubia, J.; Conn, V.M.; Salmanidis, M.; Phillips, C.A.; Roslan, S.; Schreiber, A.W.; Gregory, P.A.; Goodall, G.J. The RNA Binding Protein Quaking Regulates Formation of circRNAs. *Cell* **2015**, *160*, 1125–1134. [CrossRef] [PubMed]

71. Ng, K.W.; Anderson, C.; Marshall, E.A.; Minatel, B.C.; Enfield, K.S.S.; Saprunoff, H.L.; Lam, W.L.; Martinez, V.D. Piwi-interacting RNAs in cancer: Emerging functions and clinical utility. *Mol. Cancer* **2016**, *15*, 5. [CrossRef] [PubMed]

72. Hashim, A.; Rizzo, F.; Marchese, G.; Ravo, M.; Tarallo, R.; Nassa, G.; Giurato, G.; Santamaria, G.; Cordella, A.; Cantarella, C.; et al. RNA sequencing identifies specific PIWI-interacting small non-coding RNA expression patterns in breast cancer. *Oncotarget* **2014**, *5*, 9901–9910. [CrossRef] [PubMed]

73. Huang, G.; Hu, H.; Xue, X.; Shen, S.; Gao, E.; Guo, G.; Shen, X.; Zhang, X. Altered expression of piRNAs and their relation with clinicopathologic features of breast cancer. *Clin. Transl. Oncol.* **2013**, *15*, 563–568. [CrossRef] [PubMed]

74. Zhang, H.; Ren, Y.; Xu, H.; Pang, D.; Duan, C.; Liu, C. The expression of stem cell protein Piwil2 and piR-932 in breast cancer. *Surg. Oncol.* **2013**, *22*, 217–223. [CrossRef] [PubMed]

75. Rapisuwon, S.; Vietsch, E.E.; Wellstein, A. Circulating biomarkers to monitor cancer progression and treatment. *Comput. Struct. Biotechnol. J.* **2016**, *14*, 211–222. [CrossRef] [PubMed]

76. Su, H.; Xu, T.; Ganapathy, S.; Shadfan, M.; Long, M.; Huang, T.H.-M.; Thompson, I.; Yuan, Z.-M. Elevated snoRNA biogenesis is essential in breast cancer. *Oncogene* **2014**, *33*, 1348–1358. [CrossRef] [PubMed]

77. Ji, X.; Lu, H.; Zhou, Q.; Luo, K. LARP7 suppresses P-TEFb activity to inhibit breast cancer progression and metastasis. *Elife* **2014**, *3*, e02907. [CrossRef] [PubMed]

78. Waldron, C.; Lacroute, F. Effect of growth rate on the amounts of ribosomal and transfer ribonucleic acids in yeast. *J. Bacteriol.* **1975**, *122*, 855–865. [PubMed]

79. Goodenbour, J.M.; Pan, T. Diversity of tRNA genes in eukaryotes. *Nucleic Acids Res.* **2006**, *34*, 6137–6146. [CrossRef] [PubMed]

80. Goodarzi, H.; Nguyen, H.C.B.; Zhang, S.; Dill, B.D.; Molina, H.; Tavazoie, S.F. Modulated Expression of Specific tRNAs Drives Gene Expression and Cancer Progression. *Cell* **2016**, *165*, 1416–1427. [CrossRef] [PubMed]

81. Pavon-Eternod, M.; Gomes, S.; Geslain, R.; Dai, Q.; Rosner, M.R.; Pan, T. tRNA over-expression in breast cancer and functional consequences. *Nucleic Acids Res.* **2009**, *37*, 7268–7280. [CrossRef] [PubMed]

82. Rudolph, K.L.M.; Schmitt, B.M.; Villar, D.; White, R.J.; Marioni, J.C.; Kutter, C.; Odom, D.T. Codon-Driven Translational Efficiency Is Stable across Diverse Mammalian Cell States. *PLOS Genet.* **2016**, *12*, e1006024. [CrossRef] [PubMed]

83. Geslain, R.; Eriani, G. Regulation of translation dynamic and neoplastic conversion by tRNA and their pieces. *Transl. Austin.* **2014**, *2*, e28586. [CrossRef] [PubMed]

84. Gingold, H.; Pilpel, Y. Determinants of translation efficiency and accuracy. *Mol. Syst. Biol.* **2011**, *7*, 481. [CrossRef] [PubMed]

85. Gingold, H.; Tehler, D.; Christoffersen, N.R.; Nielsen, M.M.; Asmar, F.; Kooistra, S.M.; Christophersen, N.S.; Christensen, L.L.; Borre, M.; Sørensen, K.D.; et al. A dual program for translation regulation in cellular proliferation and differentiation. *Cell* **2014**, *158*, 1281–1292. [CrossRef] [PubMed]

86. Ruijtenberg, S.; van den Heuvel, S. Coordinating cell proliferation and differentiation: Antagonism between cell cycle regulators and cell type-specific gene expression. *Cell Cycle* **2016**, *15*, 196–212. [CrossRef] [PubMed]

87. Kumar, S.M.; Liu, S.; Lu, H.; Zhang, H.; Zhang, P.J.; Gimotty, P.A.; Guerra, M.; Guo, W.; Xu, X. Acquired cancer stem cell phenotypes through Oct4-mediated dedifferentiation. *Oncogene* **2012**, *31*, 4898–4911. [CrossRef] [PubMed]

88. Klochendler, A.; Weinberg-Corem, N.; Moran, M.; Swisa, A.; Pochet, N.; Savova, V.; Vikeså, J.; Van de Peer, Y.; Brandeis, M.; Regev, A.; et al. A Transgenic Mouse Marking Live Replicating Cells Reveals In Vivo Transcriptional Program of Proliferation. *Dev. Cell* **2012**, *23*, 681–690. [CrossRef] [PubMed]

89. Cole, C.; Sobala, A.; Lu, C.; Thatcher, S.R.; Bowman, A.; Brown, J.W.S.; Green, P.J.; Barton, G.J.; Hutvagner, G. Filtering of deep sequencing data reveals the existence of abundant Dicer-dependent small RNAs derived from tRNAs. *RNA NY* **2009**, *15*, 2147–2160. [CrossRef] [PubMed]

90. Lee, Y.S.; Shibata, Y.; Malhotra, A.; Dutta, A. A novel class of small RNAs: tRNA-derived RNA fragments (tRFs). *Genes Dev.* **2009**, *23*, 2639–2649. [CrossRef] [PubMed]

91. Saikia, M.; Jobava, R.; Parisien, M.; Putnam, A.; Krokowski, D.; Gao, X.-H.; Guan, B.-J.; Yuan, Y.; Jankowsky, E.; Feng, Z.; et al. Angiogenin-cleaved tRNA halves interact with cytochrome c, protecting cells from apoptosis during osmotic stress. *Mol. Cell. Biol.* **2014**, *34*, 2450–2463. [CrossRef] [PubMed]

92. Fu, H.; Feng, J.; Liu, Q.; Sun, F.; Tie, Y.; Zhu, J.; Xing, R.; Sun, Z.; Zheng, X. Stress induces tRNA cleavage by angiogenin in mammalian cells. *FEBS Lett.* **2009**, *583*, 437–442. [CrossRef] [PubMed]

93. Kumar, P.; Kuscu, C.; Dutta, A. Biogenesis and Function of Transfer RNA-Related Fragments (tRFs). *Trends Biochem. Sci.* **2016**, *41*, 679–689. [CrossRef] [PubMed]

94. Tang, Y.; Lin, Y.; Li, C.; Hu, X.; Liu, Y.; He, M.; Luo, J.; Sun, G.; Wang, T.; Li, W.; et al. MicroRNA-720 promotes in vitro cell migration by targeting Rab35 expression in cervical cancer cells. *Cell Biosci.* **2015**, *5*, 56. [CrossRef] [PubMed]

95. Hua, Y.; Choi, P.-W.; Trachtenberg, A.J.; Ng, A.C.; Kuo, W.P.; Ng, S.-K.; Dinulescu, D.M.; Matzuk, M.M.; Berkowitz, R.S.; Ng, S.-W. Epithelialization of mouse ovarian tumor cells originating in the fallopian tube stroma. *Oncotarget* **2016**, *7*, 66077–66086. [CrossRef] [PubMed]

96. Grelet, S.; McShane, A.; Hok, E.; Tomberlin, J.; Howe, P.H.; Geslain, R. SPOt: A novel and streamlined microarray platform for observing cellular tRNA levels. *PLOS ONE* **2017**, *12*, e0177939. [CrossRef] [PubMed]

![cancers logo] *cancers*

Review

Epigenetic Regulation of the Epithelial to Mesenchymal Transition in Lung Cancer

Joëlle Roche [1],*, Robert M. Gemmill [2] and Harry A. Drabkin [2]

[1] Laboratoire Ecologie et Biologie des Interactions, Equipe SEVE, Université de Poitiers, UMR CNRS 7267, F-86073 Poitiers, France

[2] Division of Hematology-Oncology, Medical University of South Carolina, 39 Sabin St., MSC 635, Charleston, SC 29425, USA; gemmill@musc.edu (R.M.G.); drabkinh@gmail.com (H.A.D.)

*   Correspondence: joelle.roche@univ-poitiers.fr; Tel.: +33-5-4945-3550

Academic Editor: Samuel C. Mok
Received: 29 May 2017; Accepted: 17 June 2017; Published: 24 June 2017

Abstract: Lung cancer is the leading cause of cancer deaths worldwide. It is an aggressive and devastating cancer because of metastasis triggered by enhanced migration and invasion, and resistance to cytotoxic chemotherapy. The epithelial to mesenchymal transition (EMT) is a fundamental developmental process that is reactivated in wound healing and a variety of diseases including cancer where it promotes migration/invasion and metastasis, resistance to treatment, and generation and maintenance of cancer stem cells. The induction of EMT is associated with reprogramming of the epigenome. This review focuses on major mechanisms of epigenetic regulation mainly in lung cancer with recent data on EZH2 (enhancer of zeste 2 polycomb repressive complex 2 subunit), the catalytic subunit of the PRC2 (Polycomb Group PcG), that behaves as an oncogene in lung cancer associated with gene repression, non-coding RNAs and the epitranscriptome.

Keywords: chromatin modifications; EMT; Epigenetics; Epitranscriptomics; EZH2; non-coding RNAs; NSCLC; SCLC

1. Introduction

Lung cancer is the leading cause of cancer deaths worldwide [1]. Non-small cell lung cancer (NSCLC) accounts for about 85% of cases and includes adenocarcinomas, characterized by *RAS* or *EGFR* mutations, squamous cell carcinoma with *FGFR1* amplification, *PTEN* or *PIK3*CA mutations [2,3], and large cell carcinoma, which are highly heterogeneous. The recent WHO classification [4] now distributes large cell carcinoma into either NSCLC or small cell lung cancer (SCLC) depending on its characteristics. SCLC accounts for approximately 15% of lung tumors and is among the most aggressive tumor types because of high proliferation and early metastasis. It is observed almost exclusively in heavy smokers, and is characterized by gene inactivation of p53 (*TP53*) and retinoblastoma (*RB1*), along with expression of neuroendocrine markers [5–7]. Metastasis is triggered by enhanced cellular migration and invasion, and the epithelial to mesenchymal transition (EMT) described in the following paragraph, believed to play a role in this process.

EMT is a fundamental developmental process that is reactivated in wound healing and a variety of diseases including cancer where it induces the development of metastasis, resistance to treatment, and generation and maintenance of cancer stem cells [8–13]. However, two recent studies show that EMT can be dispensable for metastasis, but nevertheless contributed substantially to chemoresistance [14,15]. These studies were recently questioned because they coincided with a time when the definition of the EMT was undergoing re-evaluation because of possible multiple partial EMT states, and the proofs that EMT did not occur during metastasis were not completely supported [13] The EMT program includes loss of intercellular adhesion with loss of tight junctions, adherens and gap junctions, increased invasion and migration, and a switch from epithelial to mesenchymal gene expression patterns. E-cadherin and ZO-1 expression is often lost, while in contrast there is a gain of N-cadherin, Vimentin and Fibronectin expression (Figure 1). In TGFβ-induced EMT in lung cancer cells, we found increased expression of Neuropilin-2 (NRP2), the receptor for class 3 semaphorins, ligands providing guidance to and control of cell movement [16]. Most recently, we found that TGFβ preferentially induced NRP2b, an understudied isoform of NRP2 that was responsible for promoting the oncogenic response in cell lines and correlated with tumor progression in patients [17]. In contrast, the level of SEMA3F (semaphorin 3F), a secreted semaphorin with potent antitumor activity that binds NRP2, was decreased in a ZEB1-induced EMT lung cancer model [18]. Of interest, neuropilins bind TGFβ suggesting a function of NRPs in TGFβ response, and additional ligands such as VEGF (Vascular Endothelial Growth Factor), HGF (Hepatocyte Growth Factor), platelet-derived growth factor (PDGF), and EGF (Epidermal Growth Factor) [19].

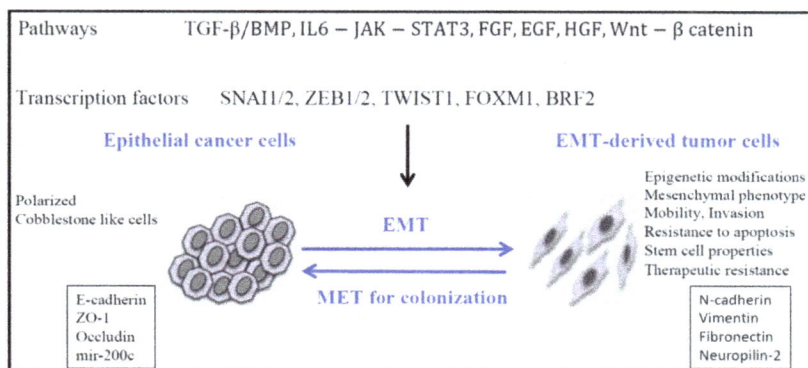

Figure 1. EMT (Epithelial to Mesenchymal transition) and epigenetic modifications. EMT is induced by different pathways that involve different transcription factors necessary to repress epithelial genes and to activate mesenchymal genes.

Many EMT studies have focused mostly on cell-based experimental models and have focused on a small number of gene promoters that are epigenetically regulated, such as E-cadherin (*CDH1*) [20–23]. However, the direct involvement of EMT in tumor progression and metastatic spread has been questioned. In fact, the difficulties to characterize EMT in vivo come, in part, from the fact that EMT can be transient and partial. Indeed, EMT is increasingly viewed as a program generating cells with a spectrum of multiple states between epithelial and mesenchymal extremes, and partial EMT would be frequent in tumors [12,13]. Our previous study on 22 NSCLC cell lines showed that they are clearly in different states with more epithelial traits such as NCI-H358 cells or with more mesenchymal traits such as A549, NCI-H661, and NCI-H460 cells [24]. Indeed, a recent integrative approach combining mRNA, miRNA, DNA methylation, and proteomic profiles of 38 cell lines representative of lung adenocarcinoma heterogeneity, and functional profiles consisting of cell invasiveness, adhesion, and motility, defined cell lines as epithelial (E), mesenchymal (M), or intermediate/hybrid (E/M) with mixed epithelial and mesenchymal characteristics. Aggressive hybrid cell lines were characterized with a signature shared with mesenchymal cell lines: upregulation of cytoskeletal and actin-binding proteins [25]. Such E/M hybrid cells have been observed among circulating tumors cells and are associated with metastasis [26].

In lung cancer, TGFβ is one of the most important physiologic EMT inducers, along with other factors such as HGF, FGF (Fibroblast Growth Factor), IGF (Insulin-like Growth factor-1)/PDGF, EGF/VEGF, each secreted by the tumor or/and its microenvironment and acting through downstream pathways including Wnt/β-catenin, TGFβ/SMAD, Notch, MAPK/ERK, and PI3K/Akt [27,28].

The induction of EMT is associated with reprogramming of the epigenome by the action of transcription factors (such as TWIST, SNAIL, SLUG, and ZEB1/2), non-coding RNAs such as micro-RNAs (miRNAs) and long non-coding RNAs (lncRNAs). Here, we focus on epigenetic regulation mainly in lung cancer, including recent data on non-coding RNAs and the epitranscriptome.

2. Epigenetics

Epigenetic regulation is normally dynamic and reversible, affecting chromatin structure for the regulation of gene expression without modification of DNA sequences [29,30]. It includes the incorporation of histone variants, covalent histone modifications, nucleosome re-positioning, DNA methylation, changes in the expression of non-coding RNAs, and RNA post-translational modifications. The impact of epigenetic changes in cancer is reflected by altered gene expression, reactivation of endogenous retro-elements and genomic instability.

The nucleosome is the basic chromatin unit consisting of a protein core formed by two copies of histones H2A, H2B, H3 and H4, encircled by 180–200 bp of DNA stabilized by histone H1 that binds DNA at its entry and exit points of the nucleosome. Histones are post-translationally modified by phosphorylation, acetylation, methylation and ubiquitylation (among others). DNA can be covalently modified by methylation, usually on cytosine residues immediately preceding guanosine (CpG), without altering the sequence of base pairs. The epigenome refers to this collective set of DNA and histone modifications, along with their precise distribution along chromatin and the dynamic changes in this pattern mediated by regulatory processes. DNA methylation at CpG is one of the best studied epigenetic marks. DNA methylation and histone modifications are added by enzymes called "writers" and removed by "erasers". These modifications are further recognized by proteins called "readers" that recruit various adapters for regulation of gene expression [31] (Table 1).

Table 1. Epigenetic modifications. Only DNA and RNA methylation, histone acetylation and some epigenetic modifications for histone H3 are shown with the corresponding players (writers, erasers and readers). For H3K79 methylation, DOT1L functions are multiple: it is involved in telomeric silencing, cellular development, cell-cycle checkpoint, DNA repair, and regulation of transcription. For RNA, only m^{5}C and m^{6}A are described but other modifications can be found in the review by Esteller and Pandolfi (2017) [32]. ** G9a weakly methylates H3K27.

	Epigenetic Modification		Function	Writer	Eraser	Reader
DNA	CpG Methylation		Transcriptional repression	DNMT1/3A/3B	TET1/2/3	MeCP, MBD1-4, UHRF1
RNA	m^{5}C		tRNA stabilization, translation, immune response	DNMT2 (=TRDMT1) NSUN family	TET	not identified
	m^{6}A		RNA splicing, export, stability, immune tolerance	METTL3/4, WTAP	FTO, ALKBH5	YTHD family, HuR HNRNPA2B1
Histones	Lysine Acetylation		Transcriptional activation	HAT	HDAC1-11, SIRT1-7	BRD bromodomain
	Lysine Methylation	HH3K4	Transcriptional activation	MLL1-5, SET1A/B, SET7/9, ASH1L	LSD1, JARID1a/b	Chromodomain, Tudor, MBT repeat, PHD finger
		HH3K9	Transcriptional repression	G9a(EHMT2) SUV39H1/2	LSD1, GASC1	
		HH3K27	Transcriptional repression	EZH1/2, G9a **	UTX, JMJD3	
		HH3K36	Transcriptional activation	SETD2, ASH1L, ASF1A, NSD1-3, SMYD2	Rph1/KDM4 Jhdm1b/Kdm2b	
		HH3K79	Transcriptional regulation	DOT1L, RE-IIBP	not known	

DNA can be methylated at the fifth position of cytosine (5mC) in the CpG dinucleotide context by DNA methyltransferases and further demethylated by TET enzymes with sequential oxidation of 5mC to 5-hydroxymethylcytosine (5hmC), 5-formylcytosine (5fC), and finally 5-carboxylcytosine (5caC). 5fC and 5caC are further cleaved by thymine-DNA glycosylases to restore unmethylated cytosine via the base-excision repair machinery [33]. In cancer, abnormal DNA methylation is observed with a global decrease of DNA methylation on repetitive sequences, and transposons, associated with chromosome anomalies and active transposition [34,35]. Of interest, gene bodies of tumor suppressor genes are undermethylated but, in contrast, their promoters are hypermethylated. Gene promoter hypermethylation is generally associated with transcriptional repression.

Histones are covalently modified primarily on their amino-terminal tails [36]. The nature and combination of histone modifications forms a "histone code" that affects transcription of nearby genes [37].

Whole genome and transcriptome sequencing established that about 70% of the genome is transcribed into non-coding RNAs. These non-coding RNAs emerged as a novel class of functional molecules with involvement in a variety of physiological processes and in tumorigenesis when misregulated. Non-coding RNAs are divided in two classes depending on their size: the small non-coding RNA group which is less than 200 nucleotides long includes microRNAs (miRNA), while the long non-coding RNA (lncRNA) group extends from 200 nucleotides to over 100 kb. More than 2000 human miRNAs have been identified and a large fraction is deregulated in cancers including lung cancer [38]. miRNAs regulate around 30% of coding genes at the post-transcriptional and translational levels and each typically targets multiple genes within a pathway. LncRNAs belong to a very heterogeneous group of transcripts (for review see [39]). Interestingly, the architecture of lncRNAs is complex: different structural domains can be combined by alternative splicing to sense or bind other RNAs, proteins and possibly DNA [40]. In addition, they undergo conformational switches that modulate their functions. Consequently, lncRNAs are considered as platforms to perform multiple functions in the cell and they act either in *cis* or in *trans*. They are involved in chromatin remodeling, transcriptional co-activation and co-repression, protein inhibition, post-transcriptional modifications (splicing), decoy binding platforms, mRNA stabilization, and more recently in nuclear organization of multichromosomal regions [41–48]. Among these functions, lncRNAs bind to chromatin-modifying proteins for their recruitment to specific sites in the genome, and lncRNAs act as competing endogenous RNAs (ceRNAs) for miRNAs. These two functions have been described in the epigenetic regulation of EMT [49–51].

To these classical epigenetic modifications, epitranscriptomics and the RNA code came into the spotlight recently [52]. Epigenetic modifications not only affect DNA and proteins, but also coding and non-coding RNAs. RNAs show a diverse spectrum of more than 100 modifications including N^7-methylguanosine (m^{7}G), N^6-methyladenosine (m^{6}A), m^{5}C, pseudouridine, and queuosine [32,53]. These modifications are reversible, dynamic, and important for gene expression. Enzymes that are writers, erasers, and protein readers of these RNA modifications, analogous to the equivalent functions for histone modification, have been identified [54] (Table 1). However, additional technology development is necessary to detect, quantify, and map these modifications [52,55]. This is a challenging and exciting field still in its early stage, with much to be discovered to understand the role of RNA epigenetic modifications.

3. Epigenetics and Lung Cancer

Mutations in genes encoding epigenetic proteins and abnormalities in epigenetic regulation are clearly linked to cancer (for review see [56]). They have been described previously for lung cancer [34,57–60].

Large-scale genomic studies have identified recurrent alterations of epigenetic regulators in lung cancer (Table 2).

Table 2. Mutations in epigenetic regulators in lung cancers. ADC: Adenocarcinoma; SC: Squamous cell carcinoma.

Lung Cancer	Gene	Function	Mutation	References
SCLC	KAT3A/CREBBP	histone acetytransferase	inactivating mutation	[7]
	KAT3B/EP300	histone acetytransferase	inactivating mutation	[7]
	KAT6B	H3K23 histone acetytransferase	genomic loss	[61]
	KMT2D/MLL2	H3K4me1/2 histone methyltransferase	frequent inactivation	[7,62]
	KDM6A/UTX	H3K27 histone demethylase	truncating mutation in a small number of SCLC patients	[62,63]
	PBRM1	chromatin remodeling factor	mutation	[62]
	ARID1A		mutation	[62]
	ARID1B		mutation	[62]
NSCLC	KMT2D/MLL2	H3K4me1/2 histone methyltransferase	mutation in 20% SC	[3]
	SETD2	H3K36 histone methyltransferase	9% ADC	[3]
	DOT1L	H3K79 histone methyltransferase	3% ADC	[64]
	ARID1A	chromatin remodeling factor	7% ADC	[3]
	ARID1B		6% ADC	
	ARID2		7% ADC	
	SMARCA4/BRG1		6% ADC	
	BRD3	Bromodomain, binds hyperacetylated chromatin		[65]

In SCLCs, mutations are found for histone acetyltransferases, histone methyltransferases and demethylases, and remodeling factors [7,62,66]. High frequency truncating mutations for the H3K4 histone methyltransferase *KMT2D/MLL2* gene have been reported in 17% of SCLC cell lines and 8% of SCLC tumors, and *KMTD2/MLL2* loss is associated with reduced H3K4me1 and impaired enhancer function [62]. Less frequent are mutations in the H3K27 histone demethylase, *KDM6A/UTX* gene, that occur in an exclusive fashion with *KMTD2/MLL2* mutations. Of interest, bivalent promoters are characterized by the presence of both H3K4me3 active mark and H3K27me3 inactive mark that poise developmental genes, enabling them to respond rapidly to suitable stimuli [67]. Therefore, *KMTD2/MLL2* or *KDM6A/UTX* mutations in addition to abnormal EZH2 expression (see below) lead to impaired expression of genes under control of these bivalent promoters.

For NSCLC, the most frequent mutations are found in the chromatin remodeling factors including SMARCA4/BRG1, and the H3K36 histone methyltransferase SETD2, in 6% and 9% of adenocarcinomas, respectively. BRG1 is one of the two ATPase subunits in the SWI/SNF chromatin-remodeling complex and has been reported to be frequently mutated or silenced in primary human NSCLC tumors and cell lines (for reviews [68,69]). To a lesser extent, the DOT1L methyltransferase for H3K79 is mutated in 3% of lung adenocarcinomas [64].

Abnormal expression of epigenetic players in lung cancer includes EZH2 overexpression in SCLC and NSCLC, where EZH2 acts as an oncogene in these tissues. EZH2 is the catalytic subunit of the PRC2 (Polycomb Group PcG) complex, which includes additional core components (SUZ12, EED, RBBP4). It mediates methylation of lysine 27 on histone H3 (H3K27), associated with gene repression. EZH2 overexpression in lung cancer is common and associated with aggressive tumor characteristics, advanced stage and poor prognosis [70–73]. In cell lines and patient-derived xenografts, EZH2 inhibition attenuated cell-cycle progression, growth and invasion [72–75]. Interestingly, tobacco smoke causes upregulation of Wnt signaling by recruiting EZH2 to suppress Dickkopf-1, a Wnt antagonist [76].

At the transcriptional level, EZH2 is suppressed by the pRB protein. At least in part, this explains the upregulation of EZH2 in SCLC, a disease with near-universal mutation of *RB* [77,78]. Of interest, the retinoblastoma protein occupies repetitive sequences that include tandem sequence repeats and interspersed repeats (endogenous retroviruses and LINE-1 elements) in somatic cells and is associated with EZH2. Consequently, H3K27 trimethylation maintains silencing of these sequences [79]. In cancer cells, these repetitive sequences can be reactivated. When pRB binding to these sequences is inhibited, the consequence is an increased susceptibility for spontaneous lymphoma in mice. Therefore, loss of pRB in SCLC or abnormal level of EZH2 would be involved in repetitive sequence expression responsible for mutagenesis and multiple possible aberrations.

Links between epitranscriptome and cancer have emerged with the discovery that FTO (Fat mass and obesity-associated protein), first identified as the m^6A eraser, but recently described as the m^6Am (N^6, 2'-*O*-dimethyladenosine) eraser, is abnormally expressed in acute myeloid leukemia [80], and affects the stability and subcellular location of mRNAs [81]. FTO is also overexpressed in breast cancer compared to adjacent breast tissues [82] and overexpression is associated with higher metastatic potential and resistance to chemotherapy in an in vitro cellular breast cancer model [83]. The ncRNA epitranscriptome is also altered in cancer [32] but more data are needed for lung cancer.

4. EMT and Epigenetics in Lung Cancer

A review about epigenetic regulation of EMT in NSCLC was recently published [28]. Additional data will be described below for lung cancer.

The involvement of the Snail and ZEB1 transcription factors strongly supports epigenetic modifications during EMT. Indeed, Snail has been described as a pseudosubstrate "hook" for the histone demethylase LSD1 that demethylates H3K4 [84,85] and H3K9 (Table 1). Through its SNAG domain that mimics a histone H3 tail, Snail binds to LSD1 and the complex is stabilized by association of the co-repressor CoREST. HDAC1/2 and PRC2 are later recruited for *CDH1* repression in cancer cells, including mammary epithelial tumor cells. SNAIL was also found to associate with G9a (EHMT2), a major histone methyltransferase responsible for creating the H3K9me2 repressive mark [86] and which also contributes to methylation of H3K27 [87] (Table 1). In addition, SNAIL interacts with SUV39H1 (another histone methyltransferase) during EMT induced by TGF-β and mediates silencing of *CDH1* by the addition of a third methyl group to H3K9 that confers a more stable and durable repressive state than H3K9me2 [88].

ZEB1 also associates with partners involved in epigenetic regulation. For gene activation, ZEB1 associates with the histone acetyltransferases (HATs) p300, PCAF, and Tip60. In contrast, as a repressor, ZEB1 interacts with CtBP [89] and recruits class I and II histone deacetylases (HDACs) [90]. Of interest, CtBP associates with several other partners including the Polycomb complex PRC2 as described above, G9a (EHMT2), the co-repressor CoREST, and the histone demethylase LSD1 (for review see [91,92]). ZEB1 can also recruit the nicotinamide adenine dinucleotide-dependent sirtuin, SIRT1, in prostate cancer cells to repress *CDH1* and to induce several EMT markers [93]. In addition, ZEB1 interacts via its N-terminal region with BRG1 to repress *CDH1* in colon cancer cells [94].

In H358 NSCLC cells where EMT was induced by ZEB1 expression, we used Western blot and immunocytochemistry to identify a global decrease in H3K27 acetylation. We found that ZEB1 binding resulted in decreased acetylation of histone H3 on residues K9 and K27 of target genes [95] (Figure 2). Our results showed that ZEB1 increased trimethylation of H3K27 on selected target genes and that H3K4me2 did not change drastically upon ZEB1 binding. These results suggested that ZEB1 recruitment of PRC2 during EMT would create bivalent domains for epithelial gene repression. Such genes would then be poised for rapid reversal of the repressed state to facilitate the mesenchymal to epithelial transition (MET) thought to be important for the subsequent growth of metastatic deposits. This hypothesis fits with the PRC2 repressive activity on promoters with the activating mark H3K4me2/3 [96].

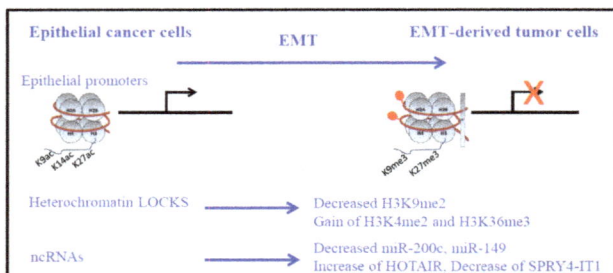

Figure 2. Epigenetic modifications during EMT. Nucleosomes are represented with DNA in red, wrapped around the histone core (2 copies of each histone H2A, H2B, H3 and H4) with the presence of histone H1 in more compact chromatin. Repression of epithelial genes (red cross) is shown with corresponding epigenetic modifications on their promoters: more compact chromatin, decreased H3K9/14/27ac and gain of H3K9me3 and H3K27me3, associated to DNA methylation (red lollipop). Of note, only major modifications for histone H3 are presented. LOCKS (long-range chromatin domains) epigenetic marks during EMT: Global reduction in the heterochromatin mark H3K9me2, increase in the euchromatin mark H3K4me3, and increase in the transcriptional mark H3K36me3. Non-coding RNAs (ncRNAs) expressions are also modified during EMT for miRNAs (loss of miR-200c and miR-149) and lncRNAs (increased HOTAIR and decreased SPRYA-IT1 in lung cancer).

Recent studies showed that specific "long-range" chromatin domains across the genome, called "LOCKs", found in non-repetitive heterochromatin domains up to several megabases, are epigenetically remodeled as a major driving force during EMT [97]. A global reduction in the heterochromatin mark H3K9me2, an increase in the euchromatin mark, H3K4me3, and an increase in the transcriptional mark, H3K36me3, were described (Figure 2). These changes depended largely on LSD1. Of interest, DNA methylation was preserved across the genome during EMT.

Epigenetic modifications also affect super-enhancers that are localized to unique relatively small subsets of genes that differ between cell states during EMT. They are often found at key oncogenes, such as *MYC*. Loss of BRD4, a bromodomain protein that binds acetylated histones, or its pharmacological inhibition, can cause super enhancer-mediated gene expression to be lost [98]. In SCLCs, MYCL, one of the three MYC family oncogenes, is often overexpressed and treatment with JQ1, a BET bromodomain inhibitor, considerably decreased cell growth, induced cell cycle arrest and apoptosis, and reduced expression of the three *MYC* genes [99].

Non-coding RNAs contribute to EMT in NSCLC. For example, micro-RNAs including miR-132 and miR-149 that target ZEB2 and FOXM1 (Forkhead box M1), respectively, and downregulation of miR-149 was inversely correlated with invasive and EMT phenotypes in NSCLC [100]. The miR-200 family (that includes miR-200a, miR-200b, miR-200c, miR-141, and miR-429) received a lot of interest because of a negative transcriptional feedback loop between miR-200c and ZEB1 [101,102]. In our ZEB1-induced EMT model in the H358 NSCLC cell line, we observed a decrease of miR-200c during EMT [95]. Moreover, in the HCC4006 NSCLC cell line that became resistant to EGFR inhibitors, acquired resistance was associated with ZEB1, an EMT phenotype, and decreased miR-200c levels [103].

Several long non-coding RNAs (lncRNAs) are also associated with an EMT signature and aggressiveness [51,60]. In NSCLC, a positive correlation with lymph node metastasis was described for HOTAIR, CARLO-5, PVT1, MVIH and ZXF1, while a negative correlation was found with MEG3, SPRY4-ITI, BANCR and GAS6-AS1 [60]. Among these lncRNAs, HOTAIR and SPRY4-ITI are of interest in EMT regulation (Figure 2). HOTAIR interacts with the PRC2 complex to induce in *trans* H3K27 trimethylation of the *HOXD* locus, and the LSD1/coREST/REST complex that catalyzes H3K4me2 demethylation. However, this model of PRC2 binding to HOTAIR for targeting specific sequences was recently challenged, since PRC2 interacts with strong affinity but weak specificity to a wide set of all RNAs. Indeed, PRC2 is dispensable for HOTAIR-mediated transcriptional repression while PRC2 recruitment and H3K27 trimethylation were proposed to occur due to HOTAIR-dependent silencing [104,105]. How this happens mechanistically will require further investigation. HOTAIR is also involved in the maintenance of stemness and EMT in colon and breast cancer cell lines [106]. SPRY4-ITI is an inhibitor of the MAPK signaling pathway, and tumor suppressive functions were described in NSCLC cell lines. SPRY4-ITI inhibits NSCLC cell migration/invasion, and suppresses metastasis. SPRY4-ITI is downregulated by EZH2 and is involved in the modulation of EMT through induction of E-cadherin expression and repression of vimentin [107].

MALAT-1 (metastasis associated lung adenocarcinoma transcript-1, also called NEAT2) is a predictive marker for metastasis and shorter survival in early stage lung adenocarcinoma [51,60]. It can act as a ceRNA for several miRNAs including miR-200c, and can recruit PRC2 subunits (EZH2 and Suz12) to the *CDH1* promoter. CCAT2 (colon cancer associated transcript 2) is also a predictive marker for metastasis in lung cancer. In addition, lncRNAs are associated with EMT, stem cell properties and drug resistance. For example, UCA1, BC087858 and GAS5 are associated with resistance to EGFR tyrosine kinase inhibitors (EGFR-TKIs) in lung cancer possibly through the Akt signaling pathway.

Lastly, the epitranscriptome is likely involved in EMT but to our knowledge data are missing in lung cancer. However, we suspect that modifications of lncRNAs MALAT1 and TUG1 are candidates. Indeed, for MALAT1, two major m^6A sites were found at consensus sites in two hairpin stem structures in several human cell lines including, MDA-MB231 breast cancer cells [108]. The consequence of this modification would destabilize these secondary structures and would modify interactions with RNA binding proteins. One can speculate that FTO overexpression would alter m^6A levels during EMT. The other lncRNA TUG1 function might be altered as well, as one site shows m^6A modification [108].

5. Therapeutic Inhibition of EMT

From a therapeutic standpoint, epigenetic processes are targets for therapeutic exploitation [11,22,31], and epigenetic regulators implicated in cancer and their corresponding inhibitors are described in recent reviews [56,109]. Some of them would be appropriate to target EMT in lung cancer. EZH2 inhibitors would be good candidates because of EZH2 oncogenic status in lung cancer, and some are in clinical development [110]. We also suspect that bromodomain inhibitors such as JQ1 (known as TEN-010 in clinical trials), the BET inhibitor for BRD4, would be appropriate because JQ1 demonstrated efficacy in blocking tumor progression in several cancer models including SCLC lung cancer [98,99]. One possible mechanism would be a decrease of C-Myc recruitment to EZH2 and consequently reduced EZH2 expression, as shown in bladder cancer [111]. Combined treatments with histone deacetylase inhibitors (such as SAHA, i.e., vorinostat, an FDA-approved drug) and bromodomain inhibitors should be considered in NSCLC for their synergistic therapeutic benefit seen in animal models [112,113]. HDAC inhibitors were also shown to induce E-cadherin in lung cancer and this restoration increased sensitivity to EGFR inhibitors [114,115]. Although the mode of action of the two drugs is multifaceted, one reason for this benefit would be their immunostimulatory effects on tumor growth arrest and prolonged survival as shown in a mouse model for lung adenocarcinoma [113].

6. Conclusions

In this review, epigenetics regulation of EMT was described in lung cancer with a particular interest in EZH2, an oncogene in lung tumors. Recent data on non-coding RNAs and epitranscriptomics were introduced. Attempts to inhibit EMT should consider combined treatments with HDAC and bromodomain inhibitors. However, caution is needed with these treatments to address the mesenchymal to epithelial transition (MET) that occurs at sites of metastasis.

Acknowledgments: Joëlle Roche was supported by Région Poitou-Charentes, France; Robert M. Gemmill and Harry A. Drabkin by DOD LC150622.

Conflicts of Interest: The authors declare no conflict of interest.

Abbreviations

The following abbreviations are used in this manuscript:

ADC	Adenocarcinoma
ceRNA	competing endogenous RNA
EGF	Epidermal Growth Factor
EGFR	Epidermal Growth Factor Recptor
EMT	Epithelial to Mesenchymal transition
EZH2	enhancer of zeste 2 polycomb repressive complex 2 subunit
HDAC	Histone deacetylase
HGF	Hepatocyte Growth Factor
lncRNA	long non-coding RNA
MET	Mesenchymal to Epithelial Transition
NSCLC	noRsmall cell lung cancer
PRC2	polycomb repressive complex 2
SC	squamous cell carcinoma
SCLC	small cell lung cancer
VEGF	Vascular Endothelial Growth Factor

References

1. Torre, L.A.; Siegel, R.L.; Jemal, A. Lung cancer statistics. *Adv. Exp. Med. Biol.* **2016**, *893*, 1–19. [CrossRef] [PubMed]
2. Heist, R.S.; Engelman, J.A. SnapShot: Non-small cell lung cancer. *Cancer Cell* **2012**, *21*, 448.e2. [CrossRef] [PubMed]
3. The Cancer Genome Atlas Research Network. Comprehensive molecular profiling of lung adenocarcinoma. *Nature* **2014**, *511*, 543–550. [CrossRef]
4. Travis, W.D.; Brambilla, E.; Nicholson, A.G.; Yatabe, Y.; Austin, J.H.; Beasley, M.B.; Chirieac, L.R.; Dacic, S.; Duhig, E.; Flieder, D.B.; et al. The 2015 world health organization classification of lung tumors: Impact of genetic, clinical and radiologic advances since the 2004 classification. *J. Thorac. Oncol.* **2015**, *10*, 1243–1260. [CrossRef] [PubMed]
5. Takahashi, T.; Nau, M.M.; Chiba, I.; Birrer, M.J.; Rosenberg, R.K.; Vinocour, M.; Levitt, M.; Pass, H.; Gazdar, A.F.; Minna, J.D. p53: A frequent target for genetic abnormalities in lung cancer. *Science* **1989**, *246*, 491–494. [CrossRef] [PubMed]
6. Wistuba, II; Gazdar, A.F.; Minna, J.D. Molecular genetics of small cell lung carcinoma. *Semin. Oncol.* **2001**, *28*, 3–13. [CrossRef]
7. George, J.; Lim, J.S.; Jang, S.J.; Cun, Y.; Ozretic, L.; Kong, G.; Leenders, F.; Lu, X.; Fernandez-Cuesta, L.; Bosco, G.; et al. Comprehensive genomic profiles of small cell lung cancer. *Nature* **2015**, *524*, 47–53. [CrossRef] [PubMed]
8. Thiery, J.P.; Acloque, H.; Huang, R.Y.; Nieto, M.A. Epithelial-mesenchymal transitions in development and disease. *Cell* **2009**, *139*, 871–890. [CrossRef] [PubMed]
9. Mani, S.A.; Guo, W.; Liao, M.J.; Eaton, E.N.; Ayyanan, A.; Zhou, A.Y.; Brooks, M.; Reinhard, F.; Zhang, C.C.; Shipitsin, M.; et al. The epithelial-mesenchymal transition generates cells with properties of stem cells. *Cell* **2008**, *133*, 704–715. [CrossRef] [PubMed]
10. Lamouille, S.; Xu, J.; Derynck, R. Molecular mechanisms of epithelial-mesenchymal transition. *Nat. Rev. Mol. Cell Biol.* **2014**, *15*, 178–196. [CrossRef] [PubMed]
11. Marcucci, F.; Stassi, G.; De Maria, R. Epithelial-mesenchymal transition: A new target in anticancer drug discovery. *Nat. Rev. Drug Discov.* **2016**, *15*, 311–325. [CrossRef] [PubMed]
12. Nieto, M.A.; Huang, R.Y.; Jackson, R.A.; Thiery, J.P. EMT: 2016. *Cell* **2016**, *166*, 21–45. [CrossRef] [PubMed]
13. Lambert, A.W.; Pattabiraman, D.R.; Weinberg, R.A. Emerging biological principles of metastasis. *Cell* **2017**, *168*, 670–691. [CrossRef] [PubMed]
14. Fischer, K.R.; Durrans, A.; Lee, S.; Sheng, J.; Li, F.; Wong, S.T.; Choi, H.; El Rayes, T.; Ryu, S.; Troeger, J.; et al. Epithelial-to-mesenchymal transition is not required for lung metastasis but contributes to chemoresistance. *Nature* **2015**, *527*, 472–476. [CrossRef] [PubMed]

15. Zheng, X.; Carstens, J.L.; Kim, J.; Scheible, M.; Kaye, J.; Sugimoto, H.; Wu, C.C.; LeBleu, V.S.; Kalluri, R. Epithelial-to-mesenchymal transition is dispensable for metastasis but induces chemoresistance in pancreatic cancer. *Nature* **2015**, *527*, 525–530. [CrossRef] [PubMed]

16. Nasarre, P.; Gemmill, R.M.; Potiron, V.A.; Roche, J.; Lu, X.; Baron, A.E.; Korch, C.; Garrett-Mayer, E.; Lagana, A.; Howe, P.H.; et al. Neuropilin-2 is upregulated in lung cancer cells during TGF-beta1-induced epithelial-mesenchymal transition. *Cancer Res.* **2013**, *73*, 7111–7121. [CrossRef] [PubMed]

17. Gemmill, R.M.; Nasarre, P.; Nair-Menon, J.; Cappuzzo, F.; Landi, L.; D'Incecco, A.; Uramoto, H.; Yoshida, T.; Haura, E.B.; Armeson, K.; et al. The neuropilin 2 isoform NRP2b uniquely supports TGFbeta-mediated progression in lung cancer. *Sci. Signal.* **2017**, *10*. [CrossRef] [PubMed]

18. Clarhaut, J.; Gemmill, R.M.; Potiron, V.A.; Ait-Si-Ali, S.; Imbert, J.; Drabkin, H.A.; Roche, J. ZEB-1, a repressor of the semaphorin 3F tumor suppressor gene in lung cancer cells. *Neoplasia* **2009**, *11*, 157–166. [CrossRef] [PubMed]

19. Nasarre, P.; Gemmill, R.M.; Drabkin, H.A. The emerging role of class-3 semaphorins and their neuropilin receptors in oncology. *Onco. Targets Ther.* **2014**, *7*, 1663–1687. [CrossRef] [PubMed]

20. Tam, W.L.; Weinberg, R.A. The epigenetics of epithelial-mesenchymal plasticity in cancer. *Nat. Med.* **2013**, *19*, 1438–1449. [CrossRef] [PubMed]

21. Cieslik, M.; Hoang, S.A.; Baranova, N.; Chodaparambil, S.; Kumar, M.; Allison, D.F.; Xu, X.; Wamsley, J.J.; Gray, L.; Jones, D.R.; et al. Epigenetic coordination of signaling pathways during the epithelial-mesenchymal transition. *Epigenet. Chromatin* **2013**, *6*, 28. [CrossRef] [PubMed]

22. Mishra, V.K.; Johnsen, S.A. Targeted therapy of epigenomic regulatory mechanisms controlling the epithelial to mesenchymal transition during tumor progression. *Cell Tissue Res.* **2014**, *356*, 617–630. [CrossRef] [PubMed]

23. Kiesslich, T.; Pichler, M.; Neureiter, D. Epigenetic control of epithelial-mesenchymal-transition in human cancer. *Mol. Clin. Oncol.* **2013**, *1*, 3–11. [CrossRef] [PubMed]

24. Gemmill, R.M.; Roche, J.; Potiron, V.A.; Nasarre, P.; Mitas, M.; Coldren, C.D.; Helfrich, B.A.; Garrett-Mayer, E.; Bunn, P.A.; Drabkin, H.A. ZEB1-responsive genes in non-small cell lung cancer. *Cancer Lett.* **2011**, *300*, 66–78. [CrossRef] [PubMed]

25. Schliekelman, M.J.; Taguchi, A.; Zhu, J.; Dai, X.; Rodriguez, J.; Celiktas, M.; Zhang, Q.; Chin, A.; Wong, C.H.; Wang, H.; et al. Molecular portraits of epithelial, mesenchymal, and hybrid States in lung adenocarcinoma and their relevance to survival. *Cancer Res.* **2015**, *75*, 1789–1800. [CrossRef] [PubMed]

26. Jolly, M.K.; Boareto, M.; Huang, B.; Jia, D.; Lu, M.; Ben-Jacob, E.; Onuchic, J.N.; Levine, H. Implications of the Hybrid Epithelial/Mesenchymal Phenotype in Metastasis. *Front. Oncol.* **2015**, *5*, 155. [CrossRef] [PubMed]

27. Dong, N.; Shi, L.; Wang, D.C.; Chen, C.; Wang, X. Role of epigenetics in lung cancer heterogeneity and clinical implication. *Semin. Cell Dev. Biol.* **2017**, *64*, 18–25. [CrossRef] [PubMed]

28. O'Leary, K.; Shia, A.; Schmid, P. Epigenetic Regulation of EMT in non-small cell lung cancer. *Curr. Cancer Drug Targets* **2017**. [CrossRef]

29. Kouzarides, T. SnapShot: Histone-modifying enzymes. *Cell* **2007**, *128*, 802. [CrossRef] [PubMed]

30. Kouzarides, T. Chromatin modifications and their function. *Cell* **2007**, *128*, 693–705. [CrossRef] [PubMed]

31. Arrowsmith, C.H.; Bountra, C.; Fish, P.V.; Lee, K.; Schapira, M. Epigenetic protein families: A new frontier for drug discovery. *Nat. Rev. Drug Discov.* **2012**, *11*, 384–400. [CrossRef] [PubMed]

32. Esteller, M.; Pandolfi, P.P. The Epitranscriptome of noncoding RNAs in cancer. *Cancer Discov.* **2017**, *7*, 359–368. [CrossRef] [PubMed]

33. Zhao, H.; Chen, T. Tet family of 5-methylcytosine dioxygenases in mammalian development. *J. Hum. Genet.* **2013**, *58*, 421–427. [CrossRef] [PubMed]

34. Mehta, A.; Dobersch, S.; Romero-Olmedo, A.J.; Barreto, G. Epigenetics in lung cancer diagnosis and therapy. *Cancer Metastasis Rev.* **2015**, *34*, 229–241. [CrossRef] [PubMed]

35. Heyn, H.; Esteller, M. DNA methylation profiling in the clinic: Applications and challenges. *Nat. Rev. Genet.* **2012**, *13*, 679–692. [CrossRef] [PubMed]

36. Dawson, M.A.; Kouzarides, T. Cancer epigenetics: From mechanism to therapy. *Cell* **2012**, *150*, 12–27. [CrossRef] [PubMed]

37. Jenuwein, T.; Allis, C.D. Translating the histone code. *Science* **2001**, *293*, 1074–1080. [CrossRef] [PubMed]

38. Inamura, K. Diagnostic and therapeutic potential of microRNAs in lung cancer. *Cancers* **2017**, *9*. [CrossRef] [PubMed]

39. Quinn, J.J.; Chang, H.Y. Unique features of long non-coding RNA biogenesis and function. *Nat. Rev. Genet.* **2016**, *17*, 47–62. [CrossRef] [PubMed]

40. Mercer, T.R.; Mattick, J.S. Structure and function of long noncoding RNAs in epigenetic regulation. *Nat. Struct. Mol. Biol.* **2013**, *20*, 300–307. [CrossRef] [PubMed]

41. Cheetham, S.W.; Gruhl, F.; Mattick, J.S.; Dinger, M.E. Long noncoding RNAs and the genetics of cancer. *Br. J. Cancer* **2013**, *108*, 2419–2425. [CrossRef] [PubMed]

42. Morriss, G.R.; Cooper, T.A. Protein sequestration as a normal function of long noncoding RNAs and a pathogenic mechanism of RNAs containing nucleotide repeat expansions. *Hum. Genet.* **2017**. [CrossRef] [PubMed]

43. Tian, H.; Zhou, C.; Yang, J.; Li, J.; Gong, Z. Long and short noncoding RNAs in lung cancer precision medicine: Opportunities and challenges. *Tumour Biol.* **2017**, *39*. [CrossRef] [PubMed]

44. Palmieri, G.; Paliogiannis, P.; Sini, M.C.; Manca, A.; Palomba, G.; Doneddu, V.; Tanda, F.; Pascale, M.R.; Cossu, A. Long non-coding RNA CASC2 in human cancer. *Crit. Rev. Oncol. Hematol.* **2017**, *111*, 31–38. [CrossRef] [PubMed]

45. Xu, Z.; Yan, Y.; Qian, L.; Gong, Z. Long non-coding RNAs act as regulators of cell autophagy in diseases (Review). *Oncol. Rep.* **2017**, *37*, 1359–1366. [CrossRef] [PubMed]

46. Wei, J.W.; Huang, K.; Yang, C.; Kang, C.S. Non-coding RNAs as regulators in epigenetics (Review). *Oncol. Rep.* **2017**, *37*, 3–9. [CrossRef] [PubMed]

47. Engreitz, J.M.; Ollikainen, N.; Guttman, M. Long non-coding RNAs: Spatial amplifiers that control nuclear structure and gene expression. *Nat. Rev. Mol. Cell Biol.* **2016**, *17*, 756–770. [CrossRef] [PubMed]

48. Wei, M.M.; Zhou, G.B. Long Non-coding RNAs and their roles in non-small-cell lung cancer. *Genom. Proteom. Bioinform.* **2016**, *14*, 280–288. [CrossRef] [PubMed]

49. Khalil, A.M.; Guttman, M.; Huarte, M.; Garber, M.; Raj, A.; Rivea Morales, D.; Thomas, K.; Presser, A.; Bernstein, B.E.; van Oudenaarden, A.; et al. Many human large intergenic noncoding RNAs associate with chromatin-modifying complexes and affect gene expression. *Proc. Natl. Acad. Sci. USA* **2009**, *106*, 11667–11672. [CrossRef] [PubMed]

50. Hendrickson, D.G.; Kelley, D.R.; Tenen, D.; Bernstein, B.; Rinn, J.L. Widespread RNA binding by chromatin-associated proteins. *Genome Biol.* **2016**, *17*, 28. [CrossRef] [PubMed]

51. Heery, R.; Finn, S.P.; Cuffe, S.; Gray, S.G. Long non-coding RNAs: Key regulators of epithelial-mesenchymal transition, tumour drug resistance and cancer stem cells. *Cancers* **2017**, *9*. [CrossRef] [PubMed]

52. Chi, K.R. The RNA code comes into focus. *Nature* **2017**, *542*, 503–506. [CrossRef] [PubMed]

53. Gilbert, W.V.; Bell, T.A.; Schaening, C. Messenger RNA modifications: Form, distribution, and function. *Science* **2016**, *352*, 1408–1412. [CrossRef] [PubMed]

54. Jia, G.; Fu, Y.; Zhao, X.; Dai, Q.; Zheng, G.; Yang, Y.; Yi, C.; Lindahl, T.; Pan, T.; Yang, Y.G.; et al. N^6-methyladenosine in nuclear RNA is a major substrate of the obesity-associated FTO. *Nat. Chem. Biol.* **2011**, *7*, 885–887. [CrossRef] [PubMed]

55. Helm, M.; Motorin, Y. Detecting RNA modifications in the epitranscriptome: Predict and validate. *Nat. Rev. Genet.* **2017**, *18*, 275–291. [CrossRef] [PubMed]

56. Pfister, S.X.; Ashworth, A. Marked for death: Targeting epigenetic changes in cancer. *Nat. Rev. Drug. Discov.* **2017**, *16*, 241–263. [CrossRef] [PubMed]

57. Balgkouranidou, I.; Liloglou, T.; Lianidou, E.S. Lung cancer epigenetics: Emerging biomarkers. *Biomark Med.* **2013**, *7*, 49–58. [CrossRef] [PubMed]

58. Liloglou, T.; Bediaga, N.G.; Brown, B.R.; Field, J.K.; Davies, M.P. Epigenetic biomarkers in lung cancer. *Cancer Lett.* **2014**, *342*, 200–212. [CrossRef] [PubMed]

59. Van Den Broeck, A.; Ozenne, P.; Eymin, B.; Gazzeri, S. Lung cancer: A modified epigenome. *Cell Adh. Migr.* **2010**, *4*, 107–113. [CrossRef] [PubMed]

60. Roth, A.; Diederichs, S. Long Noncoding RNAs in Lung Cancer. *Curr. Top. Microbiol. Immunol.* **2016**, *394*, 57–110. [CrossRef] [PubMed]

61. Simo-Riudalbas, L.; Perez-Salvia, M.; Setien, F.; Villanueva, A.; Moutinho, C.; Martinez-Cardus, A.; Moran, S.; Berdasco, M.; Gomez, A.; Vidal, E.; et al. KAT6B is a tumor suppressor histone H3 lysine 23 acetyltransferase undergoing genomic loss in small cell lung cancer. *Cancer Res.* **2015**, *75*, 3936–3945. [CrossRef] [PubMed]

62. Augert, A.; Zhang, Q.; Bates, B.; Cui, M.; Wang, X.; Wildey, G.; Dowlati, A.; MacPherson, D. Small cell lung cancer exhibits frequent inactivating mutations in the histone methyltransferase KMT2D/MLL2: CALGB 151111 (Alliance). *J. Thorac. Oncol.* **2017**, *12*, 704–713. [CrossRef] [PubMed]

63. Van Haaften, G.; Dalgliesh, G.L.; Davies, H.; Chen, L.; Bignell, G.; Greenman, C.; Edkins, S.; Hardy, C.; O'Meara, S.; Teague, J.; et al. Somatic mutations of the histone H3K27 demethylase gene UTX in human cancer. *Nat. Genet.* **2009**, *41*, 521–523. [CrossRef] [PubMed]

64. Campbell, J.D.; Alexandrov, A.; Kim, J.; Wala, J.; Berger, A.H.; Pedamallu, C.S.; Shukla, S.A.; Guo, G.; Brooks, A.N.; Murray, B.A.; et al. Distinct patterns of somatic genome alterations in lung adenocarcinomas and squamous cell carcinomas. *Nat. Genet.* **2016**, *48*, 607–616. [CrossRef] [PubMed]

65. Imielinski, M.; Berger, A.H.; Hammerman, P.S.; Hernandez, B.; Pugh, T.J.; Hodis, E.; Cho, J.; Suh, J.; Capelletti, M.; Sivachenko, A.; et al. Mapping the hallmarks of lung adenocarcinoma with massively parallel sequencing. *Cell* **2012**, *150*, 1107–1120. [CrossRef] [PubMed]

66. Gardner, E.E.; Poirier, J.T.; Rudin, C.M. Histone code aberrancies in small cell lung cancer. *J. Thorac. Oncol.* **2017**, *12*, 599–601. [CrossRef] [PubMed]

67. Voigt, P.; Tee, W.W.; Reinberg, D. A double take on bivalent promoters. *Genes Dev.* **2013**, *27*, 1318–1338. [CrossRef] [PubMed]

68. Wilson, B.G.; Roberts, C.W. SWI/SNF nucleosome remodellers and cancer. *Nat. Rev. Cancer* **2011**, *11*, 481–492. [CrossRef] [PubMed]

69. Hargreaves, D.C.; Crabtree, G.R. ATP-dependent chromatin remodeling: Genetics, genomics and mechanisms. *Cell Res.* **2011**, *21*, 396–420. [CrossRef] [PubMed]

70. Behrens, C.; Solis, L.M.; Lin, H.; Yuan, P.; Tang, X.; Kadara, H.; Riquelme, E.; Galindo, H.; Moran, C.A.; Kalhor, N.; et al. EZH2 protein expression associates with the early pathogenesis, tumor progression, and prognosis of non-small cell lung carcinoma. *Clin. Cancer Res.* **2013**, *19*, 6556–6565. [CrossRef] [PubMed]

71. Kikuchi, J.; Kinoshita, I.; Shimizu, Y.; Kikuchi, E.; Konishi, J.; Oizumi, S.; Kaga, K.; Matsuno, Y.; Nishimura, M.; Dosaka-Akita, H. Distinctive expression of the polycomb group proteins Bmi1 polycomb ring finger oncogene and enhancer of zeste homolog 2 in nonsmall cell lung cancers and their clinical and clinicopathologic significance. *Cancer* **2010**, *116*, 3015–3024. [CrossRef] [PubMed]

72. Takawa, M.; Masuda, K.; Kunizaki, M.; Daigo, Y.; Takagi, K.; Iwai, Y.; Cho, H.S.; Toyokawa, G.; Yamane, Y.; Maejima, K.; et al. Validation of the histone methyltransferase EZH2 as a therapeutic target for various types of human cancer and as a prognostic marker. *Cancer Sci.* **2011**, *102*, 1298–1305. [CrossRef] [PubMed]

73. Huqun, Ishikawa, R.; Zhang, J.; Miyazawa, H.; Shimizu, Y.; Hagiwara, K.; Koyama, N. Enhancer of zeste homolog 2 is a novel prognostic biomarker in nonsmall cell lung cancer. *Cancer* **2012**, *118*, 1599–1606. [CrossRef]

74. Cao, W.; Ribeiro Rde, O.; Liu, D.; Saintigny, P.; Xia, R.; Xue, Y.; Lin, R.; Mao, L.; Ren, H. EZH2 promotes malignant behaviors via cell cycle dysregulation and its mRNA level associates with prognosis of patient with non-small cell lung cancer. *PLoS ONE* **2012**, *7*, e52984. [CrossRef] [PubMed]

75. Poirier, J.T.; Gardner, E.E.; Connis, N.; Moreira, A.L.; de Stanchina, E.; Hann, C.L.; Rudin, C.M. DNA methylation in small cell lung cancer defines distinct disease subtypes and correlates with high expression of EZH2. *Oncogene* **2015**, *34*, 5869–5878. [CrossRef] [PubMed]

76. Hussain, M.; Rao, M.; Humphries, A.E.; Hong, J.A.; Liu, F.; Yang, M.; Caragacianu, D.; Schrump, D.S. Tobacco smoke induces polycomb-mediated repression of Dickkopf-1 in lung cancer cells. *Cancer Res.* **2009**, *69*, 3570–3578. [CrossRef] [PubMed]

77. Coe, B.P.; Thu, K.L.; Aviel-Ronen, S.; Vucic, E.A.; Gazdar, A.F.; Lam, S.; Tsao, M.S.; Lam, W.L. Genomic deregulation of the E2F/Rb pathway leads to activation of the oncogene EZH2 in small cell lung cancer. *PLoS ONE* **2013**, *8*, e71670. [CrossRef]

78. Bracken, A.P.; Pasini, D.; Capra, M.; Prosperini, E.; Colli, E.; Helin, K. EZH2 is downstream of the pRB-E2F pathway, essential for proliferation and amplified in cancer. *EMBO J.* **2003**, *22*, 5323–5335. [CrossRef] [PubMed]

79. Ishak, C.A.; Marshall, A.E.; Passos, D.T.; White, C.R.; Kim, S.J.; Cecchini, M.J.; Ferwati, S.; MacDonald, W.A.; Howlett, C.J.; Welch, I.D.; et al. An RB-EZH2 complex mediates silencing of repetitive DNA sequences. *Mol. Cell* **2016**, *64*, 1074–1087. [CrossRef] [PubMed]

80. Li, Z.; Weng, H.; Su, R.; Weng, X.; Zuo, Z.; Li, C.; Huang, H.; Nachtergaele, S.; Dong, L.; Hu, C.; et al. FTO plays an oncogenic role in acute myeloid leukemia as a N^6-methyladenosine RNA demethylase. *Cancer Cell* **2017**, *31*, 127–141. [CrossRef] [PubMed]

81. Mauer, J.; Luo, X.; Blanjoie, A.; Jiao, X.; Grozhik, A.V.; Patil, D.P.; Linder, B.; Pickering, B.F.; Vasseur, J.J.; Chen, Q.; et al. Reversible methylation of m6Am in the 5' cap controls mRNA stability. *Nature* **2017**, *541*, 371–375. [CrossRef] [PubMed]

82. Tan, A.; Dang, Y.; Chen, G.; Mo, Z. Overexpression of the fat mass and obesity associated gene (FTO) in breast cancer and its clinical implications. *Int. J. Clin. Exp. Pathol.* **2015**, *8*, 13405–13410. [PubMed]

83. Singh, B.; Kinne, H.E.; Milligan, R.D.; Washburn, L.J.L.; Olsen, M.; Lucci, A. Important role of FTO in the survival of rare panresistant triple-negative inflammatory breast cancer cells facing a severe metabolic challenge. *PLoS ONE* **2016**, *11*, e0159072. [CrossRef]

84. Christofori, G. Snail1 links transcriptional control with epigenetic regulation. *EMBO J.* **2010**, *29*, 1787–1789. [CrossRef] [PubMed]

85. Lin, Y.; Wu, Y.; Li, J.; Dong, C.; Ye, X.; Chi, Y.I.; Evers, B.M.; Zhou, B.P. The SNAG domain of Snail1 functions as a molecular hook for recruiting lysine-specific demethylase 1. *EMBO J.* **2010**, *29*, 1803–1816. [CrossRef] [PubMed]

86. Dong, C.; Wu, Y.; Yao, J.; Wang, Y.; Yu, Y.; Rychahou, P.G.; Evers, B.M.; Zhou, B.P. G9a interacts with Snail and is critical for Snail-mediated E-cadherin repression in human breast cancer. *J. Clin. Investig.* **2012**, *122*, 1469–1486. [CrossRef] [PubMed]

87. Wu, H.; Chen, X.; Xiong, J.; Li, Y.; Li, H.; Ding, X.; Liu, S.; Chen, S.; Gao, S.; Zhu, B. Histone methyltransferase G9a contributes to H3K27 methylation in vivo. *Cell Res.* **2011**, *21*, 365–367. [CrossRef] [PubMed]

88. Dong, C.; Wu, Y.; Wang, Y.; Wang, C.; Kang, T.; Rychahou, P.G.; Chi, Y.I.; Evers, B.M.; Zhou, B.P. Interaction with Suv39H1 is critical for Snail-mediated E-cadherin repression in breast cancer. *Oncogene* **2013**, *32*, 1351–1362. [CrossRef] [PubMed]

89. Postigo, A.A.; Dean, D.C. ZEB represses transcription through interaction with the corepressor CtBP. *Proc. Natl. Acad. Sci. USA* **1999**, *96*, 6683–6688. [CrossRef] [PubMed]

90. Aghdassi, A.; Sendler, M.; Guenther, A.; Mayerle, J.; Behn, C.O.; Heidecke, C.D.; Friess, H.; Buchler, M.; Evert, M.; Lerch, M.M.; et al. Recruitment of histone deacetylases HDAC1 and HDAC2 by the transcriptional repressor ZEB1 downregulates E-cadherin expression in pancreatic cancer. *Gut* **2012**, *61*, 439–448. [CrossRef] [PubMed]

91. Chinnadurai, G. CtBP, an unconventional transcriptional corepressor in development and oncogenesis. *Mol. Cell* **2002**, *9*, 213–224. [CrossRef]

92. Shi, Y.; Sawada, J.; Sui, G.; Affar el, B.; Whetstine, J.R.; Lan, F.; Ogawa, H.; Luke, M.P.; Nakatani, Y. Coordinated histone modifications mediated by a CtBP co-repressor complex. *Nature* **2003**, *422*, 735–738. [CrossRef] [PubMed]

93. Byles, V.; Zhu, L.; Lovaas, J.D.; Chmilewski, L.K.; Wang, J.; Faller, D.V.; Dai, Y. SIRT1 induces EMT by cooperating with EMT transcription factors and enhances prostate cancer cell migration and metastasis. *Oncogene* **2012**, *31*, 4619–4629. [CrossRef] [PubMed]

94. Sanchez-Tillo, E.; Lazaro, A.; Torrent, R.; Cuatrecasas, M.; Vaquero, E.C.; Castells, A.; Engel, P.; Postigo, A. ZEB1 represses E-cadherin and induces an EMT by recruiting the SWI/SNF chromatin-remodeling protein BRG1. *Oncogene* **2010**, *29*, 3490–3500. [CrossRef] [PubMed]

95. Roche, J.; Nasarre, P.; Gemmill, R.; Baldys, A.; Pontis, J.; Korch, C.; Guilhot, J.; Ait-Si-Ali, S.; Drabkin, H. Global decrease of histone H3K27 acetylation in ZEB1-induced epithelial to mesenchymal transition in lung cancer cells. *Cancers* **2013**, *5*, 334–356. [CrossRef] [PubMed]

96. Jadhav, U.; Nalapareddy, K.; Saxena, M.; O'Neill, N.K.; Pinello, L.; Yuan, G.C.; Orkin, S.H.; Shivdasani, R.A. Acquired tissue-specific promoter bivalency is a basis for prc2 necessity in adult cells. *Cell* **2016**, *165*, 1389–1400. [CrossRef] [PubMed]

97. McDonald, O.G.; Wu, H.; Timp, W.; Doi, A.; Feinberg, A.P. Genome-scale epigenetic reprogramming during epithelial-to-mesenchymal transition. *Nat. Struct. Mol. Biol.* **2011**, *18*, 867–874. [CrossRef] [PubMed]

98. Loven, J.; Hoke, H.A.; Lin, C.Y.; Lau, A.; Orlando, D.A.; Vakoc, C.R.; Bradner, J.E.; Lee, T.I.; Young, R.A. Selective inhibition of tumor oncogenes by disruption of super-enhancers. *Cell* **2013**, *153*, 320–334. [CrossRef] [PubMed]

99. Kato, F.; Fiorentino, F.P.; Alibes, A.; Perucho, M.; Sanchez-Cespedes, M.; Kohno, T.; Yokota, J. MYCL is a target of a BET bromodomain inhibitor, JQ1, on growth suppression efficacy in small cell lung cancer cells. *Oncotarget* **2016**, *7*, 77378–77388. [CrossRef] [PubMed]

100. Ke, Y.; Zhao, W.; Xiong, J.; Cao, R. miR-149 Inhibits non-small-cell lung cancer cells EMT by targeting FOXM1. *Biochem. Res. Int.* **2013**, *2013*, 506731. [CrossRef] [PubMed]

101. De Craene, B.; Berx, G. Regulatory networks defining EMT during cancer initiation and progression. *Nat. Rev. Cancer* **2013**, *13*, 97–110. [CrossRef] [PubMed]

102. Burk, U.; Schubert, J.; Wellner, U.; Schmalhofer, O.; Vincan, E.; Spaderna, S.; Brabletz, T. A reciprocal repression between ZEB1 and members of the miR-200 family promotes EMT and invasion in cancer cells. *EMBO Rep.* **2008**, *9*, 582–589. [CrossRef] [PubMed]

103. Yoshida, T.; Song, L.; Bai, Y.; Kinose, F.; Li, J.; Ohaegbulam, K.C.; Munoz-Antonia, T.; Qu, X.; Eschrich, S.; Uramoto, H.; et al. ZEB1 mediates acquired resistance to the epidermal growth factor receptor-tyrosine kinase inhibitors in non-small cell lung cancer. *PLoS ONE* **2016**, *11*, e0147344. [CrossRef] [PubMed]

104. Portoso, M.; Ragazzini, R.; Brencic, Z.; Moiani, A.; Michaud, A.; Vassilev, I.; Wassef, M.; Servant, N.; Sargueil, B.; Margueron, R. PRC2 is dispensable for HOTAIR-mediated transcriptional repression. *EMBO J.* **2017**, *36*, 981–994. [CrossRef] [PubMed]

105. Blanco, M.R.; Guttman, M. Re-evaluating the foundations of lncRNA-Polycomb function. *EMBO J.* **2017**, *36*, 964–966. [CrossRef] [PubMed]

106. Padua Alves, C.; Fonseca, A.S.; Muys, B.R.; de Barros, E.L.B.R.; Burger, M.C.; de Souza, J.E.; Valente, V.; Zago, M.A.; Silva, W.A., Jr. Brief report: The lincRNA hotair is required for epithelial-to-mesenchymal transition and stemness maintenance of cancer cell lines. *Stem Cells* **2013**, *31*, 2827–2832. [CrossRef] [PubMed]

107. Sun, M.; Liu, X.H.; Lu, K.H.; Nie, F.Q.; Xia, R.; Kong, R.; Yang, J.S.; Xu, T.P.; Liu, Y.W.; Zou, Y.F.; et al. EZH2-mediated epigenetic suppression of long noncoding RNA SPRY4-IT1 promotes NSCLC cell proliferation and metastasis by affecting the epithelial-mesenchymal transition. *Cell Death Dis.* **2014**, *5*, e1298. [CrossRef] [PubMed]

108. Liu, N.; Parisien, M.; Dai, Q.; Zheng, G.; He, C.; Pan, T. Probing N^6-methyladenosine RNA modification status at single nucleotide resolution in mRNA and long noncoding RNA. *RNA* **2013**, *19*, 1848–1856. [CrossRef] [PubMed]

109. Shortt, J.; Ott, C.J.; Johnstone, R.W.; Bradner, J.E. A chemical probe toolbox for dissecting the cancer epigenome. *Nat. Rev. Cancer* **2017**, *17*, 160–183. [CrossRef] [PubMed]

110. Comet, I.; Riising, E.M.; Leblanc, B.; Helin, K. Maintaining cell identity: PRC2-mediated regulation of transcription and cancer. *Nat. Rev. Cancer* **2016**, *16*, 803–810. [CrossRef] [PubMed]

111. Wu, X.; Liu, D.; Tao, D.; Xiang, W.; Xiao, X.; Wang, M.; Wang, L.; Luo, G.; Li, Y.; Zeng, F.; et al. BRD4 regulates EZH2 transcription through upregulation of C-MYC and represents a novel therapeutic target in bladder cancer. *Mol. Cancer Ther.* **2016**, *15*, 1029–1042. [CrossRef] [PubMed]

112. Mazur, P.K.; Herner, A.; Mello, S.S.; Wirth, M.; Hausmann, S.; Sanchez-Rivera, F.J.; Lofgren, S.M.; Kuschma, T.; Hahn, S.A.; Vangala, D.; et al. Combined inhibition of BET family proteins and histone deacetylases as a potential epigenetics-based therapy for pancreatic ductal adenocarcinoma. *Nat. Med.* **2015**, *21*, 1163–1171. [CrossRef] [PubMed]

113. Adeegbe, D.; Liu, Y.; Lizotte, P.H.; Kamihara, Y.; Aref, A.R.; Almonte, C.; Dries, R.; Li, Y.; Liu, S.; Wang, X.; et al. Synergistic immunostimulatory effects and therapeutic benefit of combined histone deacetylase and bromodomain inhibition in non-small cell lung cancer. *Cancer Discov.* **2017**. [CrossRef] [PubMed]

114. Witta, S.E.; Gemmill, R.M.; Hirsch, F.R.; Coldren, C.D.; Hedman, K.; Ravdel, L.; Helfrich, B.; Dziadziuszko, R.; Chan, D.C.; Sugita, M.; et al. Restoring E-cadherin expression increases sensitivity to epidermal growth factor receptor inhibitors in lung cancer cell lines. *Cancer Res.* **2006**, *66*, 944–950. [CrossRef] [PubMed]

115. Kakihana, M.; Ohira, T.; Chan, D.; Webster, R.B.; Kato, H.; Drabkin, H.A.; Gemmill, R.M. Induction of E-cadherin in lung cancer and interaction with growth suppression by histone deacetylase inhibition. *J. Thorac. Oncol.* **2009**, *4*, 1455–1465. [CrossRef] [PubMed]

![cancers logo]

Review

The Emerging Role of Polo-Like Kinase 1 in Epithelial-Mesenchymal Transition and Tumor Metastasis

Zheng Fu * and Donghua Wen

Department of Human and Molecular Genetics, VCU Institute of Molecular Medicine, VCU Massey Cancer Center, School of Medicine, Virginia Commonwealth University, School of Medicine, Richmond, VA 23298, USA; donghua.wen@vcuhealth.org
* Correspondence: zheng.fu@vcuhealth.org; Tel.: +1-804-628-3843

Academic Editor: Joëlle Roche
Received: 7 September 2017; Accepted: 25 September 2017; Published: 27 September 2017

Abstract: Polo-like kinase 1 (PLK1) is a serine/threonine kinase that plays a key role in the regulation of the cell cycle. PLK1 is overexpressed in a variety of human tumors, and its expression level often correlates with increased cellular proliferation and poor prognosis in cancer patients. It has been suggested that PLK1 controls cancer development through multiple mechanisms that include canonical regulation of mitosis and cytokinesis, modulation of DNA replication, and cell survival. However, emerging evidence suggests novel and previously unanticipated roles for PLK1 during tumor development. In this review, we will summarize the recent advancements in our understanding of the oncogenic functions of PLK1, with a focus on its role in epithelial-mesenchymal transition and tumor invasion. We will further discuss the therapeutic potential of these functions.

Keywords: PLK1; EMT; tumor invasion and metastasis; drug resistance; cancer therapy

---

1. Introduction

Polo-like kinases (PLKs) belong to the polo subfamily of serine/threonine kinases, which are highly evolutionarily conserved, from yeast to humans [1]. PLKs have emerged as important regulators for cell cycle progression, cell proliferation, differentiation, and adaptive responses (see reviews; [1–3]). The prototypic founding member of the family was identified in *Drosophila melanogaster* in 1988 and was named "Polo" since the knockout of the gene induced abnormal spindle poles during mitosis [4]. Only one Plk has been reported in the genomes of *Drosophila* (Polo), budding yeast (Cdc5) and fission yeast (Plo1) [2], whereas vertebrates have many PLK family members [2]. In humans, five PLK members (PLK1-PLK5) have been identified and exhibit differential tissue distributions and distinct functions with no or partial overlap in substrates [1,2,5,6] (Figure 1). Among the human PLKs, PLK1 has been most extensively studied.

Sharing a similar domain topology with other PLKs, full-length PLK1 is composed of an N-terminal serine/threonine kinase domain and the characteristic polo-box domain (PBD) in the C-terminus [7] (Figure 1). The PBD is comprised of two polo boxes, polo box 1 and polo box 2, which fold together to form a functional PBD. The PBD binds phosphorylated serine/threonine motifs in PLK1's substrates. The optimal binding motif of its substrates is Ser-[pSer/pThr]-[Pro/X], in which X represents any amino acid [8,9]. By binding with such motifs on its substrates, the PBD brings the enzyme to an array of substrates found at different subcellular structures, including centrosomes, kinetochores, the mitotic spindle, and the midbody. This confers diversity to PLK1's function and allows exquisite regulation of the cell cycle [2,10]. A PBD mutant (H538A, K540M) that is deficient in phospho-binding delocalizes PLK1 and disrupts its function [11]. PLK1 also interacts with some of its binding partners in a phospho-independent or PBD-independent manner. For instance, aurora borealis (Bora), aurora kinase A activator, was reported to be capable of binding to a PLK1 deletion mutant that lacks the PBD [12]. In addition to the role of the PBD in interacting with PLK1's substrates, the PBD also modulates PLK1's kinase activity through intramolecular interaction [13,14]. The PBD inhibits the kinase domain by reducing its flexibility. Reciprocally, the kinase domain induces a conformational alteration of the PBD that renders it less capable of interacting with its binding targets. Phosphopeptide binding or activational phosphorylation of the T210 residue of PLK1 within the kinase activation loop relieves the inhibitory intramolecular interaction [9,15].

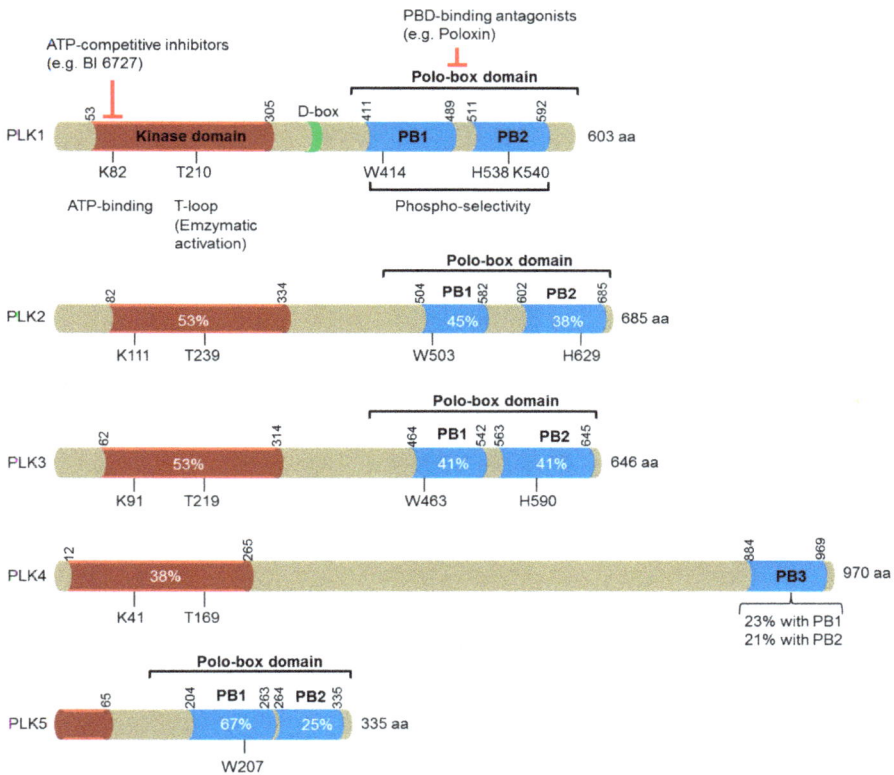

Figure 1. A schematic diagram illustrating the domain structures of the human polo-like kinase (PLK) family of proteins (PLK1-5). The number of amino acids in each family member is indicated on the right. The location of the kinase domains is shown in orange, whereas the polo-box domains (PBD), made of two polo-boxes (PB), are represented in blue. These two domains are separated by the interdomain linker, which comprises a destruction box (D-Box) indicated in green. The numbers indicate the first and the last residues of these domains in human PLKs. Residues that are essential for ATP-binding and enzymatic activation (T-loop) within the kinase domains, and for phosphoselectivity within the polo-box domains, are depicted. Sequence identities with the corresponding domains in PLK1 are provided in percentages. Two distinct strategies for targeting PLK1 are included: ATP-competitive inhibitors targeting the catalytic activity of PLK1, and PBD-binding antagonists competitively inhibiting the function of PBD.

PLK1 mediates almost every stage of cell division, including mitotic entry, centrosome maturation, bipolar spindle formation, chromosome congression and segregation, mitotic exit, and cytokinesis execution [2]. In addition to its canonical role in mitosis and cytokinesis, recent studies suggest that PLK1 may have other important functions such as regulation of microtubule dynamics, DNA replication, chromosome dynamics, p53 activity, and recovery from DNA damage-induced G2 arrest [16,17].

PLK1 is overexpressed in a variety of human tumors, and its expression level often correlates with increased cellular proliferation and poor prognosis in cancer patients [18,19]. It has been suggested that PLK1 controls cancer development through multiple mechanisms that include the canonical regulation of mitosis and cytokinesis, as well as modulation of DNA replication and cell survival [20,21]. However, emerging evidence suggests that the oncogenic functions of PLK1 extend far beyond what is currently known [21]. Here, we will discuss the recent advances in the understanding of PLK1 as an oncogene, with a focus on its role in epithelial-mesenchymal transition (EMT) and tumor invasion. We will further discuss the potential for therapeutic targeting of these newly identified oncogenic actions of PLK1.

2. PLK1 in Tumor Development

2.1. PLK1 Expression in Human Cancers

Consistent with its role in mitosis, PLK1 is highly expressed in the late G2 and M phases of the cell cycle, and enhanced PLK1 activity is observed in cells with high mitotic rates, including tumor cells [22,23]. Increasing evidence suggests that PLK1 is closely linked to human cancer development. For example, *PLK1* is overexpressed in a variety of cancers, including prostate cancer [24], non-small cell lung cancer [25], head and neck cancer [26,27], esophageal and gastric cancer [28], melanoma [29], breast cancer [30], ovarian cancer [31], endometrial cancer [32], colorectal cancer [33], glioma [34], thyroid cancer [35], and hepatocellular cancer [36]. More importantly, its expression level often correlates with poor patient prognosis [19,24,26–29,37–44], suggesting that PLK1 is essential for tumorigenesis. Indeed, emerging evidence supports the notion that PLK1 is actively involved throughout the course of human cancer development [18,35,45–49] (Figure 2).

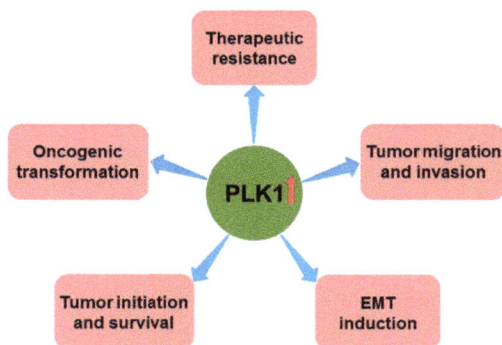

Figure 2. Role of polo-like kinase 1 (PLK1) overexpression in cancer. In addition to its role in promoting cancer cell proliferation and suppressing apoptosis, PLK1 overexpression has also been reported to have important roles in oncogenic transformation, tumor initiation and survival, epithelial-mesenchymal transition (EMT) induction, tumor migration and invasion, and therapeutic resistance.

2.2. PLK1 and Oncogenic Pathways

A defining characteristic of cancer is the uncontrolled, abnormal growth of cells [50]. Since PLK1 plays a major role in regulating the cell cycle and maintaining genomic stability, it is believed that PLK1 controls cancer development through multiple mechanisms, including classic regulation of mitosis and cytokinesis, as well as response to cellular stress and cell survival [20,21].

However, recent studies show that PLK1 is capable of contributing to carcinogenesis through interconnections with multiple cancer-associated pathways. Several interacting partners of PLK1 have been identified that are encoded by tumor suppressor genes and oncogenes. For instance, the tumor suppressor *p53* is considered to be the "guardian of the genome" and plays an important role in antiproliferation. Recent studies suggest that PLK1 has the ability to control p53's activity through multiple pathways: (1) PLK1 phosphorylates and inhibits *p53*-dependent transcriptional activation as well as p53's pro-apoptotic activity [51]; (2) PLK1 phosphorylates MDM2, an E3 ubiquitin ligase for p53, to promote p53 turnover [52,53]; (3) PLK1-mediated phosphorylation of S718 on Topors, a ubiquitin and SUMO E3 ligase, inhibits Topors-mediated SUMOylation of p53 and enhances ubiquitin-mediated degradation of p53 [53]; (4) PLK1 also phosphorylates G2 and S-phase-expressed 1 (GTSE1), resulting in GTSE1's translocation into the nucleus, where it binds to and shuttles p53 out of the nucleus for degradation [54].

Phosphatase and tensin homologue (PTEN) is one of the most commonly disrupted tumor suppressors in human cancers [55]. PLK1 has been identified as an important regulator of PTEN. PLK1 catalytic activity has been shown to phosphorylate PTEN near its C-terminal tail, which contributes to the mitotic function of PTEN [56]. Moreover, Li and colleagues showed that PLK1 phosphorylates PTEN and Nedd4-1, an E3 ubiquitin ligase of PTEN, which leads to the inactivation of PTEN and activation of the phosphatidylinositol 3-kinase (PI3K) pathway, thereby facilitating aerobic glycolysis and promoting tumorigenesis [57]. Analyses using prostate cancer cell lines, a prostate-specific PTEN-deletion mouse model, and a xenograft mouse model revealed that PLK1 is critical for PTEN-depleted cells to adapt to mitotic stress for survival, which assists the loss of PTEN-induced prostate cancer formation [58]. It has been reported that PLK1 also interacts with other tumor suppressors such as *CHK2* [59], *BRCA1/2* [60,61], *ATM* [62] and *ATR* [63], *BUB1B or BUBR1* [64,65], *CYLD* [66], *REST* [67], and *TSC1/2* [68,69]. The imbalance between these interactions and the resulting deregulation of oncogenic pathways could contribute to cancer development.

While negatively regulating tumor suppressors, PLK1 also intensively interplays with numerous oncogenes [70]. For instance, the Forkhead box protein M1 (FoxM1) transcription factor is a major mitotic transcription factor that is required for the proliferation of normal cells [71]. However, FoxM1 is frequently overexpressed in a wide spectrum of human cancers [72]. More importantly, overwhelming evidence reveals that FoxM1 is implicated in different phases of cancer development, and all major hallmarks of cancer delineated by Hanahan and Weinberg [50]. Our previous studies showed that PLK1 directly interacts with and phosphorylates FoxM1, leading to the activation of FoxM1's transcriptional activity. Activated FoxM1 then transcribes multiple mitotic regulators, including *PLK1*, which generate a positive feedback loop to further increase PLK1 levels and FoxM1 activity [73].

The *MYC* family of oncogenes contains three members (c-Myc, L-Myc, and N-Myc), which have been implicated in the genesis of specific human cancers [74]. Several studies have shown that PLK1 induces c-Myc accumulation by direct phosphorylation [75,76]. PLK1 can also stabilize N-Myc via the PLK1-Fbw7-Myc signaling circuit [77]. PLK1 binds to and phosphorylates the specificity factor Fbw7 of SCF^{Fbw7} ubiquitin ligase, which promotes Fbw7 auto-polyubiquitination and proteasomal degradation and in turn prevents SCF^{Fbw7}-mediated degradation of N-Myc. Stabilized N-Myc further activates the transcription of PLK1, resulting in a positive feed-forward regulatory loop that strengthens N-Myc-regulated oncogenic programs [77].

2.3. PLK1 and Oncogenic Transformation

The constitutive expression of *PLK1* in NIH/3T3 cells causes oncogenic foci formation and is tumorigenic in nude mice [78]. In contrast, depleting PLK1 in U2OS cells abrogates anchorage-independent growth [79]. These results highlight PLK1 as a possible driver of oncogenic transformation, although it remains unclear whether PLK1 itself is sufficient to induce tumor development. The oncogenic transformation potential of PLK1 has recently been documented in human cells. Our recent studies show that *PLK1* overexpression in human prostate epithelial cells leads to cellular transformation in vitro and promotes tumor formation in NOD/SCID/γ_c^{null} (NSG) mice, which provides convincing evidence that PLK1 is directly involved in neoplastic transformation, and that PLK1 has a tumor-promoting role in the prostate [47].

2.4. PLK1 and EMT

A recent study from our group revealed an important additional function of PLK1 [47]. We documented an interesting observation that *PLK1* overexpression in prostate epithelial cells causes the cells to change shape from an orthogonal epithelial cell morphology to a spindle-shaped fibroblast-like morphology, reminiscent of cells having undergone EMT. EMT is an important mechanism of tumor progression and metastasis [80,81]. It involves a loss of epithelial cell characteristics (cell–cell junctions, apicobasal cell polarity, and cobblestone morphology) and an acquisition of mesenchymal characteristics (fibroblast-like cell morphology, increased cell-matrix adhesions, and motility). On the molecular level, EMT can be easily recognized by the reduced expression of epithelial markers such as E-cadherin and some cytokeratin isoforms, and the elevated expression of mesenchymal markers such as N-cadherin and vimentin. Significantly, the loss of cell–cell contacts and the reorganization of the intracellular cytoskeleton during EMT result in increased cell migration and invasion [82], which allows cells to invade the surrounding stroma and vasculature, thereby leading to tumor dissemination and metastases [83]. In addition, EMT enables cancer cells to avoid apoptosis, anoikis, and oncogene addiction [84]. Indeed, forced overexpression of PLK1 in prostate epithelial cells led to the downregulation of epithelial markers (E-cadherin and cytokeratin 19) and upregulation of mesenchymal markers (N-cadherin, vimentin, fibronectin, and SM22) [47]. The switch from epithelial to mesenchymal markers did not depend on a specific stage of the cell cycle. Importantly, *PLK1* overexpression in prostate epithelial cells disrupted the localization of E-cadherin, β-catenin, and junctional adhesion molecule (JAM)-A in areas of cell-cell contacts, which are indicative of the profound disassembly of adherens and tight junctions. In addition, this was accompanied by the dramatic reorganization of the actomyosin cytoskeleton manifested by the redistribution of non-muscle myosin IIB from perijunctional F-actin bundles into basal stress fibers. A comparison of EMT induction in cells expressing wild-type, constitutively active, or kinase-defective PLK1 suggests that a PLK1-mediated phosphorylation event contributes to the induction of EMT in prostate epithelial cells [47]. The role of PLK1 in EMT induction was further substantiated by the observation that PLK1 downregulation in metastatic prostate cancer cells enhances epithelial characteristics [47]. Moreover, an androgen-refractory cancer of the prostate (ARCaP) model was adopted for further validation [47]. ARCaP cells were derived from the ascites fluid of an 83-year-old Caucasian man diagnosed with metastatic prostate cancer [85]. Epithelium-like ARCaP$_E$ cells and mesenchymal-like ARCaP$_M$ cells are sublines of ARCaP cells that were isolated by single-cell dilution cloning [86]. Interestingly, PLK1 is not only differentially expressed and activated in these two cell lines (higher in the highly metastatic ARCaP$_M$ cells and lower in the less metastatic ARCaP$_E$ cells), it also controls the switch between EMT and mesenchymal-to-epithelial transition (MET) in those two cell lines (EMT induction in ARCaP$_E$ cells upon PLK1 overexpression, and MET induction in ARCaP$_M$ cells with PLK1 downregulation). Taken together, these results convincingly established a novel function of PLK1 as a critical regulator of EMT in prostate cancer.

Subsequently, the molecular mechanism underlying PLK1-mediated EMT was investigated [47]. We demonstrated that CRAF a member of the Raf kinase family of serine/threonine-specific protein kinases, is a physiological substrate of PLK1. CRAF consists of an N-terminal regulatory domain and a C-terminal catalytic domain. PLK1 directly interacts with and phosphorylates CRAF at S338 and S339 (the critical activating phosphorylation sites), resulting in CRAF activation. The activated CRAF undergoes autophosphorylation of S621, which hinders the proteasome-mediated degradation of CRAF, and thereby generates a positive feedback loop, leading to a further increase in the level and activity of CRAF. This activation event triggers the activation of downstream MEK1/2-ERK1/2 signaling in prostate epithelial cells overexpressing PLK1. Through a series of biochemical analyses, the events between PLK1-triggered MAPK signaling and EMT induction were elucidated [47]. ERK activation stimulates Fra1 expression; Fra1 belongs to the *Fos* gene family, whose protein products can dimerize with proteins of the JUN family, thereby forming the transcription factor complex AP-1. The ectopic expression of Fra1 in epithelioid cells resulted in morphologic changes that resembled fibroblastoid conversion, and increased motility and invasiveness [87]. Fra1 has been implicated as a potent regulator of anti-apoptosis, cell motility, and invasion in a variety of tumor cell types [88,89]. Enhanced expression of Fra1 then leads to the transcriptional activation of zinc finger E-box binding homeobox (ZEB) 1 and 2, two key transcription factors in EMT that orchestrate the EMT program [47]. Later, Cai et al. documented that PLK1 promotes EMT in gastric carcinoma cells through regulation of the AKT pathway [90], which suggests that regulating EMT is a general, and not a cell-type specific, function of PLK1, and the underlying mechanisms of PLK1-dependent EMT induction may vary dramatically from one setting to another. In addition to direct regulation, PLK1 may indirectly contribute to EMT induction through its substrates. For instance, FoxM1 has been linked to EMT in various tumor types, including pancreatic cancer [91,92], breast cancer [93], prostate cancer [94], gastric cancer [95], and lung cancer [96]. Several EMT regulators, such as Snail [97], Slug [93], and Twist [98] have been documented as direct targets of FoxM1. Given the aforementioned PLK1–FoxM1 regulatory circuit [73], it is likely that PLK1 may contribute to the EMT process by directly binding to and phosphorylating FoxM1, resulting in the activation of its transcriptional activity (Figure 3).

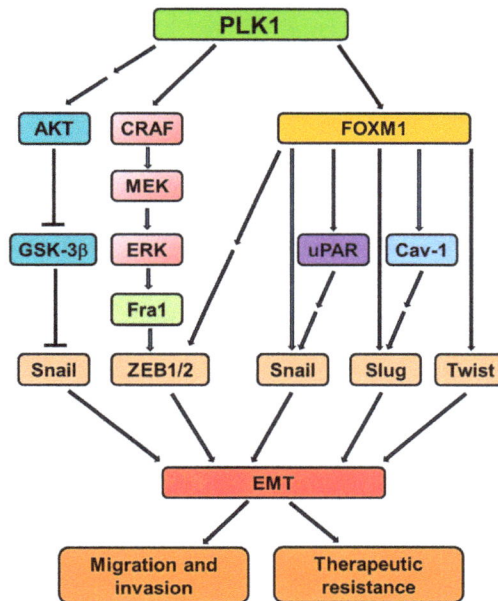

Figure 3. An overview of signaling cascades involved in PLK1-induced EMT. PLK1 activates the MAPK pathway by directly binding and phosphorylating CRAF. The activated MAPK pathway causes transcriptional upregulation of Fra1, which in turn triggers the accumulation of ZEB1/2, thus orchestrating the transcriptional network necessary for the EMT program. PLK1 also induces EMT through AKT or FoxM1-dependent pathways. Together, these signaling events contribute to EMT induction and associated events (such as invasion and therapeutic resistance) in tumor cells overexpressing PLK1.

3. PLK1 in Tumor Invasion and Metastasis

Elevated PLK1 expression has been associated with an increased invasiveness of colorectal, breast, renal, and thyroid cancer cells [45–49]. PLK1 inhibition using either siRNA or pharmacological inhibitors caused significant reductions in the invasiveness of glioblastoma, bladder carcinoma, renal cell carcinoma, anaplastic thyroid carcinoma, and colorectal cancer cells [45,48,49,99,100]. Our recent study has provided direct evidence of the pro-invasive activity of PLK1 in tumor progression [47]. PLK1 was differentially expressed and/or activated in prostate cancer cells (higher in metastatic prostate cancer cell lines and lower in non-metastatic cell lines) [47]. In addition to EMT induction, *PLK1* overexpression in prostate epithelial cells led to enhanced motility and invasiveness, as manifested by wound-healing scratch and Transwell invasion analyses [47]. The results were further validated by monitoring the random movement of the cells using time-lapse video microscopy and cell tracking, which indicated that PLK1 directly regulates the velocity of epithelial cell migration, independently of its effects on other cellular processes. Interestingly, NOD/SCID/γ_c^{null} (NSG) mice engrafted with PLK1-overexpressing prostate epithelial cells developed not only primary tumors, but also lung micrometastases, which suggests that *PLK1* overexpression not only leads to the oncogenic transformation of prostate epithelial cells, but may also drive prostate cancer metastasis [47]. Consistently, PLK1 downregulation in metastatic prostate cancer cells inhibited cell motility [47].

Both the profound disassembly of adherens and tight junctions and the dramatic reorganization of the actomyosin cytoskeleton were observed in prostate epithelial cells undergoing PLK1-mediated EMT [47]. Therefore, the following mechanisms by which EMT induction promotes prostate cancer cell motility were proposed: (1) disassembly of epithelial junctions that weaken intercellular adhesions, thereby allowing cell dissemination [101,102], and (2) rearrangement of the actomyosin cytoskeleton from epithelia-specific perijunctional bundles to basal stress fibers that are characteristic of mesenchymal cells. This rearrangement enhances cell–matrix adhesion and enables more efficient cell migration [103,104].

In line with these findings, Rizki et al. showed that PLK1 mediates invasion through vimentin and $\beta1$ integrin in breast cancer cells, which is independent of its mitotic function [45]. PLK1 phosphorylates vimentin on S82, which regulates cell surface levels of $\beta1$ integrin and thereby promotes the invasiveness of breast cancer cells [46]. In addition, it has been reported that the downregulation of PLK1 in thyroid cancer cells led to a significant decrease in CD44v6, matrix metalloproteinase (MMP)-2, and MMP-9, which are all key players in tumor invasion and metastasis [49].

4. PLK1 as a Key Target for Cancer Therapy

PLK1 has been reported as widely overexpressed in tumor samples from cancer patients, and its overexpression has been validated as a biomarker of poor prognosis in a variety of human cancers [18, 19]. Importantly, several studies have shown that inhibiting PLK1 expression or function by antibodies, RNAi, or small molecule inhibitors leads to mitotic arrest and apoptotic cell death in a wide range of human cancer cells, and is sufficient to prompt tumor regression in mouse xenograft models [105]. In contrast, toxicity modeling of PLK1-targeted therapies using primary human cells and various organs of adult Plk1 RNAi mice reveals that normal cells can tolerate up to a ~80% reduction in PLK1 level [106]. Thus, it has been repeatedly proposed that PLK1 could be a particularly attractive target for anti-cancer drug discovery [107]. Over the years, PLK1 has been the subject of an extensive effort in developing anti-mitotic agents that primarily target fast-growing mitotic cancer cells while leaving normal cells unscathed. To date, a large number of anti-PLK1 agents have been developed and tested under various preclinical and clinical settings, and several of them are currently in clinical trials, with varying degrees of success (for a comprehensive review, see [108,109]).

There are two druggable domains of PLK1 that have been pursued extensively: the catalytic domain and the PBD domain [108,109] (Figure 1). Volasertib (BI6727, a dihydropteridine derivative; Boehringer Ingelheim) is the most advanced inhibitor in the class of ATP-competitive inhibitors directed against the catalytic activity of PLK1. Its anti-cancer efficacies have been evaluated and proven to be superior in multiple nude mouse xenograft models [110]. Considerably, volasertib has also shown significant clinical efficacies against advanced solid and hematologic cancers in phase I/II clinical trials. Subsequently, a phase III clinical trial in elderly patients with acute myeloid leukemia was undertaken. The initial outcome, however, turned out to be less than satisfactory (presented at the 21st Annual Congress of the European Hematology Association, 2016 [111]). One of the major problems associated with the currently available PLK1 ATP-competitive inhibitors is their low degree of selectivity against other kinases, and their toxicity could be partly due to their interference with other kinases [108]. In the case of volasertib, it inhibits PLK2 and PLK3, with similar IC50 values as PLK1 [110,112]. A new generation of anti-PLK1 agents that target the PBD domain of PLK are currently being tested pre-clinically and have demonstrated improved specificity towards PLK1 [113]. Among all of the published PBD-interfering compounds, the most potent and selective inhibitors of the PBD have been peptide-like molecules [114–116]. However, they are often associated with poor cell membrane permeability due to their large size and the presence of charged groups [115,117]. Therefore, further efforts towards anti-PLK1 drug discovery will need to find new compounds with increased potency and specificity and improved pharmacokinetic properties to achieve better clinical outcomes.

The current rationale behind targeting PLK1 for anti-cancer therapy lies in its multifaceted functions throughout the cell cycle that target cancer's sustaining proliferative signaling. Our recent study showed that *PLK1* overexpression induces EMT and promotes cell motility and invasiveness in human prostate epithelial cells; whereas the attenuation of *PLK1* expression reduces the invasiveness of human prostate cancer cells [47]. These novel findings not only provide mechanistic insight into the important role of *PLK1* overexpression in human cancer development and metastasis, but will also aid the advancement of the prevention and treatment of advanced prostate cancer human cancers. In this regard, PLK1 can serve as a molecular biomarker to improve the stratification of cancer patients at high-, intermediate-, or low-risk of metastatic progression. In addition, PLK1 inhibition could potentially be a promising strategy to prevent prostate cancer dissemination. Consistent with this notion, a recent study reported that PLK1 depletion, mediated by PLK1 siRNA delivered by an antioxidant nanoparticle platform, inhibits lung metastasis and prolongs overall survival in a mouse model of breast cancer metastasis [118]. PLK1 inhibition by a small molecule inhibitor hindered brain metastases and prolonged survival in a mouse model of breast cancer brain metastasis [119]. These encouraging findings provide a proof of concept to substantiate the hypothesis above. Since the pro-metastatic properties of *PLK1* overexpression were also reported in other different types of cancers, including breast and thyroid cancers [46,49], this suggests that targeting cancer's ability to activate invasion and metastasis by PLK1 inhibition could be generalized to a growing list of a variety of cancers that have *PLK1* overexpression implicated in their metastatic progression.

In recent years, EMT has also emerged to be a major driver of resistance to anti-cancer therapies, manifested not only in experimental models, but also in clinical settings [120]. Remnant cancer cells that survive after different types of therapies (including chemotherapy, molecularly targeted therapy, and immunotherapy), recurrently display signs of EMT activation [120]. Gene expression profiling of tumor samples revealed a strong correlation between an EMT gene signature and resistance to chemotherapy [121,122]. However, the mechanisms by which the activation of the EMT program triggers the development of resistance to therapeutics in cancer cells remain elusive. Nonetheless, these findings suggest that targeting cancer cells that have activated the EMT program might significantly improve the efficiency of therapeutic modalities in generating durable clinical responses. In this regard, targeting signaling pathways that are critical for the activation and subsequent maintenance of the EMT program have been demonstrated to be an effective therapeutic approach to prevent and/or reverse the EMT process, thereby overcoming therapeutic resistance [120]. As a major driving force of the EMT process in prostate cancer [47], PLK1 overexpression might contribute to therapeutic resistance to anti-cancer therapies through EMT. Indeed, PLK1 has been reported to be closely associated with drug resistance in cancer cells to a number of chemotherapy drugs, including doxorubicin, paclitaxel, and gemcitabine [108]. Therefore, the inhibition of PLK1 might reverse the drug resistance and increase sensitivity to chemotherapy. In this line of thinking, PLK1 inhibition has been reported to be effective in treating EGFR-inhibitor resistant non-small cell lung cancer through EMT [123,124]. Furthermore, PLK1 inhibition has been documented to enhance anti-cancer drug efficacy in a variety of types of human cancers [125–128]. Taken together, the emerging evidence for the novel oncogenic roles of PLK1 (ranging from neoplastic transformation, EMT induction, tumor invasion and metastasis, and therapeutic resistance) further highlights PLK1 as a fascinating anti-cancer target, and may substantially aid in developing and deploying anti-PLK1 therapeutics. It is reasonable to speculate that targeting PLK1 has the potential for multi-dimensional actions against cancer, and ultimately paves the way for curative cancer treatments.

5. Conclusions and Outlook

PLK1 is a fascinating multifaceted protein that targets many binding partners to ensure proper cell cycle progression and cell proliferation, and its deregulation contributes to the genesis of a broad range of human cancers. The differential requirement of PLK1 levels in cancer versus normal cells for survival makes PLK1 a particularly attractive target for anti-cancer drug discovery. Recently, a wealth of data has shed new light on the additional biochemical functions of PLK1 proteins and on the mechanisms through which they function in neoplastic transformation, tumor progression and dissemination, and the development of therapeutic resistance. The identification of the diverse roles of PLK1 throughout the course of tumor development highlights PLK1 as one of the most appealing anti-cancer drug targets.

Anti-PLK1 drug discovery has reached an advanced stage of development. A number of PLK1 inhibitors have been developed. The major problems commonly associated with currently available PLK1 inhibitors are insufficient specificity and cancer cell-selective killing. Further studies are needed to identify new compounds with increased potency and specificity and improved pharmacokinetic properties. Furthermore, there is no doubt that additional functions of PLK1 will be uncovered in the near future. Additional studies aimed at disclosing all of the molecular mechanisms of PLK1 signaling in cancers are needed to achieve the full therapeutic potential of an anti-PLK1 drug. In other words, better understanding of the oncogenic action of PLK1 overexpression will greatly facilitate the optimization of treatment regimens targeting PLK1 signaling to significantly enhance therapeutic efficacy.

Acknowledgments: This work was supported in part by grants from the American Cancer Society (ACS Research Scholar Grant 127626-RSG-15-005-01-CCG to Zheng Fu), and the National Institutes of Health (NIH R01 CA191002 to Zheng Fu). The authors thank Heidi Sankala Bauer for editorial assistance with the manuscript.

Conflicts of Interest: The authors declare no conflict of interest.

References

1. Archambault, V.; Glover, D.M. Polo-like kinases: Conservation and divergence in their functions and regulation. *Nat. Rev. Mol. Cell Biol.* **2009**, *10*, 265–275. [CrossRef] [PubMed]
2. Barr, F.A.; Sillje, H.H.; Nigg, E.A. Polo-like kinases and the orchestration of cell division. *Nat. Rev. Mol. Cell Biol.* **2004**, *5*, 429–440. [CrossRef] [PubMed]
3. van de Weerdt, B.C.; Medema, R.H. Polo-like kinases: A team in control of the division. *Cell Cycle* **2006**, *5*, 853–864. [CrossRef] [PubMed]
4. Sunkel, C.E.; Glover, D.M. polo, a mitotic mutant of Drosophila displaying abnormal spindle poles. *J. Cell Sci.* **1988**, *89*, 25–38. [PubMed]
5. de Carcer, G.; Manning, G.; Malumbres, M. From Plk1 to Plk5: Functional evolution of polo-like kinases. *Cell Cycle* **2011**, *10*, 2255–2262. [CrossRef] [PubMed]
6. Winkles, J.A.; Alberts, G.F. Differential regulation of polo-like kinase 1, 2, 3, and 4 gene expression in mammalian cells and tissues. *Oncogene* **2005**, *24*, 260–266. [CrossRef] [PubMed]
7. Hamanaka, R.; Maloid, S.; Smith, M.R.; O'Connell, C.D.; Longo, D.L.; Ferris, D.K. Cloning and characterization of human and murine homologues of the Drosophila polo serine-threonine kinase. *Cell Growth Differ.* **1994**, *5*, 249–257. [PubMed]
8. Elia, A.E.; Cantley, L.C.; Yaffe, M.B. Proteomic screen finds pSer/pThr-binding domain localizing Plk1 to mitotic substrates. *Science* **2003**, *299*, 1228–1231. [CrossRef] [PubMed]

9.	Elia, A.E.; Rellos, P.; Haire, L.F.; Chao, J.W.; Ivins, F.J.; Hoepker, K.; Mohammad, D.; Cantley, L.C.; Smerdon, S.J.; Yaffe, M.B. The molecular basis for phosphodependent substrate targeting and regulation of Plks by the Polo-box domain. *Cell* **2003**, *115*, 83–95. [CrossRef]

10.	Donaldson, M.M.; Tavares, A.A.; Hagan, I.M.; Nigg, E.A.; Glover, D.M. The mitotic roles of Polo-like kinase. *J. Cell Sci.* **2001**, *114*, 2357–2358. [PubMed]

11.	Lee, K.S.; Grenfell, T.Z.; Yarm, F.R.; Erikson, R.L. Mutation of the polo-box disrupts localization and mitotic functions of the mammalian polo kinase Plk. *Proc. Natl. Acad. Sci. USA* **1998**, *95*, 9301–9306. [CrossRef] [PubMed]

12.	Seki, A.; Coppinger, J.A.; Jang, C.Y.; Yates, J.R.; Fang, G. Bora and the kinase Aurora a cooperatively activate the kinase Plk1 and control mitotic entry. *Science* **2008**, *320*, 1655–1658. [CrossRef] [PubMed]

13.	Park, J.E.; Soung, N.K.; Johmura, Y.; Kang, Y.H.; Liao, C.; Lee, K.H.; Park, C.H.; Nicklaus, M.C.; Lee, K.S. Polo-box domain: A versatile mediator of polo-like kinase function. *Cell Mol. Life Sci.* **2010**, *67*, 1957–1970. [CrossRef] [PubMed]

14.	Bruinsma, W.; Raaijmakers, J.A.; Medema, R.H. Switching Polo-like kinase-1 on and off in time and space. *Trends Biochem. Sci.* **2012**, *37*, 534–542. [CrossRef] [PubMed]

15.	Jang, Y.J.; Lin, C.Y.; Ma, S.; Erikson, R.L. Functional studies on the role of the C-terminal domain of mammalian polo-like kinase. *Proc. Natl. Acad. Sci. USA* **2002**, *99*, 1984–1989. [CrossRef] [PubMed]

16.	Liu, X.S.; Song, B.; Liu, X. The substrates of Plk1, beyond the functions in mitosis. *Protein Cell* **2010**, *1*, 999–1010. [CrossRef] [PubMed]

17.	Song, B.; Liu, X.S.; Davis, K.; Liu, X. Plk1 phosphorylation of Orc2 promotes, D.N.A replication under conditions of stress. *Mol. Cell. Biol.* **2011**, *31*, 4844–4856. [CrossRef] [PubMed]

18.	Cholewa, B.D.; Liu, X.; Ahmad, N. The role of polo-like kinase 1 in carcinogenesis: Cause or consequence? *Cancer Res.* **2013**, *73*, 6848–6855. [CrossRef] [PubMed]

19.	Takai, N.; Hamanaka, R.; Yoshimatsu, J.; Miyakawa, I. Polo-like kinases (Plks) and cancer. *Oncogene* **2005**, *24*, 287–291. [CrossRef] [PubMed]

20.	Deeraksa, A.; Pan, J.; Sha, Y.; Liu, X.D.; Eissa, N.T.; Lin, S.H.; Yu-Lee, L.Y. Plk1 is upregulated in androgen-insensitive prostate cancer cells and its inhibition leads to necroptosis. *Oncogene* **2012**, *32*, 2973–2983. [CrossRef] [PubMed]

21.	Luo, J.; Liu, X. Polo-like kinase 1, on the rise from cell cycle regulation to prostate cancer development. *Protein Cell* **2012**, *3*, 182–197. [CrossRef] [PubMed]

22.	Golsteyn, R.M.; Mundt, K.E.; Fry, A.M.; Nigg, E.A. Cell cycle regulation of the activity and subcellular localization of Plk1, a human protein kinase implicated in mitotic spindle function. *J. Cell Biol.* **1995**, *129*, 1617–1628. [CrossRef] [PubMed]

23.	Lee, K.S.; Yuan, Y.L.; Kuriyama, R.; Erikson, R.L. Plk is an M-phase-specific protein kinase and interacts with a kinesin-like protein, CHO1/MKLP-1. *Mol. Cell. Biol.* **1995**, *15*, 7143–7151. [CrossRef] [PubMed]

24.	Weichert, W.; Schmidt, M.; Gekeler, V.; Denkert, C.; Stephan, C.; Jung, K.; Loening, S.; Dietel, M.; Kristiansen, G. Polo-like kinase 1 is overexpressed in prostate cancer and linked to higher tumor grades. *Prostate* **2004**, *60*, 240–245. [CrossRef] [PubMed]

25.	Wolf, G.; Elez, R.; Doermer, A.; Holtrich, U.; Ackermann, H.; Stutte, H.J.; Altmannsberger, H.M.; Rubsamen-Waigmann, H.; Strebhardt, K. Prognostic significance of polo-like kinase (PLK) expression in non-small cell lung cancer. *Oncogene* **1997**, *14*, 543–549. [CrossRef] [PubMed]

26.	Knecht, R.; Elez, R.; Oechler, M.; Solbach, C.; von Ilberg, C.; Strebhardt, K. Prognostic significance of polo-like kinase (PLK) expression in squamous cell carcinomas of the head and neck. *Cancer Res.* **1999**, *59*, 2794–2797. [PubMed]

27.	Knecht, R.; Oberhauser, C.; Strebhardt, K. PLK (polo-like kinase), a new prognostic marker for oropharyngeal carcinomas. *Int. J. Cancer* **2000**, *89*, 535–536. [CrossRef]

28.	Tokumitsu, Y.; Mori, M.; Tanaka, S.; Akazawa, K.; Nakano, S.; Niho, Y. Prognostic significance of polo-like kinase expression in esophageal carcinoma. *Int. J. Oncol.* **1999**, *15*, 687–692. [CrossRef] [PubMed]

29.	Strebhardt, K.; Kneisel, L.; Linhart, C.; Bernd, A.; Kaufmann, R. Prognostic value of pololike kinase expression in melanomas. *JAMA* **2000**, *283*, 479–480. [CrossRef] [PubMed]

30.	Wolf, G.; Hildenbrand, R.; Schwar, C.; Grobholz, R.; Kaufmann, M.; Stutte, H.J.; Strebhardt, K.; Bleyl, U. Polo-like kinase: A novel marker of proliferation: Correlation with estrogen-receptor expression in human breast cancer. *Pathol. Res. Pract.* **2000**, *196*, 753–759. [PubMed]

31. Weichert, W.; Denkert, C.; Schmidt, M.; Gekeler, V.; Wolf, G.; Kobel, M.; Dietel, M.; Hauptmann, S. Polo-like kinase isoform expression is a prognostic factor in ovarian carcinoma. *Br. J. Cancer* **2004**, *90*, 815–821. [CrossRef] [PubMed]

32. Takai, N.; Miyazaki, T.; Fujisawa, K.; Nasu, K.; Hamanaka, R.; Miyakawa, I. Polo-like kinase (PLK) expression in endometrial carcinoma. *Cancer Lett.* **2001**, *169*, 41–49. [CrossRef]

33. Takahashi, T.; Sano, B.; Nagata, T.; Kato, H.; Sugiyama, Y.; Kunieda, K.; Kimura, M.; Okano, Y.; Saji, S. Polo-like kinase 1 (PLK1) is overexpressed in primary colorectal cancers. *Cancer Sci.* **2003**, *94*, 148–152. [CrossRef] [PubMed]

34. Dietzmann, K.; Kirches, E.; von, B.; Jachau, K.; Mawrin, C. Increased human polo-like kinase-1 expression in gliomas. *J. Neurooncol.* **2001**, *53*, 1–11. [CrossRef] [PubMed]

35. Ito, Y.; Miyoshi, E.; Sasaki, N.; Kakudo, K.; Yoshida, H.; Tomoda, C.; Uruno, T.; Takamura, Y.; Miya, A.; Kobayashi, K.; et al. Polo-like kinase 1 overexpression is an early event in the progression of papillary carcinoma. *Br. J. Cancer* **2004**, *90*, 414–418. [CrossRef] [PubMed]

36. Mok, W.C.; Wasser, S.; Tan, T.; Lim, S.G. Polo-like kinase 1, a new therapeutic target in hepatocellular carcinoma. *World J. Gastroenterol.* **2012**, *18*, 3527–3536. [CrossRef] [PubMed]

37. Feng, Y.B.; Lin, D.C.; Shi, Z.Z.; Wang, X.C.; Shen, X.M.; Zhang, Y.; Du, X.L.; Luo, M.L.; Xu, X.; Han, Y.L.; et al. Overexpression of PLK1 is associated with poor survival by inhibiting apoptosis via enhancement of survivin level in esophageal squamous cell carcinoma. *Int J. Cancer* **2009**, *124*, 578–588. [CrossRef] [PubMed]

38. Kanaji, S.; Saito, H.; Tsujitani, S.; Matsumoto, S.; Tatebe, S.; Kondo, A.; Ozaki, M.; Ito, H.; Ikeguchi, M. Expression of polo-like kinase 1 (PLK1) protein predicts the survival of patients with gastric carcinoma. *Oncology* **2006**, *70*, 126–133. [CrossRef] [PubMed]

39. King, S.I.; Purdie, C.A.; Bray, S.E.; Quinlan, P.R.; Jordan, L.B.; Thompson, A.M.; Meek, D.W. Immunohistochemical detection of Polo-like kinase-1 (PLK1) in primary breast cancer is associated with TP53 mutation and poor clinical outcom. *Breast Cancer Res.* **2012**, *14*, R40. [CrossRef] [PubMed]

40. Tut, T.G.; Lim, S.H.; Dissanayake, I.U.; Descallar, J.; Chua, W.; Ng, W.; de Souza, P.; Shin, J.S.; Lee, C.S. Upregulated Polo-Like Kinase 1 Expression Correlates with Inferior Survival Outcomes in Rectal Cancer. *PLoS ONE* **2015**, *10*, e0129313. [CrossRef] [PubMed]

41. Weichert, W.; Kristiansen, G.; Schmidt, M.; Gekeler, V.; Noske, A.; Niesporek, S.; Dietel, M.; Denkert, C. Polo-like kinase 1 expression is a prognostic factor in human colon cancer. *World J. Gastroenterol.* **2005**, *11*, 5644–5650. [CrossRef] [PubMed]

42. Weichert, W.; Kristiansen, G.; Winzer, K.J.; Schmidt, M.; Gekeler, V.; Noske, A.; Muller, B.M.; Niesporek, S.; Dietel, M.; Denkert, C. Polo-like kinase isoforms in breast cancer: Expression patterns and prognostic implications. *Virchows Arch.* **2005**, *446*, 442–450. [CrossRef] [PubMed]

43. Yamada, S.; Ohira, M.; Horie, H.; Ando, K.; Takayasu, H.; Suzuki, Y.; Sugano, S.; Hirata, T.; Goto, T.; Matsunaga, T.; et al. Expression profiling and differential screening between hepatoblastomas and the corresponding normal livers: Identification of high expression of the PLK1 oncogene as a poor-prognostic indicator of hepatoblastomas. *Oncogene* **2004**, *23*, 5901–5911. [CrossRef] [PubMed]

44. Zhang, Z.; Zhang, G.; Kong, C. High expression of polo-like kinase 1 is associated with the metastasis and recurrence in urothelial carcinoma of bladder. *Urol. Oncol.* **2013**, *31*, 1222–1230. [CrossRef] [PubMed]

45. Han, D.P.; Zhu, Q.L.; Cui, J.T.; Wang, P.X.; Qu, S.; Cao, Q.F.; Zong, Y.P.; Feng, B.; Zheng, M.H.; Lu, A.G. Polo-like kinase 1 is overexpressed in colorectal cancer and participates in the migration and invasion of colorectal cancer cells. *Med. Sci. Monit.* **2012**, *18*, BR237–BR246. [CrossRef] [PubMed]

46. Rizki, A.; Mott, J.D.; Bissell, M.J. Polo-like kinase 1 is involved in invasion through extracellular matrix. *Cancer Res.* **2007**, *67*, 11106–11110. [CrossRef] [PubMed]

47. Wu, J.; Ivanov, A.I.; Fisher, P.B.; Fu, Z. Polo-like kinase 1 induces epithelial-to-mesenchymal transition and promotes epithelial cell motility by activating CRAF/ERK signaling. *Elife* **2016**, *5*, e10734. [CrossRef] [PubMed]

48. Zhang, G.; Zhang, Z.; Liu, Z. Polo-like kinase 1 is overexpressed in renal cancer and participates in the proliferation and invasion of renal cancer cells. *Tumour Biol.* **2013**, *34*, 1887–1894. [CrossRef] [PubMed]

49. Zhang, X.G.; Lu, X.F.; Jiao, X.M.; Chen, B.; Wu, J.X. PLK1 gene suppresses cell invasion of undifferentiated thyroid carcinoma through the inhibition of CD44v6, MMP-2 and MMP-9. *Exp. Ther. Med.* **2012**, *4*, 1005–1009. [CrossRef] [PubMed]

50. Hanahan, D.; Weinberg, R.A. Hallmarks of cancer: The next generation. *Cell* **2011**, *144*, 646–674. [CrossRef] [PubMed]

51. Ando, K.; Ozaki, T.; Yamamoto, H.; Furuya, K.; Hosoda, M.; Hayashi, S.; Fukuzawa, M.; Nakagawara, A. Polo-like kinase 1 (Plk1) inhibits p53 function by physical interaction and phosphorylation. *J. Biol. Chem.* **2004**, *279*, 25549–25561. [CrossRef] [PubMed]

52. Dias, S.S.; Hogan, C.; Ochocka, A.M.; Meek, D.W. Polo-like kinase-1 phosphorylates MDM2 at Ser260 and stimulates MDM2-mediated p53 turnover. *FEBS Lett.* **2009**, *583*, 3543–3548. [CrossRef] [PubMed]

53. Yang, X.; Li, H.; Zhou, Z.; Wang, W.H.; Deng, A.; Andrisani, O.; Liu, X. Plk1-mediated phosphorylation of Topors regulates p53 stability. *J. Biol. Chem.* **2009**, *284*, 18588–18592. [CrossRef] [PubMed]

54. Liu, X.S.; Li, H.; Song, B.; Liu, X. Polo-like kinase 1 phosphorylation of G2 and S-phase-expressed 1 protein is essential for p53 inactivation during G2 checkpoint recovery. *EMBO Rep.* **2010**, *11*, 626–632. [CrossRef] [PubMed]

55. Song, M.S.; Salmena, L.; Pandolfi, P.P. The functions and regulation of the PTEN tumour suppressor. *Nat. Rev. Mol. Cell. Biol.* **2012**, *13*, 283–296. [CrossRef] [PubMed]

56. Choi, B.H.; Pagano, M.; Dai, W. Plk1 protein phosphorylates phosphatase and tensin homolog (PTEN) and regulates its mitotic activity during the cell cycle. *J. Biol. Chem.* **2014**, *289*, 14066–14074. [CrossRef] [PubMed]

57. Li, Z.; Li, J.; Bi, P.; Lu, Y.; Burcham, G.; Elzey, B.D.; Ratliff, T.; Konieczny, S.F.; Ahmad, N.; Kuang, S.; et al. Plk1 phosphorylation of PTEN causes a tumor-promoting metabolic state. *Mol. Cell. Biol.* **2014**, *34*, 3642–3661. [CrossRef] [PubMed]

58. Liu, X.S.; Song, B.; Elzey, B.D.; Ratliff, T.L.; Konieczny, S.F.; Cheng, L.; Ahmad, N.; Liu, X. Polo-like kinase 1 facilitates loss of Pten tumor suppressor-induced prostate cancer formation. *J. Biol. Chem.* **2011**, *286*, 35795–35800. [CrossRef] [PubMed]

59. Tsvetkov, L.; Xu, X.; Li, J.; Stern, D.F. Polo-like kinase 1 and Chk2 interact and co-localize to centrosomes and the midbody. *J. Biol. Chem.* **2003**, *278*, 8468–8475. [CrossRef] [PubMed]

60. Chabalier-Taste, C.; Brichese, L.; Racca, C.; Canitrot, Y.; Calsou, P.; Larminat, F. Polo-like kinase 1 mediates BRCA1 phosphorylation and recruitment at DNA double-strand breaks. *Oncotarget* **2016**, *7*, 2269–2283. [CrossRef] [PubMed]

61. Lee, M.; Daniels, M.J.; Venkitaraman, A.R. Phosphorylation of BRCA2 by the Polo-like kinase Plk1 is regulated by DNA damage and mitotic progression. *Oncogene* **2004**, *23*, 865–872. [CrossRef] [PubMed]

62. Lee, H.J.; Hwang, H.I.; Jang, Y.J. Mitotic DNA damage response: Polo-like kinase-1 is dephosphorylated through ATM-Chk1 pathway. *Cell Cycle* **2010**, *9*, 2389–2398. [CrossRef] [PubMed]

63. Deming, P.B.; Flores, K.G.; Downes, C.S.; Paules, R.S.; Kaufmann, W.K. ATR enforces the topoisomerase II-dependent G2 checkpoint through inhibition of Plk1 kinase. *J. Biol. Chem.* **2002**, *277*, 36832–36838. [CrossRef] [PubMed]

64. Elowe, S.; Hummer, S.; Uldschmid, A.; Li, X.; Nigg, E.A. Tension-sensitive Plk1 phosphorylation on BubR1 regulates the stability of kinetochore microtubule interactions. *Genes Dev.* **2007**, *21*, 2205–2219. [CrossRef] [PubMed]

65. Izumi, H.; Matsumoto, Y.; Ikeuchi, T.; Saya, H.; Kajii, T.; Matsuura, S. BubR1 localizes to centrosomes and suppresses centrosome amplification via regulating Plk1 activity in interphase cells. *Oncogene* **2009**, *28*, 2806–2820. [CrossRef] [PubMed]

66. Stegmeier, F.; Sowa, M.E.; Nalepa, G.; Gygi, S.P.; Harper, J.W.; Elledge, S.J. The tumor suppressor CYLD regulates entry into mitosis. *Proc. Natl. Acad. Sci. USA* **2007**, *104*, 8869–8874. [CrossRef] [PubMed]

67. Karlin, K.L.; Mondal, G.; Hartman, J.K.; Tyagi, S.; Kurley, S.J.; Bland, C.S.; Hsu, T.Y.; Renwick, A.; Fang, J.E.; Migliaccio, I.; et al. The oncogenic STP axis promotes triple-negative breast cancer via degradation of the REST tumor suppressor. *Cell Rep.* **2014**, *9*, 1318–1332. [CrossRef] [PubMed]

68. Astrinidis, A.; Senapedis, W.; Henske, E.P. Hamartin, the tuberous sclerosis complex 1 gene product, interacts with polo-like kinase 1 in a phosphorylation-dependent manner. *Hum. Mol. Genet.* **2006**, *15*, 287–297. [CrossRef] [PubMed]

69. Rosner, M.; Hanneder, M.; Siegel, N.; Valli, A.; Hengstschlager, M. The tuberous sclerosis gene products hamartin and tuberin are multifunctional proteins with a wide spectrum of interacting partners. *Mutat. Res.* **2008**, *658*, 234–246. [CrossRef] [PubMed]

70. Liu, Z.; Sun, Q.; Wang, X. PLK1, A Potential Target for Cancer Therapy. *Transl. Oncol.* **2017**, *10*, 22–32. [CrossRef] [PubMed]

71. Laoukili, J.; Kooistra, M.R.; Bras, A.; Kauw, J.; Kerkhoven, R.M.; Morrison, A.; Clevers, H.; Medema, R.H. FoxM1 is required for execution of the mitotic programme and chromosome stability. *Nat. Cell Biol.* **2005**, *7*, 126–136. [CrossRef] [PubMed]

72. Halasi, M.; Gartel, A.L. FOX(M1) news—It is cancer. *Mol. Cancer Ther.* **2013**, *12*, 245–254. [CrossRef] [PubMed]

73. Fu, Z.; Malureanu, L.; Huang, J.; Wang, W.; Li, H.; van Deursen, J.M.; Tindall, D.J.; Chen, J. Plk1-dependent phosphorylation of FoxM1 regulates a transcriptional programme required for mitotic progression. *Nat. Cell. Biol.* **2008**, *10*, 1076–1082. [CrossRef] [PubMed]

74. Adhikary, S.; Eilers, M. Transcriptional regulation and transformation by Myc proteins. *Nat. Rev. Mol. Cell Biol.* **2005**, *6*, 635–645. [CrossRef] [PubMed]

75. Padmanabhan, A.; Li, X.; Bieberich, C.J. Protein kinase A regulates MYC protein through transcriptional and post-translational mechanisms in a catalytic subunit isoform-specific manner. *J. Biol. Chem.* **2013**, *288*, 14158–14169. [CrossRef] [PubMed]

76. Tan, J.; Li, Z.; Lee, P.L.; Guan, P.; Aau, M.Y.; Lee, S.T.; Feng, M.; Lim, C.Z.; Lee, E.Y.; Wee, Z.N.; et al. PDK1 signaling toward PLK1-MYC activation confers oncogenic transformation, tumor-initiating cell activation, and resistance to mTOR-targeted therapy. *Cancer Discov.* **2013**, *3*, 1156–1171. [CrossRef] [PubMed]

77. Xiao, D.; Yue, M.; Su, H.; Ren, P.; Jiang, J.; Li, F.; Hu, Y.; Du, H.; Liu, H.; Qing, G. Polo-like Kinase-1 Regulates Myc Stabilization and Activates a Feedforward Circuit Promoting Tumor Cell Survival. *Mol. Cell* **2016**, *64*, 493–506. [CrossRef] [PubMed]

78. Smith, M.R.; Wilson, M.L.; Hamanaka, R.; Chase, D.; Kung, H.; Longo, D.L.; Ferris, D.K. Malignant transformation of mammalian cells initiated by constitutive expression of the polo-like kinase. *Biochem. Biophys. Res. Commun.* **1997**, *234*, 397–405. [CrossRef] [PubMed]

79. Eckerdt, F.; Yuan, J.; Strebhardt, K. Polo-like kinases and oncogenesis. *Oncogene* **2005**, *24*, 267–276. [CrossRef] [PubMed]

80. Kalluri, R.; Weinberg, R.A. The basics of epithelial-mesenchymal transition. *J. Clin. Investig.* **2009**, *119*, 1420–1428. [CrossRef] [PubMed]

81. Yang, J.; Weinberg, R.A. Epithelial-mesenchymal transition: At the crossroads of development and tumor metastasis. *Dev. Cell* **2008**, *14*, 818–829. [CrossRef] [PubMed]

82. Moreno-Bueno, G.; Portillo, F.; Cano, A. Transcriptional regulation of cell polarity in EMT and cancer. *Oncogene* **2008**, *27*, 6958–6969. [CrossRef] [PubMed]

83. Hugo, H.; Ackland, M.L.; Blick, T.; Lawrence, M.G.; Clements, J.A.; Williams, E.D.; Thompson, E.W. Epithelial—Mesenchymal and mesenchymal–epithelial transitions in carcinoma progression. *J. Cell Physiol.* **2007**, *213*, 374–383. [CrossRef] [PubMed]

84. Tiwari, N.; Gheldof, A.; Tatari, M.; Christofori, G. EMT as the ultimate survival mechanism of cancer cells. *Semin. Cancer Biol.* **2012**, *22*, 194–207. [CrossRef] [PubMed]

85. Zhau, H.Y.; Chang, S.M.; Chen, B.Q.; Wang, Y.; Zhang, H.; Kao, C.; Sang, Q.A.; Pathak, S.J.; Chung, L.W. Androgen-repressed phenotype in human prostate cancer. *Proc. Natl. Acad. Sci. USA* **1996**, *93*, 15152–15157. [CrossRef] [PubMed]

86. Xu, J.; Wang, R.; Xie, Z.H.; Odero-Marah, V.; Pathak, S.; Multani, A.; Chung, L.W.; Zhau, H.E. Prostate cancer metastasis: Role of the host microenvironment in promoting epithelial to mesenchymal transition and increased bone and adrenal gland metastasis. *Prostate* **2006**, *66*, 1664–1673. [CrossRef] [PubMed]

87. Kustikova, O.; Kramerov, D.; Grigorian, M.; Berezin, V.; Bock, E.; Lukanidin, E.; Tulchinsky, E. Fra-1 induces morphological transformation and increases in vitro invasiveness and motility of epithelioid adenocarcinoma cells. *Mol. Cell. Biol.* **1998**, *18*, 7095–7105. [CrossRef] [PubMed]

88. Milde-Langosch, K. The Fos family of transcription factors and their role in tumourigenesis. *Eur. J. Cancer* **2005**, *41*, 2449–2461. [CrossRef] [PubMed]

89. Young, M.R.; Colburn, N.H. Fra-1 a target for cancer prevention or intervention. *Gene* **2006**, *379*, 1–11. [CrossRef] [PubMed]

90. Cai, X.P.; Chen, L.D.; Song, H.B.; Zhang, C.X.; Yuan, Z.W.; Xiang, Z.X. PLK1 promotes epithelial-mesenchymal transition and metastasis of gastric carcinoma cells. *Am. J. Transl. Res.* **2016**, *8*, 4172–4183. [PubMed]

91. Bao, B.; Wang, Z.; Ali, S.; Kong, D.; Banerjee, S.; Ahmad, A.; Li, Y.; Azmi, A.S.; Miele, L.; Sarkar, F.H. Over-expression of FoxM1 leads to epithelial-mesenchymal transition and cancer stem cell phenotype in pancreatic cancer cells. *J. Cell Biochem.* **2011**, *112*, 2296–2306. [CrossRef] [PubMed]

92. Huang, C.; Xie, D.; Cui, J.; Li, Q.; Gao, Y.; Xie, K. FOXM1c promotes pancreatic cancer epithelial-to-mesenchymal transition and metastasis via upregulation of expression of the urokinase plasminogen activator system. *Clin. Cancer Res.* **2014**, *20*, 1477–1488. [CrossRef] [PubMed]
93. Yang, C.; Chen, H.; Tan, G.; Gao, W.; Cheng, L.; Jiang, X.; Yu, L.; Tan, Y. FOXM1 promotes the epithelial to mesenchymal transition by stimulating the transcription of Slug in human breast cancer. *Cancer Lett.* **2013**, *340*, 104–112. [CrossRef] [PubMed]
94. Wang, Y.; Yao, B.; Zhang, M.; Fu, S.; Gao, H.; Peng, R.; Zhang, L.; Tang, J. Increased FoxM1 expression is a target for metformin in the suppression of EMT in prostate cancer. *Int. J. Mol. Med.* **2014**, *33*, 1514–1522. [CrossRef] [PubMed]
95. Miao, L.; Xiong, X.; Lin, Y.; Cheng, Y.; Lu, J.; Zhang, J.; Cheng, N. Down-regulation of FoxM1 leads to the inhibition of the epithelial-mesenchymal transition in gastric cancer cells. *Cancer Genet.* **2014**, *207*, 75–82. [CrossRef] [PubMed]
96. Kong, F.F.; Qu, Z.Q.; Yuan, H.H.; Wang, J.Y.; Zhao, M.; Guo, Y.H.; Shi, J.; Gong, X.D.; Zhu, Y.L.; Liu, F.; et al. Overexpression of FOXM1 is associated with EMT and is a predictor of poor prognosis in non-small cell lung cancer. *Oncol. Rep.* **2014**, *31*, 2660–2668. [CrossRef] [PubMed]
97. Wei, P.; Zhang, N.; Wang, Y.; Li, D.; Wang, L.; Sun, X.; Shen, C.; Yang, Y.; Zhou, X.; Du, X. FOXM1 promotes lung adenocarcinoma invasion and metastasis by upregulating SNAIL. *Int. J. Biol. Sci.* **2015**, *11*, 186–198. [CrossRef] [PubMed]
98. Su, J.; Wu, S.; Wu, H.; Li, L.; Guo, T. CD44 is functionally crucial for driving lung cancer stem cells metastasis through Wnt/beta-catenin-FoxM1-Twist signaling. *Mol. Carcinog.* **2016**, *55*, 1962–1973. [CrossRef] [PubMed]
99. Brassesco, M.S.; Pezuk, J.A.; Morales, A.G.; de Oliveira, J.C.; Roberto, G.M.; da Silva, G.N.; Francisco de Oliveira, H.; Scrideli, C.A.; Tone, L.G. In vitro targeting of Polo-like kinase 1 in bladder carcinoma: Comparative effects of four potent inhibitors. *Cancer Biol. Ther.* **2013**, *14*, 648–657. [CrossRef] [PubMed]
100. Pezuk, J.A.; Brassesco, M.S.; Morales, A.G.; de Oliveira, J.C.; de Paula Queiroz, R.G.; Machado, H.R.; Carlotti, C.G., Jr.; Neder, L.; Scrideli, C.A.; Tone, L.G. Polo-like kinase 1 inhibition causes decreased proliferation by cell cycle arrest, leading to cell death in glioblastoma. *Cancer Gene Ther.* **2013**, *20*, 499–506. [CrossRef] [PubMed]
101. Godde, N.J.; Galea, R.C.; Elsum, I.A.; Humbert, P.O. Cell polarity in motion: Redefining mammary tissue organization through EMT and cell polarity transitions. *J. Mammary Gland Biol. Neoplasia* **2010**, *15*, 149–168. [CrossRef] [PubMed]
102. Le Bras, G.F.; Taubenslag, K.J.; Andl, C.D. The regulation of cell-cell adhesion during epithelial-mesenchymal transition, motility and tumor progression. *Cell Adh. Migr.* **2012**, *6*, 365–373. [CrossRef] [PubMed]
103. Martin, S.K.; Kamelgarn, M.; Kyprianou, N. Cytoskeleton targeting value in prostate cancer treatment. *Am. J. Clin. Exp. Urol.* **2014**, *2*, 15–26. [PubMed]
104. Yilmaz, M.; Christofori, G. EMT, the cytoskeleton, and cancer cell invasion. *Cancer Metastasis Rev.* **2009**, *28*, 15–33. [CrossRef] [PubMed]
105. Weiss, L.; Efferth, T. Polo-like kinase 1 as target for cancer therapy. *Exp. Hematol. Oncol.* **2012**, *1*, 38. [CrossRef] [PubMed]
106. Raab, M.; Kappel, S.; Kramer, A.; Sanhaji, M.; Matthess, Y.; Kurunci-Csacsko, E.; Calzada-Wack, J.; Rathkolb, B.; Rozman, J.; Adler, T.; et al. Toxicity modelling of Plk1-targeted therapies in genetically engineered mice and cultured primary mammalian cells. *Nat. Commun.* **2011**, *2*, 395. [CrossRef] [PubMed]
107. Degenhardt, Y.; Lampkin, T. Targeting Polo-like kinase in cancer therapy. *Clin. Cancer Res.* **2010**, *16*, 384–389. [CrossRef] [PubMed]
108. Gutteridge, R.E.; Ndiaye, M.A.; Liu, X.; Ahmad, N. Plk1 Inhibitors in Cancer Therapy: From Laboratory to Clinics. *Mol. Cancer Ther.* **2016**, *15*, 1427–1435. [CrossRef] [PubMed]
109. Park, J.E.; Hymel, D.; Burke, T.R., Jr.; Lee, K.S. Current progress and future perspectives in the development of anti-polo-like kinase 1 therapeutic agents. *F1000Res* **2017**, *6*, 1024. [CrossRef] [PubMed]
110. Rudolph, D.; Steegmaier, M.; Hoffmann, M.; Grauert, M.; Baum, A.; Quant, J.; Haslinger, C.; Garin-Chesa, P.; Adolf, G.R. BI 6727, a Polo-like kinase inhibitor with improved pharmacokinetic profile and broad antitumor activity. *Clin. Cancer Res.* **2009**, *15*, 3094–3102. [CrossRef] [PubMed]
111. Results of Phase III Study of Volasertib for the Treatment of Acute Myeloid Leukemia Presented at European Hematology Association Annual Meeting. Available online: http://www.evaluategroup.com/Universal/View.aspx?type=Story&id=649494 (accessed on September 2017).

112. Gjertsen, B.T.; Schoffski, P. Discovery and development of the Polo-like kinase inhibitor volasertib in cancer therapy. *Leukemia* **2015**, *29*, 11–19. [CrossRef] [PubMed]

113. Archambault, V.; Normandin, K. Several inhibitors of the Plk1 Polo-Box Domain turn out to be non-specific protein alkylators. *Cell Cycle* **2017**, *16*, 1220–1224. [CrossRef] [PubMed]

114. Lee, K.S.; Burke, T.R., Jr.; Park, J.E.; Bang, J.K.; Lee, E. Recent Advances and New Strategies in Targeting Plk1 for Anticancer Therapy. *Trends Pharmacol. Sci.* **2015**, *36*, 858–877. [CrossRef] [PubMed]

115. Liu, F.; Park, J.E.; Qian, W.J.; Lim, D.; Graber, M.; Berg, T.; Yaffe, M.B.; Lee, K.S.; Burke, T.R., Jr. Serendipitous alkylation of a Plk1 ligand uncovers a new binding channel. *Nat. Chem. Biol.* **2011**, *7*, 595–601. [CrossRef] [PubMed]

116. Liu, F.; Park, J.E.; Qian, W.J.; Lim, D.; Scharow, A.; Berg, T.; Yaffe, M.B.; Lee, K.S.; Burke, T.R., Jr. Identification of high affinity polo-like kinase 1 (Plk1) polo-box domain binding peptides using oxime-based diversification. *ACS Chem. Biol.* **2012**, *7*, 805–810. [CrossRef] [PubMed]

117. Qian, W.J.; Park, J.E.; Lim, D.; Lai, C.C.; Kelley, J.A.; Park, S.Y.; Lee, K.W.; Yaffe, M.B.; Lee, K.S.; Burke, T.R., Jr. Mono-anionic phosphopeptides produced by unexpected histidine alkylation exhibit high Plk1 polo-box domain-binding affinities and enhanced antiproliferative effects in HeLa cells. *Biopolymers* **2014**, *102*, 444–455. [CrossRef] [PubMed]

118. Morry, J.; Ngamcherdtrakul, W.; Gu, S.; Reda, M.; Castro, D.J.; Sangvanich, T.; Gray, J.W.; Yantasee, W. Targeted Treatment of Metastatic Breast Cancer by, PLK1 siRNA Delivered by an Antioxidant Nanoparticle Platform. *Mol. Cancer Ther.* **2017**, *16*, 763–772. [CrossRef] [PubMed]

119. Qian, Y.; Hua, E.; Bisht, K.; Woditschka, S.; Skordos, K.W.; Liewehr, D.J.; Steinberg, S.M.; Brogi, E.; Akram, M.M.; Killian, J.K.; et al. Inhibition of Polo-like kinase 1 prevents the growth of metastatic breast cancer cells in the brain. *Clin. Exp. Metastasis* **2011**, *28*, 899–908. [CrossRef] [PubMed]

120. Shibue, T.; Weinberg, R.A. EMT, CSCs, and drug resistance: The mechanistic link and clinical implications. *Nat. Rev. Clin. Oncol.* **2017**, *14*, 611–629. [CrossRef] [PubMed]

121. Byers, L.A.; Diao, L.; Wang, J.; Saintigny, P.; Girard, L.; Peyton, M.; Shen, L.; Fan, Y.; Giri, U.; Tumula, P.K.; et al. An epithelial-mesenchymal transition gene signature predicts resistance to EGFR and PI3K inhibitors and identifies Axl as a therapeutic target for overcoming EGFR inhibitor resistance. *Clin. Cancer Res.* **2013**, *19*, 279–290. [CrossRef] [PubMed]

122. Farmer, P.; Bonnefoi, H.; Anderle, P.; Cameron, D.; Wirapati, P.; Becette, V.; Andre, S.; Piccart, M.; Campone, M.; Brain, E.; et al. A stroma-related gene signature predicts resistance to neoadjuvant chemotherapy in breast cancer. *Nat. Med.* **2009**, *15*, 68–74. [CrossRef] [PubMed]

123. Crystal, A.S.; Shaw, A.T.; Sequist, L.V.; Friboulet, L.; Niederst, M.J.; Lockerman, E.L.; Frias, R.L.; Gainor, J.F.; Amzallag, A.; Greninger, P.; et al. Patient-derived models of acquired resistance can identify effective drug combinations for cancer. *Science* **2014**, *346*, 1480–1486. [CrossRef] [PubMed]

124. Wang, Y.; Singh, R.; Wang, L.; Nilsson, M.; Goonatilake, R.; Tong, P.; Li, L.; Giri, U.; Villalobos, P.; Mino, B.; et al. Polo-like kinase 1 inhibition diminishes acquired resistance to epidermal growth factor receptor inhibition in non-small cell lung cancer with T790M mutations. *Oncotarget* **2016**, *7*, 47998–48010. [CrossRef] [PubMed]

125. Jimeno, A.; Rubio-Viqueira, B.; Rajeshkumar, N.V.; Chan, A.; sSolomon, A.; Hidalgo, M. A fine-needle aspirate-based vulnerability assay identifies polo-like kinase 1 as a mediator of gemcitabine resistance in pancreatic cancer. *Mol. Cancer Ther.* **2010**, *9*, 311–318. [CrossRef] [PubMed]

126. Spankuch, B.; Heim, S.; Kurunci-Csacsko, E.; Lindenau, C.; Yuan, J.; Kaufmann, M.; Strebhardt, K. Down-regulation of Polo-like kinase 1 elevates drug sensitivity of breast cancer cells in vitro and in vivo. *Cancer Res.* **2006**, *66*, 5836–5846. [CrossRef] [PubMed]

127. Spankuch, B.; Kurunci-Csacsko, E.; Kaufmann, M.; Strebhardt, K. Rational combinations of siRNAs targeting Plk1 with breast cancer drugs. *Oncogene* **2007**, *26*, 5793–5807. [CrossRef] [PubMed]

128. Yu, C.; Zhang, X.; Sun, G.; Guo, X.; Li, H.; You, Y.; Jacobs, J.L.; Gardner, K.; Yuan, D.; Xu, Z.; et al. RNA interference-mediated silencing of the polo-like kinase 1 gene enhances chemosensitivity to gemcitabine in pancreatic adenocarcinoma cells. *J. Cell. Mol. Med.* **2008**, *12*, 2334–2349. [CrossRef] [PubMed]

cancers

MDPI

Article

Stem Cell-Like Properties of CK2β-down Regulated Mammary Cells

Eve Duchemin-Pelletier [1,2,3], Megghane Baulard [1,2,3], Elodie Spreux [1,2,3], Magali Prioux [1,2,3,4], Mithila Burute [4], Baharia Mograbi [5], Laurent Guyon [1,2,3], Manuel Théry [4], Claude Cochet [1,2,3] and Odile Filhol [1,2,3,*]

[1] Chemistry and Biology Department, Université Grenoble Alpes, F-38400 Grenoble, France;
    eve_dp87@yahoo.fr (E.D.-P.); megghanebaulard@yahoo.fr (M.B.); elodie.spreux@hotmail.fr (E.S.);
    magali.prioux@free.fr (M.P.); laurent.guyon@cea.fr (L.G.); claude.cochet@cea.fr (C.C.)
[2] Biology of Cancer and Infection, UMRS1036, Inserm, F-38054 Grenoble, France
[3] Biology of Cancer and Infection, Biosciences & Biotechnology Institute of Grenoble, CEA,
    F-38054 Grenoble, France
[4] Laboratoire de Physiologie Cellulaire et Végétale, Biosciences & Biotechnology Institute of Grenoble,
    UMR5168, CEA/INRA/CNRS, F-38054 Grenoble, France; mithila.pune@gmail.com (M.B.);
    manuel.thery@cea.fr (M.T.)
[5] Biology Department, Inserm, CNRS, IRCAN, Université Côte d'Azur, F-06000 Nice, France;
    Baharia.MOGRABI@unice.fr
*   Correspondence: odile.filhol-cochet@cea.fr; Tel.: +33-438-785-645; Fax: +33-438-785-056

Academic Editor: Joëlle Roche
Received: 21 August 2017; Accepted: 28 August 2017; Published: 31 August 2017

Abstract: The ubiquitous protein kinase CK2 has been demonstrated to be overexpressed in a number of human tumours. This enzyme is composed of two catalytic α or α' subunits and a dimer of β regulatory subunits whose expression levels are probably implicated in CK2 regulation. Several recent papers reported that unbalanced expression of CK2 subunits is sufficient to drive epithelial to mesenchymal transition, a process involved in cancer invasion and metastasis. Herein, through transcriptomic and miRNA analysis together with comparison of cellular properties between wild type and CK2β-knock-down MCF10A cells, we show that down-regulation of CK2β subunit in mammary epithelial cells induces the acquisition of stem cell-like properties associated with perturbed polarity, CD44^{high}/CD24^{low} antigenic phenotype and the ability to grow under anchorage-independent conditions. These data demonstrate that a CK2β level establishes a critical cell fate threshold in the control of epithelial cell plasticity. Thus, this regulatory subunit functions as a nodal protein to maintain an epithelial phenotype and its depletion drives breast cell stemness.

Keywords: protein kinase CK2; stem cell; breast cancer; EMT; epithelial plasticity

1. Introduction

Tumour formation is a complex process that originates from epithelial cells. A wealth of evidence supports the dogma that different events, described by Hanahan and Weinberg as "hallmarks of cancer" participate in tumour progression, from initiation to metastasis [1,2]. Metastasis remains the major cause of cancer-associated death, as cancer cells evade the primary tumour upon tumour microenvironment changes. Thus, a better understanding in the intra- and extra-cellular parameters involved in these processes would be helpful, to identify new targets to fight this scourge. A first mechanism known to allow tumour cells migration is the epithelial-to-mesenchymal transition (EMT), which leads to enhanced cell motility. Several protein kinases are involved in the different molecular signaling circuits that drive EMT. Among them, protein kinase CK2, whose expression is abnormally high in a wide range of tumours, operates as a cancer driver by creating the cellular environment favorable to neoplasia [3]. As a signaling protein, CK2 is a multi-subunit holoenzyme which has many cellular functions associated with a wide repertoire of substrates located in several of cellular compartments. The CK2α catalytic subunits possess a constitutive activity, while the homodimer of CK2β regulatory subunits operate as a regulatory component, modifying the accessibility of binding substrates to the catalytic site of the holoenzyme [4]. Using live-cell fluorescence imaging studies, we previously provided evidence of independent and rapid movement of CK2α and CK2β [5], showing that this kinase can rapidly target specific proteins in response to different stimuli [6].

In mammary epithelial cells, including MCF10A cells, TGFβ induces an EMT by driving expression of specific transcription factors, such as Snail1, Slug, and Twist. Accumulating evidences also suggest that EMT-induced cells exhibited hallmarks commonly attributed to cancer stem cells [7,8]. Recently, we provided evidence that CK2 plays a key role in EMT signaling pathways. More precisely, we showed that in the absence of its CK2β regulatory subunits, the CK2-mediated Snail1 phosphorylation is abrogated, a post-transcriptional modification that is required for Snail1 proteasomal-degradation in epithelial cells. Of note, the expression of CK2α at both transcriptional and protein levels was not affected upon CK2β depletion. Moreover, this CK2β-dependent phosphorylation had a cumulative positive effect on GSK3β-mediated Snail1 phosphorylation. As the phosphorylation of Snail1 participates in its degradation, these results suggest that both kinases can negatively regulate Snail1 stability through its hierarchal phosphorylation. In accordance, we found that in the absence of CK2β, Snail1 was no longer degraded [9]. Another EMT-transcription factor, FoxC2 is retained in the cytoplasm in its CK2-phosphorylated state. However, in the absence of CK2β, FoxC2 cannot be phosphorylated, and enters into the nucleus and activates the transcription of EMT-related genes [10]. Thus, unbalanced expression of CK2 subunits is sufficient to drive epithelial-to-mesenchymal transition, a process involved in cancer invasion and metastasis. This dysregulated expression of CK2 subunits has been observed in breast cancer, renal cancer, lung adenocarcinoma, and glioblastoma [9,11–13].

Normal and malignant stem cells share the ability to self-renew while generating differentiated cells [14]. Here, we show that CK2β-deficient MCF10A cells exhibited several properties commonly attributable to cancer stem cells [7,8]. However, we found that though they lacked tumour-initiating ability, they displayed enhanced plasticity and stemness characteristics of normal stem cells.

2. Results and Discussion

2.1. ΔCK2β-MCF10A Cells Have Reduced Expression of EMT miRNAs

Since miRNAs regulate differentiation, cell renewal and invasion [15], we looked for potential changes in miRNA expression that might occur upon CK2β depletion. Using microarrays with the miRNA probe set (Agilent miRNA human V.3), miRNA expression profiles were compared in ΔCK2β- and Mock-MCF10A cells (Table S1). These two cell lines were generated by shRNA-transduction as described in [9], in which less than 10% of the CK2β subunit remains in ΔCK2β-cells, whereas it was unchanged in Mock-cells (Figure S1A,B). We found that several microRNA families were significantly reduced in ΔCK2β-cells (accession number GSE102266, Figure 1A, green bars). The relative miRNA amounts were determined by RT-qPCR using TaqMan probes, including miR-200c, miR-205, miR-141, as well as miR 30b and miR-34b (Figure 1B). The result confirmed the strong decrease of these miRNAs in ΔCK2β-cells. The miR-200 family including hsa-miR-200c, hsa-miR-205 and hsa-miR-141 emerged as the most significantly reduced miRNAs ($p < 0.001$). Direct targets of miR-200c and miR-205 were reported to be the transcription factors ZEB1 and ZEB2 that regulate the epithelial-mesenchymal transition [16,17]. We thus analyzed the protein expression level of ZEB1 in both cell types. Accordingly, we found that ZEB1 expression level was strongly increased in ΔCK2β-cells (Figure 1C, left panel). It has been reported that a miR-30 reduction maintains self-renewal and inhibits apoptosis in breast tumour-initiating cells [18]. Of note, the expression of most members of the miR-30 family including miR-30b, -30c, and -30d, were also reduced in CK2β-depleted cells. A direct target gene of miR-30 is integrin β3 [18]. Consistently, we found the upregulation of the integrin β3 protein in ΔCK2β-cells either by Western blot or by immunofluorescence (Figure 1C). Members of the miR-34 family participate in the regulation of self-renewal and chemotherapeutic resistance of breast cancer cells [19]. When compared to Mock-cells, miR-34 was also significantly reduced in ΔCK2β-cells. Collectively, these data show that ΔCK2β-cells exhibit a decreased expression of specific miRNAs that are all known to regulate de/trans-differentiation, EMT, cell renewal, and invasion.

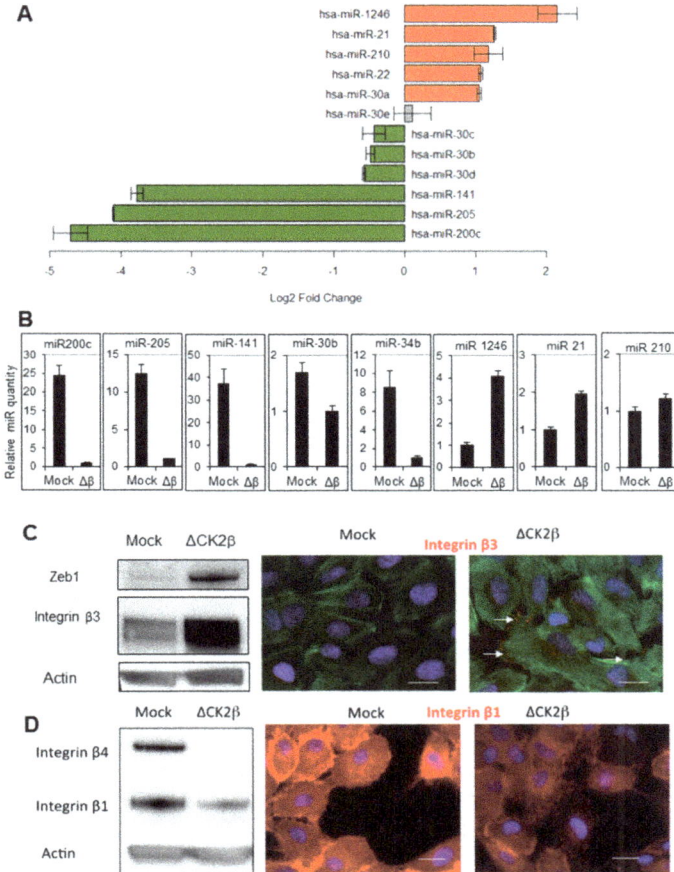

Figure 1. Modulation of miRNAs in ΔCK2β-MCF10A cells. (**A**) Log2 fold change of the main miRNAs modulated in CK2β-depleted versus parental MCF10A cells measured by miRNA array analysis; (**B**) Changes of miRNA expression between CK2β-depleted and Mock-MCF10A cells were confirmed by using the indicated TaqMan probes. The relative amount of each miRNAs was determined by cross-normalization to ΔCK2β samples using the comparative method and miR-720 as an internal reference; (**C**) Two targets of miR-200 and miR-30 families, respectively Zeb1 and integrin β3, were analyzed by Western blot and/or immunofluorescence in Mock- and CK2β-depleted cells. The ratio ΔCK2β/Mock of signal intensity in western blot was determined (3.5 and 2.3 for Zeb1 and integrin β3 respectively). Arrows indicate integrin β3 localization; (**D**) Integrin β1 and β4, targets of miR-21 were analyzed by western blot and/or immunofluorescence in Mock- and CK2β-depleted cells. The ratio ΔCK2β/Mock of signal intensity in western blot was 0.4 for integrin β1. F-actin in green, nuclei in blue, and integrin β in red. Scale bar, 10 μm.

2.2. ΔCK2β-MCF10A Cells Have Increased Expression of Specific miRNAs

We next studied the expression of miR-21, as it is one of the most frequently upregulated miRNAs in solid tumours. In addition, miR-21 is considered to be a typical "onco-miR", which acts by inhibiting the expression of phosphatases, thus limiting the activity of signaling pathways, such as AKT and MAPK [20]. When compared to Mock-cells, we found that the miR-21 expression was significantly increased in ΔCK2β-cells (Figure 1A,B). As most of the miR-21 targets are tumour suppressors, miR-21 is associated with a wide variety of cancers including breast cancers [21]. Moreover, miR-21 promotes migration and invasion through upregulation of both Sox2 and β-catenin [22], and a loss of polarity associated with an increased expression of collagen type 1 [23]. Interestingly, our transcriptomic analysis showed that different collagen types like collagen I, IV, VI, VII and XIII, were increased more than 3-fold in ΔCK2β-cells as compared to Mock-cells (Table S2). These data were confirmed in the HMEC-hTERT cell line (Figure S3). As mentioned above, integrins are also regulated by miRNAs [24]. Integrin-β3, -α4, and -αV were upregulated whereas integrin-β4 and -β1 were repressed in ΔCK2β-cells (Figure 1D and Table S2). JAG1 is another target of miR-21 that has been shown to be elevated in breast cancer [25]. By RT-qPCR we found that Jagged-1 is repressed in ΔCK2β-MCF10A cells (Figure S1B). Interestingly, miR-1246, mir-21 and miR-210 that have a link with tumour heterogeneity and tumour-initiating cell behaviour, were all induced in ΔCK2β-cells as compared to Mock-cells (Figure 1A,B) [21,26–28].

2.3. Transcriptomic Analysis

EMT in epithelial cells has been shown to be associated with stem cell traits and chemo-resistance [29]. Comparing ΔCK2β- to Mock-MCF10A cells, we looked further for individual gene expression signatures, using a transcriptomic analysis according to published profiles. Agilent microarrays were performed in duplicates, as previously described [9] (accession number GSE102265) and correlations were done with gene set collections from MSigDB 3.0, as described in Supplementary Figure S1. With this approach, we found that gene signature of Mock-MCF10A cells was correlated with the epithelial profiles while the ΔCK2β-MCF10A signature matched with mesenchymal profiles described in the Charafe, Sarrio, Gotzmann, and Jechlinger collections [30–33] (Figure S2A). These data extended and confirmed our miRNA analysis together with our previous results [9], and validated the process. Then, using the same approach, the signature of stem cell genes described as upregulated in Boquest, Lim and Pece data [34–36], showed a "good" correlation with the one expressed in ΔCK2β-MCF10A cells (Figure S2B). Moreover, this ΔCK2β-upregulated gene signature also correlates with an invasive signature, depicted by Schuetz, Wang, or Poola profiles [37–39] (Figure S2C). Altogether, these global analyses prove that ΔCK2β-cells share many genes with stem cells, and suggest that they might exhibit cancer stem cell (CSC)-like properties.

2.4. ΔCK2β-MCF10A Cells Have Hallmarks of Stem Cells

2.4.1. CD44/CD24 Stem Cell Markers

As it was reported that EMT contributes to the acquisition of cancer stem cells (CSC) traits and drug resistance [29], we further studied the cancer stem cell properties of ΔCK2β-MCF10A cells at the cellular level. Expression of a CD44^{high}/CD24^{low} configuration was associated with both human breast CSCs and normal mammary epithelial stem cells [40,41]. Therefore, we examined the consequence of CK2β silencing on CD44 and CD24 expression in ΔCK2β-MCF10A cells using flow cytometry analysis. We found an increase in the CD44^{high}/CD24^{low} population in CK2β-depleted cells (95% ± 0.1%), compared to Mock-MCF10A cells (57% ± 1.3%) (Figure 2A). CD44 can contribute to the activation of stem cell regulatory genes and can be a target of these genes [42]. Two connections between CD44 and genes that regulate stem cell characteristics have been described. First, CD44 is a target of the Wnt pathway [43]. Second, stem cells are frequently linked to secondary events such as interaction with a niche, EMT, migration, and apoptosis resistance, which have all been associated with CD44 [44,45]. Our result suggests that a decrease in CK2β expression confers stem cell-like characteristics on epithelial cells. In accordance with the subpopulation of mesenchymal breast cancer stem cells characterized by Liu et al. [46], we found that compared to Mock-cells, ΔCK2β-cells are ALDH and SSEA-1 negative (not shown), and expressed high and low mRNA levels of Vimentin and E-cadherin, respectively (Figure 2B).

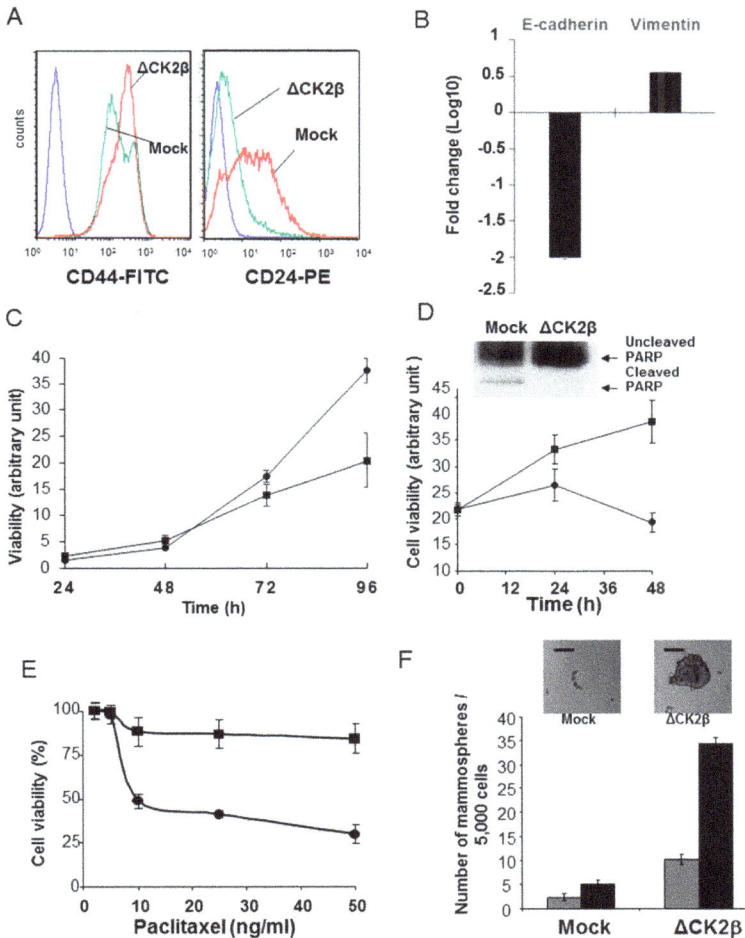

Figure 2. ΔCK2β-MCF10A cells have properties of cancer stem cells(CSCs) and are drug resistant.
(**A**) FACS analysis of CD24 and CD44 markers in in Mock- and ΔCK2β-cells (Blue line, unlabelled cells; green line, Mock-cells; red line, ΔCK2β-cells). Results are representative of three independent experiments; (**B**) E-cadherin and Vimentin expression levels measured by RT-qPCR. The fold changes compare Mock- to ΔCK2β-cells. $p < 0.05$; (**C**) Cell proliferation kinetic of ΔCK2β- (■) and Mock-MCF10A (●); (**D**) Anoikis: ΔCK2β- (■) and Mock-MCF10A (●) were grown on Poly-HEMA for 48 h. Cell viability was measured with the cell viability-GLO® assay, and apoptosis was visualized by Western blot using anti-PARP antibody; (**E**) Dose-response curves of ΔCK2β- (■) and Mock-MCF10A cells (●) treated with Paclitaxel. Bars denote the standard error ($n = 5$); (**F**) Representative images (top) and quantification (bottom) of mammosphere formation from Mock- and ΔCK2β-cells after first (grey bar) and second (black bar) dissociation steps (scale bar 50 μm).

2.4.2. Cell Proliferation and Viability

As already observed for stem cells grown in adherent conditions, cell proliferation was diminished in CK2β-depleted cells (Figure 2C). However, when ΔCK2β- and Mock-MCF10A cells were cultured on Poly-HEMA in the absence of anchorage, ΔCK2β-MCF10A cells showed evidence for increased survival without sign of cell detachment-induced apoptosis (anoikis). In contrast, a decrease of cell viability was observed in control cells, and was found associated with a PARP cleavage (Figure 2D). It has been reported that normal and cancer cell populations became resistant to chemotherapy drug treatment when experimentally induced into EMT [47]. Thus, we tested the viability of Mock- and ΔCK2β-cells treated with a commonly used chemotherapeutic drug, such as Paclitaxel [48]. Figure 2E shows that ΔCK2β-MCF10A cells were more resistant than Mock-MCF10A cells to Paclitaxel.

We further gauge the stemness of cells that have undergone EMT after CK2β depletion, using a mammosphere forming assay in serum free-medium supplemented with EGF and bFGF. We found that both control and CK2β-depleted cells developed spheres or aggregates within seven days, but after their disaggregation, most control cells died, whereas ΔCK2β-cells were still competent to form new mammospheres (Figure 2F).

2.5. Cell Positioning and Polarity

We previously investigated the cell positioning and polarity of MCF10A cells during TGFβ-induced EMT on H-shaped-micropatterned surfaces that are completely modified [49]. To gain further insight into the polarity changes that arise in ΔCK2β-cells, we first compared the localization of E-cadherin, actin, and paxillin in two-daughter-cell doublets by immunofluorescence (Figure 3A).

In Mock-MCF10A cells, the cell-cell contacts visualized by E-cadherin Figure 3A(a,c) and cortical F-actin Figure 3A(e,g) were tight. Paxillin, that links integrin to actin filaments, was restricted to the edges of the H-shaped pattern Figure 3A(i). As expected, the E-cadherin staining was lost in ΔCK2β-cells Figure 3A(b,d). Moreover, F-actin was organized as short stress filaments Figure 3A(f) that could be explained by the spatial localization of paxillin all along the adherent pattern Figure 3A(j). To study the position of the two daughter cells on stabilizing curved H-shaped patterns, the spatial coordinates of their nuclei were recorded by time-lapse microscopy during a complete cell cycle and automatically quantified [49]. From the acquired pictures, as exemplified in Figure 3(Ba), the angular distribution of the nucleus–nucleus axis was plotted (Figure 3B(b)), showing that ΔCK2β-cells were in a less stabilized state than Mock-cells. A polarity index has been evaluated, based on nucleus–nucleus axis and normalized nucleus-centrosome vector orientation [50]. Figure 3B(c) showed that this polarity index was significantly decreased in ΔCK2β-cells. These observations are in agreement with our previous data that showed a loss of polarity in CK2β-depleted cells [51]. In contrast, upon EMT induction, TGFβ-treated MCF10A cells behave differently as they underwent a polarity reversal due to both centrosomes and nuclei repositioning [50]. We next compared the behaviour of the two cell lines in 3D Matrigel culture. In these conditions, Mock-MCF10A cells generated fully polarized acinar structures whereas ΔCK2β-MCF10A cells were neither polarized, as visualized by the Golgi position, nor capable to organize in acini (Figure 3C).

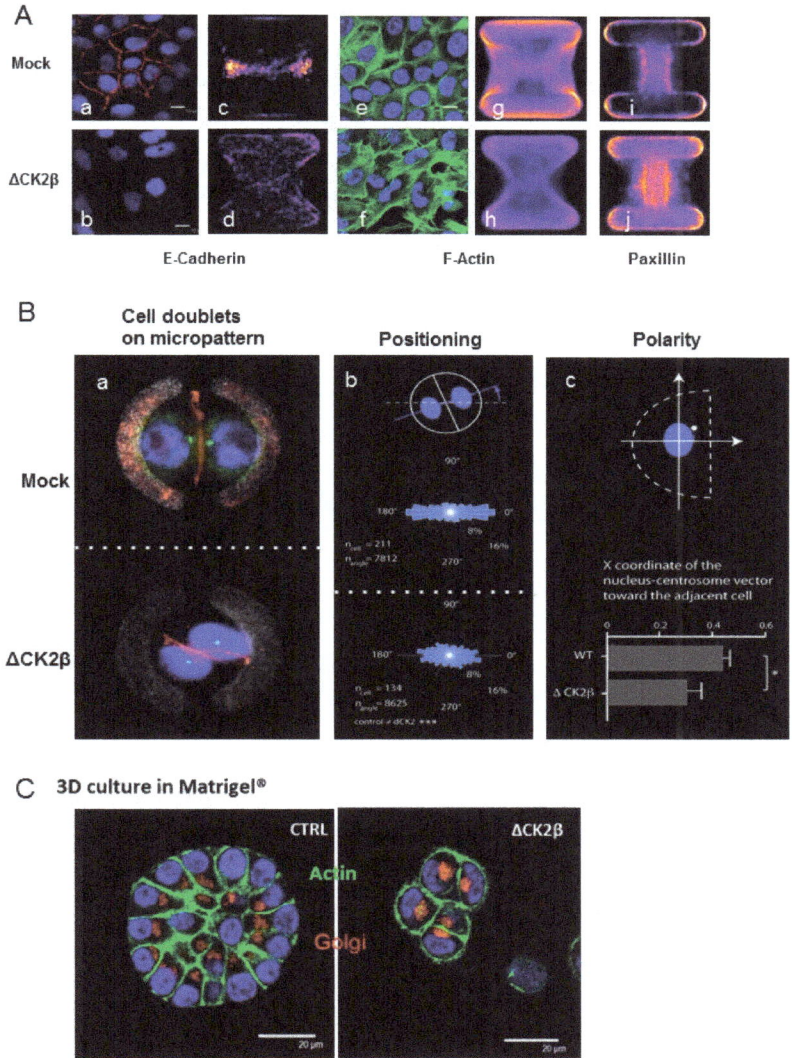

Figure 3. ΔCK2β- and Mock-MCF10A cell positioning and polarity. (**A**) Mock and ΔCK2β- MCF10A cells cultured as monolayer (a,b,e,f) or as doublets on H-shaped micropatterns (c,d,g–j) were stained for DNA (blue, a,b,e,f), E-cadherin (red, a–d), F-actin (green, e–h) or paxillin (far red, i,j). Average staining over 20 images on pattern is shown (c,d,g–j). Scale bar, 10 μm; (**B**) (a) Representative image of doublet cells stained for α-catenin (red), centrosome (green), and DNA (blue) on curved H-shaped micropattern; (b) Time-lapse acquisition of control- and ΔCK2β-MCF10A cell doublets on micropattern was performed. Automated movie analysis of Hoechst-stained cells provided the angular distribution of the nucleus–nucleus axis orientation that is represented by graph; (c) The X coordinate of the normalized nucleus-centrosome vector toward the cell-cell junction was calculated. Horizontal bar graph shows quantification of polarity index as previously described [47]; (**C**) Confocal images of 3D culture in Matrigel® for nine days, of Mock- and ΔCK2β-MCF10A cells stained for DNA in blue, F-actin in green, and Golgi apparatus in red. Scale bar, 20 μm.

Analysis of Cell Phenotypes by Orthotopic Engraftment

To analyze the cell phenotype in vivo, we injected FACS-sorted GFP-transfected wild type (WT)- or ΔCK2β-MCF10A cells into the inguinal mammary fat pads of nude mice. No tumour formation could be observed three months post-injection. However, IHC analysis of the fat pads showed that for both injected cell types, large islands of viable cells were still visible. The number of GFP-positive areas was increased between two and three months after cell injection (not shown). Compared to Mock-MCF10A cells, fat pads injected with ΔCK2β-cells showed an increased presence of gland-like structures, suggesting that these cells had an organoid-forming activity in this microenvironment (Figure 4A). Interestingly, almost no cytokeratin 5-6 (CK5-6) staining associated with basal cell differentiation, could be detected in those structures, whereas they displayed CK18 and smooth-muscle actin (SMA) staining, which are luminal and myo-epithelial characteristics, respectively. No such structures were observed in mice injected with an equal number of Mock-MCF10A cells (Figure 4B). The same protocol, performed with non GFP-labelled cells, showed that fat pads injected with ΔCK2β-MCF10A cells also displayed gland-like structures in which human mammary epithelial cells were present as visualized by human CK8/18 staining (Figure 4C).

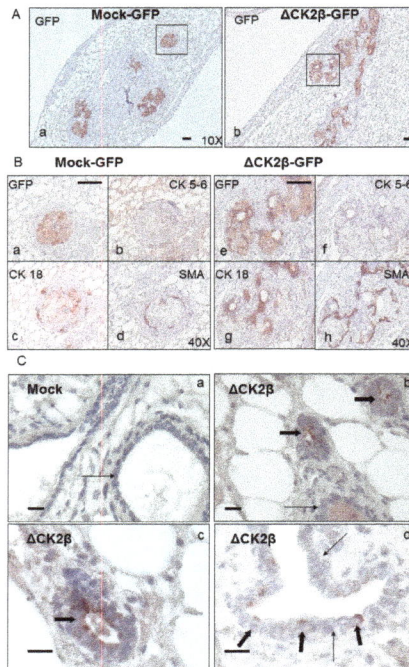

Figure 4. IHC analysis of ΔCK2β- and Mock-MCF10A cells injected in inguinal mammary fate pad.
(**A**) Three months after injection of GFP-transfected Mock- or ΔCK2β-MCF10A cells in mammary fat pad, the glands were harvested, fixed, paraffin included, and sections were stained with anti-GFP (a and b, respectively); (**B**) High magnification views (40×) of the boxed regions show the staining of GFP (a,e), cytokeratin 5-6 (b,f), cytokeratin 18 (c,g) and αSMA (d,h). Sections were counterstained with hematoxylin; (**C**) Six weeks post-injection of Mock-cells or ΔCK2β-cells, mammary gland sections were immunostained with human specific Cytokeratin 8/18. Sections were counterstained with Hematoxylin. Pictures are representative of different mammary gland sections injected with Mock-cells (a) or ΔCK2β-cells (b–d). Thin arrows indicate mouse mammary epithelial cells and thick arrows human stained luminal cells. Scale bars, 50 μm.

3. Material & Methods

3.1. Cell Culture and Retroviral Infection

MCF-10A cells from ATCC-LGS (Molsheim, France) (CRL-10317) are mammary epithelial cells derived from fibrocystic breast tissue from women with no family history of breast cancer and no evidence of disease. HMEC-hTERT were described in [8]. They were both cultured as described [52]. Stable silencing was accomplished by transduction with lentiviruses pLKO1 (Sigma-Aldrich, St. Louis, MO, USA) as described [9]. Mock-cells are MCF10A or HMEC cells transduced with an empty pLKO1 vector.

3.2. microRNA Profiling

Total RNA was isolated using Trizol Reagent (Invitrogen, Carlsbad, CA, USA) and was submitted to ProfileXpert core facility (Lyon, France) for microRNA profiling. The samples were hybridized on human v.3 miRNA Agilent array according to manufacturer instructions (Agilent protocol version 2.2), and microarray data analysis was carried out using Feature Extraction software version 10.7 (Agilent, Santa Clara, CA, USA).

For better cross-array comparison, raw data were normalized with the Genespring software version 7.3.1 (Agilent), using two LabelingSpikes-InSignal (DMR_285 and DMR_31a) as internal standard. The threshold of detection was calculated using the normalized signal intensity of negative controls ± 3 standard deviation. Spots with signal intensities below this threshold are referred to as (absent) with an arbitrary value of 0.01*, and denoted 'A' in the Tables S1 and S2. Quality of processing was evaluated by generating a scatter plot of 11 positive controls. Statistical comparison and filtering were performed using Genespring software 7.3.1 (Agilent). The average signal is averaged between two replicates, and log2 fold change is calculated between ΔCK2β and Mock conditions. The mean and standard deviation of the mean are then calculated to aggregate different probes signal of the same miRNA.

Both mRNA and miRNA datasets are available under the accession number GSE102267.

3.3. Fluorescence-Activated Cell Sorting

FITC-conjugated anti-CD44 antibody (clone G44-26) and PE-conjugated anti-CD24 antibody (clone ML5) were obtained from BD Biosciences (Grenoble, France), and used for FACS analysis in accordance with manufacturer's protocols. GFP-cells were sorted using a FACS-Calibur (BD Biosciences (Grenoble, France).

3.4. Western Blot Analysis

Primary antibodies were anti-actin (Abcam, Cambridge, UK, ab8226), anti-Zeb (Santa Cruz Biotechnology, Heidelberg, Germany, sc-25388), anti-integrin β1, anti- integrin β4 (BD Biosciences, 610467, 611232, respectively), anti- Integrin β3 (Abcam, Ab7167), and anti-PARP (Thermo Scientific, Courtaboeuf, France, 9542). Secondary antibodies were peroxidase-conjugated affinity pure Goat anti-rabbit IgG (#111035003) and peroxidase-conjugated affinity pure goat anti-mouse IgG (#115035003) from Jackson Immuno Research. Cells were lysed in RIPA buffer (10 mM Tris-HCl pH 7.4, 150 mM NaCl, 1% Triton X-100, 0.1% SDS, 0.5% DOC and 1 mM EDTA) containing both protease- and phosphatase-inhibitor cocktails (Sigma-Aldrich; P8340, P2850, P5726). Cell homogenates were quantified using BCA protein assay kit (Thermo Scientific). SDS-PAGE was performed using pre-cast 4–12% gradient gel (Bio-Rad, Hercules, CA, USA). Separated proteins at 20 µg/lane were transferred to PVDF membranes (100 V for 60 min). Blotted membranes were blocked during 1 h at room temperature with saturation buffer (1% BSA in Tris Buffer Saline 10 mM, Tween 0.1% (TBST)), and then incubated with primary antibody diluted in saturation buffer, for 2 h or overnight. After 3 washes with TBST, secondary antibodies were added for 1 h. Luminata Forte Western HRP substrate (Millipore, Billerica,

MA, USA) was added and membranes were read with Fusion Fx7 (PerkinElmer, Waltham, MA, USA). Quantification was performed using ImageJ software.

3.5. Quantitative Real-Time PCR

One microgram of total RNA prepared for the microarray hybridization was used to generate cDNAs by reverse transcription using the iScript system (Bio-Rad) as recommended by the manufacturer. Real-time PCR was performed using Bio-rad CFX96 apparatus and qPCR Master Mix (Promega, Madison, MI, USA). The values for the specific genes were normalized to the 36B4 and U6. Specific primers sequences are provided in Supplementary Table S3.

MiRNA levels were measured by RT-qPCR using TaqMan miRNA assays (Applied Biosystems, Foster City, CA, USA). Ten nanograms of tumour total RNA were reverse transcribed using the TaqMan miRNA Reverse Transcription kit and miRNA-specific stem-loop primers (Applied Biosystems). Real-time PCR was performed on the 5′-extended cDNA with Applied Biosystems TaqMan 2× Universal PCR Master Mix, and the appropriate 5 × TaqMan MicroRNA Assay Mix for each miRNA of interest (assay IDs: hsa-miR-205, 4373093; hsa-miR-141, 4373137; hsa-miR-200c, 4395411; hsa-miR-30b, 4373290; hsa-miR-34b, 4395213; hsa-miR-720, 4409111; hsa-miR-21, 000397; hsa-miR-210, 000512; hsa-miR-1246, CTRWENE). Real-time PCR was carried out on C1000 Thermal cycler (CFX96 Real Time system, Bio-Rad) at 95 °C for 10 min, followed by 40 cycles of 95 °C for 15 s and 60 °C for 1 min. Data were analyzed with CFX Manager Software version V1.5.534.0511 (Bio-Rad). The hsa-miR-720 was used as endogenous control for normalization. Normalized expression was calculated using the comparative CT method and fold changes were derived from the $2 - \Delta\Delta Ct$ values for each miRNA.

3.6. Characterization of Resistance to Cytotoxic Agents

All compounds were purchased from Sigma and dissolved in DMSO. MCF10A cells (5000/well) were plated in 100 µL per well in a 96-well plate. One day after seeding, compounds were added in five replicates per concentration. Cell viability was measured after 72 h with Cell viability Glo assay (Promega).

3.7. Colony Assay

To measure anchorage-independent growth, cells were detached with trypsin and resuspended in growth medium. Plates were prepared with a coating of 0.75% agarose (Cambrex, East Rutherford, NJ, USA) in growth medium, and then overlaid with a suspension of cells in 0.45% agarose (5×10^3 cells/well). Plates were incubated for 3 weeks at 37 °C and colonies were imaged under microscope.

3.8. Mouse Injection

Animal maintenance and experiments were performed in accordance with the animal care guidelines of the European Union and French laws. Six-week old female Athymic nude mice (Charles River laboratories) were injected with 10^6 modified MCF10A cells into a fat pad of #4 mammary gland.

3.9. Histological and Immunohistochemical Analysis

Mammary glands were fixed in 4% paraformaldehyde and embedded in paraffin. Microtome sections (5 mm thick) were stained with H&E for histological analysis. Expression of the transduced GFP was analyzed by standard immuno-histochemistry (IHC) using the anti-GFP mouse monoclonal antibody (Abcam, Ab 13970), detected with a biotin-conjugated anti-mouse IgG antibody and an avidin-biotin-peroxidase complex (Vector Laboratories). The differentiation status of tissues was determined using CK5-6 (Chemicon, Millipore, MAB 1620, Billerica, MA, USA) CK18 (Epitomics, Burlingame, CA, USA, 1433-1) and SMA (DAKO, Agilent, M0851, Santa Clara, CA, USA) antibodies.

3.10. Gene Expression Microarray Analysis

Single-sample GSEA (ssGSEA), an extension of Gene Set Enrichment Analysis (GSEA) calculates separate enrichment scores for each pairing of a sample and gene set. Each ssGSEA enrichment score represents the degree to which the genes in a particular gene set are coordinately up- or down-regulated within a sample. This analysis has been done with the GenePattern software [53]. (http://software.broadinstitute.org/cancer/software/genepattern/modules/docs/ssGSEAProjection/4).

3.11. Cell Micropatterning

Micropatterns were fabricated and cell-cell positioning was analyzed as previously described [49]. The polarity index was measured as described in [50].

3.12. Immunofluorescence

Cells were fixed as previously described [47], and incubated with anti-integrin β1, anti-integrin β4, anti-paxillin, anti-E-cadherin (BD Biosciences, 610467, 611232, 6100052, and 610181, respectively), anti-integrin β3, anti-Giantin (Abcam Ab7167 and 24586 respectively), and anti-α-catenin (B52975; Calbiochem), for 1 h, and then incubated with the corresponding secondary antibodies and FITC-phalloidin (Invitrogen) at 1 µg/mL for 30 min.

4. Conclusions

It has been reported that the EMT programs which control normal mammary stem cells and cancer stem cells are likely to differ in the activation of distinct signaling pathways [54]. Together, our data provide evidence that the downregulation of CK2β expression, observed in a subtype of breast tumours, can promote the acquisition of characteristics commonly associated with the CSC phenotype in vitro. However, since ΔCK2β-MCF10A cells were deficient in tumour-initiating ability, we suggest that under-expression of CK2β has a profound impact on both the plasticity and the stemness of breast epithelial cells. In particular, the results from Figure 4 show that micro-environmental cues are sufficient to re-direct cells that exhibit stem cell traits to acquire an organoid-forming activity with an absence of malignant transformation. Since EMT supports the induction stem-cell phenotype, identifying the CK2 substrates whose phosphorylation is modulated in CK2β-depleted cells or CK2α overexpressing cells, will improve the discovery of new specific markers furthering understanding breast cell stemness.

Supplementary Materials: The following are available online at http://www.mdpi.com/2072-6694/9/9/114/s1. Figure S1: Expression analysis of CK2β and JAG1, Figure S2: Gene expression microarray analysis, Figure S3: Modulation of miRNA-targets in ΔCK2β-HMEC-hTERT cells, Table S1: miRNA expression profile of Mock and ΔCK2β-MCF10A cells, Table S2: Normalized mRNA expression profile of Mock and ΔCK2β-MCF10A cells, Table S3: Primers used for RT-qPCR.

Acknowledgments: We thank ProfileXpert (Bron, France) for transcriptomic and miRnome analysis, Alain Puisieux for providing HMEC-hTERT cells, Caroline Roelants and Manon Gervais for qPCR studies, Véronique Collins-Faure for assistance for FACS and the animal unit staff at Biosciences and Biotechnology Institute of Grenoble (BIG) for animal husbandry. This research was supported by recurrent institutional funding from INSERM, CEA, Ligue Nationale contre le Cancer (accredited team 2010–2012) and Ligue Comité de la Loire, University Grenoble Alpes, the Espoir Foundation and grants from the French National Research Agency (PCV-08 CoCCINet).

Author Contributions: M.Bu., M.T., C.C. and O.F. designed research, E.D.-P., E.S., M.Ba., M.Bu., M.P. performed research, E.D.-P., M.Bu., M.P., B.M., L.G., M.T., C.C. and O.F. analyzed data; and C.C. and O.F. wrote the paper.

Conflicts of Interest: The authors declare no conflict of interest.

References

1. Hanahan, D.; Weinberg, R.A. Hallmarks of cancer: The next generation. *Cell* **2011**, *144*, 646–674. [CrossRef] [PubMed]
2. Hanahan, D.; Weinberg, R.A. The hallmarks of cancer. *Cell* **2000**, *100*, 57–70. [CrossRef]

3. Kreutzer, J.N.; Ruzzene, M.; Guerra, B. Enhancing chemosensitivity to gemcitabine via rna interference targeting the catalytic subunits of protein kinase ck2 in human pancreatic cancer cells. *BMC Cancer* **2010**, *10*, 440. [CrossRef] [PubMed]

4. Deshiere, A.; Duchemin-Pelletier, E.; Spreux, E.; Ciais, D.; Forcet, C.; Cochet, C.; Filhol, O. Regulation of epithelial to mesenchymal transition: Ck2β on stage. *Mol. Cell. Biochem.* **2011**, *356*, 11–20. [CrossRef] [PubMed]

5. Filhol, O.; Nueda, A.; Martel, V.; Gerber-Scokaert, D.; Benitez, M.J.; Souchier, C.; Saoudi, Y.; Cochet, C. Live-cell fluorescence imaging reveals the dynamics of protein kinase ck2 individual subunits. *Mol. Cell. Biol.* **2003**, *23*, 975–987. [CrossRef] [PubMed]

6. Filhol, O.; Cochet, C. Protein kinase ck2 in health and disease: Cellular functions of protein kinase ck2: A dynamic affair. *Cell. Mol. Life Sci.* **2009**, *66*, 1830–1839. [CrossRef] [PubMed]

7. Mani, S.A.; Guo, W.; Liao, M.J.; Eaton, E.N.; Ayyanan, A.; Zhou, A.Y.; Brooks, M.; Reinhard, F.; Zhang, C.C.; Shipitsin, M.; et al. The epithelial-mesenchymal transition generates cells with properties of stem cells. *Cell* **2008**, *133*, 704–715. [CrossRef] [PubMed]

8. Morel, A.P.; Lievre, M.; Thomas, C.; Hinkal, G.; Ansieau, S.; Puisieux, A. Generation of breast cancer stem cells through epithelial-mesenchymal transition. *PLoS ONE* **2008**, *3*, e2888. [CrossRef] [PubMed]

9. Deshiere, A.; Duchemin-Pelletier, E.; Spreux, E.; Ciais, D.; Combes, F.; Vandenbrouck, Y.; Coute, Y.; Mikaelian, I.; Giusiano, S.; Charpin, C.; et al. Unbalanced expression of ck2 kinase subunits is sufficient to drive epithelial-to-mesenchymal transition by snail1 induction. *Oncogene* **2013**, *32*, 1373–1383. [CrossRef] [PubMed]

10. Golden, D.; Cantley, L.G. Casein kinase 2 prevents mesenchymal transformation by maintaining foxc2 in the cytoplasm. *Oncogene* **2015**, *34*, 4702–4712. [CrossRef] [PubMed]

11. Kim, J.; Hwan Kim, S. Ck2 inhibitor cx-4945 blocks tgf-beta1-induced epithelial-to-mesenchymal transition in a549 human lung adenocarcinoma cells. *PLoS ONE* **2013**, *8*, e74342. [CrossRef]

12. Ferrer-Font, L.; Alcaraz, E.; Plana, M.; Candiota, A.P.; Itarte, E.; Arus, C. Protein kinase ck2 content in gl261 mouse glioblastoma. *Pathol. Oncol. Res.* **2016**, *22*, 633–637. [CrossRef] [PubMed]

13. Roelants, C.; Giacosa, S.; Duchemin-Pelletier, E.; McLeer-Florin, A.; Tisseyre, C.; Aubert, C.; Champelovier, P.; Boutonnat, J.; Descotes, J.L.; Rambeaud, J.-J.; et al. Dysregulated expression of protein kinase ck2 in renal cancer. In *Protein Kinase CK2 Cellular Function in Normal and Disease States*; Ahmed, K., Issinger, O.-G., Szyszka, R., Eds.; Springer International Publishing: Cham, Switzerland, 2015; pp. 241–257.

14. Blanpain, C.; Simons, B.D. Unravelling stem cell dynamics by lineage tracing. *Nat. Rev. Mol. Cell Biol.* **2013**, *14*, 489–502. [CrossRef] [PubMed]

15. Lu, J.; Getz, G.; Miska, E.A.; Alvarez-Saavedra, E.; Lamb, J.; Peck, D.; Sweet-Cordero, A.; Ebert, B.L.; Mak, R.H.; Ferrando, A.A.; et al. Microrna expression profiles classify human cancers. *Nature* **2005**, *435*, 834–838. [CrossRef] [PubMed]

16. Gregory, P.A.; Bert, A.G.; Paterson, E.L.; Barry, S.C.; Tsykin, A.; Farshid, G.; Vadas, M.A.; Khew-Goodall, Y.; Goodall, G.J. The mir-200 family and mir-205 regulate epithelial to mesenchymal transition by targeting zeb1 and sip1. *Nat. Cell Biol.* **2008**, *10*, 593–601. [CrossRef] [PubMed]

17. Park, S.M.; Gaur, A.B.; Lengyel, E.; Peter, M.E. The mir-200 family determines the epithelial phenotype of cancer cells by targeting the e-cadherin repressors zeb1 and zeb2. *Genes Dev.* **2008**, *22*, 894–907. [CrossRef] [PubMed]

18. Yu, F.; Deng, H.; Yao, H.; Liu, Q.; Su, F.; Song, E. Mir-30 reduction maintains self-renewal and inhibits apoptosis in breast tumor-initiating cells. *Oncogene* **2010**, *29*, 4194–4204. [CrossRef] [PubMed]

19. Li, X.J.; Ren, Z.J.; Tang, J.H. Microrna-34a: A potential therapeutic target in human cancer. *Cell Death Dis.* **2014**, *5*, e1327. [CrossRef] [PubMed]

20. Liu, L.Z.; Li, C.; Chen, Q.; Jing, Y.; Carpenter, R.; Jiang, Y.; Kung, H.F.; Lai, L.; Jiang, B.H. Mir-21 induced angiogenesis through akt and erk activation and hif-1alpha expression. *PLoS ONE* **2011**, *6*, e19139. [CrossRef]

21. Yan, L.X.; Huang, X.F.; Shao, Q.; Huang, M.Y.; Deng, L.; Wu, Q.L.; Zeng, Y.X.; Shao, J.Y. Microrna mir-21 overexpression in human breast cancer is associated with advanced clinical stage, lymph node metastasis and patient poor prognosis. *RNA* **2008**, *14*, 2348–2360. [CrossRef] [PubMed]

22. Luo, G.; Luo, W.; Sun, X.; Lin, J.; Wang, M.; Zhang, Y.; Luo, W.; Zhang, Y. Microrna21 promotes migration and invasion of glioma cells via activation of sox2 and betacatenin signaling. *Mol. Med. Rep.* **2017**, *15*, 187–193. [CrossRef] [PubMed]

23. Li, C.; Nguyen, H.T.; Zhuang, Y.; Lin, Y.; Flemington, E.K.; Guo, W.; Guenther, J.; Burow, M.E.; Morris, G.F.; Sullivan, D.; et al. Post-transcriptional up-regulation of mir-21 by type i collagen. *Mol. Carcinog.* **2011**, *50*, 563–570. [CrossRef] [PubMed]

24. Ferraro, A.; Kontos, C.K.; Boni, T.; Bantounas, I.; Siakouli, D.; Kosmidou, V.; Vlassi, M.; Spyridakis, Y.; Tsipras, I.; Zografos, G.; et al. Epigenetic regulation of mir-21 in colorectal cancer: Itgb4 as a novel mir-21 target and a three-gene network (mir-21-itgbeta4-pdcd4) as predictor of metastatic tumor potential. *Epigenetics* **2014**, *9*, 129–141. [CrossRef] [PubMed]

25. Selcuklu, S.D.; Donoghue, M.T.; Kerin, M.J.; Spillane, C. Regulatory interplay between mir-21, jag1 and 17beta-estradiol (e2) in breast cancer cells. *Biochem. Biophys. Res. Commun.* **2012**, *423*, 234–239. [CrossRef] [PubMed]

26. Zhang, W.C.; Chin, T.M.; Yang, H.; Nga, M.E.; Lunny, D.P.; Lim, E.K.; Sun, L.L.; Pang, Y.H.; Leow, Y.N.; Malusay, S.R.; et al. Tumour-initiating cell-specific mir-1246 and mir-1290 expression converge to promote non-small cell lung cancer progression. *Nat. Commun.* **2016**, *7*, 11702. [CrossRef] [PubMed]

27. Chan, Y.C.; Banerjee, J.; Choi, S.Y.; Sen, C.K. Mir-210: The master hypoxamir. *Microcirculation* **2012**, *19*, 215–223. [CrossRef] [PubMed]

28. Xu, D.; Takeshita, F.; Hino, Y.; Fukunaga, S.; Kudo, Y.; Tamaki, A.; Matsunaga, J.; Takahashi, R.U.; Takata, T.; Shimamoto, A.; et al. Mir-22 represses cancer progression by inducing cellular senescence. *J. Cell Biol.* **2011**, *193*, 409–424. [CrossRef] [PubMed]

29. Polyak, K.; Weinberg, R.A. Transitions between epithelial and mesenchymal states: Acquisition of malignant and stem cell traits. *Nat. Rev. Cancer* **2009**, *9*, 265–273. [CrossRef] [PubMed]

30. Charafe-Jauffret, E.; Ginestier, C.; Monville, F.; Finetti, P.; Adelaide, J.; Cervera, N.; Fekairi, S.; Xerri, L.; Jacquemier, J.; Birnbaum, D.; et al. Gene expression profiling of breast cell lines identifies potential new basal markers. *Oncogene* **2006**, *25*, 2273–2284. [CrossRef] [PubMed]

31. Sarrio, D.; Rodriguez-Pinilla, S.M.; Hardisson, D.; Cano, A.; Moreno-Bueno, G.; Palacios, J. Epithelial-mesenchymal transition in breast cancer relates to the basal-like phenotype. *Cancer Res.* **2008**, *68*, 989–997. [CrossRef] [PubMed]

32. Gotzmann, J.; Fischer, A.N.; Zojer, M.; Mikula, M.; Proell, V.; Huber, H.; Jechlinger, M.; Waerner, T.; Weith, A.; Beug, H.; et al. A crucial function of pdgf in tgf-beta-mediated cancer progression of hepatocytes. *Oncogene* **2006**, *25*, 3170–3185. [CrossRef] [PubMed]

33. Jechlinger, M.; Sommer, A.; Moriggl, R.; Seither, P.; Kraut, N.; Capodiecci, P.; Donovan, M.; Cordon-Cardo, C.; Beug, H.; Grunert, S. Autocrine pdgfr signaling promotes mammary cancer metastasis. *J. Clin. Investig.* **2006**, *116*, 1561–1570. [CrossRef] [PubMed]

34. Boquest, A.C.; Shahdadfar, A.; Fronsdal, K.; Sigurjonsson, O.; Tunheim, S.H.; Collas, P.; Brinchmann, J.E. Isolation and transcription profiling of purified uncultured human stromal stem cells: Alteration of gene expression after in vitro cell culture. *Mol. Biol. Cell* **2005**, *16*, 1131–1141. [CrossRef] [PubMed]

35. Lim, E.; Wu, D.; Pal, B.; Bouras, T.; Asselin-Labat, M.L.; Vaillant, F.; Yagita, H.; Lindeman, G.J.; Smyth, G.K.; Visvader, J.E. Transcriptome analyses of mouse and human mammary cell subpopulations reveal multiple conserved genes and pathways. *Breast Cancer Res.* **2010**, *12*, R21. [CrossRef] [PubMed]

36. Pece, S.; Tosoni, D.; Confalonieri, S.; Mazzarol, G.; Vecchi, M.; Ronzoni, S.; Bernard, L.; Viale, G.; Pelicci, P.G.; Di Fiore, P.P. Biological and molecular heterogeneity of breast cancers correlates with their cancer stem cell content. *Cell* **2010**, *140*, 62–73. [CrossRef] [PubMed]

37. Schuetz, C.S.; Bonin, M.; Clare, S.E.; Nieselt, K.; Sotlar, K.; Walter, M.; Fehm, T.; Solomayer, E.; Riess, O.; Wallwiener, D.; et al. Progression-specific genes identified by expression profiling of matched ductal carcinomas in situ and invasive breast tumors, combining laser capture microdissection and oligonucleotide microarray analysis. *Cancer Res.* **2006**, *66*, 5278–5286. [CrossRef] [PubMed]

38. Wang, W.; Wyckoff, J.B.; Goswami, S.; Wang, Y.; Sidani, M.; Segall, J.E.; Condeelis, J.S. Coordinated regulation of pathways for enhanced cell motility and chemotaxis is conserved in rat and mouse mammary tumors. *Cancer Res.* **2007**, *67*, 3505–3511. [CrossRef] [PubMed]

39. Poola, I.; DeWitty, R.L.; Marshalleck, J.J.; Bhatnagar, R.; Abraham, J.; Leffall, L.D. Identification of mmp-1 as a putative breast cancer predictive marker by global gene expression analysis. *Nat. Med.* **2005**, *11*, 481–483. [CrossRef] [PubMed]

40. Dontu, G.; Al-Hajj, M.; Abdallah, W.M.; Clarke, M.F.; Wicha, M.S. Stem cells in normal breast development and breast cancer. *Cell Prolif.* **2003**, *36* (Suppl. 1), 59–72. [CrossRef] [PubMed]

41. Sleeman, K.E.; Kendrick, H.; Ashworth, A.; Isacke, C.M.; Smalley, M.J. Cd24 staining of mouse mammary gland cells defines luminal epithelial, myoepithelial/basal and non-epithelial cells. *Breast Cancer Res.* **2006**, *8*, R7. [CrossRef] [PubMed]

42. Zoller, M. Cd44: Can a cancer-initiating cell profit from an abundantly expressed molecule? *Nat. Rev. Cancer* **2011**, *11*, 254–267. [CrossRef] [PubMed]

43. Zeilstra, J.; Joosten, S.P.; Dokter, M.; Verwiel, E.; Spaargaren, M.; Pals, S.T. Deletion of the wnt target and cancer stem cell marker cd44 in apc(min/+) mice attenuates intestinal tumorigenesis. *Cancer Res.* **2008**, *68*, 3655–3661. [CrossRef] [PubMed]

44. Yang, J.; Weinberg, R.A. Epithelial-mesenchymal transition: At the crossroads of development and tumor metastasis. *Dev. Cell* **2008**, *14*, 818–829. [CrossRef] [PubMed]

45. Bhat-Nakshatri, P.; Appaiah, H.; Ballas, C.; Pick-Franke, P.; Goulet, R., Jr.; Badve, S.; Srour, E.F.; Nakshatri, H. Slug/snai2 and tumor necrosis factor generate breast cells with cd44+/cd24- phenotype. *BMC Cancer* **2010**, *10*, 411. [CrossRef] [PubMed]

46. Liu, S.; Cong, Y.; Wang, D.; Sun, Y.; Deng, L.; Liu, Y.; Martin-Trevino, R.; Shang, L.; McDermott, S.P.; Landis, M.D.; et al. Breast cancer stem cells transition between epithelial and mesenchymal states reflective of their normal counterparts. *Stem Cell Rep.* **2014**, *2*, 78–91. [CrossRef] [PubMed]

47. Gupta, P.B.; Onder, T.T.; Jiang, G.; Tao, K.; Kuperwasser, C.; Weinberg, R.A.; Lander, E.S. Identification of selective inhibitors of cancer stem cells by high-throughput screening. *Cell* **2009**, *138*, 645–659. [CrossRef] [PubMed]

48. Yauch, R.L.; Januario, T.; Eberhard, D.A.; Cavet, G.; Zhu, W.; Fu, L.; Pham, T.Q.; Soriano, R.; Stinson, J.; Seshagiri, S.; et al. Epithelial versus mesenchymal phenotype determines in vitro sensitivity and predicts clinical activity of erlotinib in lung cancer patients. *Clin. Cancer Res. Off. J. Am. Assoc. Cancer Res.* **2005**, *11*, 8686–8698. [CrossRef] [PubMed]

49. Tseng, Q.; Duchemin-Pelletier, E.; Deshiere, A.; Balland, M.; Guillou, H.; Filhol, O.; Thery, M. Spatial organization of the extracellular matrix regulates cell-cell junction positioning. *Proc. Natl. Acad. Sci. USA* **2012**, *109*, 1506–1511. [CrossRef] [PubMed]

50. Burute, M.; Prioux, M.; Blin, G.; Truchet, S.; Letort, G.; Tseng, Q.; Bessy, T.; Lowell, S.; Young, J.; Filhol, O.; et al. Polarity reversal by centrosome repositioning primes cell scattering during epithelial-to-mesenchymal transition. *Dev. Cell* **2017**, *40*, 168–184. [CrossRef] [PubMed]

51. Deshiere, A.; Theis-Febvre, N.; Martel, V.; Cochet, C.; Filhol, O. Protein kinase ck2 and cell polarity. *Mol. Cell. Biochem.* **2008**, *316*, 107–113. [CrossRef] [PubMed]

52. Debnath, J.; Muthuswamy, S.K.; Brugge, J.S. Morphogenesis and oncogenesis of mcf-10a mammary epithelial acini grown in three-dimensional basement membrane cultures. *Methods* **2003**, *30*, 256–268. [CrossRef]

53. Reich, M.; Liefeld, T.; Gould, J.; Lerner, J.; Tamayo, P.; Mesirov, J.P. GenePattern 2.0. *Nat. Genet.* **2006**, *38*, 500–501. [CrossRef] [PubMed]

54. Ye, X.; Tam, W.L.; Shibue, T.; Kaygusuz, Y.; Reinhardt, F.; Ng Eaton, E.; Weinberg, R.A. Distinct emt programs control normal mammary stem cells and tumour-initiating cells. *Nature* **2015**, *525*, 256–260. [CrossRef] [PubMed]

cancers

MDPI

Review

The Role of Cancer-Derived Exosomes in Tumorigenicity & Epithelial-to-Mesenchymal Transition

Robert H. Blackwell [1], Kimberly E. Foreman [2] and Gopal N. Gupta [1],*

[1] Department of Urology, Loyola University Medical Center, 2160 S. First Ave., Maywood, IL 60153, USA; rblackwell@lumc.edu
[2] Cardinal Bernardin Cancer Center, Loyola University Medical Center, 2160 S. First Ave., Maywood, IL 60153, USA; kforema@luc.edu
* Correspondence: gogupta@lumc.edu; Tel.: +1-708-216-5098

Academic Editor: Joëlle Roche
Received: 31 May 2017; Accepted: 5 August 2017; Published: 10 August 2017

Abstract: Epithelial-to-mesenchymal transition (EMT) is a process by which epithelial cells lose their basement membrane interaction and acquire a more migratory, mesenchymal phenotype. EMT has been implicated in cancer cell progression, as cells transform and increase motility and invasiveness, induce angiogenesis, and metastasize. Exosomes are 30–100 nm membrane-bound vesicles that are formed and excreted by all cell types and released into the extracellular environment. Exosomal contents include DNA, mRNA, miRNA, as well as transmembrane- and membrane-bound proteins derived from their host cell contents. Exosomes are involved in intercellular signaling, both by membrane fusion to recipient cells with deposition of exosomal contents into the cytoplasm and by the binding of recipient cell membrane receptors. Recent work has implicated cancer-derived exosomes as an important mediator of intercellular signaling and EMT, with resultant transformation of cancer cells to a more aggressive phenotype, as well as the tropism of metastatic disease in specific cancer types with the establishment of the pre-metastatic niche.

Keywords: epithelial-mesenchymal transition; exosomes; neoplasm metastasis; neovascularization; pathologic; neoplasm invasiveness; intercellular signaling peptides and proteins

---

1. Introduction

Epithelial-to-mesenchymal transition (EMT) represents the process in which an epithelial cell, with normal basement membrane interaction, can undergo biochemical changes to lose the epithelial characteristics and develop a more migratory, mesenchymal phenotype [1,2]. While first described as a fundamental element of organ and tissue differentiation, EMT has been shown to have roles in wound healing, tumor initiation, and progression to metastatic disease [3–6].

Exosomes are membranous vesicles that are released by cells into the extracellular environment. Exosomes measure 30–100 nm in size and carry a variety of content including a lipid bilayer membrane, DNA, RNA, mRNA, miRNA, bioactive lipids (including prostaglandins, fatty acids, and leukotrienes), as well as intracellular and membrane-bound proteins [7,8]. Originally thought to function in disposal of unnecessary or waste cellular contents, exosomes have now shown involvement in cellular communication and cancer progression via target-cell: exosomal interactions as well as deposition of exosomal content into the target-cell cytoplasm [9–11]. As such, prior work has demonstrated the significant role exosomes play in cancer biology, promoting growth and survival of tumor cells distant from the primary tumor, tumor-cell apoptosis inhibition, as well as promoting angiogenesis, migration, invasiveness, and tumor cell viability [2,12–17]. Cancer-derived exosomes are now believed to be mediators of metastatic disease and contribute to the tropism of metastasis seen in different cancer cell types [18,19].

2. Exosomal Isolation

The ability to isolate exosomes has been demonstrated in cell culture media as well as nearly all bodily fluids examined (blood, urine, saliva, breast milk, etc.) [20–22]. The current standard for exosomal isolation involves differential centrifugation at increasing speeds [23]. Low speed centrifugation results in the removal of cellular debris, followed by ultracentrifugation at $100,000 \times$ *g*-force to ultimately pellet exosomes. Alternatively, kits exist to isolate exosomes that use magnetic coated beads with exosome-specific antibodies. While these kits avoid the time-intensive ultracentrifugation process, they are not intended for the isolation of large quantities of exosomes [20]. Comparison between the techniques with regard to protein, mRNA, and miRNA yields reveals that either technique is applicable for further research or clinical application; however, for protein analysis, serial ultracentrifugation resulted in a higher level of purity [24,25]. Following isolation, size determination with scanning electron microscopy or nanoparticle tracking analysis can be performed to confirm exosome isolation, with an anticipated result of vesicles ranging 30–100 nm in size [26]. In addition to size alone, the International Society for Extracellular Vesicles recommends characterization and quantification of proteins of any exosomal preparation, including those expected to be present or enriched (e.g., transmembrane or lipid-bound extracellular proteins, cytosolic proteins) and those expected to be absent or underrepresented (e.g., intracellular proteins), to properly characterize exosomes and potentially co-isolated extracellular vesicles [27]. Clinically, there are currently two exosome-based, commercially available diagnostic tests that exist for prostate and lung cancer [28].

3. Exosomal Signaling/Cargo Delivery

Exosomes have been found to carry a variety of biomolecules (including protein, DNA, mRNA, miRNA) in their lumen and/or on the surface [29]. Following release, exosomes can then interact with recipient cells to facilitate cell-cell signaling. This interaction comes in the form of direct interaction between exosomal transmembrane or membrane-associated surface proteins with recipient cell receptors resulting in cellular signaling, or by exosomal internalization with subsequent fusion and deposition of exosomal contents [30–32] (Figure 1). While not yet demonstrated in vivo, this exosomal internalization has been shown to be at least partially inhibited by treatment with heparin in vitro, as well as non-acidic conditions given alteration in membrane rigidity [32,33].

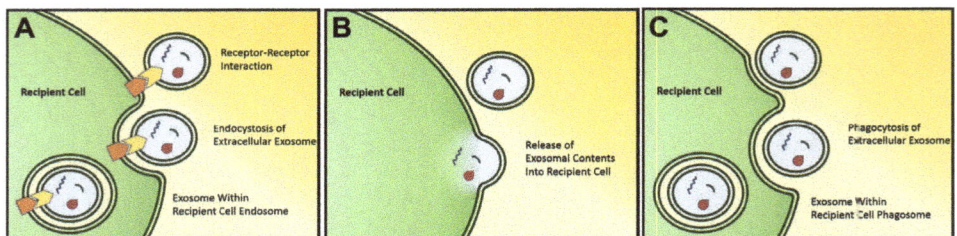

Figure 1. Several mechanisms have been found for interaction and uptake of exosomes by target cells. Most well studied mechanisms are receptor-mediated exosomal uptake (endocytosis) (**A**), direct exosomal fusion with plasma membrane (**B**), and phagocytic exosomal uptake (phagocytosis) (**C**). Used with permission [20].

4. Exosomal Proteome Alterations Following Epithelial-to-Mesenchymal Transition

Once mesenchymal transition occurs, there is an alteration in the cancer-derived exosomal proteome. While these changes are not an exact replication of the host cell, these changes affect proteins involved in junction formation, cellular adhesion, communication, and proliferation [34]. Garnier and colleagues investigated these changes to the exosomal proteome in the human squamous cell carcinoma A431 cell line. Comparisons were made between the A431 cell line vs. A431 cells treated with anti-E-cadherin antibody and TGFα to induce a mesenchymal phenotype. Using Ingenuity pathway analysis to characterize the proteome functions, cancer-derived exosomes from the treated cells, compared to untreated cells, demonstrated a marked enrichment in proteins involved in cellular growth and proliferation, cell-to-cell signaling and interaction, cellular movement, and cellular morphology. Integrin signaling, specifically, was involved in 22% of the upregulated proteins, consistent with alterations underlying EMT. This is in contrast to the cellular proteomes of the respective cells lines, which demonstrated minimal if any difference in these functions, suggesting packaging of exosomes occurs in a regulated process. A similar distribution proteome was demonstrated in hypoxia-induced A431 cells, with particular functional clustering in the domains of cellular adhesion and cell-cell junctions [35]. This suggests the exosomal proteome of cancer cells that have undergone mesenchymal transition may reflect their hypoxic nature, or that hypoxia may contribute to EMT in cancer cells [34,36,37].

5. Exosomal Stimulation of Angiogenesis

Angiogenesis naturally occurs as a response to pro-angiogenic factors during development and wound healing, but angiogenic pathways may be exploited in the setting of malignancy [38,39]. Gaining access to the host vascular system and establishing a tumor blood supply to supply nutrients, oxygen, and growth factors are rate-limiting with regard to tumor progression, with neovascularization enhancing survival, tumor growth, and dissemination [40,41]. Further, EMT-induced angiogenesis has been shown to be critical in the increase in tumorigenicity of cells undergoing EMT [42].

The role of exosomes in tumor-associated angiogenesis has been well established. In a report by Mineo and coworkers, exosomes isolated from the chronic myeloid leukemia (CML) K562 cell line were shown to induce neovascularization [43]. In vitro, upon application of cancer-derived exosomes to human umbilical endothelial cells (HUVECs), the exosomes stimulated angiotube formation compared to control, with exosomal internalization by the endothelial cells and perinuclear localization during tube formation. An in vivo murine matrigel plug model of angiogenesis demonstrated that cancer-derived exosomes stimulated vascularization to a degree greater than the negative control and equivalent to the positive control. This stimulation of angiogenesis by the cancer-derived exosomes was found to be mediated in a Src-dependent fashion, as dasatinib, a known Src-family kinase inhibitor, blocked the endothelial cell response to cancer exosomes. Further, exosome exposure caused focal adhesion kinase (FAK) phosphorylation, a process that has been shown to be required for EMT [44].

In high-grade bladder cancer, tumor-derived exosomes have similarly been shown to promote angiogenesis as demonstrated by tube formation on the HUVEC assay [14]. This was seen with both cell-line derived and patient-derived bladder cancer exosomal isolations. Beckham and coworkers further elucidated that exosomal EDIL-3, a protein with a known role in angiogenesis and abundantly present in cancer-derived exosomes of both bladder cancer cell lines and bladder cancer patient samples, was necessary to induce tube formation. When cell lines were stably transfected with silencing EDIL-3, exosomal EDIL-3 expression decreased significantly and tube formation was no different than the control. The studies by Mineo and Beckham have regulation of the Src-family of proteins in common, as EDIL-3 is known to activate the FAK-Src-AKT pathway in hepatocellular carcinoma, which may contribute to the explanation for the relationship between cancer-derived exosomes and angiogenesis [45].

Hsu et al. demonstrated similar findings with lung cancer-derived exosomes expressed under hypoxic conditions, which demonstrated enhanced angiogenesis in HUVECs via miRNA delivery [46]. Upon in vitro application of cancer-derived exosomes, increased tube formation was demonstrated compared to HUVECs exposed to normal lung exosomes under both hypoxic and normoxic conditions. In vivo murine matrigel plug models similarly demonstrated increased angiogenesis with cancer-derived exosome application. In this study, *miR-23a*, an miRNA upregulated in hypoxia in lung cancer cells, was identified in the cancer-derived exosomes. *miR-23a* levels were increased in the HUVECs in the in vitro assay following exosome application, and angiogenesis was inhibited in the in vivo model when an *miR-23a* inhibitor was applied concurrent to cancer-derived exosomes. When circulating exosomes were collected from lung cancer patients and healthy control patients' sera, the cancer patient exosomes demonstrated higher levels of *miR-23a*. When applied in vitro to HUVECs an increase in tube formation was shown. The authors demonstrated that the hypoxia-induced cancer-derived exosomes contained *miR-23a*, which when applied to cells in turn caused the downregulation of PDH1/2 and the resultant upregulation of HIF-1α, an independent promoter of EMT [47].

In renal cell carcinoma, Zhang and colleagues describe the application of cancer-derived exosomes on HUVECs with resultant increased tubularization, as well as an increased HUVEC expression of vascular endothelial growth factor (VEGF) mRNA and protein, a well-known promoter of angiogenesis [48]. Ekström and coworkers, in their work on melanoma-derived exosomes, had similar findings, but they demonstrated that the exosomes carried pro-angiogenic protein VEGF along with IL-6 as intravesical cargo to the epithelial cells [49]. With an application of cancer-derived exosomes to mouse endothelial cell line MS1, tube formation was induced, while exosome-depleted cancer cell culture supernatant did not have any effect. Of note, Ekström described the stimulation of exosome release from melanoma cells in vitro with the application of recombinant-WNT5A, as well as the suppression of exosome-release with the application of WNT5A-silencing RNA. High levels of WNT5A has been implicated in poor prognosis in advanced melanoma patients, and may hypothetically be related to stimulation of exosomal release. This exosome-release effect has been similarly documented with WNT3A treatment of rat microglial cells, and may point to the WNT-family protein regulation of exosome release and signaling [50].

Together, the above studies represent multiple avenues in which cancer-derived exosomes exert pro-angiogenic effects on epithelial cells. Cancer-derived exosomes internalize into endothelial cells to exert this effect, and carry miRNA and protein cargo to affect angiogenesis by a variety of mechanisms. Additional characterization of these effects, in particular the route of signaling (e.g., via receptor-mediated signaling or exosomal uptake and intracellular delivery of exosomal contents) and the pathways involved will elucidate the mechanisms of exosomal-mediated angiogenesis as well as possible therapeutic targets. Further, research is needed to examine the interplay between cancer-derived exosomes, EMT, and angiogenesis to better understand their relationship in promoting tumorigenicity.

6. Exosomal Enhancement of Cellular Migration and Invasion

Central to progression of malignancy to advanced disease and metastasis is development of cellular migration and invasion [51–53]. Our lab has demonstrated that not only do bladder cancer-derived exosomes increase expression of mesenchymal markers in normal uroepithelial cells, but also alter motility and invasiveness [2]. Following application of muscle-invasive bladder cancer derived exosomes to uroepithelial cells plated on Collagen IV-coated chamber glass, live cell imaging demonstrated increased distance travelled from the origin compared to negative control. Further, in a Transwell system test, bladder cancer-derived exosome application significantly enhanced both migration and invasion of uroepithelial cells.

In prostate cancer, Ramteke and colleagues demonstrated that prostate cancer-derived exosomes, specifically those under hypoxic conditions, enhance migration and invasion [17]. Using the PC3 cell line, prostate cancer cells were grown to confluence for a wound healing assay. With comparable initial wound sizes, cells grown in the presence of cancer-derived exosomes, specifically exosomes derived in the setting of hypoxia, demonstrated significantly increased cell migration and wound healing, with near complete closure of the wound. Examination of the cellular membranes revealed that exosome application decreased membrane E-cadherin expression, which is important in cell-to-cell adhesion, suggesting a pathway of the increased motility and invasiveness. Franzen et al. similarly demonstrated a decrease in cellular E-cadherin expression with the application of cancer-derived exosomes [2].

Breast cancer cell lines of increasing malignant potential have been demonstrated to release exosomes with increasing potency in promoting cell migration and invasion [54]. Harris and colleagues isolated cancer-derived exosomes from MCF-7, MCF-7 transfected with GFP-Rab27b-expressing plasmids, and MDA-MB-231 breast cancer cell lines, with low, intermediate, and high metastatic potential, respectively. When applied to several wound closure assays, the exosomes derived from moderate- and high-metastatic potential cells induced the greatest degree of cell migration and wound closure, while the exosomes from low metastatic potential cells demonstrated only a mildly increased effect compared to control. These exosomes were found to have increased protein expression of EMT promoters, including HSP90 and vimentin [55,56]. Similarly, lung cancer-derived exosomes from highly metastatic cells have also been found to increase vimentin and N-cadherin expression and decrease E-cadherin and ZO-1 expression when applied to epithelial cells, as well as increase motility and invasiveness [57]. Comparing lung-cancer derived exosomes from highly metastatic and non-highly metastatic lines, relative mRNA expression of mesenchymal markers was found to be significantly greater in the highly metastatic cancer-derived exosomes. Hsu demonstrated similar results with hypoxia-induced lung cancer-derived exosomes, which decreased cellular expression of tight junction protein ZO-1 in response to exosomal *miR-23a*, increasing permeability and transendothelial migration of HUVEC cells [46].

In breast and colon cancer, cancer-derived exosomes containing amphiregulin, an epidermal growth factor receptor ligand, have been implicated in cellular signaling and increased invasiveness [58,59]. Higginbotham et al. found that colon-cancer-derived exosomes contained several epidermal growth factor receptor ligands, including amphiregulin. Exosomes with amphiregulin were applied to either epithelial cells or breast cancer cells and demonstrated a several-fold increase in invasion on in vitro matrigel invasion assay.

Degradation of the extracellular matrix is another mechanism by which tumor cell migration and invasion is enhanced. Atay and coworkers elucidated the role of exosomes in this process in their work in gastrointestinal stromal tumors (GIST), in which cancer-derived exosomes alter nearby stromal cells to affect the tumor microenvironment [60]. In this study, oncogenic protein tyrosine kinase (KIT) was found to be packaged in cancer-derived exosomes. Upon co-culture of cancer-derived exosomes with a human uterine leiomyomatous smooth-muscle cell line, exosomes were taken up and cytosolic and membranous KIT expression was detected in 100% of observed cells within 24 h. Exposure to cancer-derived exosomes significantly increased cellular vimentin mRNA and protein expression. More impressive was the downstream KIT activity demonstrated in the previously KIT-naïve cells. Ultimately, upregulation of matrix metalloproteinase-1 was induced in the human uterine leiomyomatous smooth-muscle cells, which permitted an increased number of invasive GIST cells upon co-culture compared to cells without exosome exposure.

Finally, exosomes have been shown to transfer miRNA from an aggressive oral cancer cell line to induce cell motility and growth in the work by Sakha and colleagues [61]. Following recipient cell uptake of cancer-derived exosomes (from a highly metastatic oral cancer cell line, HOC313-LM), low malignant potential oral cancer cell lines demonstrate rapid proliferation compared to the non-exosome treated control group. When studied in a Transwell migration assay, exosome-treated recipient cells, compared to controls, demonstrated increased invasion. Analysis of cancer-derived exosomal miRNA content demonstrated upregulation of several miRNA with known oncogenic potential. When low malignant potential oral cancer cells lines were incubated directly with cancer-derived exosomes, or in a system with a 1 μm membrane to separate cell lines (thus allowing only transfer of acellular material including exosomes), an increase in miRNA expression was demonstrated on qRT-PCR.

As with the multiple examples above, it is clear that cancer-derived exosomes promote increasing cellular aggressiveness, with escalation of cellular motility, migration, and invasion. It has also been postulated that exosomes may function to concentrate proteins or RNA for signaling and transformation of nearby cells, as an explanation for the "field effect" phenomenon seen in several malignancies [59,62]. When specifically investigated, cancer-derived exosomes are demonstrated to not only carry, but also induce EMT with expression of mesenchymal markers in recipient cells, contributing to the progression to a more aggressive phenotype.

7. Exosomal Establishment of Pre-Metastatic Niche and Promotion of Metastasis

The idea that intercellular signaling is critical for the development of metastatic disease in the form of a pre-metastatic niche, as well as an organ-specific tropism of metastasis dependent on cancer type, is well established [19]. Recent development in the investigation of cancer-derived exosomes has established them as a key component of this process. Park and colleagues performed a functional and ontologic analysis of cancer-derived exosomal proteins and determined that the molecular and biological processes that occurred in the highest proportion were those associated with metastasis including cellular adhesion and extracellular matrix-receptor interaction [35].

Experimentally, this process has been eloquently described in the work of Costa-Silva and colleagues [63]. In their study, mice were treated first with injections of pancreatic ductal adenocarcinoma-derived exosomes from cell line PAN02, a cell line known for metastasis to the liver. Following three weeks of exosomal treatment, both treatment and control mice received an intrasplenic injection of PAN02 cells. At 21 days following intrasplenic injection, an increase in the incidence in liver metastasis was seen in mice treated with cancer-derived exosomes compared to PBS or normal pancreas-derived exosomal controls. To determine what cells took up the cancer-derived exosomes to promote metastasis, fluorescently labeled exosomes were then injected and liver cells examined by flow cytometry. Cancer-derived exosomes were preferentially taken up by Kupffer cells as opposed to other hepatocytes. Further, labeled normal pancreas-derived exosomes were not incorporated preferentially into any particular cell type. Finally, the authors elucidated that cancer-derived exosomal stimulation of Kupffer cells greatly upregulated TGFβ expression, a well-established inducer of EMT, resulting in increased fibronectin expression by hepatic stellate cells and ultimately development of the liver metastatic niche in this model [64]. This work indicates that cancer-derived exosomes are able to target a specific recipient cell type, as well as promote the development of metastatic disease in the target organ.

Grange and coworkers demonstrated a similar process with cancer-derived exosomes isolated from a renal cell carcinoma line expressing mesenchymal stem cell marker CD105 [65]. When cancer-derived exosomes were injected intravenously into severe combined immunodeficiency (SCID) mice for five days preceding injection of renal tumor cells, the incidence of lung metastasis was significantly greater compared to vehicle or CD105⁻ expressing tumor cell controls. Analysis of the exosomes revealed upregulation of miRNAs associated with angiogenesis. In vitro application of the CD105⁺ cancer-derived exosomes stimulated HUVEC to organize into capillary-like structures on Matrigel, while also increasing the invasiveness on Transwell assays.

In breast cancer miRNA *miR-105*, which is expressed and secreted by metastatic breast cancer cells, has been shown to be transferred by cancer-derived exosomes and to be an important factor in the development of metastasis [66]. In this study, mice were treated in vitro with cancer-derived exosomes isolated from breast cancer cell lines expressing high levels of *miR-105*. Following this pretreatment, mice were injected with breast cancer cells; subsequently, levels of lung and brain metastases were significantly increased compared to PBS and low *miR-105* exosomal controls. The mechanism for this interaction was explored, and exosomal delivery of *miR-105* to recipient cells was shown to downregulate tight-junction protein ZO-1 expression, resulting in increased permeability as measured in vitro. When cancer-derived exosomes were applied to human microvascular epithelial cells in vitro, again an increase in cellular migration as well as invasion in Transwell assay was demonstrated. It is interesting that, with the application of anti-*miR-105*, in vitro ZO-1 expression was increased and migration suppressed, while in vivo a more aggressive phenotype was inhibited, with decreased primary tumor volume and suppressed distant metastases.

Finally, Liu and colleagues detailed an investigation into the interaction between cancer-derived exosomes and Toll-like receptor 3 (TLR) in establishing a pre-metastatic niche in lung cancer [67]. It has been described that the stimulation of Toll-like receptors can be associated with multiple tumor processes, including growth and metastasis [68]. In their initial experiments, Liu et al. demonstrate that, in Tlr3⁻/⁻ mice, there was a significant reduction in spontaneous metastasis compared to wild-type mice. Further, when wild-type mice were injected with tumor-derived exosomes, there was an upregulation of TLR expression via an NF-κB- and MAPK-mediated pathways, an increase of mesenchymal markers (including fibronectin), as well as an increase in resultant metastasis. The NF-κB pathway has been demonstrated to be essential to the induction and maintenance of EMT, while the MAPK pathway appears to involve phosphorylation of ERK, JNK, or p38 MAPK, all of which occurred with the application of cancer-derived exosomes as above [67,69,70]. In contrast, when Tlr3$^{-/-}$ mice were injected with cancer-derived exosomes, the level of spontaneous metastasis was comparable to negative controls. Investigation of this interaction between TLR3 and exosomes demonstrated that, when both are present, an upregulation of cytokines is apparent; however, without exosomes, or in the setting of silencing of TLR3, the upregulation in mesenchymal markers, and ultimately metastasis, does not occur.

8. Conclusions

Epithelial-to-mesenchymal transition represents a key process in which cancer cells develop a more aggressive, motile, less adhesive, and invasive phenotype. Cancer-derived exosomes represent methods of both local and distant intercellular signaling (Figure 2). Cancer-derived exosomes induce tumor cells and normal epithelial cells to acquire a more aggressive phenotype, with increased angiogenesis, disruption of tight junctions, increased motility, increased mesenchymal markers, and, in the setting of malignancies in which organs are in contact with a lumen (e.g., bladder and colon), may represent the explanation for the "field effect" that is seen clinically. Distantly, exosomes demonstrate preferential targeting of recipient cell types via surface proteins, explaining the tropism of metastasis inherent to specific cancer types, and in this setting establish a pre-metastatic niche which has been shown in several models [71].

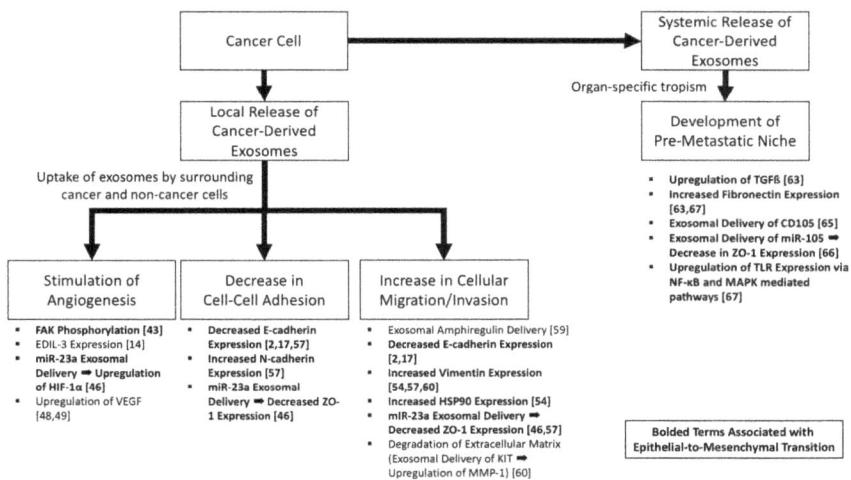

Figure 2. The role of cancer-derived exosomes in increasing tumorigenicity and epithelial-to-mesenchymal transition. Pathways associated with epithelial-to-mesenchymal transition are bolded.

The understanding of the role exosomes play in EMT remain in their infancy, and much study is needed. Directed studies into the effects of cancer-derived exosome application to target cells is needed to more fully understand the mechanisms behind migration, the loss in cell-to-cell adhesion, and the development of stem properties, which at this time are lacking. Further understanding of exosomal expression and intercellular signaling will improve the understanding of epithelial-to-mesenchymal transition as well as lead to novel diagnostic and therapeutic modalities targeting cancer progression.

Acknowledgments: No funding was received for this work.

Author Contributions: All authors contributed to the literature review, drafting, and critical revisions of this work.

Conflicts of Interest: The authors declare no conflict of interest.

References

1. Kalluri, R.; Weinberg, R.A. The basics of epithelial-mesenchymal transition. *J. Clin. Investig.* **2009**, *119*, 1420–1428. [CrossRef] [PubMed]
2. Franzen, C.A.; Blackwell, R.H.; Todorovic, V.; Greco, K.A.; Foreman, K.E.; Flanigan, R.C.; Kuo, P.C.; Gupta, G.N. Urothelial cells undergo epithelial-to-mesenchymal transition after exposure to muscle invasive bladder cancer exosomes. *Oncogenesis* **2015**, *4*, e163. [CrossRef] [PubMed]
3. Nieto, M.A. The ins and outs of the epithelial to mesenchymal transition in health and disease. *Annu. Rev. Cell Dev. Biol.* **2011**, *27*, 347–376. [CrossRef] [PubMed]
4. Weber, C.E.; Li, N.Y.; Wai, P.Y.; Kuo, P.C. Epithelial-mesenchymal transition, TGF-β, and osteopontin in wound healing and tissue remodeling after injury. *J. Burn Care Res.* **2012**, *33*, 311–318. [CrossRef] [PubMed]
5. Klymkowsky, M.W.; Savagner, P. Epithelial-mesenchymal transition. *Am. J. Pathol.* **2009**, *174*, 1588–1593. [CrossRef] [PubMed]
6. Kothari, A.N.; Mi, Z.; Zapf, M.; Kuo, P.C. Novel clinical therapeutics targeting the epithelial to mesenchymal transition. *Clin. Transl. Med.* **2014**, *3*. [CrossRef] [PubMed]
7. Akers, J.C.; Gonda, D.; Kim, R.; Carter, B.S.; Chen, C.C. Biogenesis of extracellular vesicles (EV): Exosomes, microvesicles, retrovirus-like vesicles, and apoptotic bodies. *J. Neurooncol.* **2013**, *113*, 1–11. [CrossRef] [PubMed]
8. Record, M.; Carayon, K.; Poirot, M.; Silvente-Poirot, S. Exosomes as new vesicular lipid transporters involved in cell-cell communication and various pathophysiologies. *Biochim. Biophys. Acta* **2014**, *1841*, 108–120. [CrossRef] [PubMed]
9. Henderson, M.C.; Azorsa, D.O. The genomic and proteomic content of cancer cell-derived exosomes. *Front. Oncol.* **2012**, *2*. [CrossRef] [PubMed]
10. Valadi, H.; Ekström, K.; Bossios, A.; Sjöstrand, M.; Lee, J.J.; Lötvall, J.O. Exosome-mediated transfer of mRNAs and microRNAs is a novel mechanism of genetic exchange between cells. *Nat. Cell Biol.* **2007**, *9*, 654–659. [CrossRef] [PubMed]
11. Denzer, K.; Kleijmeer, M.J.; Heijnen, H.F.; Stoorvogel, W.; Geuze, H.J. Exosome: From internal vesicle of the multivesicular body to intercellular signaling device. *J. Cell Sci.* **2000**, *113*, 3365–3374. [PubMed]
12. Iero, M.; Valenti, R.; Huber, V.; Filipazzi, P.; Parmiani, G.; Fais, S.; Rivoltini, L. Tumour-released exosomes and their implications in cancer immunity. *Cell Death Differ.* **2008**, *15*, 80–88. [CrossRef] [PubMed]
13. Valenti, R.; Huber, V.; Iero, M.; Filipazzi, P.; Parmiani, G.; Rivoltini, L. Tumor-released microvesicles as vehicles of immunosuppression. *Cancer Res.* **2007**, *67*, 2912–2915. [CrossRef] [PubMed]
14. Beckham, C.J.; Olsen, J.; Yin, P.-N.; Wu, C.-H.; Ting, H.-J.; Hagen, F.K.; Scosyrev, E.; Messing, E.M.; Lee, Y.-F. Bladder cancer exosomes contain EDIL-3/Del1 and facilitate cancer progression. *J. Urol.* **2014**, *192*, 583–592. [CrossRef] [PubMed]
15. Yang, L.; Wu, X.-H.; Wang, D.; Luo, C.-L.; Chen, L.-X. Bladder cancer cell-derived exosomes inhibit tumor cell apoptosis and induce cell proliferation in vitro. *Mol. Med. Rep.* **2013**, *8*, 1272–1278. [PubMed]
16. Jeppesen, D.K.; Nawrocki, A.; Jensen, S.G.; Thorsen, K.; Whitehead, B.; Howard, K.A.; Dyrskjøt, L.; Ørntoft, T.F.; Larsen, M.R.; Ostenfeld, M.S. Quantitative proteomics of fractionated membrane and lumen exosome proteins from isogenic metastatic and nonmetastatic bladder cancer cells reveal differential expression of EMT factors. *Proteomics* **2014**, *14*, 699–712. [CrossRef] [PubMed]
17. Ramteke, A.; Ting, H.; Agarwal, C.; Mateen, S.; Somasagara, R.; Hussain, A.; Graner, M.; Frederick, B.; Agarwal, R.; Deep, G. Exosomes secreted under hypoxia enhance invasiveness and stemness of prostate cancer cells by targeting adherens junction molecules. *Mol. Carcinog.* **2015**, *54*, 554–565. [CrossRef] [PubMed]
18. Vella, L.J. The emerging role of exosomes in epithelial-mesenchymal-transition in cancer. *Front. Oncol.* **2014**, *4*. [CrossRef] [PubMed]
19. Okuda, R.; Sekine, K.; Hisamatsu, D.; Ueno, Y.; Takebe, T.; Zheng, Y.-W.; Taniguchi, H. Tropism of cancer stem cells to a specific distant organ. *In Vivo* **2014**, *28*, 361–365. [PubMed]

20. Franzen, C.A.; Blackwell, R.H.; Foreman, K.E.; Kuo, P.C.; Flanigan, R.C.; Gupta, G.N. Urinary exosomes: The potential for biomarker utility, intercellular signaling and therapeutics in urological malignancy. *J. Urol.* **2016**, *195*, 1331–1339. [CrossRef] [PubMed]
21. Lopez-Verrilli, M.A.; Court, F.A. Exosomes: Mediators of communication in eukaryotes. *Biol. Res.* **2013**, *46*, 5–11. [CrossRef] [PubMed]
22. Qin, W.; Tsukasaki, Y.; Dasgupta, S.; Mukhopadhyay, N.; Ikebe, M.; Sauter, E.R. Exosomes in human breast milk promote EMT. *Clin. Cancer Res.* **2016**, *22*, 4517–4524. [CrossRef] [PubMed]
23. Théry, C.; Amigorena, S.; Raposo, G.; Clayton, A. Isolation and characterization of exosomes from cell culture supernatants and biological fluids. *Curr. Protoc. Cell Biol.* **2006**. [CrossRef]
24. Alvarez, M.L.; Khosroheidari, M.; Kanchi Ravi, R.; DiStefano, J.K. Comparison of protein, microRNA, and mRNA yields using different methods of urinary exosome isolation for the discovery of kidney disease biomarkers. *Kidney Int.* **2012**, *82*, 1024–1032. [CrossRef] [PubMed]
25. Blackwell, R.H.; Franzen, C.A.; Flanigan, R.C.; Kuo, P.C.; Gupta, G.N. The untapped potential of urine shed bladder cancer exosomes: Biomarkers, signaling, and therapeutics. *Bladder* **2014**, *1*, e7. [CrossRef]
26. Sokolova, V.; Ludwig, A.-K.; Hornung, S.; Rotan, O.; Horn, P.A.; Epple, M.; Giebel, B. Characterisation of exosomes derived from human cells by nanoparticle tracking analysis and scanning electron microscopy. *Colloids Surf. B Biointerfaces* **2011**, *87*, 146–150. [CrossRef] [PubMed]
27. Lötvall, J.; Hill, A.F.; Hochberg, F.; Buzás, E.I.; Di Vizio, D.; Gardiner, C.; Gho, Y.S.; Kurochkin, I.V.; Mathivanan, S.; Quesenberry, P.; et al. Minimal experimental requirements for definition of extracellular vesicles and their functions: A position statement from the International Society for Extracellular Vesicles. *J. Extracell. Vesicles* **2014**, *3*, 26913. [CrossRef] [PubMed]
28. Blackwell, R.H.; Franzen, C.A.; Gupta, G.N. Exosomes: An evolving source of urinary biomarkers and an up-and-coming therapeutic delivery vehicle. *Transl. Cancer Res.* **2017**, *6*, S226–S228. [CrossRef]
29. McGough, I.J.; Vincent, J.-P. Exosomes in developmental signalling. *Development* **2016**, *143*, 2482–2493. [CrossRef] [PubMed]
30. Montecalvo, A.; Larregina, A.T.; Shufesky, W.J.; Stolz, D.B.; Sullivan, M.L.G.; Karlsson, J.M.; Baty, C.J.; Gibson, G.A.; Erdos, G.; Wang, Z.; et al. Mechanism of transfer of functional microRNAs between mouse dendritic cells via exosomes. *Blood* **2012**, *119*, 756–766. [CrossRef] [PubMed]
31. Tian, T.; Zhu, Y.-L.; Zhou, Y.-Y.; Liang, G.-F.; Wang, Y.-Y.; Hu, F.-H.; Xiao, Z.-D. Exosome uptake through clathrin-mediated endocytosis and macropinocytosis and mediating miR-21 delivery. *J. Biol. Chem.* **2014**, *289*, 22258–22267. [CrossRef] [PubMed]
32. Franzen, C.A.; Simms, P.E.; Van Huis, A.F.; Foreman, K.E.; Kuo, P.C.; Gupta, G.N. Characterization of uptake and internalization of exosomes by bladder cancer cells. *Biomed. Res. Int.* **2014**, *2014*. [CrossRef] [PubMed]
33. Subra, C.; Grand, D.; Laulagnier, K.; Stella, A.; Lambeau, G.; Paillasse, M.; De Medina, P.; Monsarrat, B.; Perret, B.; Silvente-Poirot, S.; et al. Exosomes account for vesicle-mediated transcellular transport of activatable phospholipases and prostaglandins. *J. Lipid Res.* **2010**, *51*, 2105–2120. [CrossRef] [PubMed]
34. Garnier, D.; Magnus, N.; Meehan, B.; Kislinger, T.; Rak, J. Qualitative changes in the proteome of extracellular vesicles accompanying cancer cell transition to mesenchymal state. *Exp. Cell Res.* **2013**, *319*, 2747–2757. [CrossRef] [PubMed]
35. Park, J.E.; Tan, H.S.; Datta, A.; Lai, R.C.; Zhang, H.; Meng, W.; Lim, S.K.; Sze, S.K. Hypoxic tumor cell modulates its microenvironment to enhance angiogenic and metastatic potential by secretion of proteins and exosomes. *Mol. Cell. Proteom.* **2010**, *9*, 1085–1099. [CrossRef] [PubMed]
36. Copple, B.L. Hypoxia stimulates hepatocyte epithelial to mesenchymal transition by hypoxia-inducible factor and transforming growth factor-beta-dependent mechanisms. *Liver Int.* **2010**, *30*, 669–682. [CrossRef] [PubMed]
37. Higgins, D.F.; Kimura, K.; Bernhardt, W.M.; Shrimanker, N.; Akai, Y.; Hohenstein, B.; Saito, Y.; Johnson, R.S.; Kretzler, M.; Cohen, C.D.; et al. Hypoxia promotes fibrogenesis in vivo via HIF-1 stimulation of epithelial-to-mesenchymal transition. *J. Clin. Investig.* **2007**, *117*, 3810–3820. [CrossRef] [PubMed]
38. Weis, S.M.; Cheresh, D.A. Tumor angiogenesis: Molecular pathways and therapeutic targets. *Nat. Med.* **2011**, *17*, 1359–1370. [CrossRef] [PubMed]
39. Carmeliet, P.; Jain, R.K. Molecular mechanisms and clinical applications of angiogenesis. *Nature* **2011**, *473*, 298–307. [CrossRef] [PubMed]

40. Bergers, G.; Benjamin, L.E. Angiogenesis: Tumorigenesis and the angiogenic switch. *Nat. Rev. Cancer* **2003**, *3*, 401–410. [CrossRef] [PubMed]

41. Ribeiro, M.F.; Zhu, H.; Millard, R.W.; Fan, G.-C. Exosomes function in pro- and anti-angiogenesis. *Curr. Angiogenes.* **2013**, *2*, 54–59. [PubMed]

42. Fantozzi, A.; Gruber, D.C.; Pisarsky, L.; Heck, C.; Kunita, A.; Yilmaz, M.; Meyer-Schaller, N.; Cornille, K.; Hopfer, U.; Bentires-Alj, M.; et al. VEGF-mediated angiogenesis links EMT-induced cancer stemness to tumor initiation. *Cancer Res.* **2014**, *74*, 1566–1575. [CrossRef] [PubMed]

43. Mineo, M.; Garfield, S.H.; Taverna, S.; Flugy, A.; De Leo, G.; Alessandro, R.; Kohn, E.C. Exosomes released by K562 chronic myeloid leukemia cells promote angiogenesis in a Src-dependent fashion. *Angiogenesis* **2012**, *15*, 33–45. [CrossRef] [PubMed]

44. Cicchini, C.; Laudadio, I.; Citarella, F.; Corazzari, M.; Steindler, C.; Conigliaro, A.; Fantoni, A.; Amicone, L.; Tripodi, M. TGFbeta-induced EMT requires focal adhesion kinase (FAK) signaling. *Exp. Cell Res.* **2008**, *314*, 143–152. [CrossRef] [PubMed]

45. Feng, M.-X.; Ma, M.-Z.; Fu, Y.; Li, J.; Wang, T.; Xue, F.; Zhang, J.-J.; Qin, W.-X.; Gu, J.-R.; Zhang, Z.-G.; et al. Elevated autocrine EDIL3 protects hepatocellular carcinoma from anoikis through RGD-mediated integrin activation. *Mol. Cancer* **2014**, *13*. [CrossRef] [PubMed]

46. Hsu, Y.-L.; Hung, J.-Y.; Chang, W.-A.; Lin, Y.-S.; Pan, Y.-C.; Tsai, P.-H.; Wu, C.-Y.; Kuo, P.-L. Hypoxic lung cancer-secreted exosomal miR-23a increased angiogenesis and vascular permeability by targeting prolyl hydroxylase and tight junction protein ZO-1. *Oncogene* **2017**. [CrossRef] [PubMed]

47. Zhang, W.; Shi, X.; Peng, Y.; Wu, M.; Zhang, P.; Xie, R.; Wu, Y.; Yan, Q.; Liu, S.; Wang, J. HIF-1α promotes epithelial-mesenchymal transition and metastasis through direct regulation of ZEB1 in colorectal cancer. *PLoS ONE* **2015**, *10*. [CrossRef] [PubMed]

48. Zhang, L.; Wu, X.; Luo, C.; Chen, X.; Yang, L.; Tao, J.; Shi, J. The 786-0 renal cancer cell-derived exosomes promote angiogenesis by downregulating the expression of hepatocyte cell adhesion molecule. *Mol. Med. Rep.* **2013**, *8*, 272–276. [PubMed]

49. Ekström, E.J.; Bergenfelz, C.; von Bülow, V.; Serifler, F.; Carlemalm, E.; Jönsson, G.; Andersson, T.; Leandersson, K. WNT5A induces release of exosomes containing pro-angiogenic and immunosuppressive factors from malignant melanoma cells. *Mol. Cancer* **2014**, *13*. [CrossRef] [PubMed]

50. Hooper, C.; Sainz-Fuertes, R.; Lynham, S.; Hye, A.; Killick, R.; Warley, A.; Bolondi, C.; Pocock, J.; Lovestone, S. Wnt3a induces exosome secretion from primary cultured rat microglia. *BMC Neurosci.* **2012**, *13*. [CrossRef] [PubMed]

51. Greening, D.W.; Gopal, S.K.; Mathias, R.A.; Liu, L.; Sheng, J.; Zhu, H.-J.; Simpson, R.J. Emerging roles of exosomes during epithelial-mesenchymal transition and cancer progression. *Semin. Cell Dev. Biol.* **2015**, *40*, 60–71. [CrossRef] [PubMed]

52. Hanahan, D.; Weinberg, R.A. Hallmarks of cancer: The next generation. *Cell* **2011**, *144*, 646–674. [CrossRef] [PubMed]

53. Hendrix, A.; Westbroek, W.; Bracke, M.; De Wever, O. An ex(o)citing machinery for invasive tumor growth. *Cancer Res.* **2010**, *70*, 9533–9537. [CrossRef] [PubMed]

54. Harris, D.A.; Patel, S.H.; Gucek, M.; Hendrix, A.; Westbroek, W.; Taraska, J.W. Exosomes released from breast cancer carcinomas stimulate cell movement. *PLoS ONE* **2015**, *10*. [CrossRef] [PubMed]

55. Liu, C.-Y.; Lin, H.-H.; Tang, M.-J.; Wang, Y.-K. Vimentin contributes to epithelial-mesenchymal transition cancer cell mechanics by mediating cytoskeletal organization and focal adhesion maturation. *Oncotarget* **2015**, *6*, 15966–15983. [CrossRef] [PubMed]

56. Nagaraju, G.P.; Long, T.-E.; Park, W.; Landry, J.C.; Taliaferro-Smith, L.; Farris, A.B.; Diaz, R.; El-Rayes, B.F. Heat shock protein 90 promotes epithelial to mesenchymal transition, invasion, and migration in colorectal cancer. *Mol. Carcinog.* **2015**, *54*, 1147–1158. [CrossRef] [PubMed]

57. Rahman, M.A.; Barger, J.F.; Lovat, F.; Gao, M.; Otterson, G.A.; Nana-Sinkam, P. Lung cancer exosomes as drivers of epithelial mesenchymal transition. *Oncotarget* **2016**, *7*, 54852–54866. [CrossRef] [PubMed]

58. Busser, B.; Sancey, L.; Brambilla, E.; Coll, J.-L.; Hurbin, A. The multiple roles of amphiregulin in human cancer. *Biochim. Biophys. Acta* **2011**, *1816*, 119–131. [CrossRef] [PubMed]

59. Higginbotham, J.N.; Demory Beckler, M.; Gephart, J.D.; Franklin, J.L.; Bogatcheva, G.; Kremers, G.-J.; Piston, D.W.; Ayers, G.D.; McConnell, R.E.; Tyska, M.J.; et al. Amphiregulin exosomes increase cancer cell invasion. *Curr. Biol.* **2011**, *21*, 779–786. [CrossRef] [PubMed]

60. Atay, S.; Banskota, S.; Crow, J.; Sethi, G.; Rink, L.; Godwin, A.K. Oncogenic KIT-containing exosomes increase gastrointestinal stromal tumor cell invasion. *Proc. Natl. Acad. Sci. USA* **2014**, *111*, 711–716. [CrossRef] [PubMed]
61. Sakha, S.; Muramatsu, T.; Ueda, K.; Inazawa, J. Exosomal microRNA miR-1246 induces cell motility and invasion through the regulation of DENND2D in oral squamous cell carcinoma. *Sci. Rep.* **2016**, *6*. [CrossRef] [PubMed]
62. Chai, H.; Brown, R.E. Field effect in cancer-an update. *Ann. Clin. Lab. Sci.* **2009**, *39*, 331–337. [PubMed]
63. Costa-Silva, B.; Aiello, N.M.; Ocean, A.J.; Singh, S.; Zhang, H.; Thakur, B.K.; Becker, A.; Hoshino, A.; Mark, M.T.; Molina, H.; et al. Pancreatic cancer exosomes initiate pre-metastatic niche formation in the liver. *Nat. Cell Biol.* **2015**, *17*, 816–826. [CrossRef] [PubMed]
64. Xu, J.; Lamouille, S.; Derynck, R. TGF-beta-induced epithelial to mesenchymal transition. *Cell Res.* **2009**, *19*, 156–172. [CrossRef] [PubMed]
65. Grange, C.; Tapparo, M.; Collino, F.; Vitillo, L.; Damasco, C.; Deregibus, M.C.; Tetta, C.; Bussolati, B.; Camussi, G. Microvesicles released from human renal cancer stem cells stimulate angiogenesis and formation of lung premetastatic niche. *Cancer Res.* **2011**, *71*, 5346–5356. [CrossRef] [PubMed]
66. Zhou, W.; Fong, M.Y.; Min, Y.; Somlo, G.; Liu, L.; Palomares, M.R.; Yu, Y.; Chow, A.; O'Connor, S.T.F.; Chin, A.R.; et al. Cancer-secreted miR-105 destroys vascular endothelial barriers to promote metastasis. *Cancer Cell* **2014**, *25*, 501–515. [CrossRef] [PubMed]
67. Liu, Y.; Gu, Y.; Han, Y.; Zhang, Q.; Jiang, Z.; Zhang, X.; Huang, B.; Xu, X.; Zheng, J.; Cao, X. Tumor exosomal RNAs promote lung pre-metastatic niche formation by activating alveolar epithelial TLR3 to recruit neutrophils. *Cancer Cell* **2016**, *30*, 243–256. [CrossRef] [PubMed]
68. Pradere, J.-P.; Dapito, D.H.; Schwabe, R.F. The Yin and Yang of Toll-like receptors in cancer. *Oncogene* **2014**, *33*, 3485–3495. [CrossRef] [PubMed]
69. Huber, M.A.; Azoitei, N.; Baumann, B.; Grünert, S.; Sommer, A.; Pehamberger, H.; Kraut, N.; Beug, H.; Wirth, T. NF-kappaB is essential for epithelial-mesenchymal transition and metastasis in a model of breast cancer progression. *J. Clin. Investig.* **2004**, *114*, 569–581. [CrossRef] [PubMed]
70. Gui, T.; Sun, Y.; Shimokado, A.; Muragaki, Y. The roles of mitogen-activated protein kinase pathways in TGF-β-induced epithelial-mesenchymal transition. *J. Signal Transduct.* **2012**, *2012*. [CrossRef] [PubMed]
71. Hoshino, A.; Costa-Silva, B.; Shen, T.-L.; Rodrigues, G.; Hashimoto, A.; Tesic Mark, M.; Molina, H.; Kohsaka, S.; Di Giannatale, A.; Ceder, S.; et al. Tumour exosome integrins determine organotropic metastasis. *Nature* **2015**, *527*, 329–335. [CrossRef] [PubMed]

Review

Epithelial-to-Pericyte Transition in Cancer

Jianrong Lu [1,*] and Anitha K. Shenoy [2]

1   Department of Biochemistry and Molecular Biology, College of Medicine, University of Florida, Gainesville, FL 32610-3633, USA
2   Department of Pharmaceutics and Biomedical Sciences, California Health Sciences University, Clovis, CA 93612, USA; ashenoy@chsu.org
*   Correspondence: jrlu@ufl.edu; Tel.: +1-352-273-8200

Academic Editor: Joëlle Roche
Received: 3 May 2017; Accepted: 30 June 2017; Published: 4 July 2017

Abstract: During epithelial-to-mesenchymal transition (EMT), cells lose epithelial characteristics and acquire mesenchymal properties. These two processes are genetically separable and governed by distinct transcriptional programs, rendering the EMT outputs highly heterogeneous. Our recent study shows that the mesenchymal products generated by EMT often express multiple pericyte markers, associate with and stabilize blood vessels to fuel tumor growth, thus phenotypically and functionally resembling pericytes. Therefore, some EMT events represent epithelial-to-pericyte transition (EPT). The serum response factor (SRF) plays key roles in both EMT and differentiation of pericytes, and may inherently confer the pericyte attributes on EMT cancer cells. By impacting their intratumoral location and cell surface receptor expression, EPT may enable cancer cells to receive and respond to angiocrine factors produced by the vascular niche, and develop therapy resistance.

Keywords: EMT; EPT; SRF; myocardin-related transcription factors (MRTF); pericyte; resistance; vascular niche; angiocrine factors

---

1. Overview of Epithelial-to-Mesenchymal Transition (EMT) in Cancer

Most human cancers are malignancies of epithelial cells. Epithelial cells are tightly connected to each other by multiple types of intercellular junctions, which restrain their mobility. However, through a reprogramming process known as epithelial-to-mesenchymal transition (EMT), epithelial cells can dissolve cell–cell adhesions, reorganize the actin cytoskeleton, and transform into spindle-shaped mesenchymal cells with enhanced migratory and invasive capabilities [1,2]. During EMT, epithelial markers (e.g., the adherens junction protein E-cadherin and the tight junction proteins claudins) are downregulated, while mesenchymal markers (e.g., the adhesion protein N-cadherin, the intermediate filament protein vimentin, fibroblast-specific protein 1 [FSP1], and smooth muscle α-actin [SMA]) are upregulated [1,2]. The EMT program is executed in response to EMT-inducing signals that activate the expression of core transcription factors called EMT-TFs, such as Snail1/2, Zeb1/2, and Twist [1,3]. The EMT-TFs play a central role in driving EMT by directly or indirectly repressing epithelial genes.

The cardinal features of EMT have led to the prevailing hypothesis that EMT is crucial for cancer metastasis [4]. EMT liberates epithelial tumor cells from the surrounding tissue and thus promotes tumor invasion and metastatic spread. In addition, the EMT-TFs possess intrinsic activities to overcome apoptosis and oncogene-induced senescence; therefore, EMT may also foster tumor initiation [5]. Finally, EMT confers stemness-related properties and resistance to conventional radiation and chemotherapy, molecularly targeted therapy, and immunotherapy [4,6,7]. Overall, EMT exhibits multifaceted functions in tumor formation, disease progression, and therapy resistance.

However, the impact of EMT in cancer is still far from fully understood. In fact, recent studies have cast doubts on the actual contribution of EMT to metastasis [8]. In mouse models of breast cancer, lineage-tracing experiments showed that most metastatic cancer cells never activate the promoters of FSP1 and vimentin, two bona fide mesenchymal markers, and moreover, inhibition of EMT does not affect spontaneous lung metastasis formation, suggesting that EMT is dispensable for metastasis [9]. Genetic studies in mouse pancreatic cancer models demonstrated that metastasis development is independent of Snail and Twist [10], but partially requires Zeb1 [11]. Therefore, the functional consequences of EMT are highly context-dependent. Given the wide spectrum of intermediate phases of EMT [6], the significance of EMT in cancer remains to be elucidated. The fates and roles of epithelial tumor cells naturally transitioning to a mesenchymal state in vivo are largely elusive.

2. Epithelial-to-Pericyte Transition (EPT)

In our recent study [12], we conducted fate-mapping experiments to track cancer cells that undergo EMT in tumor xenografts in vivo. Breast epithelial cancer cells stably expressing an inducible form of Snail can be experimentally induced to initiate EMT. Such cells were fluorescently labeled and mixed with a larger number of regular epithelial cancer cells for tumor transplantation. In the resulting tumor xenografts, the majority of induced EMT cancer cells preferentially reside in the perivascular space and are associated with blood vessels, which is reminiscent of pericytes. Multiple mesenchymal cancer cell lines (including Hs578T triple-negative breast cancer cells), which are considered featuring a permanent EMT phenotype, also display similar vascular association in tumor xenografts.

Pericytes are specialized mesenchymal cells that coat and stabilize the endothelium of small blood vessels [13,14]. Pericytes are generally defined based on a combination of mesenchymal morphology, periendothelial location, and expression of multiple pericyte markers [13,14]. EMT cancer cells share a similar gene expression profile with pericytes, and indeed, EMT upregulates multiple pericyte markers in cancer cells in vitro [12]. Most Snail-induced EMT cells in tumor xenografts express the NG2 proteoglycan, which is one of the most commonly referenced pericyte markers and is required for pericyte investment of vasculature [15–17]. Experimental induction of EMT substantially increases the vascular coverage by NG2-expesing cells [12].

Mammary epithelial tumor cells undergo spontaneous EMT in vivo, which is identified by the elongated cell morphology and/or an SMA promoter-driven mesenchymal-specific fluorescent reporter [12]. The majority of spontaneous EMT cells express pericyte markers NG2 and SMA, exhibit close vascular association, and apparently constitute a great proportion of pericytes associated with tumor vasculature [12]. Importantly, depletion of such naturally occurring EMT cancer cells in transplanted tumors strongly diminishes pericyte coverage, impairs vascular integrity, and attenuates tumor growth. The results suggest that EMT enables cancer cells to phenotypically and functionally resemble pericytes, and such cancer-derived pericytes are indispensable for vascular stabilization and sustained tumor growth [12].

During blood vessel maturation, endothelial cells (ECs) secrete platelet-derived growth factor (PDGF), which chemoattracts pericytes that express its cognate receptor, PDGFRβ. This paracrine signaling plays a central role in pericyte recruitment and vascular stabilization [13,14,18]. Once recruited to the abluminal surface of endothelium, pericytes make direct peg-and-socket contact or form adhesion plaques with ECs [19]. Expressed in both ECs and pericytes, N-cadherin establishes adherens junctions that strengthen the interaction between the two cell types [19–21]. PDGFRβ and N-cadherin are critically implicated in the pericyte-endothelium association.

EMT markedly activates the expression of both PDGFRβ and N-cadherin in cancer cells [12]. In cancer, PDGFRβ expression is generally restricted to stromal cells of mesenchymal origin, and is absent in epithelial tumor cells [22]. EMT virtually universally upregulates PDGFRβ expression [12,23–28]. Therefore, unlike epithelial cells, EMT cells are able to respond to EC-secreted chemoattractant PDGF and be recruited to vasculature. On the other hand, expression of N- and E-cadherins is usually mutually exclusive, with E-cadherin primarily expressed in epithelial cells and N-cadherin in mesenchymal cells and ECs [29]. Classical cadherins exhibit preferentially homophilic interactions. Epithelial cells are thus unable to form adherens junctions with ECs. Because the

E-cadherin-to-N-cadherin switch is a hallmark of EMT [30], EMT cells acquire the new capability to associate with ECs through homotypic N-cadherin interactions. Our experimental results suggest that both PDGFRβ and N-cadherin are required for EMT cancer cells to associate with ECs in vitro and in vivo [12]. As the induction of PDGFRβ and N-cadherin is a prevalent feature of EMT, EMT may commonly confer key pericyte properties on epithelial cells, thereby often representing epithelial-to-pericyte transition (EPT).

Taken together, we propose that a small subset of epithelial cancer cells undergo EMT, which allows them to detach from adjoining neighboring cells and migrate within the tumor mass. Moreover, due to acquired expression of PDGFRβ and N-cadherin during EMT, the EMT cells are recruited to vasculature through PDGF-mediated chemotaxis, and subsequently establish intercellular adhesion with ECs through homodimerization of N-cadherin present on the plasma surface of both cell types (Figure 1). EMT cancer cells assume the identity of pericytes and perform pericyte function to stabilize tumor vasculature, thereby improving vascular support for the growth of the whole tumor. In short, the EPT program converts epithelial cancer cells into pericytes to fuel tumor growth.

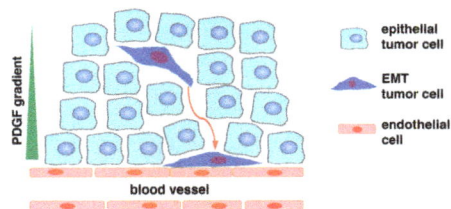

Figure 1. Schematic of epithelial-to-pericyte transition (EPT). In response to microenvironmental stimuli (e.g., hypoxia, transforming growth factor β TGFβ), a subset of carcinoma cells in the tumor mass undergo EMT, and consequently acquire increased motility and invasiveness as well as expression of PDGFRβ, N-cadherin, and other pericyte markers. As endothelial cells express N-cadherin and secrete PDGF, the EMT cancer cells are chemoattracted to vasculature via PDGF paracrine signaling and associate with endothelial cells through N-cadherin-mediated adherens junctions. The EMT cells may also upregulate CXCR4 and be recruited to endothelium in response to stromal-derived factor 1 (SDF1). Due to lack of PDGFRβ and N-cadherin, epithelial cancer cells without EMT are unable to respond to PDGF or interact with endothelium. The EMT cancer cells functionally resemble pericytes to stabilize blood vessels to fuel tumor growth.

3. SRF as a Potential Key Regulator of EPT

It remains largely elusive how EMT cells acquire pericyte properties. Pericyte markers are often mesenchymal markers. During EMT, while the repression of epithelial genes by EMT-TFs has been well understood, activation of the mesenchymal phenotype is much less clear. Loss of the epithelial characteristics and acquisition of the mesenchymal properties usually appear to be coupled and concomitant with each other during EMT in vitro; however, these two events are separable in vivo. For instance, transgenic expression of Snai1 in the mouse epidermis reduces E-cadherin expression and intercellular adhesion; however, mesenchymal markers are not ectopically induced in the epidermal cells [31,32]. In another example, FBXO11 is a ubiquitin ligase for Snai1/2 [33,34]. We generated FBXO11-deficient mutant mice, which showed aberrant Snai1/2 protein accumulation and transcriptionally downregulated E-cadherin expression in the epidermis, but no ectopic induction of mesenchymal markers in the mutant epidermal cells [34]. Furthermore, when lung epithelial cells are exposed to TGFβ, a prominent EMT-inducing signal, E-cadherin is downregulated and N-cadherin is upregulated, and cells undergo evident EMT. However, blocking E-cadherin downregulation does not affect N-cadherin upregulation [35]. Collectively, these observations suggest that suppression of the epithelial state and activation of the mesenchymal phenotype are independent and governed by different regulatory programs.

SMA, one of the reliable markers to characterize the mesenchymal products generated by EMT [1,2], is a well-established transcriptional target of serum response factor (SRF) [36–38]. SRF is a transcription factor that binds to a sequence motif known as CArG box present in many smooth muscle-specific gene promoters, and is a paramount determinant of smooth muscle differentiation [36–38]. Blood vessels are generally composed of two interacting cell types: ECs that form the inner lining of the vessel and mural cells that envelop the surface of the endothelial tube. Both vascular smooth muscle cells (SMCs) and pericytes are mural cells [13]. Vascular SMCs cover larger caliber blood vessels, whereas pericytes enwrap blood capillaries. The two cell types share strong phenotypic similarities [13,14]. Pericytes exhibit a number of characteristics consistent with muscle cell activity and express contractile SMA. There is no single molecular marker known that can be used to unequivocally distinguish pericytes from vascular SMCs, and "the field has generally adopted the view that pericytes belong to the same lineage and category of cells as vascular SMCs" [13,14].

SRF is widely expressed and its transcriptional activity is dependent on its coactivators. Myocardin-related transcription factors (MRTFs), including myocardin, MRTF-A, and MRTF-B, comprise a family of closely related transcriptional coactivators that physically associate with SRF and potently stimulate SRF-dependent gene expression [36–38]. MRTFs are regulated by actin signaling. Cytoplasmic globular actin (G-actin) retains MRTFs in the cytoplasm. Actin polymerization incorporates G-actin into filamentous actin (F-actin), thereby liberating MRTFs to enter the nucleus and interact with SRF. This activates expression of SRF-dependent genes that promote myogenic differentiation and cytoskeletal organization. Gain- and loss-of-function experiments in cultured cells and in mice have shown that MRTFs are indeed critical for vascular SMC gene activation. In addition to SMA, desmin is also a direct transcriptional target gene of SRF [39] and an established marker for pericytes and SMCs [14]. The SRF-MRTF transcriptional program is a central regulator of pericyte/vascular SMC differentiation [36–38].

Intriguingly, accumulating evidence implicates SRF and MRTFs in EMT [1]. Dynamic remodeling of actin cytoskeleton is a major event of EMT. The SRF-MRTF complex is activated by actin filament assembly [37]. The activity of SRF indeed correlates with EMT [40]. TGFβ is probably the best recognized potent inducer of EMT [1]. In renal tubular epithelial cells, TGFβ and disassembly of cell–cell junctions synergistically activate SMA expression and induce EMT [41]. MRTFs are normally localized in the cytoplasm. TGFβ triggers the nuclear translocation of MRTFs especially in epithelial cells with impaired cell–cell contacts, which subsequently act in concert with SRF to drive SMA transcription [41–43]. Ectopic expression of MRTFs in epithelial cells promotes EMT, whereas dominant-negative MRTF or knockdown of MRTF prevents TGFβ-induced EMT and impairs SMA induction [41,42]. In addition, the EMT-TF Zeb1 may also interact with SRF to transactivate the SMA promoter [44]. Overall, SRF and MRTFs critically activate EMT and SMA expression.

We postulate the following model underlying EPT regulation (Figure 2). TGFβ induces the expression of multiple EMT-TFs (primarily through the Smad signaling transducers), which in turn repress epithelial gene expression, causing the loss of epithelial features and enhancing tumor invasion and metastatic dissemination. In parallel, TGF-β signaling promotes the assembly of actin filaments from monomeric G-actin, thereby enabling nuclear import of MRTFs and their subsequent association with SRF in the nucleus to activate mesenchymal/pericyte genes. Mesenchymal acquisition is not required for metastasis [9]. Instead, because it is governed by the SRF-MRTF axis that is crucial for pericyte/SMC differentiation, mesenchymal cells resulting from EMT may inherently acquire myogenic attributes of mural cells and function like pericytes to stabilize blood vessels. Therefore, the SRF-MRTF axis may represent an intrinsic link between EMT cells and pericytes, thus supporting our discovery of EPT. Consistent with this idea, Hs578T triple-negative breast cancer cells, which behave like pericytes in our study [12], highly express multiple mesenchymal/pericyte/SMC markers, such as N-cadherin, PDGFRβ, NG2, SMA, myocardin, and smooth muscle protein 22α (SM22α or transgelin; also an established target of SRF) (Figure 3).

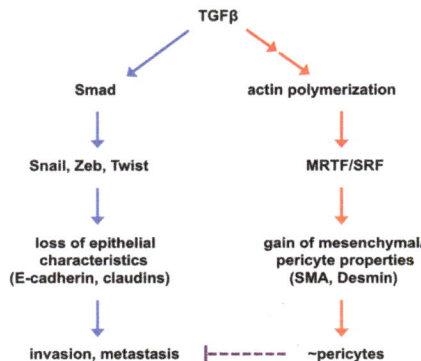

Figure 2. A simplified model of EPT regulation. EMT-inducing signaling causes epithelial cells to lose epithelial characteristics and acquire mesenchymal features. These two processes are largely independent of each other and governed by distinct transcriptional programs. The core EMT-TFs (Snail, Zeb, Twist) repress the epithelial phenotype and may promote tumor invasion and metastasis. Activation of the mesenchymal phenotype is at least in part mediated by the SRF transcription factor and its coactivators MRTFs, which are also central regulators of mural cells. Therefore, mesenchymal cells derived from EMT may inherently resemble pericytes and are able to associate with and stabilize blood vessels (this process is termed EPT). As EMT consists of a broad spectrum of intermediate phases, EPT is one of the EMT outputs. Improved pericyte coverage has been suggested to suppress tumor metastasis; therefore, EPT cancer cell-stabilized vasculature may impede metastatic spread of other cancer cells.

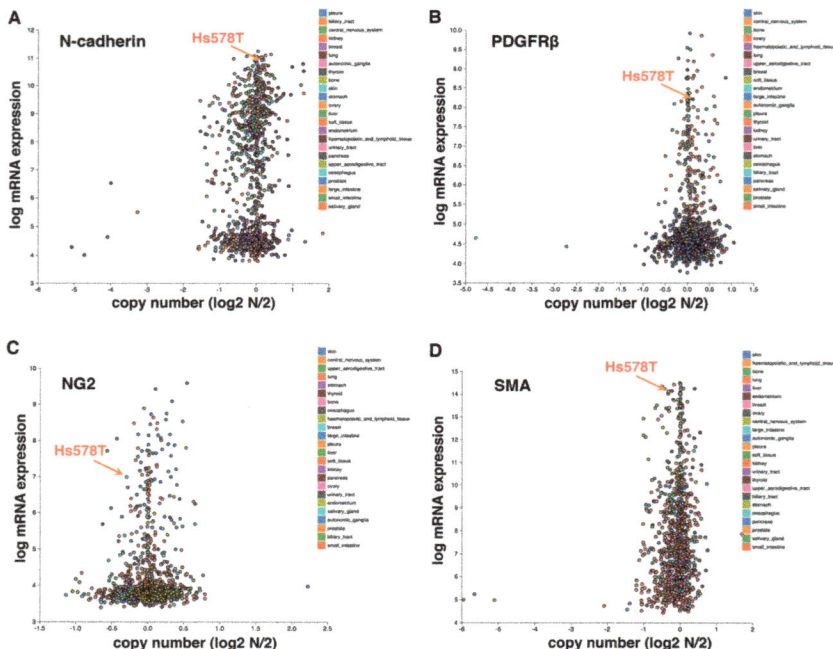

Figure 3. *Cont.*

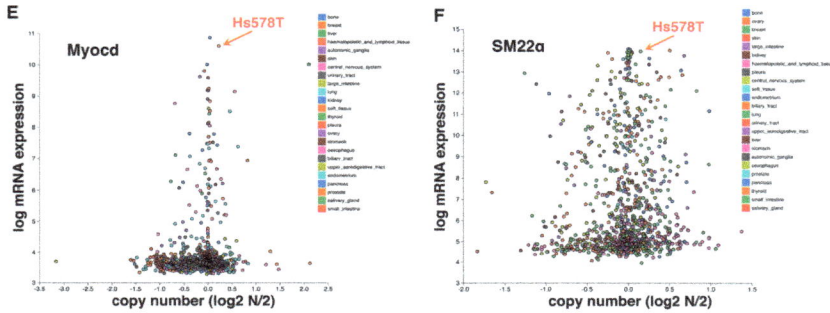

Figure 3. Hs578T triple-negative breast cancer cells, which behave like pericytes [12], express high levels of mesenchymal/mural cell markers N-cadherin (**A**), PDGFRβ (**B**), NG2 (**C**), SMA (**D**), Myocardin (Myocd) (**E**), and SM22α (**F**). Hs578T cells also highly express pericyte markers RGS5 and Angiopoietin 1 (Angpt.1) (data not shown). Gene expression is based on the Cancer Cell Line Encyclopedia (CCLE) database. Each dot represents an established human cancer cell line.

4. EPT in Development and Cancer

Pericytes are heterogeneous. Normal pericytes from different tissues may display varying morphologies, express different markers, and have diverse developmental origins [13,14,18]. Nevertheless, EMT plays a critical role in pericyte development. During embryogenesis, pericytes of the head, thymus, and outflow tract of the aorta are mostly derivatives of neural crest cells of the neuroectoderm [14,45], which is a classical model of EMT [2,4]. Pericytes in the internal viscera (such as gut, liver, lung) originate from the mesothelium that undergoes EMT [14,46–48]. A recent study showed that a subset of cardiac mural cells are derived from endocardial ECs through endothelial-to-mesenchymal transition [49]. Based on their outcomes, such EMT or EMT-like events during embryonic development are essentially EPT.

Pericytes in tumor vasculature may have malignant origins. It was previously observed that certain malignant melanoma and glioma cells occupy the perivascular location and interact with the abluminal surface of blood vessels "without any evidence of intravasation" [50]. While such cancer cells were proposed to migrate along the vascular surface to spread to distant sites (i.e., extravascular metastasis) [51], their vascular association is reminiscent of pericytes. Melanoma is derived from melanocyte transformation and tends to reactivate the EMT program that has enabled their neural crest ancestors to migrate during embryonic development. Malignant gliomas are often mesenchymal [52]. It is plausible that these vascular-associating melanoma and glioma cells resemble post-EMT cells. More recently, multipotent glioma stem cells (GSCs) were shown to be able to transdifferentiate into pericytes/SMCs [53,54]. Such GSC-derived cells are recruited to ECs through stromal-derived factor 1 (SDF1)-CXCR4 signaling [53], which is involved in pericyte recruitment [55]. Given that the GSC differentiation process is induced by TGFβ [53], it may activate the EMT program as well. In our study, we detected EPT in a HER2 breast carcinoma [12]. Currently it is unknown how frequently spontaneous EMT occurs in human cancer. Since claudin-low and metaplastic breast cancer subtypes are enriched for malignant mesenchymal cells and show the EMT core signature [56,57], EPT may occur in these tumors. A subset of perivascular soft tissue tumors, including glomus tumor, myopericytoma, angioleiomyoma, and liposarcoma, exhibit pericyte marker expression and perivascular growth [58,59]. Most of these tumors are presumed to originate from pericytes. It remains to be determined whether EPT-like transforming mechanisms may also contribute to pericyte marker adoption in some tumors. Overall, EPT may significantly contribute to the development of both normal and tumor pericytes.

5. Prospective Significance of EPT in Cancer

5.1. EPT in Tumor Vascularization and Growth

Judah Folkman proposed that all tumors are dependent on angiogenesis, the formation of new blood vessels [60]. Tumor growth requires vascular support. Avascular tumors are severely restricted in their growth due to the lack of a stable blood supply. Cancer cells are often able to stimulate angiogenesis for expansion of tumor mass. Nascent vessels consisting of only ECs are unstable and ineffective. Pericyte coverage is critical for vascular maturation and stability. Tumor vasculature is commonly portrayed as poorly organized, constantly remodeling, and lacking appreciable pericyte coverage [61]. However, microscopic studies have revealed the nearly ubiquitous presence of pericytes on tumor vessels, although such pericytes are typically less abundant and more loosely attached to the vasculature in tumors than in normal tissues [62]. Nevertheless, the existence of vessel-associated pericytes is vital to tumors, as experimental evidence indicates that pericytes critically maintain the integrity and functionality of the tumor vasculature. Pharmacological blockade of pericyte recruitment or genetic ablation of host-derived pericytes reduces pericyte coverage, destabilizes blood vessels, and decreases tumor growth [13,63,64]. Moreover, knockout of NG2 in mice causes pericyte deficiency and poor vessel functionality in transplanted tumors, leading to reduced tumor expansion [17]. Finally, in tumor xenografts derived from cancer cells prone to undergo EMT, a substantial fraction of pericytes are post-EMT cancer cells, and depletion of such EMT cells impairs pericyte coverage and vessel integrity, leading to diminished tumor growth [12]. The result suggests that EPT critically contributes to tumor vascularization and growth.

Severely deficient pericyte coverage destabilizes the vasculature, increases interstitial fluid pressure, and enables cancer cells to transit into the circulatory system, thus facilitating metastatic dissemination. Pericyte coverage affects breast cancer metastasis [65]. Accordingly, improved pericyte coverage may suppress tumor intravasation and metastasis [63,66]. As EPT cancer cells function like pericytes to stabilize blood vessels, they may prevent other cancer cells from intravasation and hence inhibit blood-borne metastasis (Figure 2).

5.2. EPT in Resistance to Anti-Angiogenesis Therapy

Vascular endothelial growth factor (VEGF) is perhaps the most important cytokine involved in tumor angiogenesis [67]. VEGF supports EC proliferation and survival. Vasculature lacking adequate pericyte coverage is vulnerable to VEGF inhibition. Anti-angiogenic therapies targeting VEGF reduce tumor vascularity and show therapeutic efficacy in human cancers, although the clinical benefits are modest and short-lived [63]. Pericytes are critical cell constituents of the tumor vasculature. Tumor pericytes express appreciable levels of VEGF and other trophic factors. Through direct support and/or paracrine interactions with ECs, pericytes mediate EC survival and protect ECs from VEGF blockade [63,68]. Tumor vessels with better pericyte coverage are less sensitive to anti-angiogenic treatment. Indeed, pericyte coverage accounts for the relative resistance of more mature vessels to VEGF withdrawal. As EPT increases pericyte coverage of the tumor vasculature, it may promote resistance to anti-angiogenic agents that target VEGF.

5.3. EPT in Resistance to Chemotherapy and Targeted Therapy

Conventional chemotherapy remains the backbone of treatment for most cancer patients, however, the effect is generally not long-lasting due to the emergence of drug resistance and tumor relapse. One major form of chemoresistance is attributed to EMT [7]. In tumor samples, population of residual cancer cells that survive after chemotherapy bear a gene signature with hallmarks of EMT [69,70]. Even in studies that EMT is dispensable for metastasis, the importance of EMT in chemoresistance is validated in mouse tumor models in vivo [9,10].

Activation of oncogenic pathways induces pro-growth and -survival signals on which tumors depend. This dependency of cancer cells on oncogenes, known as "oncogenic addiction", has been exploited in the development of targeted therapy drugs. One common means by which cancer cells resist molecularly targeted therapies involves their ability to switch to a new cell type that no longer relies on the oncogenic signaling pathway being targeted by the treatments. EMT represents such a phenotypic shift in cell state that allows cancer cells to bypass pathways targeted by therapy and survive therapeutic insult [4]. Non-small-cell lung carcinomas with activating mutations in epidermal growth factor receptor (EGFR) frequently respond to treatment with tyrosine kinase inhibitors targeting EGFR, but the responses are not durable, as tumors acquire resistance. An EMT event that switches EGFR to AXL receptor tyrosine kinase is responsible for acquired resistance to EGFR inhibition [71–73].

EMT indeed dramatically rewires signaling pathways in cancer cells [23–28]. For instance, through the EMT process, mouse mammary epithelial tumor cells downregulate HER2 and EGFR, but upregulate PDGFRs, AXL, MET, CXCR4, etc. [23]. Consistent with the receptor changes, pre- and post-EMT cells exhibited differential responsiveness to mitogenic signals and therapeutic agents [23]. PDGFRβ expression levels correlate with tumor growth, drug resistance, and poor clinical outcomes [74]. AXL is aberrantly overexpressed in mesenchymal cells and in tumor cells refractory to therapy, and potently promotes cancer cell survival and resistance to both chemotherapy and targeted therapy [75–78]. Moreover, inappropriate activation of MET and CXCR4 is frequently implicated in resistance to conventional and targeted therapies and contributes to tumor relapse [79,80].

However, altering the landscape of receptors in EMT cells alone is insufficient to confer survival and therapy resistance, the availability of cognate ligands determines whether the newly acquired receptors and downstream signaling cascades are activated. It has been recognized that capillary ECs are not just passive conduits for delivering blood. They indeed form vascular niches that produce a variety of growth factors, cytokines, and extracellular matrix components, which are defined collectively as "angiocrine factors" [81]. The angiocrine factors act in a paracrine manner to activate survival signaling and protect responsive cells in their vicinity. It has been well established that vascular niches in the bone marrow provide a sanctuary for subpopulations of leukemic cells to resist chemotherapy-induced death [82–84]. Many EMT-acquired receptors can recognize EC-derived angiocrine factors. PDGF, HGF and SDF1 can activate PDGFRβ, MET and CXCR4, respectively. GAS6 is a major ligand for AXL and is present in plasma [85].

EPT enables cancer cells to occupy the periendothelial compartments, associate with blood vessels, and express cognate receptors for angiocrine factors. Therefore, EPT cancer cells are primed to respond to pro-survival signals from blood vessels and withstand the cytotoxic effects from the treatment. EPT cancer cells may have acquired stemness-like attributes during EMT [6], which may further be sustained by EC-derived angiocrine signals in the vascular niche. By contrast, non-EPT cells do not share the same receptor repertoire and/or proximity to blood vessels, and thus fail to receive the protection by vascular niches. Through the functional interactions with vascular ECs, EPT cancer cells may gain a selective survival advantage to resist chemotherapy and targeted therapy [86].

6. Conclusions

Through the EMT reprogramming process, epithelial cells shed epithelial characteristics and/or acquire mesenchymal properties. The two events may occur independently and each to varying extents. Therefore, EMT consists of a broad spectrum of intermediate phenotypes between the completely epithelial state and the completely mesenchymal state. The outputs of EMT are heterogeneous. Cancer cells undergoing partial EMT may acquire enhanced metastatic potential, whereas cancer cells with full EMT (in particular acquiring an SRF-driven mesenchymal phenotype) may instead resemble pericytes to stabilize tumor vasculature. Such EPT cells may be protected by the vascular niche, thus gaining increased therapy resistance and contributing to tumor relapse.

Conflicts of Interest: The authors declare no conflict of interest.

References

1. Lamouille, S.; Xu, J.; Derynck, R. Molecular mechanisms of epithelial-mesenchymal transition. *Nat. Rev. Mol. Cell Biol.* **2014**, *15*, 178–196. [CrossRef] [PubMed]
2. Kalluri, R.; Weinberg, R.A. The basics of epithelial-mesenchymal transition. *J. Clin. Investig.* **2009**, *119*, 1420–1428. [CrossRef] [PubMed]
3. Peinado, H.; Olmeda, D.; Cano, A. Snail, Zeb and bHLH factors in tumour progression: An alliance against the epithelial phenotype? *Nat. Rev. Cancer* **2007**, *7*, 415–428. [CrossRef] [PubMed]
4. Thiery, J.P.; Acloque, H.; Huang, R.Y.; Nieto, M.A. Epithelial-mesenchymal transitions in development and disease. *Cell* **2009**, *139*, 871–890. [CrossRef] [PubMed]
5. Puisieux, A.; Brabletz, T.; Caramel, J. Oncogenic roles of EMT-inducing transcription factors. *Nat. Cell. Biol.* **2014**, *16*, 488–494. [CrossRef] [PubMed]
6. Nieto, M.A.; Huang, R.Y.; Jackson, R.A.; Thiery, J.P. EMT: 2016. *Cell* **2016**, *166*, 21–45. [CrossRef] [PubMed]
7. Holohan, C.; Van Schaeybroeck, S.; Longley, D.B.; Johnston, P.G. Cancer drug resistance: An evolving paradigm. *Nat. Rev. Cancer* **2013**, *13*, 714–726. [CrossRef] [PubMed]
8. Diepenbruck, M.; Christofori, G. Epithelial-mesenchymal transition (EMT) and metastasis: Yes, no, maybe? *Curr. Opin. Cell Biol.* **2016**, *43*, 7–13. [CrossRef] [PubMed]
9. Fischer, K.R.; Durrans, A.; Lee, S.; Sheng, J.T.; Li, F.H.; Wong, S.T.C.; Choi, H.; Rayes, T.E.; Ryu, S.; Troeger, J.; et al. Epithelial-to-mesenchymal transition is not required for lung metastasis but contributes to chemoresistance. *Nature* **2015**, *527*, 472–476. [CrossRef] [PubMed]
10. Zheng, X.; Carstens, J.L.; Kim, J.; Scheible, M.; Kaye, J.; Sugimoto, H.; Wu, C.C.; LeBleu, V.S.; Kalluri, R. Epithelial-to-mesenchymal transition is dispensable for metastasis but induces chemoresistance in pancreatic cancer. *Nature* **2015**, *527*, 525–530. [CrossRef] [PubMed]
11. Krebs, A.M.; Mitschke, J.; Losada, M.L.; Schmalhofer, O.; Boerries, M.; Busch, H.; Boettcher, M.; Mougiakakos, D.; Reichardt, W.; Bronsert, P.; et al. The EMT-activator Zeb1 is a key factor for cell plasticity and promotes metastasis in pancreatic cancer. *Nat. Cell Biol.* **2017**, *19*, 518–529. [CrossRef] [PubMed]
12. Shenoy, A.K.; Jin, Y.; Luo, H.C.; Tang, M.; Pampo, C.; Shao, R.; Siemann, D.W.; Wu, L.Z.; Heldermon, C.D.; Law, B.K.; et al. Epithelial-to-mesenchymal transition confers pericyte properties on cancer cells. *J. Clin. Investig.* **2016**, *126*, 4174–4186. [CrossRef] [PubMed]
13. Bergers, G.; Song, S. The role of pericytes in blood-vessel formation and maintenance. *Neuro Oncol.* **2005**, *7*, 452–464. [CrossRef] [PubMed]
14. Armulik, A.; Genove, G.; Betsholtz, C. Pericytes: Developmental, physiological, and pathological perspectives, problems, and promises. *Dev. Cell.* **2011**, *21*, 193–215. [CrossRef] [PubMed]
15. Ozerdem, U.; Grako, K.A.; Dahlin-Huppe, K.; Monosov, E.; Stallcup, W.B. NG2 proteoglycan is expressed exclusively by mural cells during vascular morphogenesis. *Dev. Dyn.* **2001**, *222*, 218–227. [CrossRef] [PubMed]
16. Ozerdem, U.; Stallcup, W.B. Pathological angiogenesis is reduced by targeting pericytes via the NG2 proteoglycan. *Angiogenesis* **2004**, *7*, 269–276. [CrossRef] [PubMed]
17. Huang, F.J.; Youa, W.K.; Bonaldob, P.; Seyfriedc, T.N.; Pasqualea, E.B.; Stallcupa, W.B. Pericyte deficiencies lead to aberrant tumor vascularizaton in the brain of the NG2 null mouse. *Dev. Biol.* **2010**, *344*, 1035–1046. [CrossRef] [PubMed]
18. Diaz-Flores, L.; Gutierrez, R.; Madrid, J.F.; Varela, H.; Valladares, F.; Acosta, E.; Diaz-Flores, L., Jr. Pericytes. Morphofunction, interactions and pathology in a quiescent and activated mesenchymal cell niche. *Histol. Histopathol.* **2009**, *24*, 909–969. [PubMed]
19. Winkler, E.A.; Bell, R.D.; Zlokovic, B.V. Central nervous system pericytes in health and disease. *Nat. Neurosci.* **2011**, *14*, 1398–1405. [CrossRef] [PubMed]
20. Li, F.; Yu, L.; Wang, Y.L.; Wang, J.; Yang, G.; Meng, F.W.; Han, H.; Meng, A.; Wang, Y.P.; Yang, X. Endothelial Smad4 maintains cerebrovascular integrity by activating N-cadherin through cooperation with Notch. *Dev. Cell* **2011**, *20*, 291–302. [CrossRef] [PubMed]
21. Gerhardt, H.; Wolburg, H.; Redies, C. N-cadherin mediates pericytic-endothelial interaction during brain angiogenesis in the chicken. *Dev. Dyn.* **2000**, *218*, 472–479. [CrossRef]
22. Heldin, C.H.; Westermark, B. Mechanism of action and in vivo role of platelet-derived growth factor. *Physiol. Rev.* **1999**, *79*, 1283–1316. [PubMed]

23. Jahn, S.C.; Law, M.E.; Corsino, P.E.; Parker, N.N.; Pham, K.; Davis, B.J.; Lu, J.R.; Law, B.K. An in vivo model of epithelial to mesenchymal transition reveals a mitogenic switch. *Cancer Lett.* **2012**, *326*, 183–190. [CrossRef] [PubMed]

24. Campbell, C.I.; Moorehead, R.A. Mammary tumors that become independent of the type I insulin-like growth factor receptor express elevated levels of platelet-derived growth factor receptors. *BMC Cancer* **2011**, *11*, 480. [CrossRef] [PubMed]

25. Jechlinger, M.; Grunert, S.; Tamir, I.H.; Janda, E.; Ludemann, S.; Waerner, T.; Seither, P.; Weith, A.; Beug, H.; Kraut, N. Expression profiling of epithelial plasticity in tumor progression. *Oncogene* **2003**, *22*, 7155–7169. [CrossRef] [PubMed]

26. Jechlinger, M.; Sommer, A.; Moriggl, R.; Seither, P.; Kraut, N.; Capodiecci, P.; Donovan, M.; Cordon-Cardo, C.; Beug, H.; Grünert, S. Autocrine PDGFR signaling promotes mammary cancer metastasis. *J. Clin. Investig.* **2006**, *116*, 1561–1570. [CrossRef] [PubMed]

27. Steller, E.J.; Raats, D.A.; Koster, J.; Rutten, B.; Govaert, K.M.; Emmink, B.L.; Snoeren, N.; van Hooff, S.R.; Holstege, F.C.; Maas, C.; et al. PDGFRB promotes liver metastasis formation of mesenchymal-like colorectal tumor cells. *Neoplasia* **2013**, *15*, 204–217. [CrossRef] [PubMed]

28. Thomson, S.; Petti, F.; Sujka-Kwok, I.; Epstein, D.; Haley, J.D. Kinase switching in mesenchymal-like non-small cell lung cancer lines contributes to EGFR inhibitor resistance through pathway redundancy. *Clin. Exp. Metastasis* **2008**, *25*, 843–854. [CrossRef] [PubMed]

29. Van Roy, F. Beyond E-cadherin: Roles of other cadherin superfamily members in cancer. *Nat. Rev. Cancer* **2014**, *14*, 121–134. [CrossRef] [PubMed]

30. Wheelock, M.J.; Shintani, Y.; Maeda, M.; Fukumoto, Y.; Johnson, K.R. Cadherin switching. *J. Cell Sci.* **2008**, *121*, 727–735. [CrossRef] [PubMed]

31. Jamora, C.; Lee, P.; Kocieniewski, P.; Azhar, M.; Hosokawa, R.; Chai, Y.; Fuchs, E. A signaling pathway involving TGF-beta2 and snail in hair follicle morphogenesis. *PLoS Biol.* **2005**, *3*, e11.

32. Du, F.; Nakamura, Y.; Tan, T.L.; Lee, P.; Lee, R.; Yu, B.; Jamoraet, C. Expression of snail in epidermal keratinocytes promotes cutaneous inflammation and hyperplasia conducive to tumor formation. *Cancer Res.* **2010**, *70*, 10080–10089. [CrossRef] [PubMed]

33. Zheng, H.; Shen, M.; Zha, Y.L.; Li, W.Y.; Wei, Y.; Blanco, M.A.; Ren, G.W.; Zhou, T.H.; Storz, P.; Wang, H.Y.; et al. PKD1 phosphorylation-dependent degradation of SNAIL by SCF-FBXO11 regulates epithelial-mesenchymal transition and metastasis. *Cancer Cell* **2014**, *26*, 358–373. [CrossRef] [PubMed]

34. Jin, Y.; Shenoy, A.K.; Doernberg, S.; Chen, H.; Luo, H.C.; Shen, H.X.; Lin, T.; Tarrash, M.; Cai, Q.S.; Hu, X.; et al. FBXO11 promotes ubiquitination of the Snail family of transcription factors in cancer progression and epidermal development. *Cancer Lett.* **2015**, *362*, 70–82. [CrossRef] [PubMed]

35. Tang, M.; Shen, H.; Jin, Y.; Lin, T.; Cai, Q.; Pinard, M.A.; Biswas, S.; Tran, Q.; Li, G.; Shenoy, A.K.; et al. The malignant brain tumor (MBT) domain protein SFMBT1 is an integral histone reader subunit of the LSD1 demethylase complex for chromatin association and epithelial-to-mesenchymal transition. *J. Biol. Chem.* **2013**, *288*, 27680–27691. [CrossRef] [PubMed]

36. Wang, D.Z.; Olson, E.N. Control of smooth muscle development by the myocardin family of transcriptional coactivators. *Curr. Opin. Genet. Dev.* **2004**, *14*, 558–566. [CrossRef] [PubMed]

37. Olson, E.N.; Nordheim, A. Linking actin dynamics and gene transcription to drive cellular motile functions. *Nat. Rev. Mol. Cell Biol.* **2010**, *11*, 353–365. [CrossRef] [PubMed]

38. Parmacek, M.S. Myocardin-related transcription factors: Critical coactivators regulating cardiovascular development and adaptation. *Circ. Res.* **2007**, *100*, 633–644. [CrossRef] [PubMed]

39. Mericskay, M.; Parlakian, A.; Porteu, A.; Dandre, F.; Bonnet, J.; Paulin, D.; Li, Z.L. An overlapping CArG/octamer element is required for regulation of desmin gene transcription in arterial smooth muscle cells. *Dev. Biol.* **2000**, *226*, 192–208. [CrossRef] [PubMed]

40. Psichari, E.; Balmain, A.; Plows, D.; Zoumpourlis, V.; Pintzas, A. High activity of serum response factor in the mesenchymal transition of epithelial tumor cells is regulated by RhoA signaling. *J. Biol. Chem.* **2002**, *277*, 29490–29495. [CrossRef] [PubMed]

41. Fan, L.; Sebe, A.; Peterfi, Z.; Masszi, A.; Thirone, A.C.; Rotstein, O.D.; Nakano, H.; McCulloch, C.A.; Szaszi, K.; Mucsi, I.; et al. Cell contact-dependent regulation of epithelial-myofibroblast transition via the rho-rho kinase-phospho-myosin pathway. *Mol. Biol. Cell* **2007**, *18*, 1083–1097. [CrossRef] [PubMed]

42. Morita, T.; Mayanagi, T.; Sobue, K. Dual roles of myocardin-related transcription factors in epithelial mesenchymal transition via slug induction and actin remodeling. *J. Cell Biol.* **2007**, *179*, 1027–1042. [CrossRef] [PubMed]

43. Busche, S.; Descot, A.; Julien, S.; Genth, H.; Posern, G. Epithelial cell-cell contacts regulate SRF-mediated transcription via Rac-actin-MAL signalling. *J. Cell Sci.* **2008**, *121*, 1025–1035. [CrossRef] [PubMed]

44. Nishimura, G.; Manabe, I.; Tsushima, K.; Fujiu, K.; Oishi, Y.; Imai, Y.; Maemura, K.; Miyagishi, M.; Higashi, Y.; Kondoh, H.; et al. DeltaEF1 mediates TGF-beta signaling in vascular smooth muscle cell differentiation. *Dev. Cell.* **2006**, *11*, 93–104. [CrossRef] [PubMed]

45. Trost, A.; Schroedl, F.; Lange, S.; Rivera, F.J.; Tempfer, H.; Korntner, S.; Stolt, C.C.; Wegner, M.; Bogner, B.; Kaser-Eichberger, A.; et al. Neural crest origin of retinal and choroidal pericytes. *Invest. Ophthalmol. Vis. Sci.* **2013**, *54*, 7910–7921. [CrossRef] [PubMed]

46. Wilm, B.; Ipenberg, A.; Hastie, N.D.; Burch, J.B.; Bader, D.M. The serosal mesothelium is a major source of smooth muscle cells of the gut vasculature. *Development* **2005**, *132*, 5317–5328. [CrossRef] [PubMed]

47. Que, J.; Wilm, B.; Hasegawa, H.; Wang, F.; Bader, D.; Hogan, B.L.M. Mesothelium contributes to vascular smooth muscle and mesenchyme during lung development. *Proc. Natl. Acad. Sci. USA* **2008**, *105*, 16626–16630. [CrossRef] [PubMed]

48. Asahina, K.; Zhou, B.; Pu, W.T.; Tsukamoto, H. Septum transversum-derived mesothelium gives rise to hepatic stellate cells and perivascular mesenchymal cells in developing mouse liver. *Hepatology* **2011**, *53*, 983–995. [CrossRef] [PubMed]

49. Chen, Q.; Zhang, H.; Liu, Y.; Adams, S.; Eilken, H.; Stehling, M.; Corada, M.; Dejana, E.; Zhou, B.; Adams, R.H.; et al. Endothelial cells are progenitors of cardiac pericytes and vascular smooth muscle cells. *Nat. Commun.* **2016**, *7*, 12422. [CrossRef] [PubMed]

50. Lugassy, C.; Haroun, R.I.; Brem, H.; Tyler, B.M.; Jones, R.V.; Fernandez, P.M.; Patierno, S.R.; Kleinman, H.K.; Barnhill, R.L. Pericytic-like angiotropism of glioma and melanoma cells. *Am. J. Dermatopathol.* **2002**, *24*, 473–478. [CrossRef] [PubMed]

51. Lugassy, C.; Peault, B.; Wadehra, M.; Kleinman, H.K.; Barnhill, R.L. Could pericytic mimicry represent another type of melanoma cell plasticity with embryonic properties? *Pigment. Cell Melanoma. Res.* **2013**, *26*, 746–754. [CrossRef] [PubMed]

52. Olar, A.; Aldape, K.D. Using the molecular classification of glioblastoma to inform personalized treatment. *J. Pathol.* **2014**, *232*, 165–177. [CrossRef] [PubMed]

53. Cheng, L.; Huang, Z.; Zhou, W.; Wu, Q.; Donnola, S.; Liu, J.K.; Fang, X.; Sloan, A.E.; Mao, Y.; Lathia, J.D. Glioblastoma stem cells generate vascular pericytes to support vessel function and tumor growth. *Cell* **2013**, *153*, 139–152. [CrossRef] [PubMed]

54. Shao, R.; Taylor, S.L.; Oh, D.S.; Schwartz, L.M. Vascular heterogeneity and targeting: The role of YKL-40 in glioblastoma vascularization. *Oncotarget* **2015**, *6*, 40507–40518. [PubMed]

55. Song, N.; Huang, Y.; Shi, H.; Yuan, S.; Ding, Y.; Song, X.; Fu, Y.; Luo, Y. Overexpression of platelet-derived growth factor-BB increases tumor pericyte content via stromal-derived factor-1alpha/CXCR4 axis. *Cancer Res.* **2009**, *69*, 6057–6064. [CrossRef] [PubMed]

56. Taube, J.H.; Herschkowitz, J.I.; Komurov, K.; Zhou, A.Y.; Gupta, S.; Yang, J.; Hartwell, K.; Onder, T.T.; Gupta, P.B.; Evans, K.W.; et al. Core epithelial-to-mesenchymal transition interactome gene-expression signature is associated with claudin-low and metaplastic breast cancer subtypes. *Proc. Natl. Acad. Sci. USA* **2010**, *107*, 15449–15454. [CrossRef] [PubMed]

57. Prat, A.; Perou, C.M. Deconstructing the molecular portraits of breast cancer. *Mol. Oncol.* **2011**, *5*, 5–23. [CrossRef] [PubMed]

58. Shen, J.; Shrestha, S.; Rao, P.N.; Asatrian, G.; Scott, M.A.; Nguyen, V.; Giacomelli, P.; Soo, C.; Ting, K.; Eilber, F.C.; et al. Pericytic mimicry in well-differentiated liposarcoma/atypical lipomatous tumor. *Hum. Pathol.* **2016**, *54*, 92–99. [CrossRef] [PubMed]

59. Shen, J.; Shrestha, S.; Yen, Y.H.; Asatrian, G.; Mravic, M.; Soo, C.; Ting, K.; Dry, S.M.; Peault, B.; James, A.W. Pericyte Antigens in Perivascular Soft Tissue Tumors. *Int. J. Surg. Pathol.* **2015**, *23*, 638–648. [CrossRef] [PubMed]

60. Folkman, J. Angiogenesis. *Annu. Rev. Med.* **2006**, *57*, 1–18. [CrossRef] [PubMed]

61. Carmeliet, P.; Jain, R.K. Molecular mechanisms and clinical applications of angiogenesis. *Nature* **2011**, *473*, 298–307. [CrossRef] [PubMed]

62. Morikawa, S.; Baluk, P.; Kaidoh, T.; Haskell, A.; Jain, R.K.; McDonald, D.M. Abnormalities in pericytes on blood vessels and endothelial sprouts in tumors. *Am. J. Pathol.* **2002**, *160*, 985–1000. [CrossRef]
63. Bergers, G.; Hanahan, D. Modes of resistance to anti-angiogenic therapy. *Nat. Rev. Cancer* **2008**, *8*, 592–603. [CrossRef] [PubMed]
64. Cooke, V.G.; LeBleu, V.S.; Keskin, D.; Khan, Z.; O'Connell, J.T.; Teng, Y.; Duncan, M.B.; Xie, L.; Maeda, G.; Vong, S.; et al. Pericyte depletion results in hypoxia-associated epithelial-to-mesenchymal transition and metastasis mediated by met signaling pathway. *Cancer Cell* **2012**, *21*, 66–81. [CrossRef] [PubMed]
65. Kim, J.; de Sampaio, P.C.; Lundy, D.M.; Peng, Q.; Evans, K.W.; Sugimoto, H.; Gagea, M.; Kienast, Y.; do Amaral, N.S.; Rocha, R.M.; et al. Heterogeneous perivascular cell coverage affects breast cancer metastasis and response to chemotherapy. *JCI Insight* **2016**, *1*, e90733. [CrossRef] [PubMed]
66. Gerhardt, H.; Semb, H. Pericytes: Gatekeepers in tumour cell metastasis? *J. Mol. Med. (Berl.)* **2008**, *86*, 135–144. [CrossRef] [PubMed]
67. Ferrara, N.; Adamis, A.P. Ten years of anti-vascular endothelial growth factor therapy. *Nat. Rev. Drug. Discov.* **2016**, *15*, 385–403. [CrossRef] [PubMed]
68. Van Beijnum, J.R.; Nowak-Sliwinska, P.; Huijbers, E.J.; Thijssen, V.L.; Griffioen, A.W. The great escape; the hallmarks of resistance to antiangiogenic therapy. *Pharmacol. Rev.* **2015**, *67*, 441–461. [CrossRef] [PubMed]
69. Dave, B.; Mittal, V.; Tan, N.M.; Chang, J.C. Epithelial-mesenchymal transition, cancer stem cells and treatment resistance. *Breast Cancer Res.* **2012**, *14*, 202. [CrossRef] [PubMed]
70. Creighton, C.J.; Li, X.; Landis, M.; Dixon, J.M.; Neumeister, V.M.; Sjolund, A.; Rimm, D.L.; Wong, H.; Rodriguez, A.; Herschkowitz, J.I.; et al. Residual breast cancers after conventional therapy display mesenchymal as well as tumor-initiating features. *Proc. Natl. Acad. Sci. USA* **2009**, *106*, 13820–13825. [CrossRef] [PubMed]
71. Zhang, Z.; Lee, J.C.; Lin, L.; Olivas, V.; Au, V.; LaFramboise, T.; Abdel-Rahman, M.; Wang, X.; Levine, A.D.; Rho, J.K.; et al. Activation of the AXL kinase causes resistance to EGFR-targeted therapy in lung cancer. *Nat. Genet.* **2012**, *44*, 852–860. [CrossRef] [PubMed]
72. Byers, L.A.; Diao, L.; Wang, J.; Saintigny, P.; Girard, L.; Peyton, M.; Shen, L.; Fan, Y.; Giri, U.; Tumula, P.K.; et al. An epithelial-mesenchymal transition gene signature predicts resistance to EGFR and PI3K inhibitors and identifies Axl as a therapeutic target for overcoming EGFR inhibitor resistance. *Clin. Cancer Res.* **2013**, *19*, 279–290. [CrossRef] [PubMed]
73. Thomson, S.; Buck, E.; Petti, F.; Griffin, G.; Brown, E.; Ramnarine, N.; Iwata, K.K.; Gibson, N.; Haley, J.D. Epithelial to mesenchymal transition is a determinant of sensitivity of non-small-cell lung carcinoma cell lines and xenografts to epidermal growth factor receptor inhibition. *Cancer Res.* **2005**, *65*, 9455–9462. [CrossRef] [PubMed]
74. Cao, Y. Multifarious functions of PDGFs and PDGFRs in tumor growth and metastasis. *Trends Mol. Med.* **2013**, *19*, 460–473. [CrossRef] [PubMed]
75. Graham, D.K.; DeRyckere, D.; Davies, K.D.; Earp, H.S. The TAM family: Phosphatidylserine sensing receptor tyrosine kinases gone awry in cancer. *Nat. Rev. Cancer* **2014**, *14*, 769–785. [CrossRef] [PubMed]
76. Wu, X.; Liu, X.; Koul, S.; Lee, C.Y.; Zhang, Z.; Halmos, B. AXL kinase as a novel target for cancer therapy. *Oncotarget* **2014**, *5*, 9546–9563. [CrossRef] [PubMed]
77. Scaltriti, M.; Elkabets, M.; Baselga, J. Molecular Pathways: AXL, a Membrane Receptor Mediator of Resistance to Therapy. *Clin. Cancer Res.* **2016**, *22*, 1313–1317. [CrossRef] [PubMed]
78. Wang, C.; Jin, H.; Wang, N.; Fan, S.; Wang, Y.; Zhang, Y.; Wei, L.; Tao, X.; Gu, D.; Zhao, F.; et al. Gas6/Axl Axis Contributes to Chemoresistance and Metastasis in Breast Cancer through Akt/GSK-3beta/beta-catenin Signaling. *Theranostics* **2016**, *6*, 1205–1219. [CrossRef] [PubMed]
79. Corso, S.; Giordano, S. Cell-autonomous and non-cell-autonomous mechanisms of HGF/MET-driven resistance to targeted therapies: From basic research to a clinical perspective. *Cancer Discov.* **2013**, *3*, 978–992. [CrossRef] [PubMed]
80. Chatterjee, S.; Behnam, A.B.; Nimmagadda, S. The intricate role of CXCR4 in cancer. *Adv. Cancer Res.* **2014**, *124*, 31–82. [PubMed]
81. Rafii, S.; Butler, J.M.; Ding, B.S. Angiocrine functions of organ-specific endothelial cells. *Nature* **2016**, *529*, 316–325. [CrossRef] [PubMed]
82. Butler, J.M.; Kobayashi, H.; Rafii, S. Instructive role of the vascular niche in promoting tumour growth and tissue repair by angiocrine factors. *Nat. Rev. Cancer* **2010**, *10*, 138–146. [CrossRef] [PubMed]

83. Tabe, Y.; Konopleva, M. Advances in understanding the leukaemia microenvironment. *Br. J. Haematol.* **2014**, *164*, 767–778. [CrossRef] [PubMed]

84. Doan, P.L.; Chute, J.P. The vascular niche: Home for normal and malignant hematopoietic stem cells. *Leukemia* **2012**, *26*, 54–62. [CrossRef] [PubMed]

85. Laurance, S.; Lemarie, C.A.; Blostein, M.D. Growth arrest-specific gene 6 (gas6) and vascular hemostasis. *Adv. Nutr.* **2012**, *3*, 196–203. [CrossRef] [PubMed]

86. Shenoy, A.K.; Lu, J. Relevance of epithelial-to-pericyte transition in cancer. *Mol. Cell. Oncol.* **2017**, *4*, e1260672. [CrossRef] [PubMed]

Review

cancers

MDPI

EMT/MET at the Crossroad of Stemness, Regeneration and Oncogenesis: The Ying-Yang Equilibrium Recapitulated in Cell Spheroids

Elvira Forte [1,*], Isotta Chimenti [2], Paolo Rosa [2], Francesco Angelini [2], Francesca Pagano [2], Antonella Calogero [2], Alessandro Giacomello [3] and Elisa Messina [4]

[1] The Jackson Laboratory, Bar Harbor, ME 04609, USA
[2] Department of Medical Surgical Sciences and Biotechnologies, "La Sapienza" University of Rome, 04100 Italy; isotta.chimenti@uniroma1.it (I.C.); p.rosa@uniroma1.it (P.R.); f.angelini@uniroma1.it (F.A.); francesca.pagano@uniroma1.it (F.P.); antonella.calogero@uniroma1.it (A.C.)
[3] Department of Molecular Medicine, "La Sapienza" University of Rome, 00195 Roma, Italy; alessandro.giacomello@uniroma1.it
[4] Department of Pediatrics and Infant Neuropsychiatry, "Umberto I" Hospital, "La Sapienza" University of Rome, 00195 Roma, Italy; elisa.messina@uniroma1.it
* Correspondence: elvira.forte@jax.org

Received: 24 June 2017; Accepted: 26 July 2017; Published: 29 July 2017

Abstract: The epithelial-to-mesenchymal transition (EMT) is an essential trans-differentiation process, which plays a critical role in embryonic development, wound healing, tissue regeneration, organ fibrosis, and cancer progression. It is the fundamental mechanism by which epithelial cells lose many of their characteristics while acquiring features typical of mesenchymal cells, such as migratory capacity and invasiveness. Depending on the contest, EMT is complemented and balanced by the reverse process, the mesenchymal-to-epithelial transition (MET). In the saving economy of the living organisms, the same (Ying-Yang) tool is integrated as a physiological strategy in embryonic development, as well as in the course of reparative or disease processes, prominently fibrosis, tumor invasion and metastasis. These mechanisms and their related signaling (e.g., TGF-β and BMPs) have been effectively studied in vitro by tissue-derived cell spheroids models. These three-dimensional (3D) cell culture systems, whose phenotype has been shown to be strongly dependent on TGF-β-regulated EMT/MET processes, present the advantage of recapitulating in vitro the hypoxic in vivo micro-environment of tissue stem cell niches and their formation. These spheroids, therefore, nicely reproduce the finely regulated Ying-Yang equilibrium, which, together with other mechanisms, can be determinant in cell fate decisions in many pathophysiological scenarios, such as differentiation, fibrosis, regeneration, and oncogenesis. In this review, current progress in the knowledge of signaling pathways affecting EMT/MET and stemness regulation will be outlined by comparing data obtained from cellular spheroids systems, as ex vivo niches of stem cells derived from normal and tumoral tissues. The mechanistic correspondence in vivo and the possible pharmacological perspective will be also explored, focusing especially on the TGF-β-related networks, as well as others, such as SNAI1, PTEN, and EGR1. This latter, in particular, for its ability to convey multiple types of stimuli into relevant changes of the cell transcriptional program, can be regarded as a heterogeneous "stress-sensor" for EMT-related inducers (growth factor, hypoxia, mechano-stress), and thus as a therapeutic target.

Keywords: spheroids; EMT/MET; TGF-β; EGR-1

Highlights:

- EMT/MET play a pivotal role in cell fate decision making for both normal and transformed cells.
- Thus, these mechanisms represent a strategic target for preclinical (basic studies, pharmacologic screening, and biotechnology advances), as well as clinical applications (personalized diagnosis and therapy).
- EMT/MET are finely reproduced within cell spheroid systems, which, as in vitro models of normal and transformed stem cell (SC) niches, represent an adequate cost/benefit biotechnological tool to investigate disease mechanisms, therapeutic targets, and related applications.

1. Introduction

Our knowledge of the shared pathways in trans-differentiation processes occurring during organogenesis, post-natal tissue repair/regeneration, and tumorigenesis has greatly expanded in the last decades, thanks also to improvement of in vitro cell culture technologies, namely three-dimensional (3D) tissue-derived spheroid systems. These 3D culture methods have been developed to recapitulate the in vivo growth, differentiation and de-differentiation conditions of normal and cancer cells, by better preserving the biological features of the original source compared to conventional 2D monolayer cultures. In particular, their hypoxic and hierarchical stem cell (SC)-supporting environment favors heterogeneous cell-cell, cell-matrix interactions and cross-talk required to mimic patho-physiological processes. Conversely, these latter are poorly represented in static 2D systems, in which cells are exposed to high O_2 and nutrient concentrations in the medium, and forced to directly interact with high-stiffness artificial substrates. This artificial condition cannot reproduce the time-course and dose-dependence of specific ligand–receptor interactions and downstream signaling induction. Furthermore, even the orthotopic transplantation of tumor cells, used to define cancer SC features, while representing the gold standard of in vivo experimental models, lacks patient-specific conditions, which are not easy to achieve in the xenograft [1], and it is less expensive and time consuming [1,2].

The reliability of 3D cell spheroid systems has allowed scientists to extend the experimental modeling of normal and malignant SC growth and differentiation. In addition to oxygen gradient, allowing a SC niche-like balance of cell quiescence/proliferation in the spheroid, the specific SC features of drug sensitivity/resistance, as well as phenotypic changes and trans-differentiation ability, can be spontaneously achieved within these systems, opening a window on the natural history of the tissue they came from. Consistently, their use for protocols of in vitro culture of normal and malignant tissue-derived SCs is now available for disease mechanism studies, drug discovery, chemoresistance and high-throughput screening, aiming at identifying molecules that inhibit cancer stem cell (CSC) proliferation, or at modulating tissue-derived stem cells (tSCs) growth and differentiation.

In this perspective, the specific trans-differentiation process of epithelial-to-mesenchymal transition (EMT), which is shared by both normal (developing/regenerating) and neoplastic tissues, can be nicely reproduced within ex vivo cultured spheroids. In these systems, a cell-migration/colonization mechanism is associated with the mechano-sensing apparatus and signaling, characterized by the reversible loss of epithelial (or endothelial) properties coupled with the acquisition of mesenchymal features.

EMT and its opposite MET (mesenchymal-to-epithelial transition) are significantly involved in stemness balance in both normal and malignant cell spheroids, and their modulation is strategic for the achievement of specific cell phenotypes [3–6]. At molecular level, EMT is mediated by the activation of several transcription factors (TFs), including those belonging to the Snail superfamily, such as SNAI1 and SNAI2 [7], by the loss of cell-junction molecules, such as E-cadherin (encoded by CDH1), and the acquisition of mesenchymal markers, such as vimentin. Activation of EMT has been particularly studied in several cancer spheroid models (e.g., mammospheres, prostaspheres, pancreatic spheres, neurospheres from nervous system tumors), as well as in normal tSCs (e.g., cardiospheres, neurospheres, retinal spheres) and embryonic SCs (e.g., blastulation, embryoid bodies, induced cell reprogramming) [8].

The controversial link between EMT and stemness in normal and neoplastic conditions has been extensively highlighted both in vitro and in vivo [9]. While a coexistence of EMT and MET is typical during development and tissue repair, allowing an intermediate phenotype which is possibly associated with stemness features, in cancer the same process is associated with invasion and progression. The two different perspectives postulated by Brabletz [10], in which both the association, as well as the separation between stemness and EMT can occur in cancer development and diffusion (depending on the time-window and microenvironment), may represent more than a realistic hypothesis, which needs to be further tested and exploited in tumor spheroid systems.

In this review, the role of EMT and its reverse process MET in the cell fate decision cross-road will be described, by taking into account the analogies and differences in the same shared signaling pathways, acting as pro-self-renewal in the context of cancer, or as self-growth limiting/differentiating in normal tissues. We will focus specifically on 3D cell spheroids as valuable SC/CSC niche models with a pivotal role in studying the role of EMT/MET related pathways in the modulation of cell stemness, differentiation and trans-differentiation. The key pathway of TGF-β, and its related network will be particularly evidenced, and potential pre-clinical/clinical application highlighted.

2. EMT and Stemness in Physiological and Transformed Tissues

EMT and MET processes have been long associated with the balance between stemness and differentiation in multiple cell models (physiologic, pathologic or transformed), with many regulatory pathways involved. This EMT/stemness relationship is often connected to cell ability for spheroid formation and growth, so that spheroid formation is indeed used as a functional SC assay in multiple systems, albeit with some limitations [11]. It is well established that EMT is a finely regulated process involving many interconnected pathways responsible for the phenotypic manifestation of epithelial versus mesenchymal features. A vast amount of data derives from embryology studies, which have identified specific properties modulated during this switch: the basement membrane structure, apical polarity and junctions, motility, and cell adhesion. Different EMT-TFs are responsible for the regulation of these properties, albeit in separate molecular systems, i.e. each cellular activity has its own control circuit made of specific TFs, so that complete EMT requires the simultaneous activation of all of them [12].

Several EMT specific TFs have been associated with stemness phenotypes through several mechanisms, such as modulation of stemness-related miRNAs. One example is the miR-200 family, which comprises members with strong epithelial-promoting effects, while concomitantly targeting multiple stem cell factors, such as Sox2 and Klf4. Zeb1 is an EMT activator which is also able to downregulate the miR-200 family, thus suppressing epithelial transcriptional programs and inducing stemness TFs in both cancer cells and embryonic SCs [13]. Conversely, miR-200c can block the physiological ability of mammary SCs to differentiate into gland tubules. Moreover, miR-200c can also inhibit clonal expansion of both adult and embryonic cancer cells through BMI1 [14], providing an interesting molecular similarity in the EMT-mediated regulation of stemness between normal tSCs and CSCs.

Indeed, pleiotropic proteins are also involved with epigenetic machineries in controlling the EMT and stemness balance. As an example, among its many guardian functions, p53 can be also considered as an "epithelium keeper", together with members of the miR-200 family, as previously mentioned, which are able to regulate EMT also by inhibiting specific E-cadherin repressors, such as Zeb1 and Zeb2 [15]. It has been shown that decreased p53 and miR-200c levels are associated with promotion of EMT and concomitant increase in the abundance of mammary epithelial and SCs [16].

Another important stemness regulating microRNA is let-7, which has been studied in multiple systems. It is downregulated in fetal neural SCs, and its expression gradually increases during postnatal life and aging, together with p16/p19, promoting the loss of neural SCs [17]. Lin28, an RNA binding protein able to regulate let-7, is also involved in SC function modulation; it is upregulated in both CSCs and induced pluripotent stem cells (iPSs), and its overexpression is able to significantly increase self-renewal and efficiency of reprogramming protocols [18]. Lin28 has been shown to be significantly expressed, particularly in more mesenchymal-like cells, while inducing EMT through let-7 downregulation. Lin28 modulates self-renewal and differentiation of mammary epithelial SCs [19], increases the efficiency of spheroid formation as mammospheres, and promotes migration in breast cancer cells [20].

The relationship between EMT and the regulation of the stemness/differentiation balance emerges also in adult tissues during wound healing when cell cycle re-entry, dedifferentiation (to some extent) and motility are needed for injury repair. In fact, during skin wound healing, basal epithelial cells are required to temporarily suppress their adherent immotile phenotype, migrate towards the wound edges, and then contribute to re-epithelialization. These cells undergo a partial EMT process activated by tissue injury [21], which is necessary for repair, and represent an example of how some dynamic features linked to EMT are required for normal tissue maintenance and healing during post-natal life. Partial EMT during repair occurs in keratinocytes as well, in an EGF/Erk5/SNAI2-regulated way [22]. Albeit historically named after epithelial cells, EGF is able to sustain SNAI2 transcription associated with intermediate EMT states and stemness features. Interestingly, EGF-signaling blocking has been related to impaired stem/progenitor cell functions also in a mesodermal tissue, such as the heart [23]. Moreover, it has been shown that modulation of the β2-adrenergic signaling pathway, which is able to affect the EMT balance in human adult cardiac progenitor cells, is associated with enhanced SC features and increased cell spheroid formation [24].

Partial EMT states associated with stem/progenitor cell functions also derive from the lung, where wound healing is again linked to the acquisition of mesenchymal traits by club and basal cells [25]. These cells are facultative SCs in lung tissue, and when activated by injury, they undergo a transient mesenchymal state, which is necessary for tissue regeneration. They activate the expression of the mesenchymal marker vimentin, while showing mixed epithelial/mesenchymal features. Interestingly, this process, occurring in vivo during tissue regeneration, has also been observed ex vivo in human lung progenitor cells when selectively cultured as spheroids, in association with features of enhanced differentiation potential, as in a SC niche-like microenvironment [26], strenghtnening the modelling potential of cell spheroids ex vivo.

Vimentin is a marker of partial EMT states also in the mammary epithelium. In fact, overexpression of SNAI1 in primary human mammary epithelial cells results in higher vimentin expression levels, with coherently reduced E-cadherin, and these mesenchymal traits are again associated with a higher efficiency of spheroid growth as mammospheres. The same mechanism can be observed in mouse mammary SCs, with mesenchymal features associated with enhanced stemness functions [27]. Moreover, other EMT inducers, such as Six1 and LBX1, can also enhance mammosphere formation through Zeb1 or SNAI1, leading again to SC expansion and mammary hyperplasia in the mouse [28].

EMT circuits have been reported to affect another important SC function, i.e. the balance between symmetric and asymmetric division, which affects the dimension of the SC population itself. In colorectal CSCs, SNAI1 has been shown to promote symmetric division through miR-146a and beta-catenin, thus amplifying the SC pool [29].

Albeit multiple similarities exist in normal versus transformed systems, it has been shown that EMT is activated to different extents and through different TFs in normal mammary epithelial SCs compared to tumor initiating cells [30]. In fact, normal SCs in the basal epithelium rely on a SNAI2-mediated signal for partial EMT activation related to their physiologic support of gland turnover, which is characterized by co-expression of epithelial and mesenchymal markers, including ZEB1 expression. In comparison, mammary tumor initiating cells with properties of CSCs are highly SNAI1 positive and display a stronger mesenchymal phenotype, including complete loss of E-cadherin expression, detachment, acquisition of motility, and lack of ZEB1 expression.

Interestingly, in another important system where the stemness/differentiation balance requires modulation, i.e. cell reprogramming to iPSs, the opposite transition is involved. In fact, fibroblast reprogramming requires MET, which by some authors has been figuratively described as "moving backwards" in their developmental program, reaching towards the more epithelial-like embryonic state. Reprogramming TFs (e.g., Sox2, Oct4, c-Myc, and Klf4) are able to downregulate multiple EMT mediators, such as SNAI1 and TGF-β, and upregulate E-cadherin [31]. Blocking this transition can significantly hamper the efficiency of reprogramming protocols. Moreover, cells at intermediate states during reprogramming closely resemble, at transcriptomic level, MET-driven developmental processes during mesendoderm formation in the primitive streak [32], providing another significant clue linking stemness and EMT.

Considering these examples, it may seem inconsistent that EMT and MET are both associated with the acquisition of stemness features in different systems. The current view is that stemness features are not simply associated with a "more epithelial" or "more mesenchymal" phenotype, but indeed to intermediate so-called "metastable" EMT states [33], which can be encountered during a transition in both directions, and have been studied in both CSCs and development [34]. Studies have compared the molecular, epigenetic, and phenotypic features of trophoblast SCs (brought to intermediate EMT states through MAP3K4 inactivation, or SNAI1 upregulation) to that of invasive breast cancer cells, finding significant similarities between their "metastable" EMT states, both characterized by enhanced stemness functions, such as self-renewal, multi-lineage potential, and motility [35]. Considering the previously mentioned epigenetic control systems, miR-200 can inhibit Lin28, thus linking an intermediate stemness state to a more mesenchymal phenotype, while if let-7 inhibits ZEB, the process is brought towards the opposite end, i.e. towards an epithelial phenotype [30]. Therefore, the Lin28/let-7 ratio seems to play a fundamental role in the balance between transitions, and in the interplay with other mediators, such as SNAI1 and Twist.

It has been proposed that miR-200, Zeb, Lin28 and let-7 are all part of a circuit that modulates the EMT-stemness network through common regulatory factors, which move the activation of stemness features between a more epithelial or more mesenchymal state [36]. This theory of a flexible "stemness window" between EMT and MET may reconcile many different studies that have described apparently contradictory results, highlighting the concept that intermediate states are the ones that may be actually associated with stemness features in both normal and transformed cells.

3. EMT-Induced Spheroids as an in Vitro Model of Stem Cell Niches and Tumors

As mentioned above, despite the importance of traditional 2D cultures, 3D systems are generally considered as more representative models of living tissues, and have been widely used in stem cell biology, cancer biology, and tissue engineering [37]. In the SC field, spheroids have been obtained from different adult organs, starting from liver [38] and brain [39] over twenty years ago, to various stromal tissues, including human and murine hearts [40], human skeletal muscle [41], human bladder [42], exocrine pancreas [43], thyroid [44], kidney [45], breast [46], lung [26,47], and bone marrow [48]. Compared to monolayer cultures, cells in spheroid cultures are more resistant to senescence and apoptosis, present better in vivo engraftment, enhanced paracrine secretion of angiogenic, pro-survival and anti-inflammatory factors, and a broader differentiation potential [26,38,45,48,49] (Figure 1).

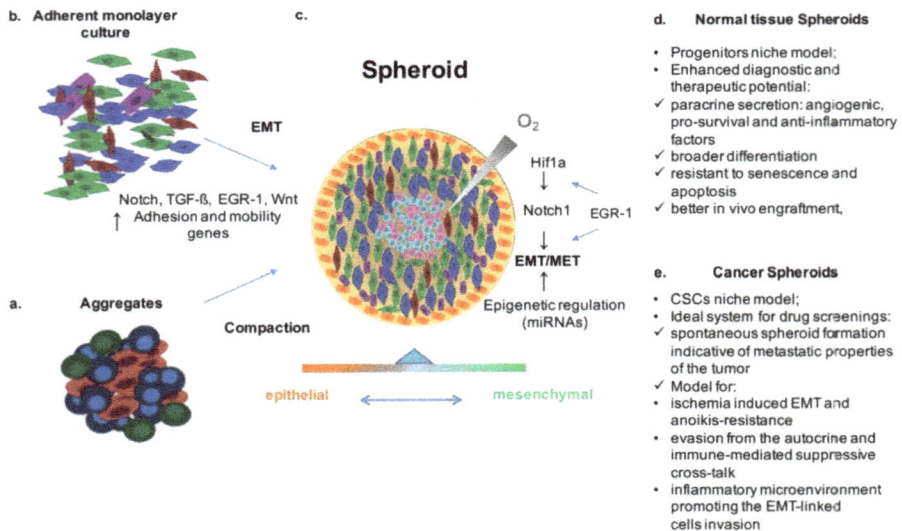

Figure 1. EMT-induced spheroids as an in vitro model of stem cell niches and tumors. Spheroids can be obtained by aggregation and compaction (**a**) or spontaneous formation from monolayer cultures, mediated by activation of EMT related pathway and gene associated with adhesion and motility (**b**). Spheroids are globular and compact structures that can be handled without causing mechanical dissociation of the cells, characterized by a hypoxic core with quiescent cells surrounded by a rim of proliferating cells. The hypoxic gradient activates Notch and other EMT associated pathways favoring the maintenance of a "metastable" state of differentiation within the spheroid (**c**). Characteristics of normal tissue spheroids (**d**) and cancer spheroids (**e**). EMT: epithelial-to-mesenchymal transition; MET: mesenchymal-to-epithelial transition; TGF-β: Transforming Growth Factor-beta; EGR-1: Early growth response protein 1.

In these conditions, cells are also exposed to more physiological cell-cell, cell-matrix interactions, and recreate a microenvironment which is thought to resemble tissue-resident SC niches [50], characterized by low-oxygen tension, which is important for the maintenance of an undifferentiated phenotype [51]. It has been proposed that low oxygen tension presents a selective advantage for SCs, which are in this way protected from the oxidative stress and DNA damages associated with aerobic metabolism [52]. In particular, oxygen levels seem to modulate the fine balance between proliferation and quiescence in SCs. Hypoxia-mediated stabilization of HIF-1a is known to activate Notch1 pathway, which has been associated with stemness maintenance, but also EMT, and may have a role in maintaining the intermediate "metastable" state previously discussed [53]. One possible drawback of this culture system is that excessive diameter of the spheroid may induce necrosis of the cells in the core. Mechanical dissociation to maintain a small diameter and spinner culture techniques can be adopted to ensure the optimal diffusion of nutrients, oxygen and waste [48,50].

Since the pioneering experiments of Sutherland et al. [54], spheroids have also been widely used in the cancer field, and have proven to be a valuable tool for drug testing, as they allow modeling in vitro the poorly vascularized microenvironment typical of solid tumors, with limited availability of blood-borne oxygen and nutrients. The low oxygen tension within tumors can lead to hypoxia-induced EMT, which is considered a landmark feature of the more aggressive and invasive cancer types [55–59] (Figure 1).

The definition of spheroids however is not always clear. Spheroids are generally described as globular and compact structures, that can be handled without causing mechanical dissociation of the cells, characterized by a hypoxic core with quiescent cells surrounded by a rim of proliferating cells [55]. Nonetheless, aggregates of loosely attached cells are sometimes inaccurately defined as spheroids, even in absence of proper spherical shape, true cell-cell and cell-matrix interactions, and hypoxic gradient [57]. Therefore, it is important to take into account how spheroids have been generated when comparing different studies.

Spheroids can be obtained by aggregation or spontaneous self-assembly. Aggregation can be induced by forcing cell-cell contact through different methods, such as hanging drop, liquid overlay/cell suspension culture, microwell array (round bottom 96 well plates), or microfluidics (i.e., in gel encapsulation) [60]. The initially loose integrins-ECM interaction within these aggregates is followed by E-cadherin accumulation and compaction. Spheroids can also be generated spontaneously from single cells in suspension, budding from monolayers or from adherent cells plated on cationic substrates, such as poly-D-lysine [40,61] or chitosan (the deacetylated derivative of chitin) [62]. Even though spheroids formed by aggregation represent a significant improvement compared to 2D cultures, their gene expression and active pathways are generally different from spontaneously formed clusters, which reflect more physiological mechanisms.

Mesenchymal/epithelial plasticity plays a central role in spontaneous formation of spheroids. Self-assembled human mesenchymal stem cell (MSC) spheroids on chitosan attach and spread on the membranes before retracting their pseudopodia and forming the multicellular spheroids [63]. This process is accompanied by activation of TGF-β, Notch, and Wnt pathways, and upregulation of genes associated with cell adhesion (e.g., integrins) and motility. Similarly, we have shown that TGF-β-mediated EMT is essential for the formation and maintenance of another model of adhesion-dependent spheroid system, that is the cardiosphere: in fact, TGF-β treatment increases cardiosphere formation, while the selective inhibitor SB431542 blocks cardiosphere formation and induces spreading of pre-existing ones [64].

EMT is important also for spontaneous formation of tumoral spheroids, as shown for example by high vimentin and lack of E-cadherin expression in spheroids spontaneously budding from monolayer cultures of ovarian cancer [65,66], which appear to be more clinically relevant models than those obtained by artificial aggregation, and an ideal system for reliable anti-cancer drug screening [67]. The ability to spontaneously form compact spheroids is reflective of an intrinsic molecular program of the parent tumor, and it is a good predictor of its progression, more than the expression of classical mesenchymal markers. While some studies have shown that expression of EMT-inducing TFs, such as Twist, is associated with the acquisition of CSC phenotype and metastatic properties [27,68], others have reported that metastatic tumors do actually retain an epithelial phenotype. Thus EMT may not be essential for the spreading of the primary tumor, but still be involved in the acquisition of chemo-resistance [69,70]. More recently, Beerling et al. [71], through high-resolution cell tracing experiments in a mouse model that spontaneously develops ductal mammary carcinoma, were able to show that the transition to a mesenchymal state is important for cell migration, but does not necessarily confer differential stemness and growth capacity, since most of the migrating cells adopt an epithelial state after the first few cell divisions. Independently from the expression of strictly mesenchymal or epithelial markers, the ability to spontaneously form spheroids is still a good predictor of metastatic potential [72]. This apparent paradox may be explained by the fact that EMT cannot be considered as a unidirectional transition between two very fixed states. As mentioned above, it is indeed a metastable process with different possible intermediate states [9], and, as such, it may not be modelled correctly by depleting or overexpressing classically EMT-associated genes, which may have oncogenic functions independently from their ability to induce EMT [71]. By phenotypic characterization of different ovarian carcinoma cell lines, Huang et al. have identified four different states along the EMT spectrum: epithelial, intermediate epithelial, intermediate mesenchymal, and mesenchymal. Cells with an intermediate mesenchymal phenotype were the most anoikis-resistant and spheroidogenic, suggesting that the most aggressive phenotype is not associated with a fully epithelial or fully mesenchymal state [73]. This Ying-Yang-like equilibrium may represent a crucial point in both normal and transformed cell spheroids in terms of changes in differentiation program and aggressiveness acquirement, respectively.

Indeed, most tumor-derived spheroids (tumorspheres) show a floating appearance, enriched in CSCs. On this basis and with the limitations linked to the method (e.g., high cell density not corresponding to clonal conditions, different spheroids size), several CSC types have been isolated and grown, often allowing the identification of the related targetable signaling pathway [8].

Interestingly, using the same method employed for the respective normal tissue, tumorspheres retaining CSC features have been isolated from brain tumors (e.g., human CD-133+ neurospheres) closely mimicking the genotype, gene-expression profile, and biology of parental tumors. Likewise, non-adherent mammospheres from human normal and transformed mammary epithelial cells have been isolated and extensively studied for their cancer initiating features, tumor heterogeneity, and pharmacologically sensible molecular networks (e.g., mouse breast cancer model of Erb-B2 receptor tyrosine kinase 2 expression and p53-deficiency [74]; study of the Wnt/β-catenin signaling pathway, and Sox2 expression [75]).

The study of colon and ovarian cancer-derived spheroids has also revealed the role of ROCK signaling inhibition in promoting cell survival and propagation, and in the acquisition of stemness features, including expression of CSC markers, capability for differentiation and tumorigenicity [8].

Consistently with their ability to reproduce the natural processes occurring in the normal or transformed tissue they came from, spheroids can also unveil epigenetic/EMT-dependent mechanisms and their related effects. In fact, as post-transcriptional gene expression regulators, several miRNAs participate in modulating self-renewal, differentiation and transformation in normal SCs and CSCs. In a spheroid model of hepatocellular tumor, miR-200a conferred a mesenchymal phenotype to oval-like progenitor cells, including an elongated cell morphology, enhanced cell migration ability, and expression of EMT-representative markers. Furthermore, several CSC-like traits and relative hepatic markers appeared in these cells, exhibiting enhanced spheroid-forming capacity and displaying superior resistance to chemotherapeutic drugs in vitro. All these miR-200-elicited effects occur by targeting the Wnt/β-catenin pathway. In addition, miR-200 participates in epigenetic modulation through a histone deacetylase 4/SP1/miR-200a regulatory network [76].

A very interesting synthesis of the potential critical step in the switch from normal multipotent mammary SCs and tumor initiating mammary CSCs has been addressed by Celia-Terrassa et al. using mammospheres as a 3D spheroid model [77]. In fact, by taking the advantage to reproduce the normal and transformed SC niche, the microenvironment-linked immuno-mediated cross-talk can be easily studied in these systems. With the hypothesis that, as CSCs can do, normal SCs (including mammary) control the immune system (for example down-regulating MHCs complex) to sustain their cellular activity, the authors addressed the role of epigenetic mechanisms, such as miRNAs, and in particular miR-199-a, in promoting both normal and transformed mammary SC properties by repression of their ability (linked to the Ligand-dependent corepressor, LCOR, nuclear receptor) of being sensitized to interferon-induced differentiation and senescence. This epigenetic mechanism, represented by the mir-199-a/LCOR/interferon axis, mediates the evasion from the autocrine and immune microenvironment-mediated suppressive cross-talk, and is conserved in normal SCs and CSCs. These finding may be both mechanistically and pharmacologically strategic, taking also into account that the inflammatory microenvironment can promote EMT-linked cell invasion [77–79].

4. Discovering Pharmacological Targets in Spheroid Model: The Case of EGR-1/TGF-β Network

Within the molecular networks strategic for cell survival, drug escape and anchorage independence, which have been studied using the spheroid models, the activation of the Early growth response protein 1 (EGR-1) and its downstream signaling components (MAPK/ERK), including its link with TGF-β, represents a milestone in the detection of critical EMT-dependent pharmacological targets. The individual role of these signaling networks, as well as their relationship with the EMT/MET process, are extensively outlined in other reports, including those collected in this issue. Here the role of 3D spheroid models to unravel their function will be better highlighted.

TGF-β signaling has been suggested to have crucial roles in several features of CSCs, such as in tumor initiation, metastasis, and resistance to anticancer drugs [80,81]. As mentioned above, it has also an important role in the spontaneous formation of tumoral spheroids, and in promoting the malignant progression of these structures [81,82].

Among other targets, TGF-β induces EGR-1, which in turn activates the transcription of several mesenchymal proteins, such as type I collagen and TGF-β itself [83,84]. EGR-1 may serve as a target regulated by TGF-β, as mediator for enhanced TGF-β gene expression and target cell responsiveness [85], as well as co-author of physiologic stress response programs [83]. EGR1 is a zinc-finger TF that binds to GC-rich recognition motifs. EGR-1 is also induced by a number of different stimuli, such as anti-cancer drugs, oxidized lipids, hyperglycemia, growth factors and ionizing radiation, and inhibits or stimulates tumor growth depending on the cellular context and the duration of EGR-1 induction [85,86]. While transient induction of EGR-1 is known to activate angiogenesis, sustained EGR-1 expression induces block of angiogenesis, growth arrest, and apoptosis [87]. This TF is able to directly regulate multiple tumor suppressors to induce apoptotic cell death [85,88], including p53 and PTEN. This latter in particular is also strongly related to the ability of CSCs to form spheres, as suggested by experiments where PTEN knockout was potentiating the invasiveness of colorectal cancer spheroidal cells through a 3D extracellular matrix [89].

In addition, EGR-1 is induced by hypoxia and plays a critical role in hypoxia-induced tumor progression, survival, and angiogenesis [90,91]. Thus, 3D spheroids, which model the hypoxic microenvironment of solid tumor, have proven to be a valuable in vitro model to study the dual role of this TF in different contexts. For example, using multicellular tumor spheroids, it has been shown that EGR-1 overexpression makes tumor cells more sensitive to necrosis induced by glucose depletion, and blocking EGR-1 with a shRNA suppresses growth of the tumorspheres [92]. On the other end, in head and neck squamous cell carcinoma, oxytocin treatment significantly reduces cell migration and spheroid formation by upregulating EGR-1 [93].

Interestingly, EGR-1 is also a target of miR-181 involved in TGF-β-mediated tumor mammosphere formation [94,95], and is upregulated in breast cancer cells expressing high level of NF-κB-induced kinase, which has been associated to a more "stemness" phenotype, promoting cancer expansion and mammosphere formation [96]. Silencing of EGR-1 with syntactic catalytic DNA has been reported to inhibit human breast carcinoma proliferation and migration [97], while on the other hand downregulation of gelsolin (indicator of breast cancer) has correlated with suppression of EGR-1 [98]. In summary, this TF appears to be among the master genes in cellular stress responses. Depending on the cell type, the duration and intensity of the stimuli, EGR-1 can act as a tumor repressor by inducing necrosis/apoptosis, block of angiogenesis and proliferation arrest, or can promote EMT-mediated cell migration, invasion, tumor growth, and acquisition of chemo-resistance [99–101].

EMT/MET processes seem to mediate adaptive responses of cancer cells and CSCs to therapy, resulting in poor chemotherapy response and negative prognosis. These developmental programs can be epitomized by oncogenically transformed cells during tumor progression [33,102]. Intriguingly, EMT can trigger reversion to a CSC-like phenotype [27,103], shedding light on a possible association between EMT, CSCs and drug resistance. Moreover, EMT/MET processes are involved in tumor spheroids formation, which have increased resistance to chemotherapeutics compared to 2D cultures, mimicking more closely in vivo tumor behavior. Therefore, standardized high-throughput ex vivo modelling of cancer with 3D cultures derived from patient biopsies can realistically provide a platform for the study of the molecular pathways involved in the evolution of the tumor, and for personalized drug screening and testing, taking into account the enormous variability among patients and within the tumor.

5. Conclusions and Future Perspectives

For several tissue-derived cells of heterogeneous origin, grown in an appropriate microenvironment, the spheroid (niche-like)-forming capacity per se is typical of stem/progenitor cells, irrespectively of their normal or neoplastic nature [1,64,104,105]. This spheroid self-building property can be considered as an EMT-dependent process, mediated by TGF-β and its network signaling. The hypoxic gradient within the spheroids might favor the maintenance of a metastable state associated with the acquisition of stem-like features.

Among its many functions, EMT supports migration in tumor (cell evasion and metastasis) and normal cells (embryonic-fetal development and adult tissue repair). Despite all the differences between normal and transformed cells, common mechanisms shared by normal and malignant SCs can be identified, such as those protecting them from suppressive immune cytokine signaling, which can be evidenced in the mammosphere model.

Moreover, spheroids offer an easy experimental tool. As an example, the molecular loops between EGR-1, TGF-β and EMT can be easily studied in 3D models as multicellular spheroid CSC compartments, where this network can be tamed under different stimuli and new drugs can be tested, such as antibodies or small inhibitors. Therefore, other than bridging the gap between in vivo and in vitro studies, the use of spheroids can accelerate the setting of protocols for a more personalized medicine, and for precision diagnostics and therapy.

Conflicts of Interest: The authors declare no conflict of interest.

References

1. Stankevicius, V.; Kunigenas, L.; Stankunas, E.; Kuodyte, K.; Strainiene, E.; Cicenas, J.; Samalavicius, N.E.; Suziedelis, K. The expression of cancer stem cell markers in human colorectal carcinoma cells in a microenvironment dependent manner. *Biochem Biophys Res Commun* **2017**, *484*, 726–733. [CrossRef] [PubMed]
2. Bartlett, R.; Everett, W.; Lim, S.; Natasha, G.; Loizidou, M.; Jell, G.; Tan, A.; Seifalian, A.M. Personalized in vitro cancer modeling—-Fantasy or reality? *Transl. Oncol.* **2014**, *7*, 657–664. [CrossRef] [PubMed]
3. Chu, J.H.; Yu, S.; Hayward, S.W.; Chan, F.L. Development of a three-dimensional culture model of prostatic epithelial cells and its use for the study of epithelial-mesenchymal transition and inhibition of PI3K pathway in prostate cancer. *Prostate* **2009**, *69*, 428–442. [CrossRef] [PubMed]
4. Shepherd, T.G.; Mujoomdar, M.L.; Nachtigal, M.W. Constitutive activation of bmp signalling abrogates experimental metastasis of ovca429 cells via reduced cell adhesion. *J. Ovarian Res.* **2010**. [CrossRef] [PubMed]
5. Sanguinetti, A.; Santini, D.; Bonafe, M.; Taffurelli, M.; Avenia, N. Interleukin-6 and pro inflammatory status in the breast tumor microenvironment. *World J. Surg. Oncol.* **2015**, *13*, 129. [CrossRef] [PubMed]
6. Seifert, A.; Posern, G. Tightly controlled mrtf-a activity regulates epithelial differentiation during formation of mammary acini. *Breast Cancer Res.* **2017**, *19*, 68. [CrossRef] [PubMed]
7. Wang, Y.; Shi, J.; Chai, K.; Ying, X.; Zhou, B.P. The role of snail in EMT and tumorigenesis. *Curr. Cancer Drug Targets* **2013**, *13*, 963–972. [CrossRef] [PubMed]
8. Ishiguro, T.; Ohata, H.; Sato, A.; Yamawaki, K.; Enomoto, T.; Okamoto, K. Tumor-derived spheroids: Relevance to cancer stem cells and clinical applications. *Cancer Sci.* **2017**, *108*, 283–289. [CrossRef] [PubMed]
9. Fabregat, I.; Malfettone, A.; Soukupova, J. New insights into the crossroads between EMT and stemness in the context of cancer. *J. Clin. Med.* **2016**. [CrossRef] [PubMed]
10. Brabletz, T. Emt and met in metastasis: Where are the cancer stem cells? *Cancer Cell* **2012**, *22*, 699–701. [CrossRef] [PubMed]
11. Pastrana, E.; Silva-Vargas, V.; Doetsch, F. Eyes wide open: A critical review of sphere-formation as an assay for stem cells. *Cell Stem Cell* **2011**, *8*, 486–498. [CrossRef] [PubMed]
12. Saunders, L.R.; McClay, D.R. Sub-circuits of a gene regulatory network control a developmental epithelial-mesenchymal transition. *Development* **2014**, *141*, 1503–1513. [CrossRef] [PubMed]
13. Wellner, U.; Schubert, J.; Burk, U.C.; Schmalhofer, O.; Zhu, F.; Sonntag, A.; Waldvogel, B.; Vannier, C.; Darling, D.; zur Hausen, A.; et al. The emt-activator zeb1 promotes tumorigenicity by repressing stemness-inhibiting micrornas. *Nat. Cell Biol.* **2009**, *11*, 1487–1495. [CrossRef] [PubMed]

14. Shimono, Y.; Zabala, M.; Cho, R.W.; Lobo, N.; Dalerba, P.; Qian, D.; Diehn, M.; Liu, H.; Panula, S.P.; Chiao, E.; et al. Downregulation of mirna-200c links breast cancer stem cells with normal stem cells. *Cell* **2009**, *138*, 592–603. [CrossRef] [PubMed]

15. Gregory, P.A.; Bert, A.G.; Paterson, E.L.; Barry, S.C.; Tsykin, A.; Farshid, G.; Vadas, M.A.; Khew-Goodall, Y.; Goodall, G.J. The mir-200 family and mir-205 regulate epithelial to mesenchymal transition by targeting zeb1 and sip1. *Nat. Cell Biol.* **2008**, *10*, 593–601. [CrossRef] [PubMed]

16. Chang, C.J.; Chao, C.H.; Xia, W.; Yang, J.Y.; Xiong, Y.; Li, C.W.; Yu, W.H.; Rehman, S.K.; Hsu, J.L.; Lee, H.H.; et al. P53 regulates epithelial-mesenchymal transition and stem cell properties through modulating mirnas. *Nat. Cell Biol.* **2011**, *13*, 317–323. [CrossRef] [PubMed]

17. Nishino, J.; Kim, I.; Chada, K.; Morrison, S.J. Hmga2 promotes neural stem cell self-renewal in young but not old mice by reducing p16ink4a and p19arf expression. *Cell* **2008**, *135*, 227–239. [CrossRef] [PubMed]

18. Hanna, J.; Saha, K.; Pando, B.; van Zon, J.; Lengner, C.J.; Creyghton, M.P.; van Oudenaarden, A.; Jaenisch, R. Direct cell reprogramming is a stochastic process amenable to acceleration. *Nature* **2009**, *462*, 595–601. [CrossRef] [PubMed]

19. Yang, X.; Lin, X.; Zhong, X.; Kaur, S.; Li, N.; Liang, S.; Lassus, H.; Wang, L.; Katsaros, D.; Montone, K.; et al. Double-negative feedback loop between reprogramming factor lin28 and microrna let-7 regulates aldehyde dehydrogenase 1-positive cancer stem cells. *Cancer Res* **2010**, *70*, 9463–9472. [CrossRef] [PubMed]

20. Liu, Y.; Li, H.; Feng, J.; Cui, X.; Huang, W.; Li, Y.; Su, F.; Liu, Q.; Zhu, J.; Lv, X.; et al. Lin28 induces epithelial-to-mesenchymal transition and stemness via downregulation of let-7a in breast cancer cells. *PLoS ONE* **2013**, *8*, e83083. [CrossRef] [PubMed]

21. Shaw, T.J.; Martin, P. Wound repair: A showcase for cell plasticity and migration. *Curr. Opin. Cell Biol.* **2016**, *42*, 29–37. [CrossRef] [PubMed]

22. Arnoux, V.; Nassour, M.; L'Helgoualc'h, A.; Hipskind, R.A.; Savagner, P. Erk5 controls slug expression and keratinocyte activation during wound healing. *Mol. Biol. Cell* **2008**, *19*, 4738–4749. [CrossRef] [PubMed]

23. Barth, A.S.; Zhang, Y.; Li, T.; Smith, R.R.; Chimenti, I.; Terrovitis, I.; Davis, D.R.; Kizana, E.; Ho, A.S.; O'Rourke, B.; et al. Functional impairment of human resident cardiac stem cells by the cardiotoxic antineoplastic agent trastuzumab. *Stem Cells Transl. Med.* **2012**, *1*, 289–297. [CrossRef] [PubMed]

24. Pagano, F.; Angelini, F.; Siciliano, C.; Tasciotti, J.; Mangino, G.; De Falco, E.; Carnevale, R.; Sciarretta, S.; Frati, G.; Chimenti, I. Beta2-adrenergic signaling affects the phenotype of human cardiac progenitor cells through emt modulation. *Pharmacol Res.* **2017**. [CrossRef] [PubMed]

25. Vaughan, A.E.; Chapman, H.A. Regenerative activity of the lung after epithelial injury. *Biochim. Biophys. Acta* **2013**, *1832*, 922–930. [CrossRef] [PubMed]

26. Chimenti, I.; Pagano, F.; Angelini, F.; Siciliano, C.; Mangino, G.; Picchio, V.; De Falco, E.; Peruzzi, M.; Carnevale, R.; Ibrahim, M.; et al. Human lung spheroids as in vitro niches of lung progenitor cells with distinctive paracrine and plasticity properties. *Stem Cells Transl. Med.* **2017**, *6*, 767–777. [CrossRef] [PubMed]

27. Mani, S.A.; Guo, W.; Liao, M.J.; Eaton, E.N.; Ayyanan, A.; Zhou, A.Y.; Brooks, M.; Reinhard, F.; Zhang, C.C.; Shipitsin, M.; et al. The epithelial-mesenchymal transition generates cells with properties of stem cells. *Cell* **2008**, *133*, 704–715. [CrossRef] [PubMed]

28. McCoy, E.L.; Iwanaga, R.; Jedlicka, P.; Abbey, N.S.; Chodosh, L.A.; Heichman, K.A.; Welm, A.L.; Ford, H.L. Six1 expands the mouse mammary epithelial stem/progenitor cell pool and induces mammary tumors that undergo epithelial-mesenchymal transition. *J. Clin. Investig.* **2009**, *119*, 2663–2677. [CrossRef] [PubMed]

29. Hwang, W.L.; Jiang, J.K.; Yang, S.H.; Huang, T.S.; Lan, H.Y.; Teng, H.W.; Yang, C.Y.; Tsai, Y.P.; Lin, C.H.; Wang, H.W.; et al. Microrna-146a directs the symmetric division of snail-dominant colorectal cancer stem cells. *Nat. Cell Biol.* **2014**, *16*, 268–280. [CrossRef] [PubMed]

30. Ye, X.; Tam, W.L.; Shibue, T.; Kaygusuz, Y.; Reinhardt, F.; Ng Eaton, E.; Weinberg, R.A. Distinct emt programs control normal mammary stem cells and tumour-initiating cells. *Nature* **2015**, *525*, 256–260. [CrossRef] [PubMed]

31. Li, R.; Liang, J.; Ni, S.; Zhou, T.; Qing, X.; Li, H.; He, W.; Chen, J.; Li, F.; Zhuang, Q.; et al. A mesenchymal-to-epithelial transition initiates and is required for the nuclear reprogramming of mouse fibroblasts. *Cell Stem Cell* **2010**, *7*, 51–63. [CrossRef] [PubMed]

32. Takahashi, K.; Tanabe, K.; Ohnuki, M.; Narita, M.; Sasaki, A.; Yamamoto, M.; Nakamura, M.; Sutou, K.; Osafune, K.; Yamanaka, S. Induction of pluripotency in human somatic cells via a transient state resembling primitive streak-like mesendoderm. *Nat Commun* **2014**, *5*, 3678. [CrossRef] [PubMed]

33. Nieto, M.A.; Huang, R.Y.; Jackson, R.A.; Thiery, J.P. Emt: 2016. *Cell* **2016**, *166*, 21–45. [CrossRef] [PubMed]
34. Abell, A.N.; Jordan, N.V.; Huang, W.; Prat, A.; Midland, A.A.; Johnson, N.L.; Granger, D.A.; Mieczkowski, P.A.; Perou, C.M.; Gomez, S.M.; et al. Map3k4/cbp-regulated h2b acetylation controls epithelial-mesenchymal transition in trophoblast stem cells. *Cell Stem Cell* **2011**, *8*, 525–537. [CrossRef] [PubMed]
35. Jordan, N.V.; Johnson, G.L.; Abell, A.N. Tracking the intermediate stages of epithelial-mesenchymal transition in epithelial stem cells and cancer. *Cell Cycle* **2011**, *10*, 2865–2873. [CrossRef] [PubMed]
36. Jolly, M.K.; Jia, D.; Boareto, M.; Mani, S.A.; Pienta, K.J.; Ben-Jacob, E.; Levine, H. Coupling the modules of emt and stemness: A tunable "stemness window" model. *Oncotarget* **2015**, *6*, 25161–25174. [CrossRef] [PubMed]
37. Gentile, C. Filling the gaps between the in vivo and in vitro microenvironment: Engineering of spheroids for stem cell technology. *Curr. Stem Cell Res. Ther.* **2016**, *11*, 652–665. [CrossRef] [PubMed]
38. Landry, J.; Bernier, D.; Ouellet, C.; Goyette, R.; Marceau, N. Spheroidal aggregate culture of rat liver cells: Histotypic reorganization, biomatrix deposition, and maintenance of functional activities. *J. Cell Biol.* **1985**, *101*, 914–923. [CrossRef] [PubMed]
39. Reynolds, B.A.; Weiss, S. Generation of neurons and astrocytes from isolated cells of the adult mammalian central nervous system. *Science* **1992**, *255*, 1707–1710. [CrossRef] [PubMed]
40. Messina, E.; De Angelis, L.; Frati, G.; Morrone, S.; Chimenti, S.; Fiordaliso, F.; Salio, M.; Battaglia, M.; Latronico, M.V.; Coletta, M.; et al. Isolation and expansion of adult cardiac stem cells from human and murine heart. *Circ. Res.* **2004**, *95*, 911–921. [CrossRef] [PubMed]
41. Alessandri, G.; Pagano, S.; Bez, A.; Benetti, A.; Pozzi, S.; Iannolo, G.; Baronio, M.; Invernici, G.; Caruso, A.; Muneretto, C.; et al. Isolation and culture of human muscle-derived stem cells able to differentiate into myogenic and neurogenic cell lineages. *Lancet* **2004**, *364*, 1872–1883. [CrossRef]
42. Fierabracci, A.; Caione, P.; Di Giovine, M.; Zavaglia, D.; Bottazzo, G.F. Identification and characterization of adult stem/progenitor cells in the human bladder (bladder spheroids): Perspectives of application in pediatric surgery. *Pediatr. Surg. Int.* **2007**, *23*, 837–839. [CrossRef] [PubMed]
43. Puglisi, M.A.; Giuliani, L.; Fierabracci, A. Identification and characterization of a novel expandable adult stem/progenitor cell population in the human exocrine pancreas. *J. Endocrinol. Investig.* **2008**, *31*, 563–572. [CrossRef] [PubMed]
44. Fierabracci, A.; Puglisi, M.A.; Giuliani, L.; Mattarocci, S.; Gallinella-Muzi, M. Identification of an adult stem/progenitor cell-like population in the human thyroid. *J. Endocrinol.* **2008**, *198*, 471–487. [CrossRef] [PubMed]
45. Huang, Y.; Johnston, P.; Zhang, B.; Zakari, A.; Chowdhry, T.; Smith, R.R.; Marban, E.; Rabb, H.; Womer, K.L. Kidney-derived stromal cells modulate dendritic and t cell responses. *J. Am. Soc. Nephrol.* **2009**, *20*, 831–841. [CrossRef] [PubMed]
46. Farnie, G.; Clarke, R.B. Mammary stem cells and breast cancer—Role of notch signalling. *Stem Cell Rev.* **2007**, *3*, 169–175. [CrossRef] [PubMed]
47. Henry, E.; Cores, J.; Hensley, M.T.; Anthony, S.; Vandergriff, A.; de Andrade, J.B.; Allen, T.; Caranasos, T.G.; Lobo, L.J.; Cheng, K. Adult lung spheroid cells contain progenitor cells and mediate regeneration in rodents with bleomycin-induced pulmonary fibrosis. *Stem Cells Transl. Med.* **2015**, *4*, 1265–1274. [CrossRef] [PubMed]
48. Cesarz, Z.; Tamama, K. Spheroid culture of mesenchymal stem cells. *Stem Cells Int.* **2016**, *2016*, 9176357. [CrossRef] [PubMed]
49. Li, T.S.; Cheng, K.; Lee, S.T.; Matsushita, S.; Davis, D.; Malliaras, K.; Zhang, Y.; Matsushita, N.; Smith, R.R.; Marban, E. Cardiospheres recapitulate a niche-like microenvironment rich in stemness and cell-matrix interactions, rationalizing their enhanced functional potency for myocardial repair. *Stem Cells* **2010**, *28*, 2088–2098. [CrossRef] [PubMed]
50. Chimenti, I.; Massai, D.; Morbiducci, U.; Beltrami, A.P.; Pesce, M.; Messina, E. Stem cell spheroids and ex vivo niche modeling: Rationalization and scaling-up. *J. Cardiovasc. Transl. Res.* **2017**, *10*, 150–166. [CrossRef] [PubMed]
51. Mohyeldin, A.; Garzon-Muvdi, T.; Quinones-Hinojosa, A. Oxygen in stem cell biology: A critical component of the stem cell niche. *Cell Stem Cell* **2010**, *7*, 150–161. [CrossRef] [PubMed]
52. Cipolleschi, M.G.; Dello Sbarba, P.; Olivotto, M. The role of hypoxia in the maintenance of hematopoietic stem cells. *Blood* **1993**, *82*, 2031–2037. [PubMed]

53. Kluppel, M.; Wrana, J.L. Turning it up a notch: Cross-talk between TGF beta and notch signaling. *Bioessays* **2005**, *27*, 115–118. [CrossRef] [PubMed]

54. Sutherland, R.M.; Inch, W.R.; McCredie, J.A.; Kruuv, J. A multi-component radiation survival curve using an in vitro tumour model. *Int J Radiat Biol Relat Stud Phys Chem Med* **1970**, *18*, 491–495. [CrossRef] [PubMed]

55. Nagelkerke, A.; Bussink, J.; Sweep, F.C.; Span, P.N. Generation of multicellular tumor spheroids of breast cancer cells: How to go three-dimensional. *Anal Biochem* **2013**, *437*, 17–19. [CrossRef] [PubMed]

56. Lee, C.; Siu, A.; Ramos, D.M. Multicellular spheroids as a model for hypoxia-induced EMT. *Anticancer Res.* **2016**, *36*, 6259–6263. [CrossRef] [PubMed]

57. Hirschhaeuser, F.; Menne, H.; Dittfeld, C.; West, J.; Mueller-Klieser, W.; Kunz-Schughart, L.A. Multicellular tumor spheroids: An underestimated tool is catching up again. *J. Biotechnol.* **2010**, *148*, 3–15. [CrossRef] [PubMed]

58. Huang, Y.J.; Hsu, S.H. Acquisition of epithelial-mesenchymal transition and cancer stem-like phenotypes within chitosan-hyaluronan membrane-derived 3d tumor spheroids. *Biomaterials* **2014**, *35*, 10070–10079. [CrossRef] [PubMed]

59. Li, Q.; Chen, C.; Kapadia, A.; Zhou, Q.; Harper, M.K.; Schaack, J.; LaBarbera, D.V. 3d models of epithelial-mesenchymal transition in breast cancer metastasis: High-throughput screening assay development, validation, and pilot screen. *J. Biomol. Screen.* **2011**, *16*, 141–154. [CrossRef] [PubMed]

60. Fennema, E.; Rivron, N.; Rouwkema, J.; van Blitterswijk, C.; de Boer, J. Spheroid culture as a tool for creating 3d complex tissues. *Trends Biotechnol.* **2013**, *31*, 108–115. [CrossRef] [PubMed]

61. Chimenti, I.; Gaetani, R.; Barile, L.; Forte, E.; Ionta, V.; Angelini, F.; Frati, G.; Messina, E.; Giacomello, A. Isolation and expansion of adult cardiac stem/progenitor cells in the form of cardiospheres from human cardiac biopsies and murine hearts. *Methods Mol. Biol.* **2012**, *879*, 327–338. [PubMed]

62. Huang, G.S.; Dai, L.G.; Yen, B.L.; Hsu, S.H. Spheroid formation of mesenchymal stem cells on chitosan and chitosan-hyaluronan membranes. *Biomaterials* **2011**, *32*, 6929–6945. [CrossRef] [PubMed]

63. Yeh, H.Y.; Liu, B.H.; Sieber, M.; Hsu, S.H. Substrate-dependent gene regulation of self-assembled human msc spheroids on chitosan membranes. *BMC Genom.* **2014**, *15*, 10. [CrossRef] [PubMed]

64. Forte, E.; Miraldi, F.; Chimenti, I.; Angelini, F.; Zeuner, A.; Giacomello, A.; Mercola, M.; Messina, E. Tgfbeta-dependent epithelial-to-mesenchymal transition is required to generate cardiospheres from human adult heart biopsies. *Stem Cells Dev.* **2012**, *21*, 3081–3090. [CrossRef] [PubMed]

65. Pease, J.C.; Brewer, M.; Tirnauer, J.S. Spontaneous spheroid budding from monolayers: A potential contribution to ovarian cancer dissemination. *Biol. Open* **2012**, *1*, 622–628. [CrossRef] [PubMed]

66. Rafehi, S.; Ramos Valdes, Y.; Bertrand, M.; McGee, J.; Prefontaine, M.; Sugimoto, A.; DiMattia, G.E.; Shepherd, T.G. TGFbeta signaling regulates epithelial-mesenchymal plasticity in ovarian cancer ascites-derived spheroids. *Endocr. Relat. Cancer* **2016**, *23*, 147–159. [CrossRef] [PubMed]

67. Theodoraki, M.A.; Rezende, C.O., Jr.; Chantarasriwong, O.; Corben, A.D.; Theodorakis, E.A.; Alpaugh, M.L. Spontaneously-forming spheroids as an in vitro cancer cell model for anticancer drug screening. *Oncotarget* **2015**, *6*, 21255–21267. [CrossRef] [PubMed]

68. Yang, J.; Mani, S.A.; Donaher, J.L.; Ramaswamy, S.; Itzykson, R.A.; Come, C.; Savagner, P.; Gitelman, I.; Richardson, A.; Weinberg, R.A. Twist, a master regulator of morphogenesis, plays an essential role in tumor metastasis. *Cell* **2004**, *117*, 927–939. [CrossRef] [PubMed]

69. Fischer, K.R.; Durrans, A.; Lee, S.; Sheng, J.; Li, F.; Wong, S.T.; Choi, H.; El Rayes, T.; Ryu, S.; Troeger, J.; et al. Epithelial-to-mesenchymal transition is not required for lung metastasis but contributes to chemoresistance. *Nature* **2015**, *527*, 472–476. [CrossRef] [PubMed]

70. Zheng, X.; Carstens, J.L.; Kim, J.; Scheible, M.; Kaye, J.; Sugimoto, H.; Wu, C.C.; LeBleu, V.S.; Kalluri, R. Epithelial-to-mesenchymal transition is dispensable for metastasis but induces chemoresistance in pancreatic cancer. *Nature* **2015**, *527*, 525–530. [CrossRef] [PubMed]

71. Beerling, E.; Seinstra, D.; de Wit, E.; Kester, L.; van der Velden, D.; Maynard, C.; Schafer, R.; van Diest, P.; Voest, E.; van Oudenaarden, A.; et al. Plasticity between epithelial and mesenchymal states unlinks emt from metastasis-enhancing stem cell capacity. *Cell Rep.* **2016**, *14*, 2281–2288. [CrossRef] [PubMed]

72. Lou, Y.; Preobrazhenska, O.; auf dem Keller, U.; Sutcliffe, M.; Barclay, L.; McDonald, P.C.; Roskelley, C.; Overall, C.M.; Dedhar, S. Epithelial-mesenchymal transition (emt) is not sufficient for spontaneous murine breast cancer metastasis. *Dev. Dyn.* **2008**, *237*, 2755–2768. [CrossRef] [PubMed]

73. Huang, R.Y.; Wong, M.K.; Tan, T.Z.; Kuay, K.T.; Ng, A.H.; Chung, V.Y.; Chu, Y.S.; Matsumura, N.; Lai, H.C.; Lee, Y.F.; et al. An emt spectrum defines an anoikis-resistant and spheroidogenic intermediate mesenchymal state that is sensitive to e-cadherin restoration by a src-kinase inhibitor, saracatinib (azd0530). *Cell Death Dis.* **2013**, *4*, e915. [CrossRef] [PubMed]

74. Li, Y.; Zhang, T.; Korkaya, H.; Liu, S.; Lee, H.F.; Newman, B.; Yu, Y.; Clouthier, S.G.; Schwartz, S.J.; Wicha, M.S.; et al. Sulforaphane, a dietary component of broccoli/broccoli sprouts, inhibits breast cancer stem cells. *Clin. Cancer Res.* **2010**, *16*, 2580–2590. [CrossRef] [PubMed]

75. Kondratyev, M.; Kreso, A.; Hallett, R.M.; Girgis-Gabardo, A.; Barcelon, M.E.; Ilieva, D.; Ware, C.; Majumder, P.K.; Hassell, J.A. Gamma-secretase inhibitors target tumor-initiating cells in a mouse model of erbb2 breast cancer. *Oncogene* **2012**, *31*, 93–103. [CrossRef] [PubMed]

76. Yuan, J.H.; Yang, F.; Chen, B.F.; Lu, Z.; Huo, X.S.; Zhou, W.P.; Wang, F.; Sun, S.H. The histone deacetylase 4/sp1/microrna-200a regulatory network contributes to aberrant histone acetylation in hepatocellular carcinoma. *Hepatology* **2011**, *54*, 2025–2035. [CrossRef] [PubMed]

77. Celia-Terrassa, T.; Liu, D.D.; Choudhury, A.; Hang, X.; Wei, Y.; Zamalloa, J.; Alfaro-Aco, R.; Chakrabarti, R.; Jiang, Y.Z.; Koh, B.I.; et al. Normal and cancerous mammary stem cells evade interferon-induced constraint through the mir-199a-lcor axis. *Nat. Cell Biol.* **2017**, *19*, 711–723. [CrossRef] [PubMed]

78. Zhou, H.S.; Su, X.F.; Fu, X.L.; Wu, G.Z.; Luo, K.L.; Fang, Z.; Yu, F.; Liu, H.; Hu, H.J.; Chen, L.S.; et al. Mesenchymal stem cells promote pancreatic adenocarcinoma cells invasion by transforming growth factor-beta1 induced epithelial-mesenchymal transition. *Oncotarget* **2016**, *7*, 41294–41305. [CrossRef] [PubMed]

79. Trivanovic, D.; Jaukovic, A.; Krstic, J.; Nikolic, S.; Okic Djordjevic, I.; Kukolj, T.; Obradovic, H.; Mojsilovic, S.; Ilic, V.; Santibanez, J.F.; et al. Inflammatory cytokines prime adipose tissue mesenchymal stem cells to enhance malignancy of mcf-7 breast cancer cells via transforming growth factor-beta1. *IUBMB Life* **2016**, *68*, 190–200. [CrossRef] [PubMed]

80. Dancea, H.C.; Shareef, M.M.; Ahmed, M.M. Role of radiation-induced tgf-beta signaling in cancer therapy. *Mol. Cell Pharmacol.* **2009**, *1*, 44–56. [CrossRef] [PubMed]

81. Chen, Z.; Wang, X.; Jin, T.; Wang, Y.; Hong, C.S.; Tan, L.; Dai, T.; Wu, L.; Zhuang, Z.; Shi, C. Increase in the radioresistance of normal skin fibroblasts but not tumor cells by mechanical injury. *Cell Death Dis.* **2017**, *8*, e2573. [CrossRef] [PubMed]

82. Bhola, N.E.; Balko, J.M.; Dugger, T.C.; Kuba, M.G.; Sanchez, V.; Sanders, M.; Stanford, J.; Cook, R.S.; Arteaga, C.L. Tgf-beta inhibition enhances chemotherapy action against triple-negative breast cancer. *J. Clin. Investig.* **2013**, *123*, 1348–1358. [CrossRef] [PubMed]

83. Chen, S.J.; Ning, H.; Ishida, W.; Sodin-Semrl, S.; Takagawa, S.; Mori, Y.; Varga, J. The early-immediate gene egr-1 is induced by transforming growth factor-beta and mediates stimulation of collagen gene expression. *J. Biol. Chem.* **2006**, *281*, 21183–21197. [CrossRef] [PubMed]

84. Bastianelli, D.; Siciliano, C.; Puca, R.; Coccia, A.; Murdoch, C.; Bordin, A.; Mangino, G.; Pompilio, G.; Calogero, A.; De Falco, E. Influence of egr-1 in cardiac tissue-derived mesenchymal stem cells in response to glucose variations. *Biomed. Res. Int.* **2014**, *2014*, 254793. [CrossRef] [PubMed]

85. Baron, V.; Adamson, E.D.; Calogero, A.; Ragona, G.; Mercola, D. The transcription factor egr1 is a direct regulator of multiple tumor suppressors including tgfbeta1, pten, p53, and fibronectin. *Cancer Gene Ther.* **2006**, *13*, 115–124. [CrossRef] [PubMed]

86. Calogero, A.; Arcella, A.; De Gregorio, G.; Porcellini, A.; Mercola, D.; Liu, C.; Lombari, V.; Zani, M.; Giannini, G.; Gagliardi, F.M.; et al. The early growth response gene egr-1 behaves as a suppressor gene that is down-regulated independent of arf/mdm2 but not p53 alterations in fresh human gliomas. *Clin. Cancer Res.* **2001**, *7*, 2788–2796. [PubMed]

87. Calogero, A.; Lombari, V.; De Gregorio, G.; Porcellini, A.; Ucci, S.; Arcella, A.; Caruso, R.; Gagliardi, F.M.; Gulino, A.; Lanzetta, G.; et al. Inhibition of cell growth by egr-1 in human primary cultures from malignant glioma. *Cancer Cell Int.* **2004**. [CrossRef] [PubMed]

88. Yu, J.; Zhang, S.S.; Saito, K.; Williams, S.; Arimura, Y.; Ma, Y.; Ke, Y.; Baron, V.; Mercola, D.; Feng, G.S.; et al. Pten regulation by akt-egr1-arf-pten axis. *EMBO J.* **2009**, *28*, 21–33. [CrossRef] [PubMed]

89. Chandrasekaran, S.; Deng, H.; Fang, Y. Pten deletion potentiates invasion of colorectal cancer spheroidal cells through 3d matrigel. *Integr. Biol. (Camb.)* **2015**, *7*, 324–334. [CrossRef] [PubMed]

90. Sperandio, S.; Fortin, J.; Sasik, R.; Robitaille, L.; Corbeil, J.; de Belle, I. The transcription factor egr1 regulates the hif-1alpha gene during hypoxia. *Mol. Carcinog.* **2009**, *48*, 38–44. [CrossRef] [PubMed]

91. Shimoyamada, H.; Yazawa, T.; Sato, H.; Okudela, K.; Ishii, J.; Sakaeda, M.; Kashiwagi, K.; Suzuki, T.; Mitsui, H.; Woo, T.; et al. Early growth response-1 induces and enhances vascular endothelial growth factor-a expression in lung cancer cells. *Am. J. Pathol.* **2010**, *177*, 70–83. [CrossRef] [PubMed]

92. Jeon, H.M.; Lee, S.Y.; Ju, M.K.; Kim, C.H.; Park, H.G.; Kang, H.S. Early growth response 1 regulates glucose deprivation-induced necrosis. *Oncol. Rep.* **2013**, *29*, 669–675. [PubMed]

93. Kim, J.; Kang, S.M.; Lee, H.J.; Choi, S.Y.; Hong, S.H. Oxytocin inhibits head and neck squamous cell carcinoma cell migration by early growth response-1 upregulation. *Anticancer Drugs* **2017**, *28*, 613–622. [CrossRef] [PubMed]

94. Sarver, A.L.; Li, L.; Subramanian, S. Microrna mir-183 functions as an oncogene by targeting the transcription factor egr1 and promoting tumor cell migration. *Cancer Res.* **2010**, *70*, 9570–9580. [CrossRef] [PubMed]

95. Verduci, L.; Azzalin, G.; Gioiosa, S.; Carissimi, C.; Laudadio, I.; Fulci, V.; Macino, G. Microrna-181a enhances cell proliferation in acute lymphoblastic leukemia by targeting egr1. *Leuk Res.* **2015**, *39*, 479–485. [CrossRef] [PubMed]

96. Vazquez-Santillan, K.; Melendez-Zajgla, J.; Jimenez-Hernandez, L.E.; Gaytan-Cervantes, J.; Munoz-Galindo, L.; Pina-Sanchez, P.; Martinez-Ruiz, G.; Torres, J.; Garcia-Lopez, P.; Gonzalez-Torres, C.; et al. Nf-kappabeta-inducing kinase regulates stem cell phenotype in breast cancer. *Sci. Rep.* **2016**, *6*, 37340. [CrossRef] [PubMed]

97. Mitchell, A.; Dass, C.R.; Sun, L.Q.; Khachigian, L.M. Inhibition of human breast carcinoma proliferation, migration, chemoinvasion and solid tumour growth by dnazymes targeting the zinc finger transcription factor egr-1. *Nucleic Acids Res.* **2004**, *32*, 3065–3069. [CrossRef] [PubMed]

98. Liu, J.; Liu, Y.G.; Huang, R.; Yao, C.; Li, S.; Yang, W.; Yang, D.; Huang, R.P. Concurrent down-regulation of egr-1 and gelsolin in the majority of human breast cancer cells. *Cancer Genom. Proteom.* **2007**, *4*, 377–385. [PubMed]

99. Irwin, M.E.; Johnson, B.P.; Manshouri, R.; Amin, H.M.; Chandra, J. A nox2/egr-1/fyn pathway delineates new targets for tki-resistant malignancies. *Oncotarget* **2015**, *6*, 23631–23646. [CrossRef] [PubMed]

100. Grotegut, S.; von Schweinitz, D.; Christofori, G.; Lehembre, F. Hepatocyte growth factor induces cell scattering through mapk/egr-1-mediated upregulation of snail. *EMBO J.* **2006**, *25*, 3534–3545. [CrossRef] [PubMed]

101. Housman, G.; Byler, S.; Heerboth, S.; Lapinska, K.; Longacre, M.; Snyder, N.; Sarkar, S. Drug resistance in cancer: An overview. *Cancers (Basel)* **2014**, *6*, 1769–1792. [CrossRef] [PubMed]

102. Thiery, J.P.; Acloque, H.; Huang, R.Y.; Nieto, M.A. Epithelial-mesenchymal transitions in development and disease. *Cell* **2009**, *139*, 871–890. [CrossRef] [PubMed]

103. Polyak, K.; Weinberg, R.A. Transitions between epithelial and mesenchymal states: Acquisition of malignant and stem cell traits. *Nat. Rev. Cancer* **2009**, *9*, 265–273. [CrossRef] [PubMed]

104. Ayinde, O.; Wang, Z.; Griffin, M. Tissue transglutaminase induces epithelial-mesenchymal-transition and the acquisition of stem cell like characteristics in colorectal cancer cells. *Oncotarget* **2017**, *8*, 20025–20041. [CrossRef] [PubMed]

105. Yoon, C.; Cho, S.J.; Chang, K.K.; Park, D.J.; Ryeom, S.W.; Yoon, S.S. Role of rac1 pathway in epithelial-to-mesenchymal transition and cancer stem-like cell phenotypes in gastric adenocarcinoma. *Mol. Cancer Res.* **2017**. [CrossRef] [PubMed]

![cancers logo] *cancers*

MDPI

Review

Phenotypic Plasticity and Cell Fate Decisions in Cancer: Insights from Dynamical Systems Theory

Dongya Jia [1,2], **Mohit Kumar Jolly** [1], **Prakash Kulkarni** [3] and **Herbert Levine** [1,4,5,6,*]

1   Center for Theoretical Biological Physics, Rice University, Houston, TX 77005, USA;
    dyajia@gmail.com (D.J.); mkjolly.15@gmail.com (M.K.J.)
2   Graduate Program in Systems, Synthetic and Physical Biology, Rice University, Houston, TX 77005, USA
3   Institute for Bioscience and Biotechnology Research, University of Maryland, Rockville, MD 20850, USA;
    pkulkar4@ibbr.umd.edu
4   Department of Bioengineering, Rice University, Houston, TX 77005, USA
5   Department of Physics and Astronomy, Rice University, Houston, TX 77005, USA
6   Department of Biosciences, Rice University, Houston, TX 77005, USA
*   Correspondence: herbert.levine@rice.edu; Tel.: +1-713-348-8122

Academic Editor: Joëlle Roche
Received: 15 May 2017; Accepted: 13 June 2017; Published: 22 June 2017

Abstract: Waddington's epigenetic landscape, a famous metaphor in developmental biology, depicts how a stem cell progresses from an undifferentiated phenotype to a differentiated one. The concept of "landscape" in the context of dynamical systems theory represents a high-dimensional space, in which each cell phenotype is considered as an "attractor" that is determined by interactions between multiple molecular players, and is buffered against environmental fluctuations. In addition, biological noise is thought to play an important role during these cell-fate decisions and in fact controls transitions between different phenotypes. Here, we discuss the phenotypic transitions in cancer from a dynamical systems perspective and invoke the concept of "cancer attractors"—hidden stable states of the underlying regulatory network that are not occupied by normal cells. Phenotypic transitions in cancer occur at varying levels depending on the context. Using epithelial-to-mesenchymal transition (EMT), cancer stem-like properties, metabolic reprogramming and the emergence of therapy resistance as examples, we illustrate how phenotypic plasticity in cancer cells enables them to acquire hybrid phenotypes (such as hybrid epithelial/mesenchymal and hybrid metabolic phenotypes) that tend to be more aggressive and notoriously resilient to therapies such as chemotherapy and androgen-deprivation therapy. Furthermore, we highlight multiple factors that may give rise to phenotypic plasticity in cancer cells, such as (a) multi-stability or oscillatory behaviors governed by underlying regulatory networks involved in cell-fate decisions in cancer cells, and (b) network rewiring due to conformational dynamics of intrinsically disordered proteins (IDPs) that are highly enriched in cancer cells. We conclude by discussing why a therapeutic approach that promotes "recanalization", i.e., the exit from "cancer attractors" and re-entry into "normal attractors", is more likely to succeed rather than a conventional approach that targets individual molecules/pathways.

Keywords: cell fate decision; cancer attractors; gene network dynamics; EMT; therapy resistance; intrinsically disordered proteins

1. Introduction

> *"The woods are lovely, dark and deep, but I have promises to keep, and miles to go before I sleep, and miles to go before I sleep."*
>
> —*Robert Frost*

Waddington's epigenetic landscape [1] depicting how a stem cell progresses from an undifferentiated phenotype to a differentiated one is one of the most famous and powerful metaphors in developmental biology. Conceptually, the differentiation of a stem cell is represented by a ball rolling downhill through a rugged landscape of bifurcating valleys, each representing a possible cell fate (Figure 1A). The valleys continue bifurcating and the ball finally enters one of many sub-valleys at the foot of the hill. These sub-valleys represent terminally differentiated states, i.e., cell fates. The cell is held permanently, unless perturbed significantly, in the terminally differentiated state by high ridges, i.e., valley walls. The deeper the valley, the more canalized the cell fate. The epigenetic landscape in the context of dynamical systems theory represents a high-dimensional state space in which each cell fate is an "attractor" shaped by the architecture of its regulatory interaction network [2]. It is generally held that cell fate is essentially irreversible; it follows the "arrow of time". However, recent developments in cellular reprogramming have illustrated that a terminally differentiated cell can be forced to switch states (phenotypes) and acquire an undifferentiated state by supraphysiological overexpression of a cocktail of transcription factors (TFs) [3]. Similarly, cancer has been also shown to be 'reversed' to a non-malignant phenotype, thereby raising questions about the sufficient and necessary role of mutations in cancer progression [4].

In a dynamical system, an "attractor" (steady state) represents a set of values of the variables towards which the system evolves from a wide variety of starting conditions, and is robust to slight perturbations. Cell phenotypes are regulated by underlying gene regulatory networks (GRNs) (Figure 1B). GRNs are dynamical systems that start from context-dependent conditions, develop temporally due to the mutual interactions between molecular regulators (genes, proteins, microRNAs etc.) and later settle down into "attractors" (stable cell phenotype), each of which is characterized by a unique gene expression pattern (Figure 1C). Different possible steady states ("attractors") of a given GRN can be identified by mathematically modeling its dynamics; each attractor is associated with a steady-state probability of finding the system in that particular configuration. Together, this set of attractors—with their relative probabilities of being realized by the system—define a "landscape". Representing a stable cell phenotype as an "attractor" has helped realize the basic concepts of both single-cell stochasticity and population determinism, i.e., single cells can shift from one attractor to another due to noise, without altering the overall population structure. This perspective facilitates viewing biological systems from the perspective of statistical mechanics, where a macrostate (a cell population structure) can correspond to multiple microstates (phenotypic heterogeneity at a single-cell level) [5].

The concept of an "attractor" representing a cell phenotype (cell fate) has been widely used to understand lineage specifications during development. Usually, lineage commitment between sister cell-fates (i.e., sharing a common progenitor) is a binary branching process that is governed by a decision-making circuit consisting of two transcription factors X and Y that mutually inhibit each other and can also self-activate [2], referred to as a "self-activating toggle switch" [6]. X and Y are usually the master regulators of the two sister cell-fates. Such a "self-activating toggle switch" usually generates three stable "attractors' that are characterized by X^{high}/Y^{low}, X^{low}/Y^{high} and X^{medium}/Y^{medium} corresponding to two differentiated cell fates and an undifferentiated progenitor state respectively [2,6,7] (Figure 1A). Such "self-activating toggle switches" governing lineage commitments have been studied in various scenarios, such as the Gata1/PU.1 switch in the lineage commitment of multipotent progenitor cells [8], the Cdx2/Oct4 switch in the differentiation of a totipotent embryo [9], the Gata6/Nanog switch in the branching process of inner cell mass [10] and the T-bet/Gata3 switch in the lineage specification of the T-helper cells [11].

The concept of an "attractor" representing a cell phenotype is used not only in understanding embryonic development, but also in elucidating cancer initiation and progression. Cancer cells are regarded as abnormal cell phenotypes, i.e., "cancer attractors", and are believed to be the "hidden stable states" enabled by the regulatory networks that are not commonly occupied by normal cells [10]. Accesses to "cancer attractors" can be facilitated by genetic events (mutations) and/or non-genetic events (contextual signals and biological noise). For example, loss-of-function mutations in tumor suppressor genes such as TP53 and BRCA and/or gain-of-function mutations in proto-oncogenes such as MYC and RAS facilitate oncogenic properties of cells [12]. In addition to genetic events, the microenvironment surrounding cells can also promote tumorigenesis. For instance, overexpression of a stromal proteinase-matrix metalloproteinase-3 (MMP3) in both mouse phenotypically normal mammary epithelial cells (Scp2) and the mammary glands of transgenic mice, results in a reactive stroma and eventually leads to infiltrative mammary tumors [13]. Similarly, overexpression of the platelet-derived growth factor subunit B (PDGF-B) in the non-tumorigenic immortalized human keratinocytes (HaCaT) leads to a conversion to epithelial tumor cells through stromal cell activation [14]. These examples suggest that the probability to get access to "cancer attractors" can be enhanced due to gene mutations and/or contextual signals in the microenvironment. Furthermore, transitions can happen among "cancer attractors" to benefit cancer cells for survival and progression, referred to as phenotypic plasticity in cancer [15].

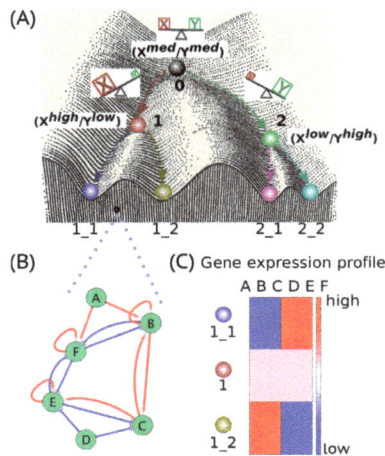

Figure 1. Schematic illustration of Waddington's epigenetic landscape. (**A**) Waddington's epigenetic landscape (adopted and revised from [1]). The balls with different colors on the landscape represent different cell phenotypes, each settles steadily in one of the sub-valleys at the foot of the hill. X and Y are the master regulators driving a cell to attain the phenotypes "1" and "2" respectively. The phenotype "0", characterized by the co-expression of both X and Y at a medium level X^{med}/Y^{med}, represents the progenitor state of the two differentiated states "1" and "2" which are characterized by X^{high}/Y^{low} and X^{low}/Y^{high} respectively. Due to inherent stochasticity in the progenitor cell "0", the level of X (Y) becomes higher than that of Y (X). This asymmetry can trigger a cascade of events where the levels of X (Y) continually increase and those of Y (X) continually decrease, because X (Y) can progressively repress its repressor Y (X) strongly, rendering its own inhibition by Y (X) ineffective. Consequently, the cell attains the differentiated state X^{high}/Y^{low} (X^{low}/Y^{high}). (**B**) Schematic illustration of a gene regulatory network (GRN) governing the differentiation of "1" to two lineages "1_1" and "1_2". The nodes A–F represent different genes whose regulatory behaviors usually can be approximated by the interplay between two master regulators X and Y as aforementioned. Various kinds of regulation can be found in the GRN, such as transcriptional activation, represented by red arrows, transcriptional inhibition, represented by blue bar-headed arrows, and self-activation, represented by circled arrows. (**C**) Schematic illustration of a heatmap that depicts the gene expression patterns of different cell phenotypes. The two sister lineages "1_1" and "1_2" are characterized by different gene expression patterns, i.e., relatively high expression of one gene set and low of another. The progenitor of "1_1" and "1_2", i.e., "1", usually co-expresses both sets of genes at some intermediate level.

In this review, we invoke the concept of "cancer attractors" and discuss the phenotypic plasticity of cancer cells from a dynamical systems perspective. Using epithelial-to-mesenchymal transition (EMT) and the acquisition of stem-like properties, metabolic reprogramming and the emergence of drug/hormone resistance in cancer as examples, we illustrate how non-genetic heterogeneity regulates phenotypic plasticity of cancer cells that enables them to acquire phenotypes that are notoriously aggressive and resilient to drug/hormone treatment. With enhanced plasticity, cancer cells can potentially rewire the regulatory network to access latent "attractors" suggesting that cancer initiation and progression may, at least in part, be due to a "de-canalization" of normal cell fates. Finally, we highlight the potential role of intrinsically disordered proteins (IDPs) that comprise a vast majority of the proteins over-expressed in cancer, and how biological noise due to IDP conformational dynamics may further enhance phenotypic plasticity of cancer cells. Since the perspective is intended to encourage cross pollination of ideas between biologists, especially cancer biologists, and physicists interested in exploring the physics of biology, technical jargon is limited to its minimum and equations are omitted.

2. Cancer Cell States: The Hidden "Attractors"

Cell phenotypes manifested during embryonic development are governed by specific gene regulatory networks (GRNs) (Figure 1B). The GRNs give rise to an epigenetic landscape consisting of multiple stable gene expression patterns (Figure 1C) characterizing various "attractors", i.e., "stable states" or "phenotypes" [16,17]. The "attractors" are usually self-stabilized and robust to local perturbations [18]. However, certain transitions between "attractors", i.e., phenotypic switching, can be triggered by regulatory signals, such as cytokines and noise due to gene expression as well as IDP conformational dynamics in addition to mutational events [19,20].

Cancer cells are viewed as abnormal cell types that are characterized by hallmarks such as sustained proliferation, invasion and metabolic reprogramming [21]. Extensive inherent heterogeneity of cancer cells has been shown at both the genetic level due to genomic instability [22], and the non-genetic level, resulting from cellular plasticity, i.e., the ability of cells to switch between phenotypes [23,24]. The examples of non-genetic heterogeneity in cancer include, but are not restricted to, epithelial-to-mesenchymal transition (EMT) [25], acquiring "stem-like" properties [26], and metabolic plasticity [27,28]. In certain cases, these processes have been shown to be coupled. For instance, cells undergoing EMT can acquire stem-like properties [29], stem-like properties associate with metabolic changes [30], and metabolic programming involves changes in EMT [27,31,32].

This extensive plasticity of cancer cells may enable the occupancy of the "attractors" that are unpopulated or inaccessible during embryonic development, or equivalently, acquire phenotypes not usually observed during development or homeostasis. The concept of "cancer attractors" representing abnormal cell types was first proposed by Stuart Kauffman in 1971 [33] and recently revisited by Huang, Ao and colleagues [34,35]. In the following sections, we will review progress in elucidating the phenotypic plasticity of cancer cells from the dynamical systems perspective, namely, by viewing cancer cell phenotypes as different "cancer attractors" in the state space determined by the underlying regulatory networks.

3. Cell Fate Decision-Making during Epithelial-to-Mesenchymal Transition

Epithelial-to-Mesenchymal Transition (EMT) is a trans-differentiation program by which epithelial cells lose their cell-cell adhesion and gain migratory property to become mesenchymal cells. Both EMT and its reverse—Mesenchymal-to-Epithelial Transition (MET)—play crucial roles during embryogenesis (during processes such as gastrulation, neural crest delamination and myogenesis) and tissue repair (during wound healing and fibrosis) [36]. However, EMT may sometimes be "hijacked" by carcinoma cells to acquire enhanced migratory properties that can contribute to metastasis and/or acquired therapy resistance [37,38]. Moreover, the EMT transcription factors (EMT-TFs), such as ZEB (zinc finger E-box-binding homeobox) and SNAIL (zinc finger protein SNAI1), have even been shown to play an important role in tumor progression in non-carcinomas, such as melanoma [39,40] and glioblastoma [41,42].

During metastasis, cancer cells do not always undergo a complete EMT, instead a partial EMT (leading to a hybrid epithelial/mesenchymal (E/M)) phenotype, in which cells exhibit both epithelial (cell-cell adhesion) and mesenchymal (migration and/or invasion) traits, has often been observed [43–45] (Figure 2A). Cells in a hybrid E/M phenotype can migrate collectively as a cluster instead of migrating individually like a cell that has undergone a complete EMT. These clusters of circulating tumor cells (CTCs) associate with up to 50-fold higher metastasis potential and higher tumor-initiating potential compared with single CTCs [43,46], thus being proposed as the primary "bad actors" of metastasis [44].

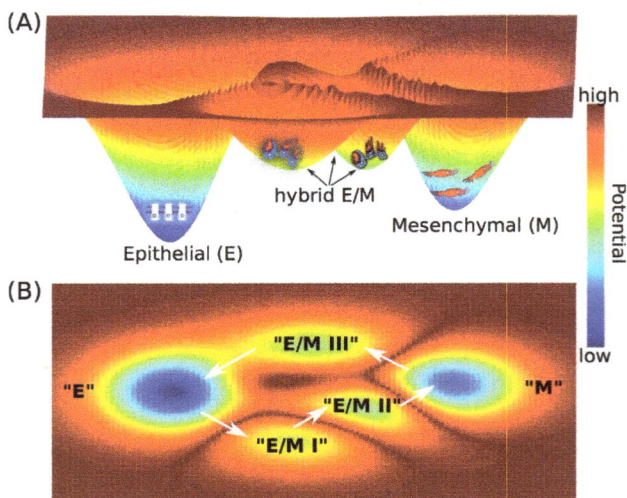

Figure 2. Schematic illustration of the quasi-potential landscape for epithelial-to-mesenchymal transition (EMT) in 3-dimensional space (**A**) and 2-dimensional projection (**B**). In (**A**), the basins of attraction depicting the attractors "E", "E/M" and "M" are labeled respectively along with the cartoons representing the epithelial (tight cell-cell adhesion, cobblestone shaped), hybrid E/M (some cell-cell adhesion and invasive) and mesenchymal (no cell-cell adhesion, invasive and spindle-shaped) phenotypes. The quasi-potential of "attractors", i.e., stability of "attractors", is derived from the probability of finding cells in that "attractors". Lower potential here represents more stable "attractor" in the landscape. The "potential well" depicted here is an analog of "valleys" in Waddington epigenetic landscape.

To understand the epithelial-mesenchymal plasticity, i.e., transitions among epithelial (E), hybrid E/M and mesenchymal (M) phenotypes, a core EMT regulatory circuit consisting of two transcription factor families—ZEB and SNAIL and two microRNA families—miR-200 and miR-34, has been characterized. High expression of the transcription factors ZEB and SNAIL promotes a mesenchymal phenotype while high expression of microRNAs miR-200 and miR-34 maintains an epithelial phenotype. Two mathematical models [47–49] that were independently proposed have been applied to analyze the dynamics of the core EMT circuit. Both models elucidate that (1) the core EMT decision-making circuit functions as a "three-way" switch, that can give rise to three stable states—"E" characterized by (E marker^{high}/M marker^{low}), "M" characterized by (E marker^{low}/M marker^{high}) and "E/M" characterized by (E marker^{medium}/M marker^{medium}). (2) EMT is a two-step processes—from "E" to "E/M" to "M" [47,48]. Once the cells transition into a mesenchymal phenotype, the stable state or phenotype "M" can be self-stabilized, by feedback loops such as increased inhibition of ZEB on miR-34 [50], and/or the decreased inhibition of miR-200 on the endogenous TGF-β [48,50]. The landscape approach has been utilized to quantify the transition processes among these three stable states, i.e., "attractors"—"E", "E/M" and "M" [51]. This study suggested that attainment of a hybrid E/M state often decreases the required strength of EMT-inducing signals to initiate EMT, i.e., pulling cells out of the stable state "E", thus enabling cancer cells to be more plastic [51].

The hybrid E/M phenotype has been observed in circulating tumor cells (CTCs), primary tumors, metastases, and 3D reconstructions of 2D histological sections [44,52], but it has tacitly been largely assumed as a "metastable" or transient phenotype [53]. However, recently, in part driven by these mathematical models, a stable hybrid E/M phenotype has been observed in the non-small cell lung cancer (NSCLC) cell line—H1975, in which individual cells co-express an epithelial marker—E-cadherin and a mesenchymal marker—Vimentin [54]. These cells can maintain their hybrid E/M phenotype for over two months after multiple passages, thus being characterized as a stable phenotype [54]. Moreover, such an integrated computational-experimental analysis has also helped identify two transcription factors GRHL2 (grainyhead like transcription factor 2) and OVOL2 (ovo-like zinc finger 2) that can stabilize the hybrid E/M phenotype [54–56]. Knockdown of either GRHL2 or OVOL2 in H1975 cells destabilized the hybrid E/M phenotype and cells progressing to a complete EMT state [54]. Thus, these "phenotypic stability factors" (PSFs) GRHL2 and OVOL [57] act as "critical molecular brakes" by preventing "cells that have gained partial plasticity from crossing the line to undergo complete EMT" [58]. Of note, there may exist multiple hybrid E/M phenotypes characterized by different gene expression profiles [56,59], and other players such as JAG1 (ligand of cell-cell communication pathway—Notch signaling) and ΔNP63α can also act as PSFs [60,61]. EMT and MET need not be symmetric [47], i.e., EMT and MET could potentially proceed via different hybrid E/M phenotypes, that enables cancer cells to have more phenotypic plasticity (Figure 2).

4. EMT and Stemness

Cancer cells undergoing EMT can acquire stemness, i.e., stem-like properties or tumor-initiation potential [29], and thus behave operationally as Cancer Stem Cells (CSCs) as observed in multiple solid tumors [62]. The coupling between EMT and stemness is finely regulated. On the one hand, EMT promotes the acquirement of stemness in breast [29,63] hepatocellular [64], pancreatic [65] and colorectal [66] carcinomas; on the other hand, repression of EMT is required for tumor initiation and metastatic colonization [67–69].

As the first step to understand the coupled decision-making of EMT and stemness, Jolly et al. [70] formulated a mathematical model to analyze the dynamics of the coupled decision-making circuits of EMT-ZEB/miR-200 and stemness-LIN28/let-7 [71]. It suggests that the "stemness window" is most likely to lie at an intermediate position on the "EMT axis" with E and M phenotypes as the two ends. Further, this positioning of "stemness window" can be adjusted and the phenotypic stability factors such as OVOL promote the association of a hybrid E/M phenotype with stemness, a prediction that has been supported by recent experimental work. For instance, HMLER breast cancer cells co-expressing both epithelial and mesenchymal genes, thus being characterized as hybrid E/M cells, exhibited highest mammosphere formation potential compared with epithelial and mesenchymal HMLER cells [72]. Besides, Cancer Stem Cell (CSC)-enriched population resides in a hybrid E/M phenotype of triple-negative breast cancer cells [73]. Last but not least, a subpopulation of normal mammary cells, accompanied by both epithelial-like and mesenchymal-like characteristics, i.e., hybrid E/M phenotype, displays the highest mammosphere-formation capacity [74]. Thus, a biphasic relationship between stemness and EMT—stemness increases initially during EMT progression, but then subsides as cells complete EMT—seems to be the emerging notion [43,75,76].

CSCs have also been observed to display enriched drug resistance [77]. For example, a hybrid E/M phenotype has been reported to be resistant to paclitaxel and salinomycin [78]. Moreover, adaptive drug resistance involves transitioning to a $CD24^{high}CD44^{high}$ state [79]—a proposed signature for hybrid E/M phenotype [72]. Future work on quantifying the landscape [80] for the coupled circuits of EMT and stemness, along with a better mechanistic understanding of drug resistance pathways, are required to generate valuable insights into the EMT-stemness interplay.

5. Metabolic Reprogramming and EMT

Abnormal metabolism is an emerging hallmark of cancer [21,81]. Unlike normal cells, cancer cells mainly utilize glycolysis for ATP production even in presence of oxygen, a phenomenon referred to as aerobic glycolysis or the Warburg effect [82]. Although aerobic glycolysis has been proposed to be the dominant metabolism phenotype in cancer cells [83,84], emerging evidence shows that mitochondria in cancer cells are actively functioning and oxidative phosphorylation (OXPHOS) can enhance metastasis [85–90].

As the first step to understand the metabolic plasticity in cancer, Yu et al. [91] constructed a core metabolism regulatory network consisting of AMPK and HIF-1—master regulators for OXPHOS and glycolysis, respectively—and ROS (reactive oxygen species) that mediates the interplay between AMPK and HIF-1. This AMPK:HIF-1:ROS regulatory network enables three stable states—(pAMPK^{high}/HIF-1^{low}), (pAMPK^{low}/HIF-1^{high}) and (pAMPK^{medium}/HIF-1^{medium})—corresponding to an OXPHOS, a glycolysis and a hybrid OXPHOS/glycolysis metabolic phenotype respectively (pAMPK denotes phosphorylated AMPK, i.e., the active form of AMPK). The hybrid metabolic state, in which cancer cells can utilize both glycolysis and OXPHOS, facilitates relatively high plasticity for ATP production and proliferation for cancer cells. The hybrid metabolism phenotype can be stabilized by increased HIF-1 activity, high oncogene (MYC, RAS, c-SRC) activity and high mitochondria ROS production in cancer cells compared with that in normal cells [91].

The hybrid metabolism phenotype proposed by the aforementioned modeling work has been observed in many experimental studies to be associated with metastatic potential. The supermetastatic human tumor cells SiHa-F3 by in vitro selection and the mouse melanoma cells B16F10, B16-M1 to M5 by in vivo selection have an increased OXPHOS activity together with an enhanced invasive activity [92]. The non-small cell lung carcinoma A549 cells undergoing EMT induced by TGF-β show elevated respiration [27]. The metastatic breast cancer cells 66cl4 and 4T1 have both enhanced oxidative as well as glycolytic metabolism accompanied by increased extracellular acidification rate and oxygen consumption rate compared with non-metastatic 67NR cells [93]. In addition, cells in the hybrid metabolism phenotype can maintain ROS at a moderate level [91], thus avoiding excessive DNA damage [94] while using ROS signaling to promote metastasis [95]. Moreover, cells in the hybrid phenotype can simultaneously produce energy and generate biomass for proliferation [30]. Therefore, a combination therapy that target the hybrid metabolism phenotype, i.e., blocking both glycolysis and OXPHOS in cancer cells, could be relatively more effective [30,91] than the therapy targeting only one metabolic pathway.

Of note, regulation of metabolic plasticity has been shown to be coupled with the EMT decision-making [31]. EMT enhances glycolysis in MCF-7 and BT-474 cells [96] while shifts metabolism from glycolysis to OXPHOS in MCF10 cells [97]. Fatty acid oxidation is more utilized in the mesenchymal breast cancer cells D492M than that in epithelial cells D492 (D492M cells are isolated following a spontaneous EMT in D492 cells) [32]. Blocking fatty acid oxidation in MDA-MB-231 cells decreases their migratory and colony-formation properties, suggesting multiple feedback loops between regulatory circuit of metabolism, EMT and stemness [90]. This situation remains to be clarified on the basis of models.

Metabolic plasticity has also been observed in CSCs. Epithelial-like CSCs, characterized by ALDH^{high}, have higher oxygen consumption rate and lower glycolytic activity compared with the mesenchymal-like breast CSCs, characterized by CD44^{high}CD24^{low} [98,99]. Recent work highlighted that ALDH^{high} cells may exhibit a hybrid E/M state [74]. Future work to analyze the coupled decision-making of metabolism, EMT and stemness needs to be done to comprehensively chart the stable states characterized by varied EMT, stem-like property and metabolism traits, while taking into consideration the direct coupling between gene expression and metabolites, at least partly through epigenetic mechanisms [100].

6. EMT and Therapy Resistance

EMT has been associated with both *de novo* and acquired resistance. *De novo* resistance implies intrinsic refractory response of patients, whereas acquired resistance refers to cases where patients first respond to therapy but later relapse. A relationship between EMT and *de novo* resistance has been well studied in cases of targeted therapy. For instance, increased levels of E-cadherin were associated with sensitivity to EGFR kinase inhibitors such as gefitinib in non-small-cell lung cancer (NSCLC) cell lines, and pre-treatment of resistant cell lines to induce E-cadherin levels improved their sensitivity [101]. Similarly, knockdown of the levels of SLUG, an EMT-TF, in *de novo* trastuzumab-resistant HER2+ breast cancer cells can drive them to being sensitive to trastuzumab [102]. Besides, recent in vivo reports that questioned an indispensable role of EMT in metastasis only strengthened a potential causal role of EMT in driving chemoresistance. For example, knocking down TWIST or SNAIL sensitized tumors to gemcitabine in pancreatic cancer mouse models [103], and miR-200 overexpression abrogated resistance to cyclophosphamide, a drug commonly used in breast cancer [104]. Taken together, these studies suggest that cellular plasticity mediated by EMT can act as a switch enabling cells to "enter" and "exit" a drug-resistant cell state dynamically. Recent mathematical modeling attempts that investigate the crosstalk among signaling players have highlighted that non-genetic heterogeneity can drive this dynamic "entry" into and "exit" from a stem-like therapy-resistant state [70,71,80,105,106].

This dynamic "entry" and "exit" may also underlie acquired or adaptive drug resistance, where different therapies may induce cells to access the "cancer attractors" which are relatively inaccessible otherwise, but can be used to play "hide-and-seek" with different therapies. For instance, in ovarian cancer, treatment with chemotherapeutic drugs such as cisplatin, doxorubicin, and paclitaxel can reversibly increase a small population of CXCR4^{high} cells that is drug-resistant, mesenchymal-like, and has enhanced tumor-initiation potential [107]. Other examples of adaptive resistance include melanoma cells switching to a NGFR^{high} state upon exposure to RAF/MEK inhibitors [108], NSCLC cells upregulating ZEB1 on prolonged exposure to increasing concentrations of erlotinib [109], vemurafenib driving epigenetic reprogramming to a drug-resistant state in melanoma [110] and chemotherapy enriching a CD24^{high}CD4^{high} drug-resistant population in breast cancer cells [79].

Mechanism-based mathematical models have helped tease out that this adaptive enrichment of a drug-resistant cancer subpopulation can result from phenotypic plasticity, for instance, the emergence of a drug-resistant CD24^{high}/CD44^{high} state [79]. The CD24^{high}/CD44^{high} state was also suggested to associate with an elevated Notch-Jagged signaling, a prediction that has been validated experimentally at least preliminarily [60]. Similarly, in an attempt to understand the experimentally observed correlation between EMT and immune evasion, a mathematical model involving the transcription factors STAT1, STAT3, and the microRNA miR-200 predicted and guided the experimental design for how inhibiting STAT3 activation altered the levels of a set of immune-evasion mediators PSMB8 and PSMB9 in the mesenchymal NSCLC cells [111]. Therefore, mathematical models can be valuable tools in elucidating the principles of phenotypic plasticity governing both *de novo* and acquired resistance to various therapies.

7. Role of Intrinsically Disordered Proteins in Phenotypic Plasticity

From the foregoing, it is obvious that cancer cells retain high plasticity which facilitates phenotypic transitions among various phenotypes to adjust to microenvironments. A hallmark of many master regulators that regulate cancer phenotypic plasticity such as, oncoproteins that cause cellular transformation, factors that induce reprograming of somatic cells to pluripotent stem (iPS) cells, and several EMT-TFs that play a critical role in EMT/MET is that, they are intrinsically disordered proteins (IDPs) [20,112–114].

IDPs are proteins, or large regions within ordered proteins, that lack three-dimensional structure. They exist as ensembles instead but can transition from disorder to order upon interacting with a biological target (reviewed in [115,116]). However, there are several cases where IDPs stochastically sample the conformational state space *a priori* [117,118] or are functional even when remaining highly disordered [119–122]. Regardless however, because IDPs populate multiple conformational states albeit transiently, and display rapid conformational dynamics, they are prone to stochastically engage in myriad "promiscuous" interactions, especially when they are overexpressed [123,124].

In an attempt to understand the roles of IDPs in cancer phenotypic plasticity, Mahmoudabadi et al. [125] have suggested that these promiscuous interactions result in "noise" in the system. Further, to distinguish this noise from the widely recognized "transcriptional noise" that stems from gene expression, the authors coined the term "conformational noise". This new source of biological noise stems from IDP conformational dynamics and is an inherent characteristic of IDP interactions. However, notwithstanding the distinction, the authors postulated that just like transcriptional noise which plays an important role in generating phenotypic heterogeneity [126,127], the collective effect of conformational noise is an ensemble of protein regulatory network configurations, from which the most suitable configuration can be explored by the cancer cell to "make" appropriate decisions, thus conferring it with remarkable phenotypic plasticity. Moreover, the ubiquitous presence of intrinsic disorder in transcription factors and, more generally, in proteins that occupy hub positions in regulatory networks is thought to be indicative of the role of IDPs in propagation and amplification of transcriptional as well as other types of noise (e.g., noise in signaling pathways) in the system. Therefore, as effectors of conformational and transcriptional noise, IDPs can rewire regulatory networks unmasking latent regulatory circuits in response to perturbations and switch phenotypes to generate phenotypic heterogeneity [125]. Thus, from Waddington's epigenetic landscape perspective, conformational noise-driven rewiring results in the system exploring the high-dimensional state space and homing to attractor basins that harbor "cancer attractors". Implicit in the model proposed by Mahmoudabadi et al. [125], phenotypic switching can result from stochastic (non-genetic) rather than by deterministic events alone (genetic), and the regulatory network configuration contains information that can aid cell fate decisions.

In a recent paper, Mooney et al. [20] reviewed the role of IDPs in EMT and discussed how IDP conformational dynamics can contribute to phenotypic plasticity using prostate cancer (PCa) as an example. In addition, Kulkarni et al. [106] discussed the role of IDPs in the emergence of androgen resistance (independence), yet another paradigm of phenotypic plasticity in PCa. Here, we highlight their role in the emergence of androgen resistance.

The onset of androgen resistance in patients treated with androgen-deprivation therapy (ADT) is a major impediment in PCa. However, the underlying molecular mechanisms are not fully understood. To gain new insight, Kulkarni et al. [106] recently employed multiple biophysical approaches that report conformational preferences of Prostate-Associated Gene 4 (PAGE4). PAGE4 is an IDP that acts as a potentiator of the Activator Protein-1 (AP-1) transcription factor [128,129]. PAGE4 is phosphorylated by Homeodomain-Interacting Protein Kinase 1 (HIPK1) predominantly at T51 which is critical for its transcriptional activity [130]. However, PAGE4 is also hyperphosphorylated by CDC-Like Kinase2 (CLK2) at multiple S/T residues including T51. Further, while HIPK1 is expressed in both androgen-dependent and androgen-independent PCa cells, CLK2 and PAGE4 are expressed only in androgen-dependent cells. Cell-based reporter assays indicated that PAGE4 interaction with the two kinases leads to opposing functions. Thus, whereas HIPK1-phosphorylated PAGE4 (HIPK1-PAGE4) potentiates c-Jun, CLK2-phosphorylated PAGE4 (CLK2-PAGE4) attenuates c-Jun activity. Consistent with the cellular data, biophysical measurements employing small-angle X-ray scattering, single-molecule fluorescence resonance energy transfer, and multidimensional NMR indicated that HIPK1-PAGE4 exhibits a relatively compact conformational ensemble that binds AP-1, whereas CLK2-PAGE4 is more expanded and resembles a random coil with diminished affinity for AP-1 [106,128].

AP-1 can negatively regulate androgen receptor (AR) activity [131,132], and AR can transcriptionally inhibit CLK2 expression [106]. Furthermore, cells resistant to ADT often have enhanced AR activity (AR protein expression can increase >25 fold) suggesting a positive correlation between ADT resistance and AR activity [133]. These observations combined with the data [106] allowed the construction of a circuit representing the PAGE4/AP-1/AR interactions and the development of a mathematical model that represents the dynamics of this circuit.

The model predicts that the circuit can display sustained or damped oscillations suggesting that androgen dependence of a cell need not be a fixed state and can vary temporally. Thus, contrary to the prevailing deterministic model that tacitly assumes PCa cells to acquire an androgen-dependent or an independent state (mutually exclusive "binary" model driven by genetic events), cells can enter or exit the androgen-independent state or phenotype (it is reversible) (Figure 3). Even in the case of damped oscillations that eventually settle to one state, the system can revert to displaying sustained oscillations under the effect of biological "noise". Such noise can originate from multiple sources such as, limited quantities of PAGE4, HIPK1, or CLK2, and/or the conformational dynamics of PAGE4. Furthermore, the model also predicts that the intracellular CLK2, HIPK1-PAGE4, and CLK2-PAGE4 oscillations need not be synchronized across cells. Thus, individual cells in an isogenic population would have varying levels of androgen dependence or independence at a given point in time consequently giving rise to non-genetic phenotypic heterogeneity observed in a seemingly homogenous population of PCa cells [134]. In other words, androgen dependence represents a trait whose values can display a broad distribution across the population. Indeed, this predicted heterogeneity in the levels of HIPK1, CLK2 and PAGE4 is corroborated by quantitative immunohistochemistry and qRT-PCR data [106]. Thus, the model that is developed using the tools of nonlinear dynamics demonstrates how differential phosphorylation of PAGE4 can lead to transitions between androgen-dependent and androgen-independent phenotypes by altering the AP-1/androgen receptor regulatory circuit in PCa cells. Although additional work needs to be done, the study underscores IDPs can stochastically orchestrate phenotypic heterogeneity in PCa due to their conformational dynamics when overexpressed or aberrantly expressed.

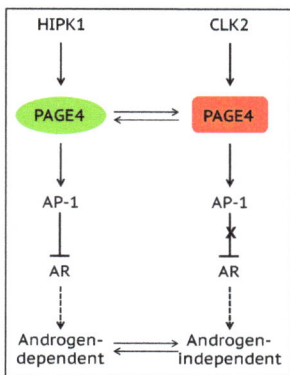

Figure 3. IDP conformational dynamics and phenotypic heterogeneity in prostate cancer cells. The stress-response kinase HIPK1 phosphorylates the IDP PAGE4 resulting in a relatively compact PAGE4 ensemble (HIPK1-PAGE4) that can potentiate AP-1 in androgen-dependent cells. In contrast, the dual-specificity kinase CLK2 hyperphosphorylates PAGE4 leading to a more random-like PAGE4 ensemble (CLK2-PAGE4) that attenuates AP-1 function. Mathematical modeling suggests that the oscillatory dynamics of HIPK1-PAGE4, CLK2-PAGE4, and CLK2 in the circuit enable the cells to transition from an androgen-dependent to an androgen-independent phenotype. This prediction is supported by the experimentally observed heterogeneity in a population of isogenic PCa cells (see [106] for details).

8. Conclusions and Future Vision

Waddington's epigenetic landscape initially depicting the differentiation process of stem cells now have been used to understand the phenotypic plasticity in cancer cells. The regulatory network underlying the landscape can give rise to various "attractors", i.e., "stable states" corresponding to different cell phenotypes, each of which is characterized by a unique gene expression pattern. Emerging insights demonstrate that cancer cells are often behaving as "moving targets" and often find new adaptive ways to resist therapeutic attacks. This search for "cancer attractors" that increase their fitness and/or survival likelihood can be considered akin to "de-canalization". "Canalization" refers to buffering of biological noise during development, such that cellular phenotypes are stabilized against genetic and/or environmental perturbations, and their variability is decreased [135]. Thus, "de-canalization" would imply supraphysiological plasticity that can make the "valleys" in Waddington's landscape more shallow (by decreasing the height of the ridge between "valleys") [136], thereby enabling stochastic sampling of the landscape by cells, hence disrupting the stable cellular phenotypes obtained and maintained in specific niches.

"De-canalization" into "cancer attractors" can be facilitated by gene mutations that rewires the underlying regulatory network. For example, both gain-of-function mutations in proto-oncogenes RAS and MYC and loss-of-function mutations in tumor suppressor genes TP53 and BRCA1 can trigger abnormal cell growth and provoke cancer formation [12]. Once cells enter "cancer attractors", they acquire high cellular plasticity that allows phenotypic transitions to adjust to the microenvironment. The high plasticity in cancer can be promoted by (a) increased biological noise due to the intrinsic variability in gene expression [137] and the conformational dynamics of intrinsically disordered proteins, such as oncoproteins, reprogramming TFs and EMT-TFs in cancer cells [20,112–114]; (b) the changed physiological parameters for cancer cells due to the modified microenvironment [138]. For example, cancer cells usually face hypoxia condition due to their rapid proliferation and the hypoxia condition stabilizes HIF-1. This can then promote cancer cells to acquire a hybrid OXPHOS/glycolysis phenotype that has been shown to be associated with higher metastatic potential as compared with only OXPHOS or glycolysis phenotypes [91]. The high phenotypic plasticity of cancer cells can contribute to metastasis and therapeutic failure.

Recent studies have also highlighted that phenotypic transitions do not have to be cell-autonomous events. Instead, the microenvironment of a cell can often modulate such phenotypic switching, for instance, (a) the lineage commitment of naïve mesenchymal stem cells can be directed by the matrix elasticity and soft matrices generate nerve-like cells, stiff matrices generate muscle-like cells and rigid matrices generate bone-like cells [139], (b) simulated microgravity can dramatically alter the cytoskeletal architecture of MDA-MB-231 cells with consequent effects on proliferation and apoptosis [140], (c) parallel microgrooves on the surface of cell-adhesive substrates can mechanically modulate a cell's epigenetic state and induce an MET, thereby increasing the efficiency of cellular reprogramming [141], and (d) signals from mammary microenvironment can overrule the 'terminal commitment' of a stem cell belonging to a "foreign" tissue [142]. Together, these studies highlight the need to revisit whether a cell is ever "terminally differentiated", and how much cell-autonomy there is in a cell-fate [143].

9. Therapeutic Approach That Promotes "Re-canalization"

Can cells transition from "cancer attractors" back to "normal attractors", i.e., "re-canalization"? The answer seems to be yes based on some existing data. First, inactivation of the oncogene MYC in hepatocellular carcinoma cells leads to the formation of normal hepatic structures [144]. Second, decreasing the intracellular levels of TCTP (transcriptionally controlled tumor protein) is sufficient to revert the malignancy of MCF7 or T47D cells (breast cancer), U937 cells (histiocytic lymphoma) [145] and v-Src-transformed NIH3T3 cells (fibroblasts) [146], partially through recovering the function of the P53/MDM2 axis [147]. Third, replacement of mitochondria in metastatic triple negative breast cancer cells SUM159 with mitochondria from benign breast cancer cells MCF10A or A1N4 abolish cell migration potential and in vivo tumor formation potential [90]. Forth, modification of the surface integrins of human breast cancer cells in 3-dimensional culture results in a reversion to a normal cell phenotype both morphologically and functionally despite the malignant genome [148]. Therefore, we believe that targeting the sources for phenotypic plasticity in cancer cells, for instance, deactivation of oncoproteins and/or modification of tumor microenvironment can contribute to the "re-canalization".

Even though it may be difficult to revert cancer cells directly to normal cells, we can still help cancers cells transition out from highly aggressive "attractors". One possible approach is to perturb factors that help maintain the aggressive "cancer attractors". For example, knockdown of the phenotypic stability factors OVOL and GRHL2 in H1975 cells can destabilize the hybrid E/M phenotype [54], the "primary bad actors" of metastasis [43,44,46]. Therefore, instead of targeting individual signaling pathways with insufficient knowledge of how they impinge on the epigenetic landscape for each cell, future therapeutic approaches might consider a stepwise approach from the dynamical systems perspective, start with the destabilization of the "cancer attractors", followed by transitions into "normal attractors", then deepening the basin of attraction of "normal attractor" to prevent future tumor relapse. As attractive as it may seem, the proposed approach remains to be clarified on the basis of combined modeling and experimental work.

Acknowledgments: We would like to thank Xuefei Li (Center for Theoretical Biological Physics, Rice University, Houston, TX 77005, USA) and Min-Yeh Tsai (Center for Theoretical Biological Physics, Rice University, Houston, TX 77005, USA) for a critical reading and helpful discussion of the manuscript. Herbert Levine was supported by the Physics Frontiers Center National Science Foundation (NSF) grant PHY-1427654 and the NSF grants DMS-1361411 and PHY-1605817. Herbert Levine was also supported by the Cancer Prevention and Research Institute of Texas (CPRIT) grants R1111. Mohit Kumar Jolly has a training fellowship from the Keck Center for Interdisciplinary Bioscience Training of the Gulf Coast Consortia (CPRIT Grant RP170593). Prakash Kulkarni would like to dedicate this article to Prof. Vidyanand Nanjundiah on the occasion of his 70th birthday.

Conflicts of Interest: The authors declare that they have no conflict of interest.

References

1. Waddington, C.H. *The Strategy of the Genes*; George Allen & Unwin: London, UK, 1957.
2. Zhou, J.X.; Huang, S. Understanding gene circuits at cell-fate branch points for rational cell reprogramming. *Trends Genet.* **2011**, *27*, 55–62. [CrossRef] [PubMed]
3. Takahashi, K.; Yamanaka, S. Induction of pluripotent stem cells from mouse embryonic and adult fibroblast cultures by defined factors. *Cell* **2006**, *126*, 663–676. [CrossRef] [PubMed]
4. Amson, R.; Karp, J.E.; Telerman, A. Lessons from tumor reversion for cancer treatment. *Curr. Opin. Oncol.* **2013**, *25*, 59–65. [CrossRef] [PubMed]
5. Giuliani, A. Collective motions and specific effectors: A statistical mechanics perspective on biological regulation. *BMC Genomics* **2010**, *11*, S2. [CrossRef] [PubMed]
6. Lu, M.; Jolly, M.K.; Gomoto, R.; Huang, B.; Onuchic, J.; Ben-Jacob, E. Tristability in cancer-associated microRNA-TF chimera toggle switch. *J. Phys. Chem. B* **2013**, *117*, 13164–13174. [CrossRef] [PubMed]
7. Jia, D.; Jolly, M.K.; Harrison, W.; Boareto, M.; Ben-Jacob, E.; Levine, H. Operating principles of tristable circuits regulating cellular differentiation. *Phys. Biol.* **2017**, *14*, 035007. [CrossRef] [PubMed]
8. Huang, S.; Guo, Y.-P.; May, G.; Enver, T. Bifurcation dynamics in lineage-commitment in bipotent progenitor cells. *Dev. Biol.* **2007**, *305*, 695–713. [CrossRef] [PubMed]
9. Niwa, H.; Toyooka, Y.; Shimosato, D.; Strumpf, D.; Takahashi, K.; Yagi, R.; Rossant, J. Interaction between Oct3/4 and Cdx2 determines trophectoderm differentiation. *Cell* **2005**, *123*, 917–929. [CrossRef] [PubMed]
10. Yamanaka, Y.; Lanner, F.; Rossant, J. FGF signal-dependent segregation of primitive endoderm and epiblast in the mouse blastocyst. *Development* **2010**, *137*, 715–724. [CrossRef] [PubMed]
11. Huang, S. Hybrid T-helper cells: Stabilizing the moderate center in a polarized system. *PLoS Biol.* **2013**, *11*, e1001632. [CrossRef] [PubMed]
12. Lee, E.Y.H.P.; Muller, W.J. Oncogenes and tumor suppressor genes. *Cold Spring Harb. Perspect. Biol.* **2010**, *2*, a003236. [CrossRef] [PubMed]

13. Sternlicht, M.D.; Lochter, A.; Sympson, C.J.; Huey, B.; Rougier, J.-P.; Gray, J.W.; Pinkel, D.; Bissell, M.J.; Werb, Z. The stromal proteinase MMP3/Stromelysin-1 promotes mammary carcinogenesis. *Cell* **1999**, *98*, 137–146. [CrossRef]
14. Skobe, M.; Fusenig, N.E. Tumorigenic conversion of immortal human keratinocytes through stromal cell activation. *Proc. Natl. Acad. Sci. USA* **1998**, *95*, 1050–1055. [CrossRef] [PubMed]
15. Ferrao, P.T.; Behren, A.; Anderson, R.L.; Thompson, E.W. Cellular and phenotypic plasticity in cancer. *Front. Oncol.* **2015**, *5*, 171. [CrossRef] [PubMed]
16. Kauffman, S.A. Metabolic stability and epigenesis in randomly constructed genetic nets. *J. Theor. Biol.* **1969**, *22*, 437–467. [CrossRef]
17. Kauffman, S.A. *The Origins of Order: Self-Organization and Selection in Evolution*; Oxford University Press: New York, NY, USA, 1993.
18. Huang, S.; Eichler, G.; Bar-Yam, Y.; Ingber, D.E. Cell fates as high-dimensional attractor states of a complex gene regulatory network. *Phys. Rev. Lett.* **2005**, *94*, 128701. [CrossRef] [PubMed]
19. Raj, A.; van Oudenaarden, A. Nature, nurture, or chance: Stochastic gene expression and its consequences. *Cell* **2008**, *135*, 216–226. [CrossRef] [PubMed]
20. Mooney, S.M.; Jolly, M.K.; Levine, H.; Kulkarni, P. Phenotypic plasticity in prostate cancer: Role of intrinsically disordered proteins. *Asian J. Androl.* **2016**, *18*, 704–710. [CrossRef] [PubMed]
21. Hanahan, D.; Weinberg, R.A. Hallmarks of cancer: The next generation. *Cell* **2011**, *144*, 646–674. [CrossRef] [PubMed]
22. Burrell, R.A.; McGranahan, N.; Bartek, J.; Swanton, C. The causes and consequences of genetic heterogeneity in cancer evolution. *Nature* **2013**, *501*, 338–345. [CrossRef] [PubMed]
23. Brock, A.; Chang, H.; Huang, S. Non-genetic heterogeneity—A mutation-independent driving force for the somatic evolution of tumours. *Nat. Rev. Genet.* **2009**, *10*, 336–342. [CrossRef] [PubMed]
24. Huang, S. Non-genetic heterogeneity of cells in development: More than just noise. *Development* **2009**, *136*, 3853–3862. [CrossRef] [PubMed]
25. Polyak, K.; Weinberg, R.A. Transitions between epithelial and mesenchymal states: Acquisition of malignant and stem cell traits. *Nat. Rev. Cancer* **2009**, *9*, 265–273. [CrossRef] [PubMed]
26. Wicha, M.S.; Liu, S.; Dontu, G. Cancer stem cells: An old idea—A paradigm shift. *Cancer Res.* **2006**, *66*, 1883–1890. [CrossRef] [PubMed]
27. Jiang, L.; Xiao, L.; Sugiura, H.; Huang, X.; Ali, A.; Kuro-o, M.; Deberardinis, R.J.; Boothman, D.A. Metabolic reprogramming during TGFβ1-induced epithelial-to-mesenchymal transition. *Oncogene* **2015**, *34*, 3908–3916. [CrossRef] [PubMed]
28. Viale, A.; Corti, D.; Draetta, G.F. Tumors and mitochondrial respiration: A neglected connection. *Cancer Res.* **2015**, *75*, 3685–3686. [CrossRef] [PubMed]
29. Mani, S.A.; Guo, W.; Liao, M.-J.; Eaton, E.N.; Ayyanan, A.; Zhou, A.Y.; Brooks, M.; Reinhard, F.; Zhang, C.C.; Shipitsin, M.; et al. The epithelial-mesenchymal transition generates cells with properties of stem cells. *Cell* **2008**, *133*, 704–715. [CrossRef] [PubMed]
30. Peiris-Pagès, M.; Martinez-Outschoorn, U.E.; Pestell, R.G.; Sotgia, F.; Lisanti, M.P. Cancer stem cell metabolism. *Breast Cancer Res.* **2016**, *18*, 55. [CrossRef] [PubMed]
31. Morandi, A.; Taddei, M.L.; Chiarugi, P.; Giannoni, E. Targeting the metabolic reprogramming that controls epithelial-to-mesenchymal transition in aggressive Tumors. *Front. Oncol.* **2017**, *7*, 40. [CrossRef] [PubMed]
32. Halldorsson, S.; Rohatgi, N.; Magnusdottir, M.; Choudhary, K.S.; Gudjonsson, T.; Knutsen, E.; Barkovskaya, A.; Hilmarsdottir, B.; Perander, M.; Mælandsmo, G.M.; et al. Metabolic re-wiring of isogenic breast epithelial cell lines following epithelial to mesenchymal transition. *Cancer Lett.* **2017**, *396*. 117–129. [CrossRef] [PubMed]
33. Kauffman, S. Differentiation of malignant to benign cells. *J. Theor. Biol.* **1971**, *31*, 429–451. [CrossRef]
34. Huang, S.; Ernberg, I.; Kauffman, S. Cancer attractors: A systems view of tumors from a gene network dynamics and developmental perspective. *Semin. Cell Dev. Biol.* **2009**, *20*, 869–876. [CrossRef] [PubMed]
35. Zhu, X.; Yuan, R.; Hood, L.; Ao, P. Endogenous molecular-cellular hierarchical modeling of prostate carcinogenesis uncovers robust structure. *Prog. Biophys. Mol. Biol.* **2015**, *117*, 30–42. [CrossRef] [PubMed]
36. Thiery, J.P.; Acloque, H.; Huang, R.Y.J.; Nieto, M.A. Epithelial-mesenchymal transitions in development and disease. *Cell* **2009**, *139*, 871–890. [CrossRef] [PubMed]

37. Ye, X.; Weinberg, R.A. Epithelial-mesenchymal plasticity: A central regulator of cancer progression. *Trends Cell Biol.* **2015**, *25*, 675–686. [CrossRef] [PubMed]

38. Jolly, M.K.; Ware, K.E.; Gilja, S.; Somarelli, J.A.; Levine, H. EMT and MET: Necessary or permissive for metastasis? *Mol. Oncol.* **2017**. [CrossRef] [PubMed]

39. Caramel, J.; Papadogeorgakis, E.; Hill, L.; Browne, G.J.; Richard, G.; Wierinckx, A.; Saldanha, G.; Osborne, J.; Hutchinson, P.; Tse, G.; et al. A switch in the expression of embryonic EMT-inducers drives the development of malignant melanoma. *Cancer Cell* **2013**, *24*, 466–480. [CrossRef] [PubMed]

40. Li, F.Z.; Dhillon, A.S.; Anderson, R.L.; McArthur, G.; Ferrao, P.T. Phenotype switching in melanoma: Implications for progression and therapy. *Front. Oncol.* **2015**, *13*, 5. [CrossRef] [PubMed]

41. Lee, J.-K.; Joo, K.M.; Lee, J.; Yoon, Y.; Nam, D.-H. Targeting the epithelial to mesenchymal transition in glioblastoma: The emerging role of MET signaling. *Oncol. Targets Ther.* **2014**, *7*, 1933–1944. [CrossRef] [PubMed]

42. Iwadate, Y. Epithelial-mesenchymal transition in glioblastoma progression. *Oncol. Lett.* **2016**, *11*, 1615–1620. [CrossRef] [PubMed]

43. Shibue, T.; Weinberg, R.A. EMT, CSCs, and drug resistance: The mechanistic link and clinical implications. *Nat. Rev. Clin. Oncol.* **2017**. [CrossRef] [PubMed]

44. Jolly, M.K.; Boareto, M.; Huang, B.; Jia, D.; Lu, M.; Ben-Jacob, E.; José, N.O.; Herbert, L. Implications of the hybrid epithelial/mesenchymal phenotype in metastasis. *Front. Oncol.* **2015**, *20*, 5. [CrossRef] [PubMed]

45. Yu, M.; Bardia, A.; Wittner, B.S.; Stott, S.L.; Smas, M.E.; Ting, D.T.; Isakoff, S.J.; Ciciliano, J.C.; Wells, M.N.; Shah, A.M.; et al. Circulating breast tumor cells exhibit dynamic changes in epithelial and mesenchymal composition. *Science* **2013**, *339*, 580–584. [CrossRef] [PubMed]

46. Aceto, N.; Bardia, A.; Miyamoto, D.T.; Donaldson, M.C.; Wittner, B.S.; Spencer, J.A.; Yu, M.; Pely, A.; Engstrom, A.; Zhu, H.; et al. Circulating tumor cell clusters are oligoclonal precursors of breast cancer metastasis. *Cell* **2014**, *158*, 1110–1122. [CrossRef] [PubMed]

47. Lu, M.; Jolly, M.K.; Levine, H.; Onuchic, J.N.; Ben-Jacob, E. MicroRNA-based regulation of epithelial-hybrid-mesenchymal fate determination. *Proc. Natl. Acad. Sci. USA* **2013**, *110*, 18144–18149. [CrossRef] [PubMed]

48. Zhang, J.; Tian, X.-J.; Zhang, H.; Teng, Y.; Li, R.; Bai, F.; Elankumaran, S.; Xing, J. TGF-β-induced epithelial-to-mesenchymal transition proceeds through stepwise activation of multiple feedback loops. *Sci. Signal.* **2014**, *7*, ra91. [CrossRef] [PubMed]

49. Tian, X.-J.; Zhang, H.; Xing, J. Coupled reversible and irreversible bistable switches underlying tgfβ-induced epithelial to mesenchymal transition. *Biophys. J.* **2013**, *105*, 1079–1089. [CrossRef] [PubMed]

50. Jia, D.; Jolly, M.K.; Tripathi, S.C.; Hollander, P.D.; Huang, B.; Lu, M.; Celiktas, M.; Ramirez-Pena, E.; Ben-Jacob, E.; Onuchic, J.N.; et al. Distinguishing mechanisms underlying EMT tristability. *ArXiv* **2017**, arXiv:1701.01746.

51. Li, C.; Hong, T.; Nie, Q. Quantifying the landscape and kinetic paths for epithelial-mesenchymal transition from a core circuit. *Phys. Chem. Chem. Phys.* **2016**, *18*, 17949–17956. [CrossRef] [PubMed]

52. Grigore, A.D.; Jolly, M.K.; Jia, D.; Farach-Carson, M.C.; Levine, H. Tumor budding: The name is EMT. prtial EMT. *J. Clin. Med.* **2016**, *5*, 51. [CrossRef] [PubMed]

53. Savagner, P. Epithelial-mesenchymal transitions: From cell plasticity to concept elasticity. *Curr. Top. Dev. Biol.* **2015**, *112*, 273–300. [CrossRef] [PubMed]

54. Jolly, M.K.; Tripathi, S.C.; Jia, D.; Mooney, S.M.; Celiktas, M.; Hanash, S.M.; Mani, S.A.; Pienta, K.J.; Ben-Jacob, E.; Levine, H. Stability of the hybrid epithelial/mesenchymal phenotype. *Oncotarget* **2016**, *7*, 27067–27084. [CrossRef] [PubMed]

55. Jia, D.; Jolly, M.K.; Boareto, M.; Parsana, P.; Mooney, S.M.; Pienta, K.J.; Levine, H.; Ben-Jacob, E. OVOL guides the epithelial-hybrid-mesenchymal transition. *Oncotarget* **2015**, *6*, 15436–15448. [CrossRef] [PubMed]

56. Hong, T.; Watanabe, K.; Ta, C.H.; Villarreal-Ponce, A.; Nie, Q.; Dai, X. An Ovol2-Zeb1 mutual inhibitory circuit governs bidirectional and multi-step transition between epithelial and mesenchymal states. *PLoS Comput. Biol.* **2015**, *11*, e1004569. [CrossRef] [PubMed]

57. Yaswen, P. Reinforcing targeted therapeutics with phenotypic stability factors. *Cell Cycle* **2014**, *13*, 3818–3822. [CrossRef] [PubMed]

58. Watanabe, K.; Villarreal-Ponce, A.; Sun, P.; Salmans, M.L.; Fallahi, M.; Andersen, B.; Dai, X. Mammary morphogenesis and regeneration require the inhibition of EMT at terminal end buds by Ovol2 transcriptional repressor. *Dev. Cell* **2014**, *29*, 59–74. [CrossRef] [PubMed]

59. Huang, B.; Lu, M.; Jia, D.; Ben-Jacob, E.; Levine, H.; Onuchic, J.N. Interrogating the topological robustness of gene regulatory circuits by randomization. *PLoS Comput. Biol.* **2017**, *13*, e1005456. [CrossRef] [PubMed]

60. Boareto, M.; Jolly, M.K.; Goldman, A.; Pietilä, M.; Mani, S.A.; Sengupta, S.; Ben-Jacob, E.; Levine, H.; Onuchic, J.N. Notch-jagged signalling can give rise to clusters of cells exhibiting a hybrid epithelial/mesenchymal phenotype. *J. R. Soc. Interface* **2016**, *13*. [CrossRef] [PubMed]

61. Jolly, M.K.; Boareto, M.; Debeb, B.G.; Aceto, N.; Farach-Carson, M.C.; Woodward, W.A.; Levine, H. Inflammatory breast cancer: A model for investigating cluster-based dissemination. *NPJ Breast Cancer* **2017**, *3*, 21. [CrossRef]

62. Scheel, C.; Weinberg, R.A. Cancer stem cells and epithelial-mesenchymal transition: Concepts and molecular links. *Semin. Cancer Biol.* **2012**, *22*, 396–403. [CrossRef] [PubMed]

63. Morel, A.-P.; Lièvre, M.; Thomas, C.; Hinkal, G.; Ansieau, S.; Puisieux, A. Generation of breast cancer stem cells through epithelial-mesenchymal transition. *PLoS ONE* **2008**, *3*. [CrossRef] [PubMed]

64. Niu, R.F.; Zhang, L.; Xi, G.M.; Wei, X.Y.; Yang, Y.; Shi, Y.R.; Hao, X.S. Up-regulation of twist induces angiogenesis and correlates with metastasis in hepatocellular carcinoma. *J. Exp. Clin. Cancer Res. CR* **2007**, *26*, 385–394. [PubMed]

65. Wang, Z.; Li, Y.; Kong, D.; Banerjee, S.; Ahmad, A.; Azmi, A.S.; Ali, S.; Abbruzzese, J.L.; Gallick, G.E.; Sarkar, F.H. Acquisition of epithelial-mesenchymal transition phenotype of gemcitabine-resistant pancreatic cancer cells is linked with activation of the notch signaling pathway. *Cancer Res.* **2009**, *69*, 2400–2407. [CrossRef] [PubMed]

66. Brabletz, T.; Hlubek, F.; Spaderna, S.; Schmalhofer, O.; Hiendlmeyer, E.; Jung, A.; Kirchner, T. Invasion and metastasis in colorectal cancer: Epithelial-mesenchymal transition, mesenchymal-epithelial transition, stem cells and beta-catenin. *Cells Tissues Organs* **2005**, *179*, 56–65. [CrossRef] [PubMed]

67. Celià-Terrassa, T.; Meca-Cortés, O.; Mateo, F.; Martínez de Paz, A.; Rubio, N.; Arnal-Estapé, A.; Ell, B.J.; Bermudo, R.; Díaz, A.; Guerra-Rebollo, M.; et al. Epithelial-mesenchymal transition can suppress major attributes of human epithelial tumor-initiating cells. *J. Clin. Investig.* **2012**, *122*, 1849–1868. [CrossRef] [PubMed]

68. Ocaña, O.H.; Córcoles, R.; Fabra, A.; Moreno-Bueno, G.; Acloque, H.; Vega, S.; Barrallo-Gimeno, A.; Cano, A.; Nieto, M.A. Metastatic colonization requires the repression of the epithelial-mesenchymal transition inducer Prrx1. *Cancer Cell* **2012**, *22*, 709–724. [CrossRef] [PubMed]

69. Li, R.; Liang, J.; Ni, S.; Zhou, T.; Qing, X.; Li, H.; He, W.; Chen, J.; Li, F.; Zhuang, Q.; et al. A mesenchymal-to-epithelial transition initiates and is required for the nuclear reprogramming of mouse fibroblasts. *Cell Stem Cell* **2010**, *7*, 51–63. [CrossRef] [PubMed]

70. Jolly, M.K.; Jia, D.; Boareto, M.; Mani, S.A.; Pienta, K.J.; Ben-Jacob, E.; Levine, H. Coupling the modules of EMT and stemness: A tunable "stemness window" model. *Oncotarget* **2015**, *6*, 25161–25174. [CrossRef] [PubMed]

71. Jolly, M.K.; Huang, B.; Lu, M.; Mani, S.A.; Levine, H.; Ben-Jacob, E. Towards elucidating the connection between epithelial-mesenchymal transitions and stemness. *J. R. Soc. Interface* **2014**, *11*. [CrossRef] [PubMed]

72. Grosse-Wilde, A.; Fouquier d'Hérouël, A.; McIntosh, E.; Ertaylan, G.; Skupin, A.; Kuestner, R.E.; del Sol, A.; Walters, K.A.; Huang, S. Stemness of the hybrid epithelial/mesenchymal state in breast cancer and its association with poor survival. *PLoS ONE* **2015**, *10*. [CrossRef] [PubMed]

73. Bierie, B.; Pierce, S.E.; Kroeger, C.; Stover, D.G.; Pattabiraman, D.R.; Thiru, P.; Donaher, J.L.; Reinhardt, F.; Chaffer, C.L.; Keckesova, Z.; et al. Integrin-β4 identifies cancer stem cell-enriched populations of partially mesenchymal carcinoma cells. *Proc. Natl. Acad. Sci. USA* **2017**, *114*, E2337–E2346. [CrossRef] [PubMed]

74. Colacino, J.; Azizi, E.; Brooks, M.; Fouladdel, S.; McDermott, S.P.; Lee, M.; Hill, D.; Sartor, M.; Rozek, L.; Wicha, M. Heterogeneity of normal human breast stem and progenitor cells as revealed by transcriptional profiling. *BioRxiv* **2017**. [CrossRef]

75. Li, W.; Kang, Y. Probing the fifty shades of EMT in metastasis. *Trends Cancer* **2016**, *2*, 65–67. [CrossRef] [PubMed]

76. Ombrato, L.; Malanchi, I. The EMT universe: Space between cancer cell dissemination and metastasis initiation. *Crit. Rev. Oncog.* **2014**, *19*, 349–361. [CrossRef] [PubMed]

77. Singh, A.; Settleman, J. EMT, cancer stem cells and drug resistance: An emerging axis of evil in the war on cancer. *Oncogene* **2010**, *29*, 4741–4751. [CrossRef] [PubMed]

78. Biddle, A.; Gammon, L.; Liang, X.; Costea, D.E.; Mackenzie, I.C. Phenotypic plasticity determines cancer stem cell therapeutic resistance in oral squamous cell carcinoma. *EBioMedicine* **2016**, *4*, 138–145. [CrossRef] [PubMed]

79. Goldman, A.; Majumder, B.; Dhawan, A.; Ravi, S.; Goldman, D.; Kohandel, M.; Majumder, P.K.; Sengupta, S. Temporally sequenced anticancer drugs overcome adaptive resistance by targeting a vulnerable chemotherapy-induced phenotypic transition. *Nat. Commun.* **2015**, *6*, 6139. [CrossRef] [PubMed]

80. Li, C.; Wang, J. Quantifying the landscape for development and cancer from a core cancer stem cell circuit. *Cancer Res.* **2015**, *75*, 2607–2618. [CrossRef] [PubMed]

81. Pavlova, N.N.; Thompson, C.B. The emerging hallmarks of cancer metabolism. *Cell Metab.* **2016**, *23*, 27–47. [CrossRef] [PubMed]

82. Warburg, O. On the origin of cancer cells. *Science* **1956**, *123*, 309–314. [CrossRef] [PubMed]

83. Vander Heiden, M.G.; Cantley, L.C.; Thompson, C.B. Understanding the warburg effect: The metabolic requirements of cell proliferation. *Science* **2009**, *324*, 1029–1033. [CrossRef] [PubMed]

84. Hsu, P.P.; Sabatini, D.M. Cancer cell metabolism: Warburg and beyond. *Cell* **2008**, *134*, 703–707. [CrossRef] [PubMed]

85. Viale, A.; Pettazzoni, P.; Lyssiotis, C.A.; Ying, H.; Sánchez, N.; Marchesini, M.; Carugo, A.; Green, T.; Seth, S.; Giuliani, V.; et al. Oncogene ablation-resistant pancreatic cancer cells depend on mitochondrial function. *Nature* **2014**, *514*, 628–632. [CrossRef] [PubMed]

86. Maiuri, M.C.; Kroemer, G. Essential role for oxidative phosphorylation in cancer progression. *Cell Metab.* **2015**, *21*, 11–12. [CrossRef] [PubMed]

87. Strohecker, A.M.; White, E. Targeting mitochondrial metabolism by inhibiting autophagy in BRAF-driven cancers. *Cancer Discov.* **2014**, *4*, 766–772. [CrossRef] [PubMed]

88. Lu, C.-L.; Qin, L.; Liu, H.-C.; Candas, D.; Fan, M.; Li, J.J. Tumor cells switch to mitochondrial oxidative phosphorylation under radiation via mTOR-mediated hexokinase II inhibition—A Warburg-reversing effect. *PLoS ONE* **2015**, *10*, e0121046. [CrossRef] [PubMed]

89. Huang, D.; Li, T.; Li, X.; Zhang, L.; Sun, L.; He, X.; Zhong, X.; Jia, D.; Song, L.; Semenza, G.L.; et al. HIF-1-mediated suppression of Acyl-CoA dehydrogenases and fatty acid oxidation is critical for cancer progression. *Cell Rep.* **2014**, *8*, 1930–1942. [CrossRef] [PubMed]

90. Park, J.H.; Vithayathil, S.; Kumar, S.; Sung, P.-L.; Dobrolecki, L.E.; Putluri, V.; Bhat, V.B.; Bhowmik, S.K.; Gupta, V.; Arora, K.; et al. Fatty acid oxidation-driven src links mitochondrial energy reprogramming and oncogenic properties in triple-negative breast cancer. *Cell Rep.* **2016**, *14*, 2154–2165. [CrossRef] [PubMed]

91. Yu, L.; Lu, M.; Jia, D.; Ma, J.; Ben-Jacob, E.; Levine, H.; Kaipparettu, B.A.; Onuchic, J.N. Modeling the genetic regulation of cancer metabolism: Interplay between glycolysis and oxidative phosphorylation. *Cancer Res.* **2017**, *77*, 1564–1574. [CrossRef] [PubMed]

92. Porporato, P.E.; Payen, V.L.; Pérez-Escuredo, J.; De Saedeleer, C.J.; Danhier, P.; Copetti, T.; Dhup, S.; Tardy, M.; Vazeille, T.; Bouzin, C.; et al. A mitochondrial switch promotes tumor metastasis. *Cell Rep.* **2014**, *8*, 754–766. [CrossRef] [PubMed]

93. Dupuy, F.; Tabariès, S.; Andrzejewski, S.; Dong, Z.; Blagih, J.; Annis, M.G.; Omeroglu, A.; Gao, D.; Leung, S.; Amir, E.; et al. PDK1-dependent metabolic reprogramming dictates metastatic potential in breast cancer. *Cell Metab.* **2015**, *22*, 577–589. [CrossRef] [PubMed]

94. Piskounova, E.; Agathocleous, M.; Murphy, M.M.; Hu, Z.; Huddlestun, S.E.; Zhao, Z.; Leitch, A.M.; Johnson, T.M.; DeBerardinis, R.J.; et al. Oxidative stress inhibits distant metastasis by human melanoma cells. *Nature* **2015**, *527*, 186–191. [CrossRef] [PubMed]

95. Ishikawa, K.; Takenaga, K.; Akimoto, M.; Koshikawa, N.; Yamaguchi, A.; Imanishi, H.; Nakada, K.; Honma, Y.; Hayashi, J.-I. ROS-generating mitochondrial DNA mutations can regulate tumor cell metastasis. *Science* **2008**, *320*, 661–664. [CrossRef] [PubMed]

96. Kondaveeti, Y.; Guttilla Reed, I.K.; White, B.A. Epithelial-mesenchymal transition induces similar metabolic alterations in two independent breast cancer cell lines. *Cancer Lett.* **2015**, *364*, 44–58. [CrossRef] [PubMed]

97. Farris, J.C.; Pifer, P.M.; Zheng, L.; Gottlieb, E.; Denvir, J.; Frisch, S.M. Grainyhead-like 2 Reverses the Metabolic Changes Induced by the Oncogenic Epithelial-Mesenchymal Transition: Effects on Anoikis. *Mol. Cancer Res. MCR* **2016**, *14*, 528–538. [CrossRef] [PubMed]

98. Gammon, L.; Biddle, A.; Heywood, H.K.; Johannessen, A.C.; Mackenzie, I.C. Sub-sets of cancer stem cells differ intrinsically in their patterns of oxygen metabolism. *PLoS ONE* **2013**, *8*, e62493. [CrossRef] [PubMed]
99. Diehn, M.; Cho, R.W.; Lobo, N.A.; Kalisky, T.; Dorie, M.J.; Kulp, A.N.; Qian, D.; Lam, J.S.; Ailles, L.E.; Wong, M.; et al. Association of reactive oxygen species levels and radioresistance in cancer stem cells. *Nature* **2009**, *458*, 780–783. [CrossRef] [PubMed]
100. Paldi, A. What makes the cell differentiate? *Prog. Biophys. Mol. Biol.* **2012**, *110*, 41–43. [CrossRef] [PubMed]
101. Witta, S.E.; Gemmill, R.M.; Hirsch, F.R.; Coldren, C.D.; Hedman, K.; Ravdel, L.; Helfrich, B.; Dziadziuszko, R.; Chan, D.C.; Sugita, M.; et al. Restoring E-cadherin expression increases sensitivity to epidermal growth factor receptor inhibitors in lung cancer cell lines. *Cancer Res.* **2006**, *66*, 944–950. [CrossRef] [PubMed]
102. Oliveras-Ferraros, C.; Corominas-Faja, B.; Vazquez-Martin, S.A.; Martin-Castillo, B.; Iglesias, J.M.; López-Bonet, E.; Martin, Á.G.; Menendez, J.A. Epithelial-to-mesenchymal transition (EMT) confers primary resistance to trastuzumab (Herceptin). *Cell Cycle* **2012**, *11*, 4020–4032. [CrossRef] [PubMed]
103. Zheng, X.; Carstens, J.L.; Kim, J.; Scheible, M.; Kaye, J.; Sugimoto, H.; Wu, C.-C.; LeBleu, V.S.; Kalluri, R. Epithelial-to-mesenchymal transition is dispensable for metastasis but induces chemoresistance in pancreatic cancer. *Nature* **2015**, *527*, 525–530. [CrossRef] [PubMed]
104. Fischer, K.R.; Durrans, A.; Lee, S.; Sheng, J.; Li, F.; Wong, S.T.C.; Choi, H.; El Rayes, T.; Ryu, S.; Troeger, J.; et al. Epithelial-to-mesenchymal transition is not required for lung metastasis but contributes to chemoresistance. *Nature* **2015**, *527*, 472–476. [CrossRef] [PubMed]
105. Chen, C.; Baumann, W.T.; Xing, J.; Xu, L.; Clarke, R.; Tyson, J.J. Mathematical models of the transitions between endocrine therapy responsive and resistant states in breast cancer. *J. R. Soc. Interface* **2014**, *11*, 20140206. [CrossRef] [PubMed]
106. Kulkarni, P.; Jolly, M.K.; Jia, D.; Mooney, S.M.; Bhargava, A.; Kagohara, L.T.; Chen, Y.; Hao, P.; He, Y.; Veltri, R.W.; et al. Phosphorylation-induced conformational dynamics in an intrinsically disordered protein and potential role in phenotypic heterogeneity. *Proc. Natl. Acad. Sci. USA* **2017**, *114*, E2644–E2653. [CrossRef] [PubMed]
107. Lee, H.H.; Bellat, V.; Law, B. Chemotherapy induces adaptive drug resistance and metastatic potentials via phenotypic CXCR4-expressing cell state transition in ovarian cancer. *PLoS ONE* **2017**, *12*, e0171044. [CrossRef] [PubMed]
108. Fallahi-Sichani, M.; Becker, V.; Izar, B.; Baker, G.J.; Lin, J.-R.; Boswell, S.A.; Shah, P.; Rotem, A.; Garraway, L.A.; Sorger, P.K. Adaptive resistance of melanoma cells to RAF inhibition via reversible induction of a slowly dividing de-differentiated state. *Mol. Syst. Biol.* **2017**, *13*, 905. [CrossRef] [PubMed]
109. Yoshida, T.; Song, L.; Bai, Y.; Kinose, F.; Li, J.; Ohaegbulam, K.C.; Muñoz-Antonia, T.; Qu, X.; Eschrich, S.; Uramoto, H.; Tanaka, F.; et al. ZEB1 mediates acquired resistance to the epidermal growth factor receptor-tyrosine kinase inhibitors in non-small cell lung cancer. *PLoS ONE* **2016**, *11*, 1–22. [CrossRef] [PubMed]
110. Shaffer, S.M.; Dunagin, M.C.; Torborg, S.R.; Torre, E.A.; Emert, B.; Krepler, C.; Beqiri, M.; Sproesser, K.; Brafford, P.A.; Xiao, M.; et al. Rare cell variability and drug-induced reprogramming as a mode of cancer drug resistance. *Nature* **2017**, *546*, 431–435. [CrossRef] [PubMed]
111. Tripathi, S.C.; Peters, H.L.; Taguchi, A.; Katayama, H.; Wang, H.; Momin, A.; Jolly, M.K.; Celiktas, M.; Rodriguez-Canales, J.; Liu, H.; et al. Immunoproteasome deficiency is a feature of non-small cell lung cancer with a mesenchymal phenotype and is associated with a poor outcome. *Proc. Natl. Acad. Sci. USA* **2016**, *113*, E1555–E1564. [CrossRef] [PubMed]
112. Iakoucheva, L.M.; Brown, C.J.; Lawson, J.D.; Obradović, Z.; Dunker, A.K. Intrinsic disorder in cell-signaling and cancer-associated proteins. *J. Mol. Biol.* **2002**, *323*, 573–584. [CrossRef]
113. Uversky, V.N.; Oldfield, C.J.; Dunker, A.K. Intrinsically disordered proteins in human diseases: Introducing the D2 concept. *Annu. Rev. Biophys.* **2008**, *37*, 215–246. [CrossRef] [PubMed]
114. Xue, B.; Oldfield, C.J.; Van, Y.-Y.; Dunker, A.K.; Uversky, V.N. Protein intrinsic disorder and induced pluripotent stem cells. *Mol. Biosyst.* **2012**, *8*, 134–150. [CrossRef] [PubMed]
115. Wright, P.E.; Dyson, H.J. Intrinsically disordered proteins in cellular signalling and regulation. *Nat. Rev. Mol. Cell Biol.* **2015**, *16*, 18–29. [CrossRef] [PubMed]
116. DeForte, S.; Uversky, V.N. Order, disorder, and everything in between. *Molecules* **2016**, *21*, 1090. [CrossRef]
117. Choi, U.B.; McCann, J.J.; Weninger, K.R.; Bowen, M.E. Beyond the random coil: Stochastic conformational switching in intrinsically disordered proteins. *Structure* **2011**, *19*, 566–576. [CrossRef] [PubMed]

118. Liu, J.; Dai, J.; He, J.; Niemi, A.J.; Ilieva, N. Multistage modeling of protein dynamics with monomeric Myc oncoprotein as an example. *Phys. Rev. E* **2017**, *95*, 32406. [CrossRef] [PubMed]
119. Chakrabortee, S.; Meersman, F.; Kaminski Schierle, G.S.; Bertoncini, C.W.; McGee, B.; Kaminski, C.F.; Tunnacliffe, A. Catalytic and chaperone-like functions in an intrinsically disordered protein associated with desiccation tolerance. *Proc. Natl. Acad. Sci. USA* **2010**, *107*, 16084–16089. [CrossRef] [PubMed]
120. Andresen, C.; Helander, S.; Lemak, A.; Farès, C.; Csizmok, V.; Carlsson, J.; Penn, L.Z.; Forman-Kay, J.D.; Arrowsmith, C.H.; Lundström, P.; et al. Transient structure and dynamics in the disordered c-Myc transactivation domain affect Bin1 binding. *Nucleic Acids Res.* **2012**, *40*, 6353–6366. [CrossRef] [PubMed]
121. Luna Maldonado, A.; Jimenez-Rios, G.; Villanueva Canadas, E. Aminopeptidase and cathepsin a activity in vitreous humour in relation to causes of death. *Acta Med. Leg. Soc. (Liege)* **1984**, *34*, 207–216. [PubMed]
122. Neira, J.L.; Bintz, J.; Arruebo, M.; Rizzuti, B.; Bonacci, T.; Vega, S.; Lanas, A.; Velázquez-Campoy, A.; Iovanna, J.L.; Abián, O. Identification of a drug targeting an intrinsically disordered protein involved in pancreatic adenocarcinoma. *Sci. Rep.* **2017**, *7*, 39732. [CrossRef] [PubMed]
123. Vavouri, T.; Semple, J.I.; Garcia-Verdugo, R.; Lehner, B. Intrinsic protein disorder and interaction promiscuity are widely associated with dosage sensitivity. *Cell* **2009**, *138*, 198–208. [CrossRef]
124. Marcotte, E.M.; Tschansky, M. Disorder, promiscuity, and toxic partnerships. *Cell* **2009**, *138*, 16–18. [CrossRef] [PubMed]
125. Mahmoudabadi, G.; Rajagopalan, K.; Getzenberg, R.H.; Hannenhalli, S.; Rangarajan, G.; Kulkarni, P. Intrinsically disordered proteins and conformational noise: Implications in cancer. *Cell Cycle* **2013**, *12*, 26–31. [CrossRef] [PubMed]
126. Chang, H.H.; Hemberg, M.; Barahona, M.; Ingber, D.E.; Huang, S. Transcriptome-wide noise controls lineage choice in mammalian progenitor cells. *Nature* **2008**, *453*, 544–547. [CrossRef]
127. Eldar, A.; Elowitz, M.B. Functional roles for noise in genetic circuits. *Nature* **2010**, *467*, 167–173. [CrossRef] [PubMed]
128. Zeng, Y.; He, Y.; Yang, F.; Mooney, S.M.; Getzenberg, R.H.; Orban, J.; Kulkarni, P. The cancer/testis antigen prostate-associated gene 4 (PAGE4) is a highly intrinsically disordered protein. *J. Biol. Chem.* **2011**, *286*, 13985–13994. [CrossRef] [PubMed]
129. Rajagopalan, K.; Qiu, R.; Mooney, S.M.; Rao, S.; Shiraishi, T.; Sacho, E.; Huang, H.; Shapiro, E.; Weninger, K.R.; Kulkarni, P. The Stress-response protein prostate-associated gene 4, interacts with c-Jun and potentiates its transactivation. *Biochim. Biophys. Acta* **2014**, *1842*, 154–163. [CrossRef] [PubMed]
130. Mooney, S.M.; Qiu, R.; Kim, J.J.; Sacho, E.J.; Rajagopalan, K.; Johng, D.; Shiraishi, T.; Kulkarni, P.; Weninger, K.R. Cancer/testis antigen PAGE4, a regulator of c-Jun transactivation, is phosphorylated by homeodomain-interacting protein kinase 1, a component of the stress-response pathway. *Biochemistry* **2014**, *53*, 1670–1679. [CrossRef] [PubMed]
131. Sato, M.; Sadar, M.; Bruchovsky, N.; Saatcioglu, F.; Rennie, P.; Sato, S.; Lange, P.; Gleave, M. Androgenic induction of prostate-specific antigen gene is repressed by protein-protein interaction between the androgen receptor and AP-1/c-Jun in the human prostate cancer cell line LNCaP. *J. Biol. Chem.* **1997**, *272*, 17485–17494. [CrossRef] [PubMed]
132. Tillman, K.; Oberfield, J.L.; Shen, X.Q.; Bubulya, A.; Shemshedini, L. c-Fos dimerization with c-Jun represses c-Jun enhancement of androgen receptor transactivation. *Endocrine* **1998**, *9*, 193–200. [CrossRef]
133. Isaacs, J.T.; D'Antonio, J.M.; Chen, S.; Antony, L.; Dalrymple, S.P.; Ndikuyeze, G.H.; Luo, J.; Denmeade, S.R. Adpative auto-regultion of androgen receptor provides a paradigm shifting rationale for bipolar androgen therapy (BAT) for castrate resistant human prostate cancer. *Prostate* **2012**, *72*, 1491–1505. [CrossRef] [PubMed]
134. Terada, N.; Shiraishi, T.; Zeng, Y.; Aw-Yong, K.-M.; Mooney, S.M.; Liu, Z.; Takahashi, S.; Luo, J.; Lupold, S.E.; Kulkarni, P.; et al. Correlation of Sprouty1 and Jagged1 with aggressive prostate cancer cells with different sensitivities to androgen deprivation. *J. Cell. Biochem.* **2014**, *115*, 1505–1515. [CrossRef] [PubMed]
135. Hornstein, E.; Shomron, N. Canalization of development by microRNAs. *Nat. Genet.* **2006**, *38*, S20–S24. [CrossRef] [PubMed]
136. Ferrell, J.E. Bistability, bifurcations, and waddington's epigenetic landscape. *Curr. Biol.* **2012**, *22*, R458–R466. [CrossRef] [PubMed]
137. Brock, A.; Krause, S.; Ingber, D.E. Control of cancer formation by intrinsic genetic noise and microenvironmental cues. *Nat. Rev. Cancer* **2015**, *15*, 499–509. [CrossRef] [PubMed]

138. Quail, D.; Joyce, J. Microenvironmental regulation of tumor progression and metastasis. *Nat. Med.* **2013**, *19*, 1423–1437. [CrossRef] [PubMed]
139. Engler, A.J.; Sen, S.; Sweeney, H.L.; Discher, D.E. Matrix elasticity directs stem cell lineage specification. *Cell* **2006**, *126*, 677–689. [CrossRef] [PubMed]
140. Masiello, M.G.; Cucina, A.; Proietti, S.; Palombo, A.; Coluccia, P.; Anselmi, F.; Dinicola, S.; Pasqualato, A.; Morini, V.; Bizzarri, M. Phenotypic Switch Induced by Simulated Microgravity on MDA-MB-231 Breast Cancer Cells. *BioMed Res. Int.* **2014**, *2014*, e652434. [CrossRef] [PubMed]
141. Downing, T.L.; Soto, J.; Morez, C.; Houssin, T.; Fritz, A.; Yuan, F.; Chu, J.; Patel, S.; Schaffer, D.V.; Li, S. Biophysical regulation of epigenetic state and cell reprogramming. *Nat. Mater.* **2013**, *12*, 1154–1162. [CrossRef] [PubMed]
142. Bissell, M.J.; Inman, J. Reprogramming stem cells is a microenvironmental task. *Proc. Natl. Acad. Sci. USA* **2008**, *105*, 15637–15638. [CrossRef] [PubMed]
143. Bizzarri, M.; Pasqualato, A.; Cucina, A.; Pasta, V. Physical forces and non linear dynamics mould fractal cell shape: Quantitative morphological parameters and cell phenotype. *Histol. Histopathol.* **2013**, *28*, 155–174. [CrossRef] [PubMed]
144. Shachaf, C.M.; Kopelman, A.M.; Arvanitis, C.; Karlsson, Å.; Beer, S.; Mandl, S.; Bachmann, M.H.; Borowsky, A.D.; Ruebner, B.; Cardiff, R.D.; et al. MYC inactivation uncovers pluripotent differentiation and tumour dormancy in hepatocellular cancer. *Nature* **2004**, *431*, 1112–1117. [CrossRef] [PubMed]
145. Tuynder, M.; Susini, L.; Prieur, S.; Besse, S.; Fiucci, G.; Amson, R.; Telerman, A. Biological models and genes of tumor reversion: cellular reprogramming through tpt1/TCTP and SIAH-1. *Proc. Natl. Acad. Sci. USA* **2002**, *99*, 14976–14981. [CrossRef] [PubMed]
146. Tuynder, M.; Fiucci, G.; Prieur, S.; Lespagnol, A.; Géant, A.; Beaucourt, S.; Duflaut, D.; Besse, S.; Susini, L.; Cavarelli, J.; et al. Translationally controlled tumor protein is a target of tumor reversion. *Proc. Natl. Acad. Sci. USA* **2004**, *101*, 15364–15369. [CrossRef] [PubMed]
147. Amson, R.; Pece, S.; Lespagnol, A.; Vyas, R.; Mazzarol, G.; Tosoni, D.; Colaluca, I.; Viale, G.; Rodrigues-Ferreira, S.; Wynendaele, J.; et al. Reciprocal repression between P53 and TCTP. *Nat. Med.* **2011**, *18*, 91–99. [CrossRef] [PubMed]
148. Weaver, V.M.; Petersen, O.W.; Wang, F.; Larabell, C.A.; Briand, P.; Damsky, C.; Bissell, M.J. Reversion of the malignant phenotype of human breast cells in three-dimensional culture and in vivo by integrin blocking antibodies. *J. Cell Biol.* **1997**, *137*, 231–245. [CrossRef] [PubMed]

MDPI AG

St. Alban-Anlage 66

4052 Basel, Switzerland

Tel. +41 61 683 77 34

Fax +41 61 302 89 18

http://www.mdpi.com

Cancers Editorial Office

E-mail: cancers@mdpi.com

http://www.mdpi.com/journal/cancers

www.ingramcontent.com/pod-product-compliance
Lightning Source LLC
Chambersburg PA
CBHW051725210326
41597CB00032B/5616